Orogeny Through Time

Geological Society Special Publications
Series Editor A.J. FLEET

GEOLOGICAL SOCIETY SPECIAL PUBLICATION NO. 121

Orogeny Through Time

EDITED BY

J.-P. BURG and M. FORD
Geologisches Institute, ETH, Zurich, Switzerland

1997
Published by
The Geological Society
London

THE GEOLOGICAL SOCIETY

The Society was founded in 1807 as The Geological Society of London and is the oldest geological society in the world. It received its Royal Charter in 1825 for the purpose of 'investigating the mineral structure of the Earth'. The Society is Britain's national society for geology with a membership of around 8000. It has countrywide coverage and approximately 1000 members reside overseas. The Society is responsible for all aspects of the geological sciences including professional matters. The Society has its own publishing house, which produces the Society's international journals, books and maps, and which acts as the European distributor for publications of the American Association of Petroleum Geologists, SEPM and the Geological Society of America.

Fellowship is open to those holding a recognized honours degree in geology or cognate subject and who have at least two years' relevant postgraduate experience, or who have not less than six years' relevant experience in geology or a cognate subject. A Fellow who has not less than five years' relevant postgraduate experience in the practice of geology may apply for validation and, subject to approval, may be able to use the designatory letters C Geol (Chartered Geologist).

Further information about the Society is available from the Membership Manager, The Geological Society, Burlington House, Piccadilly, London W1V 0JU, UK. The Society is a Registered Charity, No. 210161.

Published by The Geological Society from:
The Geological Society Publishing House
Unit 7,
Brassmill Enterprise Centre
Brassmill Lane
Bath BA1 3JN
UK
(*Orders*: Tel. 01225 445046
Fax 01225 442836)

First published 1997

British Library Cataloguing in Publication Data

A catalogue record for this book is available from the British Library.

ISBN 1–897799–75–6

Typeset by Type Study, Scarborough, UK.

Printed by The Alden Press, Osney Mead, Oxford, UK.

Distributors

USA
AAPG Bookstore
PO Box 979
Tulsa
OK 74101–0979
USA
(*Orders*: Tel. (918) 584–2555
Fax (918) 560–2652)

Australia
Australian Mineral Foundation
63 Conyngham Street
Glenside
South Australia 5065
Australia
(*Orders*: Tel. (08) 379–0444
Fax (08) 379–4634)

India
Affiliated East-West Press PVT Ltd
G-1/16 Ansari Road
New Delhi 110 002
India
(*Orders*: Tel. (11) 327–9113
Fax (11) 326–0538)

Japan
Kanda Book Trading Co.
Tanikawa Building
3–2 Kanda Surugadai
Chiyoda-Ku
Tokyo 101
Japan
(*Orders*: Tel. (03) 3255–3497
Fax (03) 3255–3495)

Contents

Orogeny through time: an overview

JEAN-PIERRE BURG & MARY FORD

Geologisches Institut, ETH-Zentrum, Sonneggstrasse 5, CH-8092 Zurich, Switzerland

Abstract: As an introduction to a diverse set of papers on orogenic studies we present a personal overview of the most notable developments in orogenic studies since the 1960s. The impact of new techniques such as deep reflection seismic profiling, geochronology, analogue and numerical modelling is discussed. The major geodynamic models and concepts which have stimulated orogenic studies in recent years are also considered.

Orogenesis is the most complex of tectonic processes and interpreting ancient mountain belts is one of the greatest challenges geologists face today. As originally defined by Gilbert (1890), an orogeny (from the classical Greek *óros* meaning 'mountain' and *genés* meaning 'stemming from') is simply a period of mountain building. To field geologists the term orogeny represents a penetrative deformation of the Earth's crust associated with phases of metamorphism and igneous activity along restricted, commonly linear zones and within a limited time interval (Dennis 1967). However our increasing understanding of the rheology of lithospheres allows geologists today to view orogenesis on a larger scale as the interaction of a series of geodynamic processes. This volume has arisen from a seminar series given by invited speakers in 1994 at the Geological Institute, ETH, Zurich. The aim of this lecture series was to provoke discussion on, and greater awareness of, the larger issues of orogenesis. In particular, is there a change in style or mode of orogeny through geological time, or is variety of orogenic features rather reflecting different rheologies and boundary conditions in space. The wealth of data and ideas presented in this lecture series are compiled here as a series of review-type papers.

This book can provide only scattered examples of orogens through time. Figure 1 shows the global distribution of orogenies of different ages and those covered by this book are marked. Many of the Cenozoic orogens which have received considerable attention in recent years (Alps: Roure *et al.* 1990; Pfiffner *et al.* 1996; Schmid *et al.* 1996; Himalayas: Treloar & Searle 1993; Pyrenees: Choukroune *et al.* 1990; Oman: Robertson *et al.* 1990) are not covered in this volume.

Changing approach to the study of orogens

The major controversies that have arisen from the study of looking at orogens in the past are documented in several works (e.g. Cady 1950; Condie 1982; Miyashiro *et al.* 1982). Based on a mixture of spiritual contemplation and observation, many nineteenth century theories on the origin of mountain belts could not conceive of any large movements in the Earth to produce orogenic belts (fixist theories). The important concept of lateral compression gained credence as late as the middle nineteenth century. It was only in the twentieth century that Earth scientists first suggested (e.g. Argand 1924; Wegener 1912; Wilson 1966) and then established with the plate tectonic paradigm (e.g. McKenzie & Parker 1967; Isacks *et al.* 1968; Le Pichon 1968; Morgan 1968) that large horizontal movements were responsible for Cenozoic orogens. As elaborated elsewhere (e.g. Condie 1982) plate tectonics unified several long-lived theories such as those of geosynclines and continental drift and rendered obsolete notions such as world-wide orogenic cycles and a contracting Earth. Shortly after the acceptance of the plate tectonic theory, modern and ancient mountain belts were analysed in terms of global tectonics (e.g. Dewey & Bird 1970; Dickinson 1971). However, partisans of primary vertical tectonics resisted the plate tectonics paradigm and maintained that ancient orogens were better explained by contraction of intracontinental mobile zones (evolved from ensialic rift zones i.e. aulacogens or geosynclines) between stable regions (cratons, e.g. Weber 1984). Zwart (1967) emphasized differences between the Alpine and the Hercynian orogens, leading to the widely used classification of orogens as either Alpinotype or

From Burg, J.-P. & Ford, M. (eds), 1997, *Orogeny Through Time*,
Geological Society Special Publication No. 121, pp. 1–17.

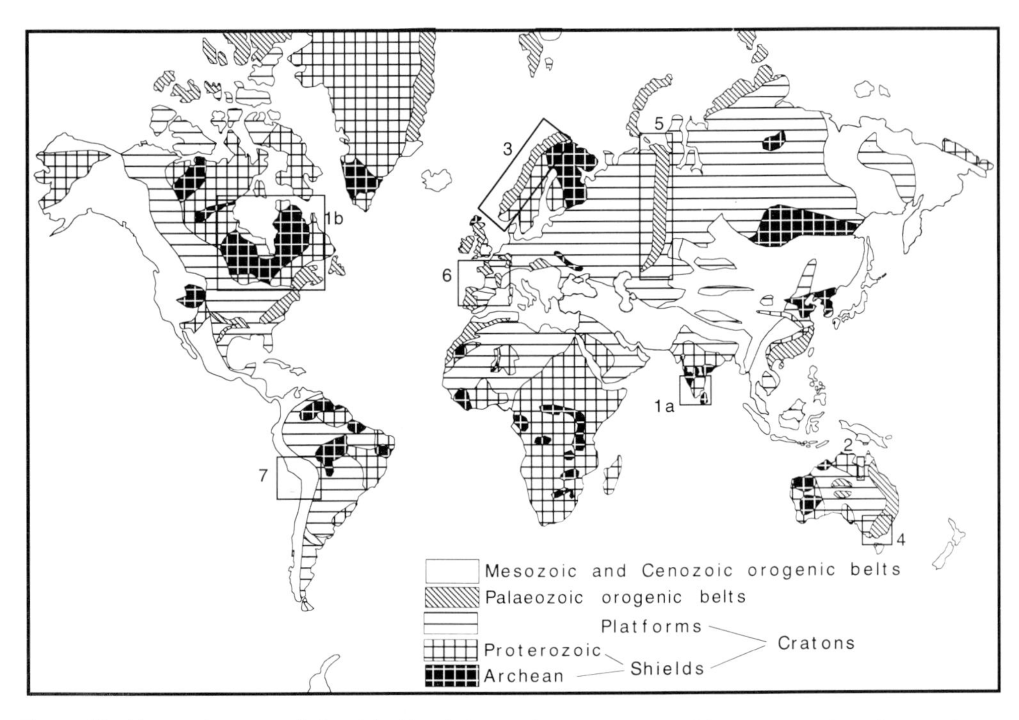

Fig. 1. World map of orogens distinguished by their age. Orogens coved in this volume are boxed. 1, Archaean orogens of Choukroune *et al.* a, Dharwar craton and b, Superior province; 2, Mount Isa Terrain, O'Dea *et al*; 3, the Scandinavian Caledonides, Milnes *et al.* and Rey *et al.*; 4, the Lachlan fold belt, Gray; 5, the Urals, Puchkov; 6, the Variscides, Rey *et al.*; 7, the Central Andes, Lamb *et al.* Adapted from Miyashiro *et al.* (1979) and Condie (1982).

Hercynotype depending principally on the amount of ophiolites, high-pressure metamorphic rocks and granites. Application of the plate tectonic concept was however more fruitful in that geologists could show that major characteristics of an orogen (namely deformation, metamorphism and igneous activity) record stages of the plate tectonic history of the orogen, i.e. successively subduction, obduction, collision and eventually post-collisional intra-continental deformation. Many articles and books, dedicated to the link between plate tectonics and mountain belts (e.g. Mitchell & Reading 1969; Coney 1970; Dewey & Bird 1970; Dietz 1972; Gilluly 1973), have convincingly revealed that modern orogenic belts occur principally at convergent plate boundaries and result from collision between continental, arc-derived or oceanic crustal blocks.

Today, it is generally agreed that plate tectonics were acting throughout Phanerozoic time, although deduction of relative directions and rates of continental drift before 200 Ma remains a problem. Some authors extrapolate the theory to the whole Precambrian (e.g. Dewey & Burke 1973; Kröner 1981; Hoffman 1989). Other researchers doubt whether Middle Proterozoic (1000 Ma) and older orogens could result from a plate tectonic regime (e.g. Hargraves 1976; Wynne-Edwards 1976; Reed *et al.* 1993). It is often argued that geothermal gradients were much higher in the Archaean than at present because heat-producing elements were much more abundant. Geological evidence for this comes from very high-temperature terrains, komatiitic lavas in greenstones and large plutonic bodies. The lithospheric plates would then have been thin and their density too low to cause buoyancy-driven subduction. However, vigorous mantle convection could have resulted in folding and faulting of the thin lithosphere (reminiscent of the contraction theory). This heat- and gravity-driven activity would indeed have triggered intraplate orogenesis. However, the argument may simply imply a secular variation in style of intraplate deformation which does not exclude plate boundary deformation, because if plates did exist in the Archaean, they would have been smaller and their motion twice as rapid as Phanerozoic rates (Sleep & Windley 1982).

Recent techniques

Since the 1960s the impact of geodynamics and geophysical data on the study of orogens has been enormous. Rapid advances in technology and increasingly powerful computers have (1) generated completely new data sets (e.g. geochronology and deep seismic profiling) which must be reconciled with more traditional field observations and (2) allowed numerical modelling of complex Earth systems and processes whose results can be compared with and constrained by factual data. Field geologists can increasingly interpret their data in terms of large scale lithospheric or crustal processes. We summarize below what we feel are the more important of these modern techniques and models and the impact they have had so far on the study of orogens.

Geochronology. Isotopic techniques permit the dating of crystalline rocks and therefore have become a prerequisite to understanding the crystalline axes of all orogens. Our knowledge, particularly in the Precambrian regions, owes much to these methods. For example, many terranes, initially recognized as Archaean from high-grade metamorphic and abundant plutonic rocks (a rock association long used to identify the Archaean) were revealed to be much younger by absolute dating techniques. A spectacular example is the so called 'Tibetan slab' believed to be Precambrian basement until its granites were dated as Miocene (e.g. Gansser 1964, 1983; Deniel *et al.* 1987) intruded into rocks that have undergone an extensive Cenozoic metamorphism (e.g. Le Fort 1975; Brunel 1986; Brunel & Kiénast 1986). In this case, the advent of absolute dating has dramatically changed the interpretation of the orogenic system and further examples can be expected from future dating of so-called microcontinents or 'Zwischengebirge' (Kober 1928) within Cenozoic belts. However different radiometric methods yield ages with different meanings (e.g. Fowler 1990, chapter 6) so that the true significance of some isotopic data is still questioned. Do isotopic ages constrain the sequential development of growing or waning stages of an orogen?

Radiometric dating has also resulted in the division of large cratons into provinces surrounded by younger orogenic belts with an age progression away from the oldest central craton (e.g. Hoffman 1988). This distribution is interpreted as recording the coalescence of arc systems, hence implying subduction and eventual collision between the provinces (e.g. DePaolo 1981). The additional information gained from dating lithospheric plates facilitates the definition of mineralogical and physical differences that may develop with time.

Seismicity and deep reflection seismic profiling: the layered lithosphere. The layered configuration of stable lithospheric plates verified by seismic studies stands as a fundamental concept that rules the mode in which an orogen may evolve. Seismic studies have used S- and P-wave velocities to recognize a layered Earth with a thermally conductive lithosphere overlying the seismic low velocity zone of the convective mantle. Thus a plate is colder and therefore more rigid than the underlying asthenosphere. Lithospheres are themselves seismically layered and consist of an upper rigid layer and a lower viscous thermal boundary layer (Parsons & McKenzie 1978). The crust is the upper part of the 100–150 km thick mechanical lithosphere. Seismic velocity–depth models simplify as one level of bulk tonalitic composition the 20–35 km thick continental crust, although geologists know it to vary greatly in lithological content. Lithospheric layering varies in young and old regions, which is particularly important for the mechanical behaviour of the plate, therefore the resistance to long term stress. It also varies in time since the crust may reach thicknesses of 60–70 km under high mountains. Less is known on the thickening processes and lower limits of the mantle lithosphere that can be subducted or thickened to 200 km such as under the Alps (Panza & Mueller 1979).

The advent of deep seismic reflection and refraction profiles has better delineated two-dimensional structural details of the crust. Since the 1970s orogens of all ages have been imaged by deep seismic reflection profiling (e.g. BIRPS in the British Isles, ECORS in France, DEKORP in Germany, COCORP in the USA, *Lithoprobe* in Canada and NFP20 in Switzerland), which has again highlighted the layered nature of the lithosphere. The bilateral symmetry of orogenic structures, often emphasized in the past, is now considered unreal. Particularly illustrative in this respect is the manner in which the Pyrenees have been portrayed before and after the ECORS deep seismic profile (Choukroune *et al.* 1990). In the Central Alps the symmetrical 'Verschluckung' models (Laubscher 1974) have been replaced by a strongly asymmetrical model as imaged on the NFP20 profiles (Fig. 2) showing subduction of the lower European crust southward into the mantle (Valasek 1992; Schmid *et al.* 1996).

Low-angle reflectors are conspicuous features

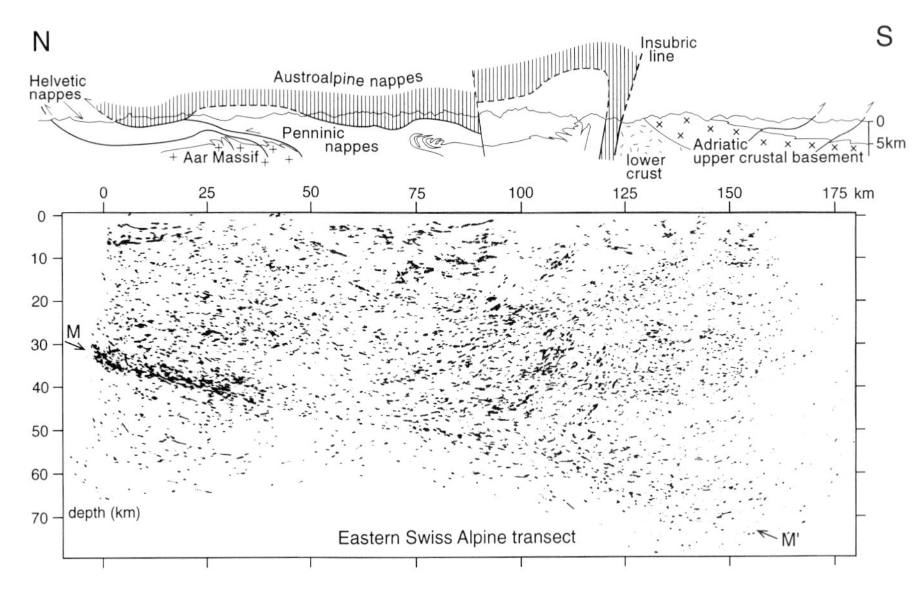

Fig. 2. Depth converted line drawing of the NFP20 deep seismic profile through the Central Alps. Adapted from Valasek (1992). The corresponding surface geology is shown above the profile. M–M′ marks the trace of the European Moho which clearly plunges down below the Adriatic plate.

identified in most orogens. They are interpreted as thrust zones possibly connected with crustal-scale décollement zones (Cook & Varsek 1994) that show the bulk asymmetry of orogens. Upper crustal faults cannot be traced downward into the ductile lower crust (e.g. Meissner 1989). Sharp irregularities of the Moho support the idea that the upper mantle is often brittle (Hirn 1988; Goleby *et al.* 1990).

Rheology: strength of the lithosphere. The mechanical properties of lithospheres have been explored from inferred mineralogical stratification, temperature gradient and pressure conditions in the plate (Kusznir & Park 1984; Ranalli & Murphy 1987). The estimated rheologies are primarily based on extrapolations of mechanical properties of minerals and rocks obtained in high-temperature high-pressure laboratory experiments (e.g. Kirby 1985; Carter & Tsenn 1987). They suggest that the cold upper continental crust (quartz–feldspar dominated) is brittle and deforms dominantly by faulting. Below 10–20 km (300–400°C), the continental crust flows and deforms by solid-state creep rather than by fracture. The uppermost part of the olivine-dominated mantle is also brittle and may undergo faulting. The lower levels of the lithospheres are ductile. Corresponding strength profiles have several maxima and minima (Carter & Tsenn 1987), which image the lithosphere as a rheological multilayer medium (Ranalli 1995, this volume; Ranalli & Murphy 1987). Temperature is the main controlling parameter, which is itself controlled by the thermo-tectonic age and composition of the crust, in particular the amount of radiogenic elements. Therefore, strength profiles differ for lithospheres of different ages (old lithospheres are 'cooler' than young lithospheres) and origin (i.e. oceanic or continental). The character of an orogen as depicted by large scale structures will obviously be dictated largely by the rheology of the lithospheres involved. It is worth noting that undisturbed Moho is known in obducted lithospheres (Nicolas 1990), which suggests that rheological layering is not necessarily coincident with this seismically defined boundary.

It is generally accepted that the lithosphere has a mechanical strength which can evolve over geological time. This strength is defined using an elastic layer whose thickness depends on thermal gradient and crustal thickness (hence age, e.g. Kusznir & Karner 1985). The effective elastic thickness of continental lithospheres has no geological reality (Burov & Diament 1995).

Lithospheric modelling of orogenesis. Two modelling approaches have been used to investigate mountain building processes: numerical and analogue modelling. Both point to the fundamental control exerted on the deformation style by the strength of the crust and its coupling with the rigid or ductile mantle at its base.

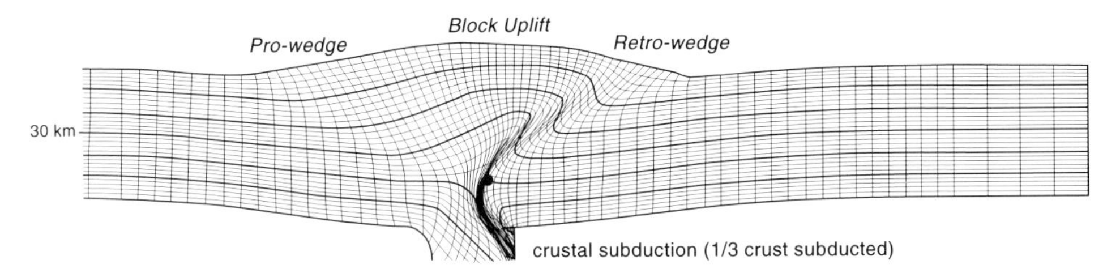

Fig. 3. Lagrangian grid showing deformation of the crust generated by a two dimensional plane strain finite element model of a small compressional orogen. Redrawn from Beaumont & Quinlan (1994, fig. 6) by kind permission of Blackwell Science, Oxford. This model represents a cold, single layer crust where one third of the crust is subducted. The crust is modelled with the rheology of wet feldspar. Decoupling of the lower one third of the left-hand crust occurs at the singularlty (black spot). The model is shown after 2 Ma and 40 km of convergence. These geometries are comparable to those seen on deep seismic lines through the Pyrenees (ECORS) and the Gulf of Bothnia, Svenofennide transects (Beaumont & Quinlan 1994) and possibly with the Central Alps (compare with Fig. 2).

Finite element models use continuum mechanics to calculate crustal deformation patterns (England & McKenzie 1983; England & Houseman 1986; Vilotte *et al.* 1986; Cloetingh *et al.* 1989). The crust is considered to have plastic and viscous rheologies. The continuum deformation calculated incorporates rheological layering and coupling between laterally uniform layers. The method has been particularly successful in understanding the dynamics of lithospheric processes in two-dimensions. Results commonly illustrate that the style of crustal deformation depends strongly on geometrical boundary conditions such as symmetry or asymmetry of mantle shortening to produce symmetrical or asymmetrical orogens, on properties of the crust and on the amount of crust that can be subducted with the mantle (Fig. 3).

Every description of collisional orogens points to folding and thrusting as the principal mechanisms of crustal thickening coeval with horizontal shortening. These structures record regional deformation but are too small to be taken into consideration in terms of thin-sheet approximation, i.e. these structures can be smoothed into a bulk homogeneous strain of the lithosphere (England & Houseman 1986; Houseman & England 1986). It frustrates geologists to know that their observations and measurements become local, mechanical anecdotes and noise when it comes to understanding collisional orogeny in terms of lithospheric deformation. First-order approximation is however the best means of appreciating how much a continental crust may thicken. Calculations show that gravitational forces increase 'non-linearly' as crustal thickness and elevation increase (England & McKenzie 1983). Both elevation and crustal thickness are buffered by the ratio of the force needed to drive convergence and gravitational forces. As a consequence, with the present-day accepted lithospheric viscosity, the crust may hardly thicken homogeneously more than 60–80 km (Dewey 1988; Molnar *et al.* 1993). Therefore, metamorphic pressures equivalent to 100 km in collisional belts (e.g. Chopin 1984; Harley & Carswell 1995; Schreyer 1995) are likely to be related to subduction of continental material below all possible Moho levels.

Lithospheric analogue models of continental shortening generally involve a strong, brittle upper crust, a weak ductile lower crust and a strong mantle floating on a fourth, low-viscosity layer simulating the asthenosphere (Davy & Cobbold 1991; Cobbold & Jackson 1992; Shemenda & Grocholsky 1992). Analogue materials are scaled and layered to reproduce the strength variability of lithospheres. They are particularly useful in investigating in three dimensions the development of large-scale lithospheric structures (Fig. 4). Relevant conclusions are (1) the thickening style depends mainly on upper mantle behaviour. Lithospheres with a ductile upper mantle (thus reduced to two-layer systems) thicken more symmetrically and homogeneously than those with a brittle upper mantle (four-layer systems), (2) the lithospheric thickness controls the width and distribution of deformation zones, (3) non-isostatically compensated buckling may occur in the initial stages of shortening and controls the strain localization with further shortening and coeval thickening, which is strongly controlled by the coupling of brittle layers through the ductile intermediate layers (Davy & Cobbold 1991).

Recent geodynamic concepts

Some new geodynamic concepts have been developed in recent years that are not fully

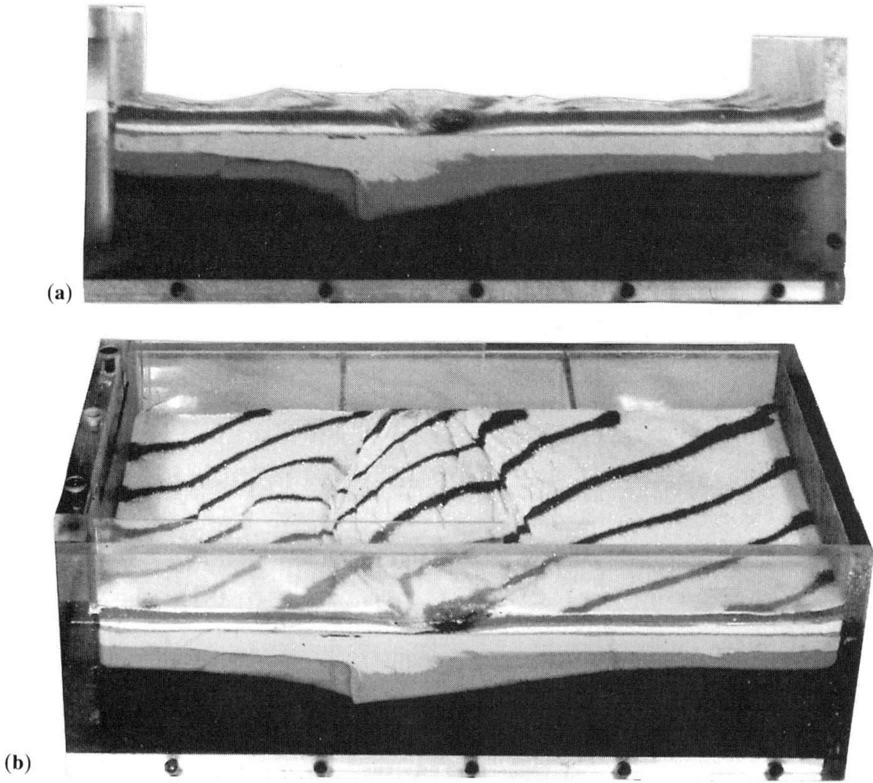

Fig. 4. An analogue model of a mountain belt after *c*. 30% shortening using a four-layer lithosphere (from Davy & Cobbold 1988). (**a**) side view of the shortened belt with a high topography and a root zone floating on a weak 'mantle'. (**b**) perspective view of the top surface of the model showing topography. Oblique stripes are passive markers to highlight thrust discontinuities and relative displacements between crustal segments.

integrated or addressed in the articles that make up this volume. These concepts deal with very large scale evolution of orogens, usually far larger than the structures recognised and mapped by geologists. We summarise some of these concepts because we believe that they have an important impact on orogenic studies and should inspire future interpretations.

Critical wedge theory and the exhumation of high pressure metamorphic rocks. The presence of very high-pressure rock units in orogens is one of the principal geodynamic problems in orogenic studies (Platt 1993). Ultrahigh-pressure crustal rocks have been found in many orogens indicating that crustal segments have been subducted to depths of 70–100 km and have then been exhumed rapidly enough to preserve the high-pressure mineral associations (Coleman & Wang 1995). It is now accepted that crustal rocks can be subducted on the downgoing slab to great depth; however the exhumation process is more problematical. Several large-scale geodynamic models for orogenic processes have been proposed in recent years that, while addressing this problem, also have a wider impact on orogenic studies. These are the critical wedge model, the doubly vergent critical wedge model, the corner flow model and the subduction channel model.

Chapple (1978) equates the deformation and displacement of a thrust sheet to a plastic layer which will deform to become a wedge-shaped continuum ahead of a rigid buttress (or backstop) and above a non-deforming, subducting slab. Davis *et al.* (1983), Dahlen *et al.* (1984) and Dahlen (1984) extend this analysis to materials with Coulomb-type (brittle) behaviour, applying their models to accretionary prisms and external fold and thrust belts of orogens. The wedge (Fig. 5) must constantly adjust in order to achieve dynamic equilibrium, i.e. to reach and maintain a stable shape (critical wedge) for which the gravitational forces balance the

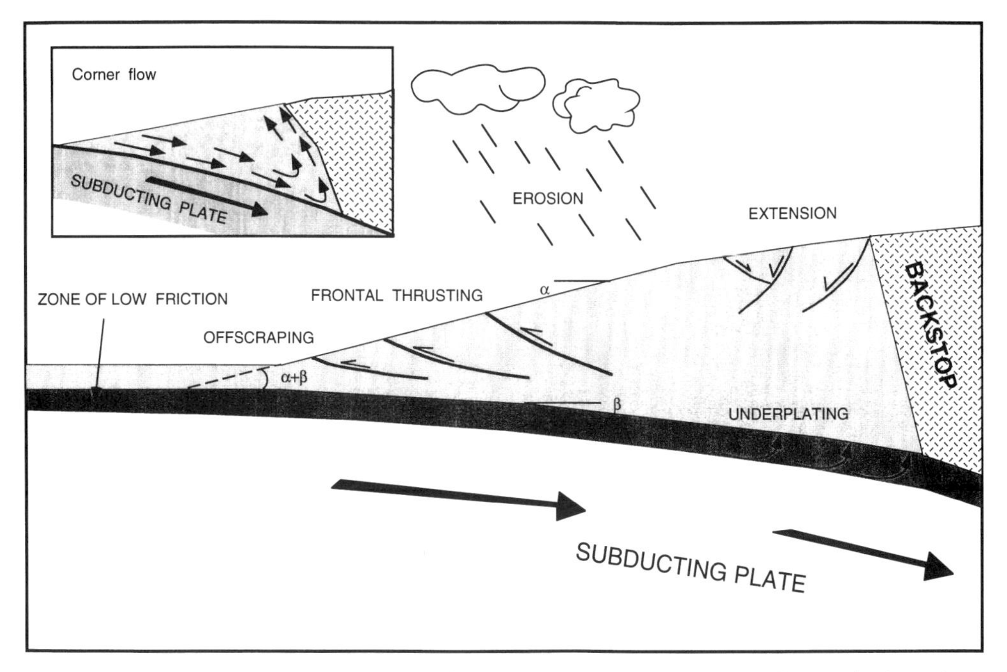

Fig. 5. Sketch representing the critical wedge model and the main processes that contribute to the dynamic system. This model can be applied to a single thrust sheet, an accretionary prism, a fold and thrust belt or to a whole orogen. α is the dip of the upper surface which will be affected by all the processes noted (see text for discussion). β is the dip of the subducting plate. Inset shows the corner flow model for a subduction complex. Arrows indicate the path followed by exhumed high pressure rocks.

traction exerted on its base by the subducting slab across a zone of low friction. The basal angle is maintained by the dip β of the subducting slab. The stable geometry generates a surface slope α in the direction of relative displacement. The 'critical taper' defines the angle at the apex of the wedge ($\alpha + \beta$), which relates the wedge shape to the compressive and gravitational forces during subduction (Davis *et al.* 1983). Material is added to the wedge by frontal offscraping (reducing α) and by underplating, the addition of material to the underside of the wedge (increasing α). Internal horizontal shortening is achieved by thrusting and folding, which produces thickening (α increases). Then the wedge may extend (decreasing α), through listric normal faulting, to regain stability. Erosion reduces α.

Platt (1986) applied the critical wedge concept to whole orogens in order to investigate the geodynamic processes by which high pressure metamorphic rocks can be exhumed during orogenic convergence. This application allows consideration of seemingly diverse processes such as the intricate association of synmetamorphic convergent deformation, decompressional extension and exhumation of high-grade rocks within a single geodynamic system. Thus the critical wedge model has come to play an important role in tectonic studies of orogens (Platt 1993).

A mechanical model for doubly-vergent orogens (such as the Alps, the Pyrenees) has been developed (Willett *et al.* 1993) which removes the need for the poorly defined backstop or buttress (Fig. 3). The model is based on numerical modelling backed up by sandbox modelling. Shortening and deformation of the crust occurs above two convergent, nearly rigid, mantle plates, one of which subducts into the asthenosphere. Many of the large scale geological processes active in orogens can be reproduced and deep seismic profiles through many orogens show the asymmetry predicted by this work (Beaumont & Quinlan 1994).

Movement of material within a subduction complex has been modelled using viscous rheologies for material caught between the subducting plate and the hanging wall buttress (Fig. 5, inset; Cowen & Silling 1978; England & Holland 1979; Emerman & Turcotte 1983). This 'corner flow' model has been proposed as a mechanism for the exhumation of high-pressure rocks although

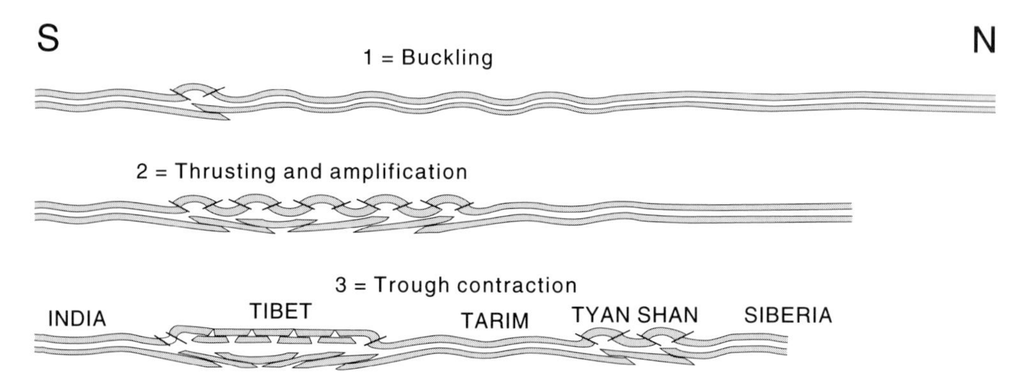

Fig. 6. Successive stages of lithospheric buckling inferred from analogue modeling used to interpret the Cenozoic intracontinental shortening/thickening of Asia after collision with India (adapted from Burg *et al.* 1994*a*). Shaded layers indicate the brittle upper crust and upper mantle. Decoupling at mid crustal level would take place along a partially molten horizon and gives rise to mechanical instabilities (buckles) with unrelated characteristic wavelengths in the lower and upper brittle layers. Conjugate thrusts, 'crocodile' wedges and Moho topography are formed.

significant incompatibilities have been found between the model and nature (Platt 1993).

The subduction channel or flow-channel model (Cloos 1982) assumes that material contained in a low viscosity mud matrix is squeezed in the wedge-like zone between a subducting plate and the overlying rigid accretionary complex. Fragments are plucked from the underside of the wedge and carried to great depth in the lower part of the flow-channel by the descending plate. In low-angle (corner) wedges, a forced and upward return flow carries this material slowly back to the surface, where it is deposited at the toe of the accretionary prism. This model can be applied to areas where high pressure metamorphic fragments are preserved within a tectonic mélange such as the Franciscan complex in California but does not explain larger scale high pressure metamorphic nappe complexes within orogenic belts (Platt 1993).

Lithospheric buckling followed by failure and thrusting. Calculations assuming an elastic plate suggest that lithospheric buckling is impossible (Turcotte & Schubert 1982). However, if the lithosphere behaves as a viscous, ductile layer floating on a weak asthenosphere, lithospheric buckling seems possible (Biot 1961; Ramberg 1970). In fact, growth of periodic lithospheric folds seems related to the plastic rheology of the 'brittle' layers of the lithosphere, with strong coupling between these brittle lithospheric layers resulting in buckling of the whole lithosphere (Stephenson & Cloetingh 1991). In compression, the wavelength of the lithospheric folds is approximately four times the thickness of the mechanical lithosphere (Martinod & Davy 1992, 1994).

Lithospheric buckling has seldom been formulated in the geological literature (Lambeck 1983; Hoffman *et al.* 1988; Stephenson *et al.* 1990; Nikishin *et al.* 1993). However it could be very important during the early stages of intraplate shortening. In particular, buckling may geometrically control the location of the crustal scale thrusts that form in the inflexion zones of the initial buckles to accommodate strong shortening (Fig. 6), and may trigger conjugate thrust faults and full ramp compressional basins in an overall asymmetric system (Cobbold *et al.* 1993; Burg *et al.* 1994a). Numerical modeling indicates that lithospheric-scale folds are expected when the lithosphere possesses a brittle-ductile stratification (Stephenson & Cloetingh 1991).

Decoupling within the lithosphere. Decoupling due to elevated pore pressure remains the best explanation for weakly deformed thrust sheets and accretionary prisms that have been displaced over long distances relative to their foot wall. The process is obviously important during subduction in separating sediments from their basement (von Huene & Lee 1983). On a larger scale, several analogue experiments show that decoupling between the upper and lower layers of continental lithosphere may be an important and underestimated feature of crustal thickening (Shemenda & Grocholsky 1992). Decoupling within the crust would indeed facilitate different structural styles in the upper, brittle layers of an orogen and in the lower, ductile root zone, even

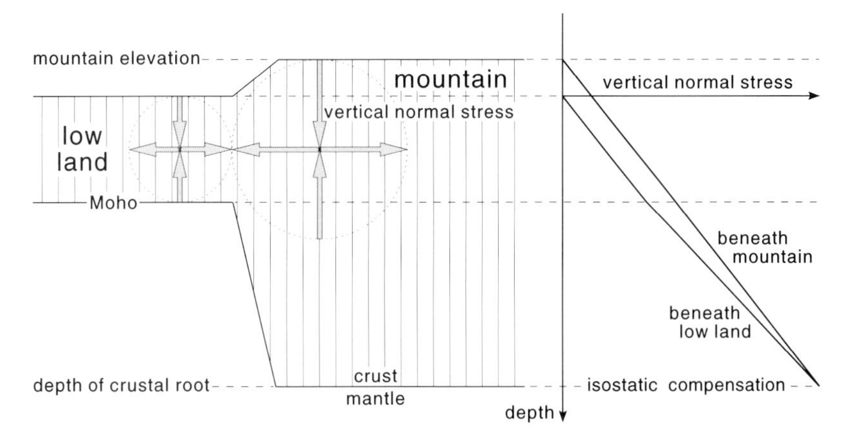

Fig. 7. Qualitative plot of vertical normal stress with depth (adapted from Bird 1991, England 1992, Tapponnier & Molnar 1976). The averaged vertical stress, equated with lithostatic pressure, is larger beneath the mountains than beneath the neighbouring lowland at any given depth in the crust. The topographic load causes collapse and expansion of elevated mountains towards low altitude forelands.

permitting antagonistic structures such as the 'crocodiles' at mid crustal level (Meissner 1989; Burg *et al.* 1994a). Subduction of the decoupled lower lithospheric layers into the mantle as imaged on deep seismic lines (e.g. Fig. 2) help to explain the balance of crustal material within collisional orogens. Décollement can occur at different levels within the lithosphere leading to variation in large-scale characteristics of orogens (Willett *et al.* 1993; Beaumont & Quinlan 1994). Direct observation of oceanic Mohos suggests that this interface is not a decoupling surface along which obduction can take place.

Late-orogenic extension – detachment of lithospheric roots. Late orogenic extension or collapse has been detected in Phanerozoic mountain belts (e.g. Wernicke 1981; Coney 1987; Dewey 1988; Andersen *et al.* 1991; Burg *et al.* 1994b) and has been suggested as an important process in the formation and exhumation of Precambrian granulite-facies terrains (Sandiford 1989). Several theoretical and mechanical considerations have shown that within a collisional plate setting, thickening of the continental lithosphere generates local internal forces and changes in horizontal forces that can produce extension (Molnar & Tapponnier 1975; England & Houseman 1989; England & Molnar 1993). In other words, shortened and consequently thickened lithospheres may undergo post-orogenic collapse to achieve equilibrium thickness (Dewey 1988). Body forces generated in high elevation regions (Fig. 7) can be important enough to develop extension in a convergent regime (Artyushkov 1973; Tapponnier & Molnar 1976; Bird 1991). However, in uniform Newtonian sheets, a marked decrease in convergence rate seems necessary before horizontal boundary forces are overcome by the buoyancy forces (England & Houseman 1989). There are regions where evidence clearly shows that convergence continued during the extensional phase and therefore decrease in rate of convergence can be eliminated as a sole cause for orogenic collapse. In these areas an increase in the potential energy of the mountain belt can be achieved by the removal of the thickened lithospheric root. There are several numerical models for the process by which the mountain root can be removed (Houseman *et al.* 1981; England & Houseman 1989; Molnar *et al.* 1993; Platt & England 1994); the convective removal of the lithospheric root and delamination of the mantle lithosphere are the two main mechanisms proposed (Fig. 8). The basic concept is that the mantle lithosphere is denser than the underlying asthenosphere. Thus a thickened lithosphere can become gravitationally unstable causing the root to detach and sink into the asthenosphere, allowing the hot asthenospheric material to rise (Fig. 8). Isostatic readjustment leads to extension and the increased geothermal gradient can lead to partial melting and the production of K_2O-rich magmas (Platt & England 1994). Lithospheric delamination below the Alboran Sea and Rif–Betic mountains is inferred from teleseismic P wave residuals (Seber *et al.* 1996).

Indenter tectonics. The Cenozoic history of Asia shows that plates are not quite rigid and

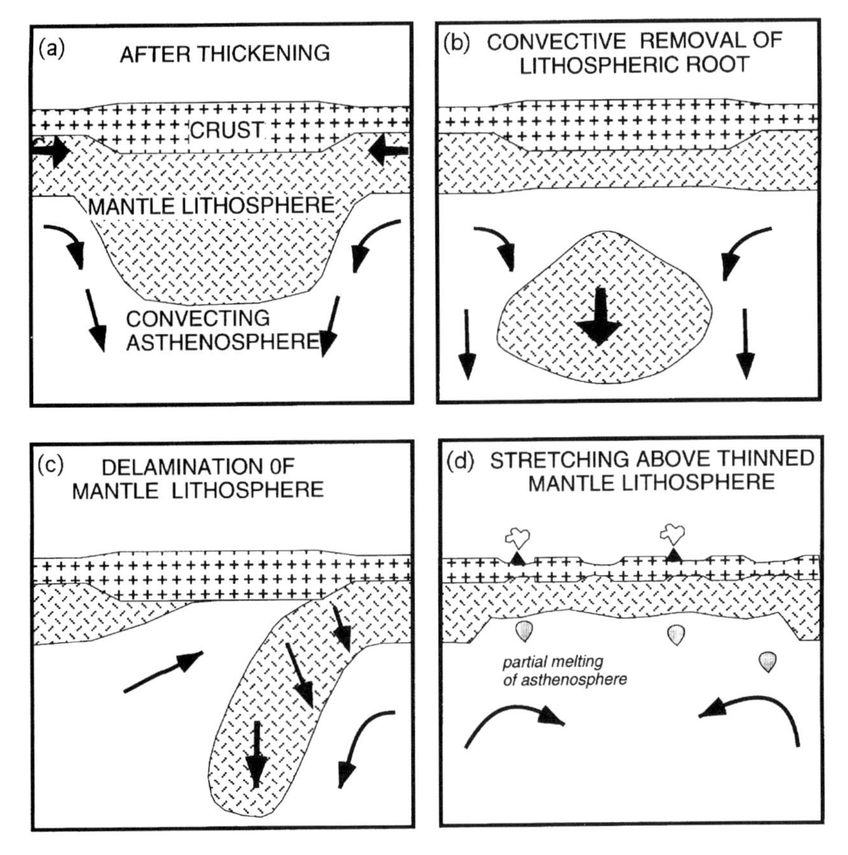

Fig. 8. Models proposed to generate orogenic collapse by removal of the lithospheric root. (**a**) Symmetrically thickened lithosphere whose root can be removed by either (**b**) convection of the underlying asthenosphere or (**c**) delamination. After the root has detached, the asthenosphere rises and re-equilibration leads to (**d**) stretching of the thinned lithosphere and possibly partial melting of the asthenosphere and K_2O-rich volcanism.

that shortening is not located only at their boundaries. Widespread deformation, and particularly wrenching within the continents continues well after plates have collided (Molnar & Tapponnier 1975; Tapponnier & Molnar 1976) and it is geologically difficult to distinguish between orogenic deformation due to active tectonics at plate boundaries and persistent intracontinental deformation. The model of a rigid block indenting into a plastic body (Molnar & Tapponnier 1975) is quite successful in describing and predicting the far stress and strain fields resulting from collision. The deformation field is expressed by slip lines whose pattern is adapted to the geometry of the deforming system (Tapponnier & Molnar 1976), a versatility that may find application to orogenic systems as complex as that of the peri-Mediterranean (Tapponnier 1977). Long lasting indentation is accommodated by large wrench faults along which lateral extrusion (escape tectonics) of continental blocks takes place towards free plate boundaries (Tapponnier *et al.* 1982). However neither the indenter nor the extrusion models address the orogenic problem proper, namely lithospheric thickening and subsequent thinning to equilibrium thickness.

Crustal thickening as a cause for topography. Geological and theoretical considerations as well as analogue modelling suggest that collision proper, that is the time of contact between the continental plates and the early deformation that follows it, does not produce high topography. For example, Eocene marine sediments as old as or younger than collision in Tibet show that the Indus–Tsangpo collision zone was below sea level (e.g. Burg & Chen 1984) at collision time. This is certainly true today in the Caribbean realm (Dercourt *et al.* 1993) and for the present day collision between Timor and Australia (Karig *et al.* 1987).

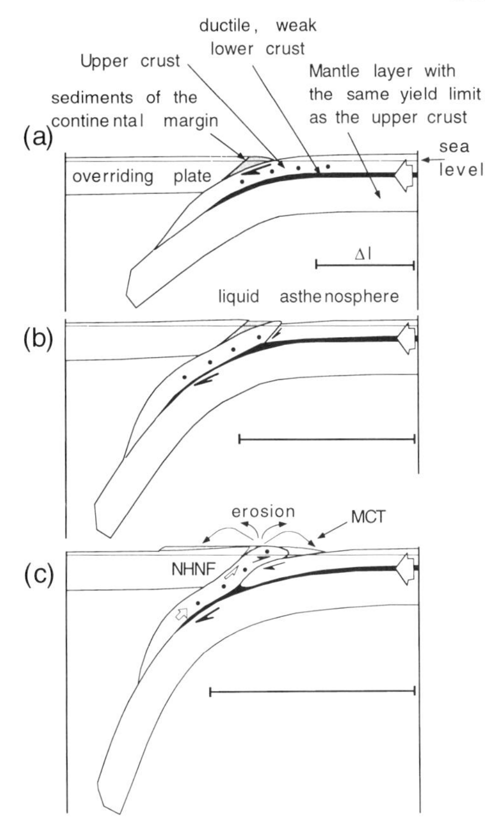

Fig. 9. Line drawings of an analogue experiment to simulate subduction of the continental lithosphere. The model is driven by a piston on the right hand side and by the drag force of the subducting lithospheric mantle. Δl represents the amount of convergence. The buoyancy force grows as the amount of continental crust subducted increases. Finally the upper crust fails and is uplifted. Uplift continues until the buoyancy forces are balanced by friction against the adjacent plates and by the weight of the part of the sheet raised above sea-level. Erosion generates further uplift. A major normal fault is formed between the uplifting block and the overriding plate. A comparison can be made with the Himalayas where the frontal thrust could be equated to the Main Central Thrust and the normal fault with the North Himalayan Normal Fault (adapted from Chemenda *et al.* 1995).

Theoretical consideration seems to account for this feature: cold roots of thickened lithospheres are denser than the warm asthenosphere at the same depth. The weight of this root causes a negative buoyancy that counters the isostatic uplift of the crust, hence diminishes the average elevation (Molnar *et al.* 1987). At this stage of collision, slab pull may rule the system. Little topography is also seen in analogue modelling of collisional zones until intracontinental thrusting appears (Fig. 9, Chemenda *et al.* 1995). Buoyancy forces may become dominant after enough continental crust has been subducted causing exhumation of high pressure metamorphic rocks (Fig. 9). Thus high mountains seem to result from intracontinental deformation that follows collision rather than collision itself. Notably, the Himalayas overlie an intracontinental thrust, the Main Central Thrust, rather than the India–Asia suture as marked by the ophiolites (Gansser 1964) and Tibet is nearly 40 Ma younger than collision (Molnar *et al.* 1993).

Orogenesis: a geodynamic overview

The question of whether orogenic processes have remained the same through geological time is directly dependant on the answer to the question: have the rheological and thermal properties of the continental lithosphere changed through geological time?

Lithospheres: their rheological variations

In this volume, **Giorgio Ranalli** discusses the thermal and rheological properties of the layered lithosphere and their importance to orogenic processes, including gravitational collapse of softened lithospheres and detachment of lithospheric roots. Thermal considerations suggest that the lithosphere has significantly changed through time, yet not enough to discard plate tectonics for the Archaean. Instead, plate tectonics probably operated at different rates. Even if plate tectonics were responsible for orogenesis, body forces exerted on hot lithospheres were also important parameters that controlled the development of Archaean orogenic structures. In other terms, there are some rheological grounds to infer that ancient orogens were not fundamentally different from modern mountain belts, varying more in their rates and structures than in the tectonic processes from which they derive.

A new dimension has been given to terrestrial geodynamics by the burgeoning research on planetary tectonics. An example of the importance of such work is given in the paper of **Pierre Thomas** and colleagues who approach the mechanics of lithospheres through the geometrical characterisitcs of impact craters. Interestingly, they show, among other points, that (1) there is a secular variation of the mechanical and thermal state of the outer layer (? lithosphere) of silicate and icy bodies, an evolution which therefore should be relevant to the Earth and (2) parameters other than lithospheric thickness should control plate tectonics. These conclusions therefore support the idea that Archaean orogeneses probably deformed lithospheres that had properties different from those measured today, which should be of some consequence for orogen distribution, duration and size through geological time.

Orogenic case histories

With the rheological approach in hand, the discussion now requires the description and comparison of orogenic systems using case studies. Conflicting interpretations for geodynamic processes in the Archaean arise partly from the techniques used to study these orogens as illustrated by the paper by **Pierre Choukroune** and colleagues, jointly written by two independant research groups. Pierre Choukroune and his team have used basic structural field techniques to study the Archaean of SW India (the Dharwar craton) while John Ludden and his group have combined surface data with deep seismic reflection data from the *Lithoprobe* project in the Superior Province of Canada. This paper perhaps demonstrates that problems in reconciling differing interpretations may in part be due to the differing approaches and data sets used.

The Proterozoic Mount Isa terrain of NE Australia (**Mark O'Dea** and collaborators) experienced repeated phases of extension and compression. This is an intracontinental orogeny generated without subduction. None of its rifting events proceeded to the breakup stage. Instead, extensive rifting accompanied by magmatism was followed by shortening and later wrench faulting associated with LP–HT metasomatism of the Isan Orogeny. There is no lateral migration of deformation and the most intense deformation, metamorphism and metasomatism are concentrated in shear zones. The rift system appears to have provided a weak lithospheric zone in which later compression and wrenching could concentrate. The source of stresses is not clear. Either they were transmitted through continental lithosphere from plate boundaries or they were generated by the underlying convective mantle.

The mid-Palaeozoic Lachlan fold belt in SE Australia (**David Gray**) developed above an accretionary plate boundary. To the east lay the long-lived active eastern plate margin of Gondwana. The Andean type orogen developed by accretion of sediments and magmatic complexes to the plate overriding a subduction zone. No lower crustal rocks seem to have been exhumed during this protracted orogeny, where a minimum of 50% shortening by a complex sequence of folding and thrusting was associated with LP–HT metamorphism and prolific granite intrusion. High-grade rocks are concentrated in a crustal block bounded by shear zones. Thrusts show no cratonic vergence and there is no metamorphic hinterland.

The geodynamic processes which contributed to the Norwegian Caledonides, an Early Palaeozoic collisional orogen, are discussed by **Geoff Milnes** and coworkers using a transect through the southern part of the orogen. Extreme crustal shortening was accommodated along low angle detachments in the upper crust while an overdeepened crustal root generated eclogites at depth. These authors argue that significant late orogenic extension which in part exhumed these high and very high pressure rocks was not due to orogenic collapse but instead due to changes in Devonian plate motions. **Patrice Rey** and associates concur with this conclusion by comparing the characteristics of late orogenic extension in the Scandinavian Caledonides and the European Variscides. They argue that while the Variscan extension was due to horizontal buoyancy forces acting on a thermally softened and thickened crust (orogenic collapse), the E–W late orogenic extension in the Scandinavian Caledonides was generated by a far distant strain field due to N–S Variscan collision.

The Late Palaeozoic Urals, an orogen which has long been poorly known in the English literature is described by **Victor Puchkov**. This orogen records a Wilson-type cycle terminating in the collision of the east European passive margin with the active Kazakhstan margin and involving intervening island arcs and crustal blocks. The asymmetrical orogen encompasses the typical features of a collisional orogen: a foreland basin (Permian), a fold and thrust belt, an ophiolitic suture and a complex metamorphic hinterland, fragments of volcanic arcs, a granitic axis.

Lamb *et al*. present a study of the Cenozoic central Andes, a subduction orogen active since the Cretaceous. The crustal thickness in places exceeds 70 km but the lithosphere is not thickened. On the contrary, the authors show, using helium isotopes, active mantle melting and thinned mantle lithosphere above the subducting plate and directly below the main body of the Andes. Thus these mountains (the Altiplano lies at *c*. 4000 m) exist in dynamic equilibrium above an active subduction zone. The crust has thickened by extensive sedimentation, volcanism and minor intrusion and shortening by thrusting and folding in the upper levels, with synchronous magmatic underplating and ductile deformation at depth. The area has gradually risen due to these processes since the Cretaceous. This case study documents extensive sedimentation within a compressional and uplifting orogen, not only in foreland basins but also in substanial intramontane basins such as the Altiplano.

Conclusions

From a geodynamic point of view, orogenesis encompasses all processes that result in crustal thickening, causing uplift and hence high topography. Whether the term implies thickening of the whole lithosphere is an obvious and crucial point of contention. Although plate tectonics is the main agent of orogenesis, continental crust can thicken by horizontal shortening or by vertical processes such as magmatic underplating and granite intrusion (e.g. Dewey & Bird 1970; Wells 1980; White & McKenzie 1989).

It is clear from these contributions that geodynamic concepts, modelling and new techniques such as deep seismic profiling and isotopic methods are important but must always be combined with field observations. Such interdisciplinary work defines the new stage in orogenic studies which has succeeded the plate tectonic revolution. It is also clear that orogenesis is not only equated with the Wilson cycle (where a subduction orogen may precede obduction, followed by full closure of the ocean and collision), but can also occur within continental plates.

Accretion that results from a collage of island arcs and accretionary prisms juxtaposed to an active continental margin (Irving *et al.* 1980; Ben-Avraham *et al.* 1981) also illustrates non-Wilson-cycle orogenesis. Large horizontal strike slip between geological units (suspect terranes) does not produce regional thickening. 'Orogeny' must result from the partitioned compressional component of oblique convergence.

The inference from both observation and modelling is that there must be subduction involved in orogens formed at plate boundaries. Therefore, these orogenic structures are fundamentally asymmetric, usually synthetic with the subducting plate. The two contrasting subduction types, A-type for continental subduction and B-type for oceanic subduction (Oxburgh 1972) may control, for understandable reasons of buoyancy, two types and/or stages of orogenesis (e.g. Tibet and Oman respectively). Theoretically, the continental lithosphere is not dense enough to be significantly subducted into the mantle. However, geological restorations are too controversial to estimate the amount that has been truly subducted in an orogen. This basic problem has received little attention from geophysicists and the presently accepted values range from a few tens to a few hundred kilometres (Molnar & Gray 1979). The amount of subducted continental lithosphere remains an important research direction for students of orogenesis because gravity, expressed through the negative buoyancy of subducted material, causes topographically high mountains and their eventual collapse.

There seems to be no rule concerning the time spanned by an orogeny. Modern examples teach us that plate motions that produce mountain ranges may last for several hundred million years. This is shown by the, at least, 200 Ma old subduction zones along the western margins of North and South America. Alternatively, several cumulative short-lived phases of motion may also produce mountains. This seems to be illustrated by the Alps. Hence, differences in orogenic structures may be a question of duration as much as a classification problem.

Did orogenic processes evolve and change through time? Weaker bulk rheology of lithospheric plates was probably responsible for the wider distribution of intraplate deformation in Archaean and Lower Proterozoic times. The abundance of anatectic granites in Archaean and Early Proterozoic orogens may have been due to the large width/length aspect ratio of these orogens. As demonstrated in younger orogens, horizontal heat transfer leading to anatexis in the thickened crust is more efficient in wider belts (Gaudemer *et al.* 1988). There is no reason for excluding orogenesis at plate boundaries during early Earth history. Younger orogens probably evolved over longer periods and formed narrower belts along longer plate margins. The larger scale of tectonic plates in the Phanerozoic thus led to orogenic belts of more dramatic dimensions. However, this difference in scale probably had little effect on the major geodynamic processes involved in orogenesis.

We thank the participants of the 'Orogeny through time' seminar series in 1994 and our colleagues at ETH for stimulating discussions on orogenesis. We also thank N. Kusznir for helpful discussion on the topic. Reviewers are thanked for their contributions.

References

ANDERSEN, T. B., JAMTVEIT, B., DEWEY, J. F. & SWENSSON, E. 1991. Subduction and eduction of continental crust : major mechanisms during continent-continent collision and orogenic ex-

tensional collapse, a model based on the south Norwegian Caledonides. *Terra Nova* **3**, 303–310.
ARGAND, E. 1924. La tectonique de l'Asie. *Compte Rendu de la 13ème Congres Géologique International, Bruxelles*, **1**, 171–372.
ARTYUSHKOV, E. V. 1973. Stresses in the lithosphere caused by crustal thickness inhomogeneities. *Journal of Geophysical Research*, **78**, 7675–7708.
BEAUMONT, C. & QUINLAN, G. 1994. A geodynamic framework for interpreting crustal-scale seismic-reflectivity patterns in compressional orogens. *Geophysical Journal International*, **116**, 754–783.
BEN-AVRAHAM, Z., NUR, A., JONES, D. & COX, A. 1981. Continental accretion and orogeny : From oceanic plateaus to allochthonous terranes. *Science*, **213**, 47–54.
BIOT, M. A. 1961. Theory of folding of stratified viscoelastic media and its implications in tectonics and orogenesis. *Geological Society of America Bulletin*, **72**, 1595–1620.
BIRD, P. 1991. Lateral extrusion of lower crust from under high topography, in the isostatic limit. *Journal of Geophysical Research*, **96**, 10,275–10,286.
BRUNEL, M. 1986. Ductile thrusting in the Himalayas: shear sense criteria and stretching lineations. *Tectonics*, **5**, *2*, 247–265.
—— & KIÉNAST, J.-R. 1986. Etude pétro-structurale des chevauchements ductiles himalayens sur la transversale de l'Everest-Makalu (Népal oriental). *Canadian Journal of Earth Sciences*, **23**, 1117–1137.
BURG, J.-P. & CHEN, G. M. 1984. Tectonics and structural zonation of southern Tibet, China. *Nature* **311**, 219–223.
——, DAVY, P. & MARTINOD, J. 1994*a*. Shortening of analogue models of the continental lithosphere: New hypothesis for the formation of the Tibetan plateau. *Tectonics*, **13**, 475–483.
——, VAN DEN DRIESSCHE, J. & BRUN, J.-P. 1994*b*. Syn- to post-thickening extension in the Variscan Belt of Western Europe: Mode and structural consequences. *Géologie de la France*, **3**, 33–51.
BUROV, E. B. & DIAMENT, M. 1995. The effective elastic thickness (T*e*) of continental lithosphere: What does it really mean? *Journal of Geophysical Research*, **100**, 3905–3927.
CADY, W. M. 1950. Classification of geotectonic elements. *American Geophysical Union Transactions*, **31**, 780–785.
CARTER, N. L. & TSENN, M. C. 1987. Flow properties of continental lithosphere. *Tectonophysics* **136**, 27–63.
CHAPPLE, W. M. 1978. Mechanics of thin-skinned fold-and-thrust belts. *Geological Society of America Bulletin*, **89**, 1189–1198.
CHEMENDA, A. I., MATTAUER, M., MALAVIEILLE, J. & BOKUN, A. N. 1995. A mechanism for syn-collisional rock exhumation and associated normal faulting: Results from physical modelling. *Earth and Planetary Sciences Letters*, **132**, 225–232.
CHOPIN, C. 1984. Coesite and pure pyrope in high-grade pelitic blueschists of the Western Alps: a first record and some consequences. *Contribution to Mineralogy and Petrology*, **86**, 107–118.
CHOUKROUNE, P. & ECORS PYRENEES TEAM 1990. The ECORS Pyrenean deep seismic profile. Reflection data and the overall structure of an orogenic belt. *Tectonics*, **8**, 23–39.
CLOETINGH, S., WORTEL, R. & VLAAR, N. J. 1989. On the initiation of subduction zones. *Pure and Applied Geophysics*, **129**, 7–25.
CLOOS, M. 1982. Flow melanges: Numerical modeling and geologic constraints on their origin in the Franciscan subduction complex, California. *Geological Society of America Bulletin*, **93**, 330–345.
COBBOLD, P. R. & JACKSON, M. P. A. 1992. Gum resin (colophony): a suitable material for thermomechanical modelling of the lithosphere. *Tectonophysics*, **210**, 255–271.
——, DAVY, P., GAPAIS, D., ROSSELLO, E. A., SADYBAKASOV, E., THOMAS, J. C., TONDJI BIYO, J. J. & DE URREIZTIETA, M. 1993. Sedimentary basins and crustal thickening. *Sedimentary Geology*, **86**, 77–89.
COLEMAN, R. G. & WANG, X. 1995. Overview of the geology and tectonics of UHPM. *In*: COLEMAN, R. G. & WANG, X. (eds) *Ultrahigh pressure metamorphism*. Cambridge University Press, Cambridge, 1–32.
CONDIE, K. C. 1982. *Plate tectonics and crustal evolution*. Pergamon Press.
CONEY, P. J. 1970. The geotectonic cycle and the new global tectonics. *Geological Society of America Bulletin*, **81**, 739–748.
—— 1987. The regional tectonic setting and possible causes of Cenozoic extension in the North American Cordillera. *Geological Society Special Publication*, **28**, 177–186.
COOK, F. A. & VARSEK, J. L. 1994. Orogen-scale décollements. *Review of Geophysics*, **32**, 37–60.
COWEN, D. D. & SILLING, R. M. 1978. A dynamic scaled model of accretion at trenches and its implications for the tectonic evolution of subduction complexes. *Journal of Geophysical Research*, **83**, 5389–5396.
DAHLEN, F. A. 1984. Non-cohesive critical Coulomb wedges: an exact solution. *Journal of Geophysical Research*, **89**, 10,125–10,133.
——, SUPPE, J. & DAVIS, D. 1984. Mechanics of fold-and-thrust belts and accretionary wedges: Cohesive Coulomb theory. *Journal of Geophysical Research*, **89**, 10,087–10,101.
DAVIS, D., SUPPÉ, J. & DAHLEN, F. A. 1983. Mechanics of fold-and-thrust belts and accretionary wedges. *Journal of Geophysical Research*, **88**, 1153–1172.
DAVY, P. & COBBOLD, P. 1991. Experiments on shortening of a 4-layer model of the continental lithosphere. *Tectonophysics*, **188**, 1–25.
—— & —— 1988. Indentation tectonics in nature and experiment, 1. Experiments scaled for gravity. *Bulletin of the Geological Institute, University of Uppsala, New Series*, **14**, 129–141.
DENIEL, C., VIDAL, P., FERNANDEZ, A., LE FORT, P. & PEUCAT, J.-J. 1987. Isotopic study of the Manaslu granite (Himalaya, Nepal); inference on the age

and source of Himalayan leucogranites. *Contributions to Mineralogy and Petrology,* **96**, 78–92.
DENNIS, J. G. 1967. International tectonic dictionary. *American Association of Petroleum Geologists Memoirs,* **7**.
DEPAOLO, D. J. 1981. Nd isotope studies; Some new perspectives on earth structure and evolution. *EOS Transaction of the American Geophysical Union,* **62**, 137–140.
DERCOURT, J., RICOU, L.-E. & VRIELYNCK, B. 1993. *Atlas Tethys palaeoenvironmental maps.* Gauthier-Villars.
DEWEY, J. F. 1988. Extensional collapse of orogens. *Tectonics,* **7**, 1123–1139.
—— & BIRD, J. M. 1970. Mountain belts and the new global tectonics. *Journal of Geophysical Research,* **75**, 2625–2647.
—— & BURKE, K. 1973. Tibetan, Variscan and Precambrian basement reactivation: Products of continental collision. *Journal of Geology,* **81**, 683–692.
DICKINSON, W. R. 1971. Plate tectonic models of geosynclines. *Earth and Planetary Sciences Letters,* **10**, 165–174.
DIETZ, R. S. 1972. Geosynclines, mountains and continent-building. *Scientific American,* **226**, 30–38.
EMERMAN, S. H. & TURCOTTE, D. L. 1983. A fluid model for the shape of accretionary prisms. *Earth and Planetary Sciences Letters,* **56**, 387–397.
ENGLAND, P. 1992. Deformation of the continental crust. *In*: BROWN, G., HAWKESWORTH, C. & WILSON, C. (eds) *Understanding the Earth. A new synthesis.* Cambridge University Press, Cambridge, 275–300.
—— & HOLLAND, T. J. B. 1979. Archimedes and the Tauern eclogites: The role of buoyancy in the preservation of exotic eclogite blocks. *Earth and Planetary Sciences Letters,* **44**, 287–294.
—— & HOUSEMAN, G. A. 1986. Finite strain calculations of continental deformation 2. Comparison with the India-Asia collision zone. *Journal of Geophysical Research,* **91**, 3664–3676.
—— & —— 1989. Extension during continental convergence, with application to the Tibetan plateau. *Journal of Geophysical Research,* **94**, 17,561–17,579.
—— & MCKENZIE, D. P. 1983. Correction to: A thin viscous sheet model for continental deformation. *Geophysical Journal of the Royal Astronomical Society,* **73**, 523–532.
—— & MOLNAR, P. 1993. Cause and effect among thrust and normal faulting, anatectic melting and exhumation in the Himalaya. *In*: TRELOAR, P. J. & SEARLE, N. P. (eds) *qv*. 401–411.
FOWLER, C. M. R. 1990. *The solid Earth.* Cambridge University Press.
GANSSER, A. 1964. *Geology of the Himalayas.* Interscience.
—— 1983. Geology of the Bhutan Himalaya. *Denkschriften der Schweizerischen Naturforschenden Gesellschaft,* **96**, 1–181.
GAUDEMER, Y., JAUPART, C. & TAPPONNIER, P. 1988. Thermal control on post-orogenic extension in collision belts. *Earth and Planetary Science Letters,* **89**, 48–62.
GILBERT, G. K. 1890. *Lake Bonneville.* US Geological Survey Monographs, **1**.
GILLULY, J. 1973. Steady plate motions and episodic orogeny and magmatism. *Geological Society of America Bulletin,* **84**, 499–514.
GOLEBY, B. R., KENNETT, B. L. N., WRIGHT, C., SHAW, R. D. & LAMBECK, K. 1990. Seismic reflection profiling in the Proterozoic Arunta Block, central Australia: processing for testing models of tectonic evolution. *Tectonophysics,* **173**, 257–268.
HARGRAVES, R. B. 1976. Precambrian geologic history. *Science,* **193**, 363–371.
HARLEY, S. L. & CARSWELL, D. A. 1995. Ultradeep crustal metamorphism: A prospective view. *Journal of Geophysical Research,* **100**, 8367–8380.
HIRN, A. 1988. Features of the crust-mantle structure of Himalayas-Tibet: a comparison with seismic traverses of Alpine, Pyrenean and Variscan orogenic belts. *Philosophical Transactions of the Royal Society of London,* **A 326**, 17–32.
HOFFMAN, P. F. 1988. United plates of America, the birth of a craton: early Proterozoic assembly and growth of Laurentia. *Annual Review of Earth and Planetary Sciences,* **16**, 543–603.
—— 1989. Precambrian geology and tectonic history of North America. *In*: BALLY, A. B. & PALMER, A. R. (eds) *The geology of North America; An overview.* Geological Society of America, Boulder, 447–512.
——, TIRRUL, R., KING, J. E., ST-ONGE, M. R. & LUCAS, S. B. 1988. Axial projections and modes of crustal thickening, eastern Wopmay orogen, northwest Canadian shield. *In*: CLARK, S. P. JR (ed.) *Processes in Continental Lithospheric Deformation.* Geological Society of America, Special Papers, **218**, 1–29.
HOUSEMAN, G. A. & ENGLAND, P. C. 1986. Finite strain calculations of continental deformation 1. Method and general results for convergent zones. *Journal of Geophysical Research,* **91**, 3651–3663.
——, MCKENZIE, D. P. & MOLNAR, P. 1981. Convective instability of a thickened boundary layer and its relevance for the thermal evolution of continental convergent belts. *Journal of Geophysical Research,* **86**, 6115–6132.
IRVING, E. J., MONGER, W. H. & YOLE, R. W. 1980. New paleomagnetic evidence for displaced terranes in British Columbia. *Special Papers of the Geological Association of Canada,* **20**, 441–456.
ISACKS, B., OLIVER, J. & SYKES, L. R. 1968. Seismology and the new global tectonics. *Journal of Geophysical Research,* **73**, 5855–5899.
KARIG, D. E., BARBER, A. J., CHARLTON, T. R., KLEMPERER, S. & HUSSONG, D. M. 1987. Nature and distribution of deformation across the banda-Arc-Australian collision zone at Timor. *Geological Society of America Bulletin,* **98**, 18–32.
KIRBY, S. H. 1985. Rock mechanics observations pertinent to the rheology of the continental lithosphere and the localization of strain along shear zones. *Tectonophysics,* **119**, 1–27.

KOBER, L. 1928. *Der Bau der Erde. Zweite Auflage*. Borntraeger.
KRÖNER, A. 1981. Precambrian plate tectonics. *In*: KRÖNER, A. (ed.) *Precambrian plate tectonics*. Elsevier, Amsterdam, 56–90.
KUSZNIR, N. & KARNER, G. 1985. Dependence of the flexural rigidity of the continental lithosphere on rheology and temperature. *Nature,* **316**, 139–142.
—— & PARK, R. G. 1984. Intraplate lithosphere strength and heat flow. *Geophysical Journal of the Royal Astronomical Society,* **79**, 513–538.
LAMBECK, K. 1983. Structure and evolution of the intracratonic basins of central Australia. *Geophysical Journal of the Royal Astronomical Society,* **74**, 843–886.
LAUBSCHER, H. 1974. The tectonics of subduction in the Alpine system. *Memoire delta Società geologica Italiana* **13**, suppl. 2, 275–283.
LE FORT, P. 1975. The collided range. Present knowledge of the continental arc. *American Journal of Science,* **275-A**, 1–44.
LE PICHON, X. 1968. Sea-floor spreading and continental drift. *Journal of Geophysical Research,* **73**, 3661–3697.
MARTINOD, J. & DAVY, P. 1992. Periodic instabilities during compression or extension of the lithosphere 1. Deformation modes from an analytical perturbation method. *Journal of Geophysical Research,* **97**, 1999–2014.
—— & —— 1994. Periodic instabilities during compression of the lithosphere 2. Analogue experiments. *Journal of Geophysical Research,* **99**, 12,057–12,069.
MCKENZIE, D. P. & PARKER, R. L. 1967. The North Pacific: an example of tectonics on a sphere. *Nature,* **216**, 1276–1280.
MEISSNER, R. 1989. Rupture, creep lamellae and crocodiles: happenings in the continental crust. *Terra Nova,* **1**, 17–28.
MITCHELL, A. H. & READING, H. G. 1969. Continental margins, geosynclines and ocean floor spreading. *Journal of Geology,* **77**, 629–646.
MIYASHIRO, A., AKI, K. & SENGÖR, A. M. C. 1982. *Orogeny*. John Wiley & Sons.
MOLNAR, P. & GRAY, D. 1979. Subduction of continental lithosphere: Some constraints and uncertainties. *Geology,* **7**, 58–62.
—— & TAPPONNIER, P. 1975. Cenozoic tectonics of Asia: Effects of a continental collision. *Science,* **189**, 419–426.
——, BURCHFIEL, B. C., LIANG, K. & ZHAO, Z. 1987. Geomorphic evidence for active faulting in the Altyn Tagh and northern Tibet and qualitative estimates of its contribution to the convergence of India and Eurasia. *Geology,* **15**, 249–253.
——, ENGLAND, P. & MARTINOD, J. 1993. Mantle dynamics, uplift of the Tibetan plateau, and the Indian monsoon. *Review of Geophysics,* **31**, 357–396.
MORGAN, W. J. 1968. Rises, trenches, great faults, and crustal blocks. *Journal of Geophysical Research,* **73**, 1959–1982.
NICOLAS, A. 1990. *Les montagnes sous la mer*. Editions du BRGM.
NIKISHIN, A. M., CLOETINGH, S., LOBKOVSKY, L. I., BUROV, E. B. & LANKREIJER, A. C. 1993. Continental lithosphere folding in Central Asia (Part I): constraints from geological observations. *Tectonophysics,* **226**, 59–72.
OXBURGH, E. R. 1972. Flake tectonics and continental collision. *Nature,* **239**, 202–215.
PANZA, G. F. & MUELLER, S. 1979. The plate boundary between Eurasia and Africa in the Alpine area. *Memorie de Scienze Geologiche,* **33**, 43–50.
PARSONS, B. & MCKENZIE, D. 1978. Mantle convection and the thermal structure of the plates. *Journal of Geophysical Research,* **83**, 4485–4496.
PFIFFNER, O. A., LEHNER, P., HEITZMANN, P., MUELLER, S. & STECK, A. (eds) 1996. *Deep structure of the Swiss Alps – results of the National Research Program 20 (NFP20)*. Birkhäuser, Basel.
PLATT, J. P. 1986. Dynamics of orogenic wedges and the uplift of high-pressure metamorphic rocks. *Geological Society of America Bulletin,* **97**, 1037–1053.
—— 1993. Exhumation of high-pressure rocks: a review of concepts and processes. *Terra Nova,* **5**, 119–133.
—— & ENGLAND, P. C. 1994. Convective removal of lithosphere beneath mountain belts: thermal and mechanical consequences. *American Journal of Science,* **294**, 307–336.
RAMBERG, H. 1970. Folding of laterally compressed multilayers in the field of gravity, II, Numerical examples. *Physics of the Earth and Planetary Interiors* **4**, 83–120.
RANALLI, G. 1995. *Rheology of the Earth*. Chapman & Hall.
—— & MURPHY, D. C. 1987. Rheological stratification of the lithosphere. *Tectonophysics,* **132**, 281–295.
REED, J. C. J., BALL, T. T., LANG FARMER, G. & HAMILTON, W. B. 1993. A broader view. *In*: REED, J. C. JR., BICKFORD, M. E., HOUSTON, R. S., LINK, P. K., RANKIN, D. W., SIMS, P. K., VAN SCHMUS, W. R. (eds) *The Geology of North America. Precambrian: Conterminous US*. Geological Society of America, Boulder, 597–636.
ROBERTSON, A. H. F., SEARLE, M. P. & RIES, A. C. (eds) 1990. *The geology and tectonics of the Oman region*. Geological Society, London, Special Publications, **49**.
ROURE, F., HEITZMANN, P. & POLINO, R. (eds) 1990. *Deep structure of the Alps*, Mémoires de la Société géologique de Suisse, **1**, Zürich.
SANDIFORD, M. 1989. Horizontal structures in granulite terrains: A record of mountain building or mountain collapse? *Geology* **17**, 449–452.
SCHMID, S. M., FROITZHEIM, N., PFIFFNER, A., SCHÖNBORN, G. & KISSLING, E. 1996. Geophysical-geological transect and tectonic evolution of the Swiss-Italian Alps. *Tectonics*, in press.
SCHREYER, W. 1995. Ultradeep metamorphic rocks: The retrospective viewpoint. *Journal of Geophysical Research,* **100**, 8353–8366.
SEBER, D., BARAZANGI, M., IBENBRAHIM, A. & DEMNATI, A. 1996. Geophysical evidence for lithospheric delamination beneath the Alboran

Sea and Rif-Betic mountains. *Nature,* **379**, 785–790.

SHEMENDA, A. I. & GROCHOLSKY, A. L. 1992. Physical modelling of lithosphere subduction in collision zones. *Tectonophysics,* **216**, 273–290.

SLEEP, N. H. & WINDLEY, B. F. 1982. Archean plate tectonics; Constraints and inferences. *Journal of Geology,* **90**, 363–379.

STEPHENSON, R. A. & CLOETINGH, S. A. P. L. 1991. Some examples and mechanical aspects of continental lithospheric folding. *Tectonophysics,* **188**, 27–37.

——, RICKETTS, B. D., CLOETINGH, S. A. P. L. & BEEKMAN, F. 1990. Lithosphere folds in the Eurekan Orogen, Arctic Canada? *Geology,* **18**, 103–106.

TAPPONNIER, P. 1977. Evolution tectonique du système alpin en Méditerranée: poinçonnement et écrasement rigide-plastique. *Bulletin de la Société géologique de France,* **7–19**, 437–460.

—— & MOLNAR, P. 1976. Slip-line field theory and large-scale continental tectonics. *Nature,* **264**, 319–324.

——, PELTZER, G., LE DAIN, A. Y., ARMIJO, R. & COBBOLD, P. 1982. Propagating extrusion tectonics in Asia: New insights from simple experiments with plasticine. *Geology,* **10**, 611–616.

TRELOAR, P. J. & SEARLE, M. P. (eds) 1993. *Himalayan Tectonics.* Geological Society, London, Special Publications **74**.

TURCOTTE, D. L. & SCHUBERT, G. 1982. *Geodynamics: applications of continuum physics to geological problems.* John Wiley & Sons.

VALASEK, P. 1992. *The tectonic structure of the Swiss Alpine crust interpreted from a 2D network of deep crustal seismic profiles and an evaluation of 3D effects.* PhD, ETH, Zürich.

VILOTTE, J. P., MADARIAGA, R., DAIGNIÈRES, M. & ZINKIEWICZ, O. 1986. Numerical study of continental collision: influence of buoyancy forces and an initial stiff inclusion. *Geophysical Journal of the Royal Astronomical Society,* **84**, 279–310.

VON HUENE, R. & LEE, H. 1983. The possible significance of pore fluid pressures in subduction zones. *In*: WATKINS, J. S. & DRACE, C. L. (eds) *Continential Margin Geology.* American Association of Petroleum Geologists Memoirs, **34**, 781–791.

WEBER, K. 1984. Variation in tectonic style with time (Variscan and Proterozoic systems). *In*: HOLLAND, H. D. & TRENDALL, A. F. (eds) *Patterns of change in Earth evolution.* Springer-Verlag, Berlin, 371–386.

WEGENER, A. 1912. Die Entstehung der Kontinente. *Geologische Rundschau,* **3**, 276–292.

WELLS, P. R. A. 1980. Thermal models for the magmatic accretion and subsequent metamorphism of continental crust. *Earth and Planetary Science Letters,* **46**, 253–265.

WERNICKE, B. 1981. Low-angle normal faults in the Basin and Range Province: nappe tectonics in an extending orogen. *Nature,* **291**, 645–648.

WHITE, R. S. & MCKENZIE, D. P. 1989. Magmatism at rift zones: The generation of volcanic continental margins and flood basalts. *Journal of Geophysical Research,* **94**, 7685–7729.

WILLETT, S., BEAUMONT, C. & FULLSACK, P. 1993. Mechanical model for the tectonics of doubly vergent compressional orogens. *Geology,* **21**, 371–374.

WILSON, J. T. 1966. Did the Atlantic close and then reopen? *Nature,* **211**, 676–681.

WYNNE-EDWARDS, H. R. 1976. Proterozoic ensialic orogenesis: the millipede model of ductile plate tectonics. *American Journal of Science,* **276**, 927–953.

ZWART, H. J. 1967. The duality of orogenic belts. *Geologie en Mijnbouw,* **46**, 283–309.

Rheology of the lithosphere in space and time

GIORGIO RANALLI
Department of Earth Sciences and Ottawa-Carleton Geoscience Centre, Carleton University, Ottawa K1S 5B6, Canada

Abstract: The rheology of the lithosphere is a factor of primary importance in the kinematics and dynamics of mountain belts. This paper attempts to clarify the role of rheology in orogenesis, by applying simple physical principles to the analysis of tectonic processes. The emphasis is on broad generalizations leading to order-of-magnitude estimates. Since the rheology of lithospheric materials is strongly dependent on temperature, the discussion opens with a review of continental and oceanic geotherms and an assessment of their reliability. Then the brittle (frictional) and ductile (high-temperature creep) properties of the lithosphere are considered. In the brittle field, particular attention is paid to the problem of fault reactivation, which is shown to be more likely in extensional than in compressional regimes. In the ductile field, a summary of creep parameters for the most common lithospheric materials is presented. The central concept of rheological profiles (strength envelopes), essential to the estimation of the depth variations of lithospheric rheology, is discussed with reference not only to its applicability but also to its limitations (a two-dimensional example from the Canadian Cordillera is given). Processes related to the rheological properties and layering of the lithosphere – gravitational collapse in thickened and softened crust, tectonic inversion following the detachment of a lithospheric root, lower crustal ductile flow with consequent relaxation of Moho topography – are analysed semi-quantitatively, mainly to show that they are indeed likely to be important players in geodynamics. Finally, in a brief Archaean detour, some possible important differences between present-day and Archaean oceanic lithosphere are examined, and the conclusion is reached that in all likelihood plate tectonics, although at different rates, has been the main agent of orogenesis during most of the history of the planet.

Orogeny is the result of tectonic forces (ultimately related to planetary processes such as mantle convection) acting upon the lithosphere that, from the material science viewpoint, consists of polycrystalline aggregates with position- and time-dependent rheological properties. In principle, therefore, analysis of dynamic processes in orogenic belts requires the specification of a *rheological function*

$$R(\sigma, \dot{\sigma}, \ldots, \epsilon, \dot{\epsilon}, \ldots, \{M_i\}, \{S_i\}) = 0 \quad (1)$$

where σ and ϵ are stress and strain (tensor indices are omitted), an overdot denotes time derivative, and $\{M_i\}$, $\{S_i\}$, $i = 1, 2, \ldots, n$ are material parameters (e.g. elastic moduli, frictional properties, viscosity) and state variables (e.g. grain size, subgrain size, dislocation density), respectively. Temperature and pressure enter equation (1) both explicitly in the function R and implicitly through their effect on material parameters and state variables. While the pressure dependence can be neglected to the first order in the lithosphere (the pressure P is small compared to the bulk modulus k, $P/k \lesssim 0.02$), the temperature dependence is one of the fundamental factors, along with type of material and deformation rate, in determining the rheological behaviour.

This paper discusses some of the interactions between tectonic forces and rheological properties of the lithosphere. The aim is to clarify the role of rheology in orogeny, with some examples, rather than considering in detail the deformation mechanisms. Since temperature exerts a controlling influence on rheology, the first section gives a brief account of the thermal state of the continental and oceanic lithospheres. Then the low-temperature (brittle–frictional) rheology of the lithosphere is discussed, with special attention to the problem of fault reactivation. The third section summarizes the present knowledge on the ductile rheology of crustal and upper mantle materials. The concept of strength envelopes, or rheological profiles, is discussed next, with emphasis not only on its first-order applicability but also on its limi-

From Burg, J.-P. & Ford, M. (eds), 1997, *Orogeny Through Time*,
Geological Society Special Publication No. 121, pp. 19–37.

tations, with some examples from the Canadian Cordillera. Some processes related to tectonic inversion and to delamination are analysed in the following section. Finally, possible differences between early Precambrian and present lithosphere rheology are examined, to assess whether plate tectonics (in particular, negative-buoyancy subduction) was operative in the Archaean.

Temperature in the lithosphere

The conductive heat transfer equation, neglecting hydrothermal fluid convection, which is important only in specific areas, is (see Carslaw & Jaeger 1959, Turcotte & Schubert 1982 and Ranalli 1995 for detailed derivation)

$$\frac{\partial}{\partial x}(\kappa\frac{\partial T}{\partial x}) + \frac{\partial}{\partial z}(\kappa\frac{\partial T}{\partial z}) + \frac{A}{\rho c} = \frac{\partial T}{\partial t} + v\frac{\partial T}{\partial x} \quad (2)$$

where T is temperature, x and z spatial coordinates (x horizontal, z positive downwards), t time, v the velocity of the material in the x-direction (assumed to be the only non-vanishing component of velocity), and A, κ, ρ, and c heat generation rate per unit volume, thermal diffusivity, density, and specific heat at constant pressure, respectively.

Thermal equilibrium is determined by different processes in continental and oceanic lithosphere. In most cases in continental lithosphere, the advective term $v\partial T/\partial x$ in equation (2) vanishes. If furthermore the situation is steady-state ($\partial T/\partial t = 0$, which applies to regions where the latest tectonothermal episode has an age $\gtrsim$ 100 Ma), equation (2) has simple analytical solutions. Assuming constant thermal conductivity and a crustal radioactive heat generation which decreases exponentially with depth

$$A = A_0\exp(-z/D) \quad (3)$$

the geotherm can be expressed as a function of the observed surface heat flow as

$$T = T_0 + \frac{(q_0 - q_r)D}{K}[1 - \exp(-z/D)] + \frac{q_r}{K}z \quad (4)$$

where K is the thermal conductivity ($K = \kappa\rho c$), T_0 the temperature at the surface, and q_0, q_r surface heat flow and heat flow from below the layer of thickness D, respectively. The choice of a heat generation rate as in equation (3) is justified by the observation that A is highest in felsic rocks (1.0–2.5 μW m^{-3}), intermediate in mafic rocks (0.05–0.5 μW m^{-3}), and lowest in ultramafic rocks (0.001–0.01 μW m^{-3}). The characteristic thickness D varies from region to region, but is usually 10 $\pm$ 5 km (see Ranalli 1995 and Stein 1995 for compilations and discussion).

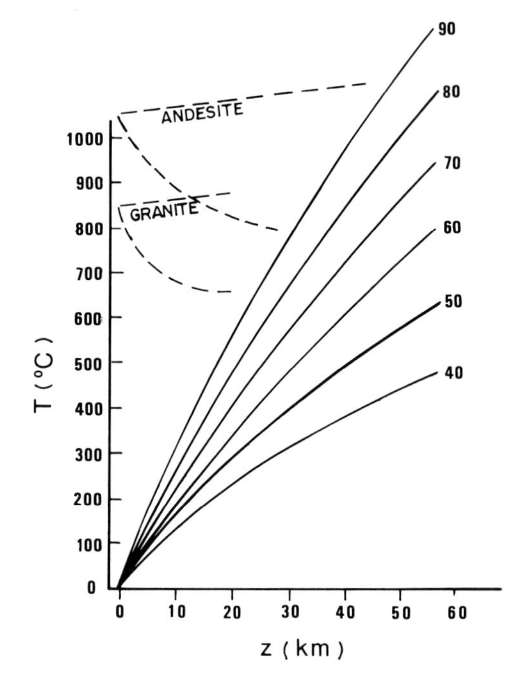

Fig. 1. Continental geotherms in terms of surface heat flow (mW m^{-2}). Dashed lines are dry (top) and wet (bottom) solidi for granite and andesite (from Chapman & Furlong 1992).

The above solution assumes the thermal conductivity to be constant. Another approximation is to take the lithosphere as composed of n layers, each with a given heat generation A_i and thermal conductivity K_i. The temperature distribution in the i-th layer is

$$T_i = T_{oi} + \frac{q_{oi}}{K_i}z - \frac{A_i}{2K_i}z^2 \quad (5)$$

where T_{oi} and q_{oi} are temperature and heat flow at the top of the layer, and z is measured from the top of the layer. Continental geotherms computed according to equation (5) are shown in Fig. 1 (Chapman & Furlong 1992), parameterized in terms of surface heat flow.

It is difficult to assign quantitative confidence limits to computed geotherms. Assuming the validity of the steady-state approximation, the effects of uncertainties in thermal parameters, heat generation and surface heat flow have been assessed by Chapman & Furlong (1992). The largest source of error is the uncertainty in surface heat flow. This is shown in Fig. 2, where geotherms for different parts of the East African Rift System have been computed assuming a $\pm$20% uncertainty in average surface heat flow (Fadaie & Ranalli 1990). As an order of

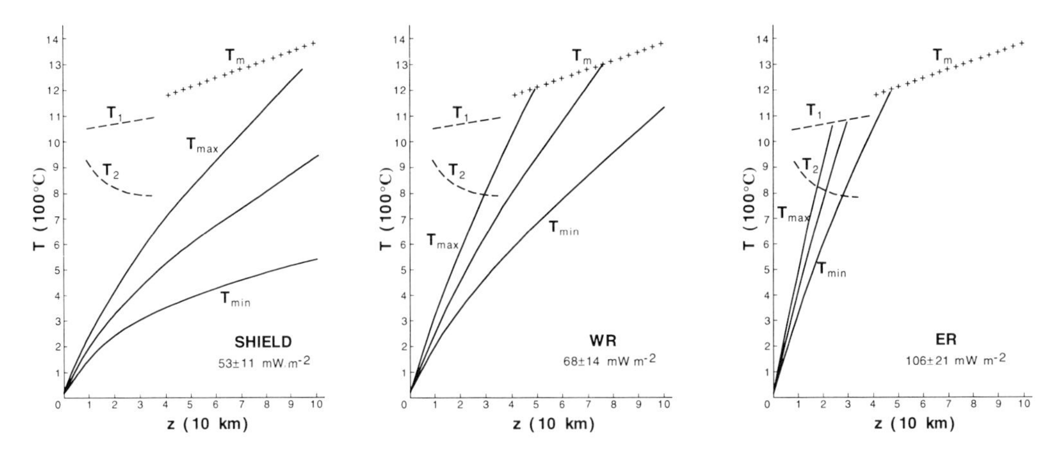

Fig. 2. Geotherms for Precambrian shield, western rift (WR), and eastern rift (ER) in the East African Rift System. T_{min} and T_{max} are the geotherms for minimum and maximum values of surface heat flow, respectively; T_1 and T_2 are dry and wet solidi for the lower crust; T_m is the solidus for a slightly hydrated upper mantle (from Fadaie & Ranalli 1990).

magnitude, temperature uncertainties at the Moho are ±100°C. Similar results have been obtained in the Charlevoix region of the Precambrian Grenville Province in eastern Canada, where 90% of 22 000 computed geotherms, for different combinations of parameters and a surface heat flow of 41 ± 10 mW m^{-2}, fall between 215 and 355°C at 25 km of depth (Lamontagne & Ranalli 1996).

What is apparent from the above considerations is the great lateral variability of temperature at a reference depth, for instance the Moho. Taking this to be at $z = 35$ km, its temperature varies from less than 400°C in regions of low surface heat flow ($q_0 \approx 40$ mW m^{-2}; Precambrian shields), to about 600°C for intermediate values of heat flow ($q_0 \approx 60$ mW m^{-2}), to more than 1000°C for very high heat flow ($q_0 \approx 100$ mW m^{-2}; continental extension zones). When compared with the variability of the Moho depth itself, these results show conclusively that the Moho is not an isotherm.

In oceanic lithosphere, heat generation is practically negligible, but the lithosphere moves away at speed v from the mid-oceanic ridge (refer to equation 2). The geotherms can be modelled by transient half-space cooling (see Turcotte & Schubert 1982 and Ranalli 1995 for reviews) which gives

$$T = T_o + (T_m - T_0) erf[\frac{z}{2(\kappa t)^{1/2}}] \qquad (6)$$

where T_m is the temperature at the base of the lithosphere (T_m = 1300°C corresponds to the solidus of an asthenosphere containing some volatiles), and t is the time since formation at the mid-oceanic ridge (age of the lithosphere). From equation (6), it follows that the thickness of the oceanic lithosphere increases as the square root of the age. (Typical oceanic geotherms are shown in Fig. 10, where they are compared to their Archaean equivalents.)

The above considerations neglect the thermal effects of tectonic processes (e.g., frictional heat related to overthrusting, metamorphic reactions, transient cooling of magmatic intrusions, etc). These phenomena can be very important at the local and regional scale, and should be included in detailed temperature calculations. Formally, from the general viewpoint of this paper, they can be considered as incorporated in the heat generation term, the transient term, and the advective term of equation (2).

Calculation of geotherms allows the definition of the thermal thickness of the lithosphere. The thermal lithosphere approximately coincides with the seismological lithosphere (material above the low-velocity zone). Its thermal thickness varies according to age in oceanic areas up to about 80 Ma, after which it becomes roughly constant (80–100 km for mature oceanic basins). In continental areas, the thickness varies from 30 km or less in rift systems and some areas that have undergone Cenozoic delamination and tectonic inversion to 250–300 km beneath Precambrian shields.

Frictional rheology: faulting and fault reactivation

Once the geotherm is known, it is possible to estimate the thickness of the brittle layers, which is approximately controlled by the condition $T \leqslant 1/2\ T_m$, where T_m is the solidus temperature of the relevant material. As a first approximation, the distribution of crustal and (when present) upper mantle seismicity outside subduction zones can be taken as an indication of the extent of brittle layers. The maximum depth of crustal seismicity correlates negatively with surface heat flow (Sibson 1982; Chen & Molnar 1983). Critical temperatures above which seismicity does not occur are approximately 350 ± 50°C in the crust and 700 ± 100°C in the upper mantle (Ranalli 1993).

To the first order, it may be assumed that the depth extent of seismicity coincides with that of faulting. Brittleness and ductility, however, refer to bulk rheological properties of rocks, while stability of sliding refers to properties of surfaces. It is therefore more appropriate to regard the lower boundary of seismicity as governed by the transition between velocity weakening (where the dynamic frictional resistance decreases with increasing sliding velocity, thereby making an instability possible), and velocity strengthening (where the frictional resistance increases with increasing velocity, leading to stable sliding). This transition seems to coincide with the onset of ductility for the softest phase (usually quartz; Scholz 1990). It should therefore occur at slightly shallower depth than the brittle-ductile transition, and this is confirmed by detailed studies of intraplate seismic zones (Lamontagne & Ranalli 1996). However, the difference is not large, as shown by the approximate coincidence of seismogenic and brittle layers.

Although some localized diffuse deformation (for instance, pressure solution) can occur at relatively low temperatures, the predominant mode of deformation in the brittle parts of the lithosphere is faulting. Slippage is governed by the Coulomb-Navier criterion, modified for pore fluid pressure (see e.g. Byerlee 1967; Jaeger & Cook 1979), which is usually referred to as 'Byerlee's law' in the geological literature. Since the medium is thoroughly pre-fractured, faults are often reactivated. Assuming one principal stress axis vertical and pre-existing planes of weakness of all orientations, and further that the planes of weakness are cohesionless, the failure criterion can be recast in terms of the critical stress differences necessary to cause thrust, normal and transcurrent faulting along the most favourably oriented planes. They are given respectively by (Sibson 1974)

$$\sigma_1 - \sigma_3 = (R-1)\rho g z(1-\lambda) \quad (7a)$$

$$\sigma_1 - \sigma_3 = \frac{R-1}{R}\rho g z(1-\lambda) \quad (7b)$$

$$\sigma_1 - \sigma_3 = \frac{R-1}{1+\delta(R-1)}\rho g z(1-\lambda) \quad (7c)$$

where σ_1 and σ_3 are maximum and minimum principal stresses (compression positive), $\rho g z$ is the overburden pressure, λ is the pore fluid factor (ratio of pore fluid pressure to overburden pressure), $\delta = (\sigma_2 - \sigma_3)/(\sigma_1 - \sigma_3)$, $0 < \delta < 1$, determines the value of the intermediate principal stress, and $R = \sigma_1/\sigma_3$ is the stress ratio. For cohesionless fracture surfaces, the latter can be expressed in terms of the static coefficient of friction μ_0 as

$$R = [(1 + \mu_0^2)^{1/2} - \mu_0]^{-2}. \quad (8)$$

The critical stress differences are therefore different for the various classes of faults, because of their dissimilar orientation with respect to the principal stress directions.

Equations (7) and (8) give the lower-bound critical stress difference necessary for faulting, since the pre-existing failure surfaces are assumed to be cohesionless and ideally oriented with respect to the stress field (that is, making an angle $\theta = (1/2)\tan^{-1}(1/\mu_0) \approx 27°$ with the σ_1 axis and containing the σ_2 axis). Even so, for low pore fluid pressures, stress differences of the order of hundreds of megapascals are required to cause slippage at mid- and lower-crustal depths ($z \gtrsim 10$ km). These stress differences are greatly reduced with increasing pore fluid pressure, and can tend to zero if the pore fluid pressure approaches the lithostatic pressure. Consequently, high pore fluid pressures seem to be necessary for faulting below the upper crust. The existence of fluids in the lower crust is supported by seismological and electrical conductivity data, at least in some regions (Hyndman & Shearer 1989). Alternatively, friction may decrease with increasing pressure, or a different high-pressure type of failure may apply (a hypothesis for which there is some experimental evidence; Hirth & Tullis 1989; Shimada & Cho 1990).

The general case where none of the principal stress axes is vertical, the pre-existing planes of weakness have finite cohesion and are not necessarily oriented at the most favourable angle to the maximum principal stress, has been considered by Ranalli & Yin (1990) and Yin & Ranalli (1992). Two critical stress differences must be considered: that necessary to cause a new fault in ideal orientation with respect to the stress system, and that necessary to cause slippage along a pre-existing fault. Denoting by

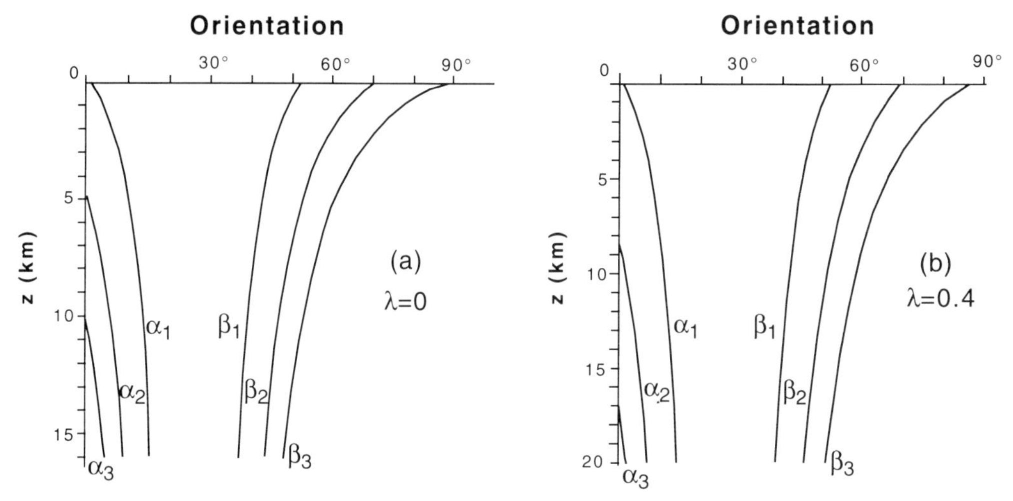

Fig. 3. Limiting angles for slippage along pre-existing failure surfaces *vs* depth for reactivation as thrust (α_1, β_1), transcurrent (α_2, β_2), and normal (α_3, β_3) faults in the two dimensional case (see text). Material parameters: $S = 75\,\text{MPa}$, $S_0 = 5\,\text{MPa}$, $\mu = \mu_0 = 0.75$, $\rho = 2700\,\text{kg m}^{-3}$, $\delta = 1/2$, $\lambda = 0$ (case a), and $\lambda = 0.4$ (case b) (from Ranalli & Yin 1990).

S, μ and S_0, μ_0 cohesion and friction coefficient, respectively, along a new and a pre-existing fault, and by m_i and n_i, $i = 1,2,3$, the direction cosines of the vertical unit vector and of the unit normal to the failure plane, respectively (both with respect to the principal stress axes), the stress difference necessary to cause a new fault is

$$\sigma_1 - \sigma_3 = \frac{2\mu\rho g z(1-\lambda) + 2S}{(\mu^2+1)^{1/2} - \mu + 2\mu(m_1^2 + \delta m_2^2)} \tag{9a}$$

and the stress difference necessary to cause slip along a pre-existing plane of weakness is

$$\sigma_1 - \sigma_3 = \frac{\mu_0\rho g z(1-\lambda) + S_0}{[(n_1^2+\delta^2 n_2^2) - (n_1^2+\delta n_2^2)^2]^{1/2} + \mu_0[(m_1^2+\delta m_2^2) - (n_1^2+\delta n_1^2)]} \tag{9b}$$

These expressions, which reduce to equations (7) with the simplifying assumptions adopted by Sibson (1974), allow the prediction of whether, for any given orientation of stress field and plane of weakness, a new fault will form, or whether deformation will be accommodated by slip along a pre-existing weakness. The process requiring the lesser stress difference occurs. Formation of a new fault takes place when the pre-existing faults are so unfavourably oriented that slip along them would require a larger stress difference than formation of a new failure surface.

These results, although strictly valid for the ideal case where the rock is homogeneous (except for the existence of planes of weakness) and the stress field is uniform, are nevertheless of some relevance to the problem of reactivation of pre-existing faults. Figure 3 gives a simple example, where the weak surface contains the intermediate stress axis and one of the principal stresses is vertical. Even if material parameters are assumed not to depend on depth, the range of orientations for which reactivation is possible is a function of depth, and increases with increasing pore fluid pressure. The favourable reactivation range, other things being equal, is largest for normal faulting and smallest for thrust faulting: consequently, reactivation of pre-existing planes of weakness as normal faults can occur more commonly than reactivation as thrust faults. Also, since the reactivation range increases with decreasing depth, erosion and tectonic denudation should favour reactivation of pre-existing faults.

The more general three-dimensional case is shown in Fig. 4, on triangular diagrams with coordinates (n_1^2, n_2^2, n_3^2), for three cases of vertical principal stress, corresponding to normal, transcurrent, and thrust faulting regimes. Each point on the diagrams corresponds to an orientation of the unit normal to the plane of weakness with respect to the principal stress directions. If the orientation is unfavourable, new faults form (shaded areas), of orientation and type appropriate to the stress field. If the orientation is within the favourable range (unshaded areas), pre-existing faults are reactivated, with one of the types of slip shown (generally oblique-slip as the fault is inclined to all three principal stress directions). Since the reactivation range is a function of depth, these

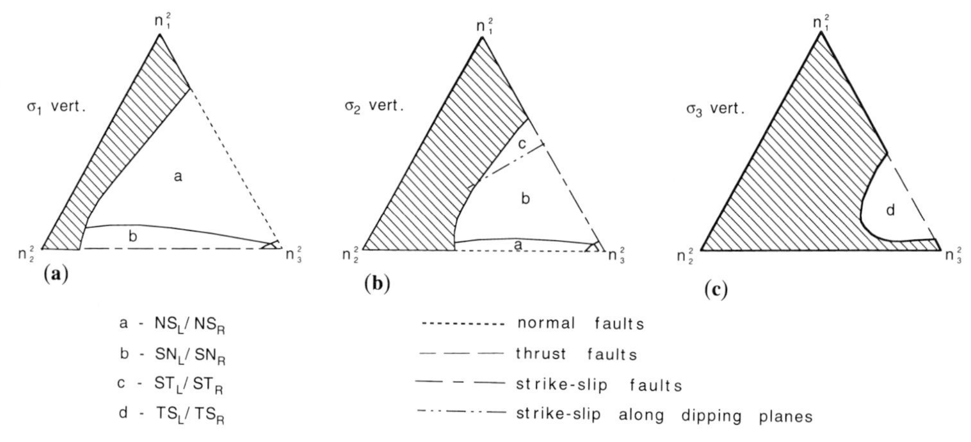

Fig. 4. Faulting regimes in rocks as a function of orientation n_i of pre-existing faults with respect to principal stress directions. Material parameters as in Fig. 3, $\lambda = 0.4$, $z = 10$ km. (**a**), (**b**) and (**c**) refer to different cases of vertical principal stress, as shown. Shaded areas denote unfavourable orientations of pre-existing faults, where new faults are formed, of type appropriate to the stress field. Unshaded areas denote orientations for which reactivation occurs. Areas a, b, c, d show the two possible slips along reactivated faults (N, normal; S, strike-slip; T, thrust; R, dextral; L, sinistral; e.g. ST_L represents oblique slip consisting of a combination of sinistral strike-slip and thrust displacement). Pure dip-slip and strike-slip reactivated faults occur along lines denoted by different symbols (from Yin & Ranalli 1992).

diagrams are valid for a given depth ($z = 10$ km in the example chosen). The much wider reactivation range for normal faulting is evident.

In both Figs 3 and 4, internal friction has been taken equal to sliding friction. The reactivation range would be larger if $\mu_0 < \mu$. Reactivation of previous thrust faults as normal faults (at an angle of 60° or more to the vertical) is possible in the upper and middle crust for moderate to high pore fluid pressures.

Ductile rheology: creep parameters

At temperatures higher than about one-half of their solidus, all rocks are ductile. Instantaneous elastic deformation and transient creep (where the strain rate decreases with increasing time) usually precede steady-state creep, where the strain rate is constant under constant stress and temperature. Although transient creep is important in some geodynamic processes (e.g. damping of seismic waves and possibly postglacial rebound), in tectonic deformation it is safe to assume that elastic and transient strains are negligible compared to steady-state strains.

The microphysics of high-temperature creep in crystalline materials is reviewed in a number of books (Frost & Ashby 1982; Poirier 1985; Ranalli 1995). The rate equation for practically all crystalline aggregates (both single phase and polyphase) occurring in the lithosphere can be written in the general form (which facilitates comparison with experimental results)

$$\dot{\epsilon} = Ad^{-m}(\sigma_1 - \sigma_3)^n \exp\left(-\frac{E}{RT}\right) \qquad (10)$$

where $\dot{\epsilon}$ is the strain rate (elongation in the usual experimental setup), d the average grain size, $\sigma_1 - \sigma_3$ the stress difference (referred to a triaxial test), T the absolute temperature, R the gas constant, and A, m, n, and E material parameters, approximately independent of stress and temperature in the range considered. Activation energy E, rather than activation enthalpy $H = E + PV$ (where P is pressure and V is the activation volume), has been used in the exponential because the pressure effect is negligible under lithospheric conditions.

It can be shown (see for instance Ranalli 1995 for a review) that if creep is controlled by diffusion-related dislocation climb, the expected value of the stress exponent n varies from 2.0 to 4.5 (power-law creep) according to the details of the recovery mechanism (most typically, $n = 3$), and there is no dependence of strain rate on grain size ($m = 0$). On the other hand, if creep occurs by vacancy diffusion between grain boundaries, the dependence on stress is linear ($n = 1$), and the grain size exponent is 2 or 3 according to whether diffusion occurs in the bulk of the crystal or along grain boundaries. The activation energy for this linear creep is usually

Table 1 *Creep parameters of lithospheric rocks and minerals*

Material	$A(\text{MPa}^{-n}\,\text{s}^{-1})$	n	$E(\text{kJ mol}^{-1})$
Quartz	1.0×10^{-3}	2.0	167
Plagioclase (An_{75})	3.3×10^{-4}	3.2	238
Orthopyroxene	3.2×10^{-1}	2.4	293
Clinopyroxene	15.7	2.6	335
Granite	1.8×10^{-9}	3.2	123
Granite (wet)	2.0×10^{-4}	1.9	137
Quartzite	6.7×10^{-6}	2.4	156
Quartzite (wet)	3.2×10^{-4}	2.3	154
Quartz diorite	1.3×10^{-3}	2.4	219
Diabase	2.0×10^{-4}	3.4	260
Anorthosite	3.2×10^{-4}	3.2	238
Felsic granulite	8.0×10^{-3}	3.1	243
Mafic granulite	1.4×10^{4}	4.2	445
Peridotite (dry)	2.5×10^{4}	3.5	532
Peridotite (wet)	2.0×10^{3}	4.0	471

Compiled from various sources; from Ranalli (1995).

less than the activation energy for nonlinear dislocation creep.

The two types of creep are found to apply for $\sigma_1-\sigma_3\leqslant 100$ MPa, above which other creep mechanisms, not relevant to geodynamics, take over. The two mechanisms are concurrent, that is to say, both are acting at the same time, and the macroscopic strain rate is the sum of the two contributions. However, since the stress and temperature dependences of the two mechanisms are different, fields in σ, T-space can be defined where either linear or nonlinear creep are predominant (this is the basis for the construction of deformation mechanism maps; Frost & Ashby 1982). The transition stress between the two, above which nonlinear creep is predominant, increases with decreasing grain size. These considerations are of general validity, although values of parameters and consequently the transition stress vary according to material. For lithospheric constituent materials, transition stresses are in the range 0.1–10 MPa for $d\geqslant 10$–100 μm. Therefore, except in fine-grained shear zones and at very low stress, nonlinear creep appears to be the relevant steady-state rheology of the bulk of the ductile lithosphere.

Several compilations of experimentally determined creep parameters are available (Frost & Ashby 1982; Kirby 1983; Kirby & Kronenberg 1987; Wilks & Carter 1990; Ji & Zhao 1993; Ranalli 1995). Values for selected lithospheric rocks and minerals are shown in Table 1. It is not possible to assign statistical error limits; however, comparison of experiments on similar materials indicates that margins of $\pm10\%$ are of the right order of magnitude. (Note that creep parameters for some materials – e.g., quartz and olivine – are better known than those for others – e.g., feldspars.) If the petrological composition of the various layers of the lithosphere is approximately known, therefore, it is possible to predict approximate variations of rheology with depth. When creep data are not available for a given rock but the main constituent phases are known, procedures are available to estimate the bulk rheology from that of the constituent phases (Tullis *et al.* 1991; Ji & Zhao 1993; Handy 1994).

There are important limitations to the use of experimental creep data to estimate the ductile properties of the lithosphere. However, the most basic ones are not those that are often mentioned. The difference between experimental ($10^{-4}-10^{-8}$ s^{-1}) and geological (10^{-12}–10^{-16} s^{-1}) strain rates is compensated by higher homologous temperatures (T/T_m, where T_m is solidus temperature) under laboratory conditions. Confidence in the validity of equation (10) is strengthened by the coincidence of solid-state theoretical models and laboratory observations. Uncertainties in temperature, strain rate, and rheological parameters can be assessed. The presence of minor phases (by volume percent) usually does not noticeably affect the rheology. The most serious limitation is the assumption of constant and uniform strain rate to calculate 'creep strength', i.e., the stress difference necessary to maintain a given strain rate. This presupposes homogeneous deformation, while ductile flow in the lower crust is highly heterogeneous (see e.g. Knipe & Rutter 1990; Fountain *et al.* 1992; Rutter & Brodie 1992). When this heterogeneity is related to compositional variations, the use of creep data is still justified, provided that the composition is correctly chosen. When, however, the heterogeneity is a consequence of dynamic factors (for instance, shear concentration along relatively soft zones), the estimated creep strength is probably an upper limit. Consequently, it is unwise to regard inferences on lithosphere rheology as giving more than order-of-magnitude estimates of the *in-situ* properties of the crust and upper mantle. However, the most important conclusion (that the lithosphere is likely to be rheologically stratified; see next section) is certainly valid.

Rheological profiles (strength envelopes)

The most important rheological transition in a solid material is the brittle–ductile transition. The brittle–ductile transition is gradual even in single-phase polycrystalline aggregates, and all the more so in polyphase aggregates in the lithosphere. Consequently, the concept of a sharp transition, occurring at a critical temperature, is only an approximation for a gradual transition probably covering a few kilometres. Also, since the lithosphere is compositionally layered and each layer has a different solidus, there can be more than one transition with increasing depth in any given region.

The variations of strength with depth in the lithosphere can be modelled, as a first approximation, by means of *rheological profiles* (or *strength envelopes*), which allow the identification of brittle and ductile layers and the estimation of strength. Rheological profiles are constructed by comparing frictional strength (which increases linearly with pressure; see equations 7 and 9) and creep strength at a given strain rate (which decreases exponentially with increasing temperature; see equation 10). In the simple case of cohesionless ideally oriented pre-existing faults, the critical stress difference for frictional sliding is

$$\sigma_1 - \sigma_3 = \alpha\rho g z(1 - \lambda) \qquad (11)$$

where $\alpha = 3.0, 1.2, 0.75$ for thrust, transcurrent, and normal faulting, respectively (for $\mu_0 = 0.75$ and $\delta = 1/2$; see equations 7 and 8). The ductile strength in the case of power-law creep (equation 10 with $m = 0$) is

$$\sigma_1 - \sigma_3 = (\frac{\dot{\epsilon}}{A})^{1/n}\exp(\frac{E}{nRT}) \qquad (12)$$

The 'strength' at any given depth is the lower of the brittle and ductile stress differences.

In the construction of rheological profiles (see e.g. Ranalli & Murphy 1987; Ranalli 1991) it is first necessary to establish the tectonic regime, since it affects frictional strength. In the ductile regime, a strain rate must be chosen, usually in the range 10^{-14}–10^{-16} s^{-1}. Information on the structure of the lithosphere (depth of the Moho, composition of crustal layers) comes from geological, gravity and seismological data. The geotherm is estimated on the basis of surface heat flow or palaeothermometric and palaeobarometric data.

Several examples of typical rheological profiles are given in Ranalli (1995). Since as a rule the creep strength of silicate rocks in the ductile field increases with decreasing silica content (see Table 1), the rheological structure of the lithosphere is layered, with one or more brittle layers separated by ductile layers. 'Type' profiles are shown in Fig. 5, for linearized geotherms and different types of lithosphere. (Note that the stress scale here is logarithmic, while in Figs 6 and 11 it is linear.) Except for cold geotherms (corresponding to surface heat flows of the order of 40 mW m^{-2}), the Moho is a strength discontinuity in the continental lithosphere (but not in the oceanic lithosphere). The lower continental crust is softer than the uppermost mantle, which is brittle for $q_0 \leqslant 60$ mW m^{-2} (see also Fadaie & Ranalli 1990). If the crust is compositionally layered, there may be more ductile layers in addition to the lower crust. Only for very high geothermal gradients ($q_0 \geqslant 100$ mW m^{-2}, not shown in the figure) does the strength discontinuity at the Moho become negligible (Fadaie & Ranalli 1990). For all but the most extreme values of surface heat flow, therefore, the lithosphere has two strong load-bearing layers: the middle crust (brittle) and the uppermost mantle (brittle or ductile). This stratification affects both the response of the lithosphere to tectonic forces and its flexural response to vertical loads (see discussions in Kusznir 1991 and Ranalli 1995).

An example of two-dimensional estimation of temperature and rheology is shown in Fig. 6, for a 350 km long cross-section of the southeastern Canadian Cordillera (Lowe & Ranalli 1993). The cross-section includes both thinned continental lithosphere (the Omineca Belt, expressed at the surface by tectonically inverted metamorphic core complexes) and cratonic lithosphere (the Foreland Belt and basinal sedimentary cover). The temperature effect on the rheology stands out quite clearly. (It should be noted that the strength of the brittle part of the upper mantle, when applicable, is estimated on the basis of extrapolation of frictional failure, which is almost certainly an overestimation; see e.g. Shimada 1993.)

The variation of rheology with position as a function of geothermal gradient, composition and structure has been studied in a variety of tectonic environments (see for instance Cloet-

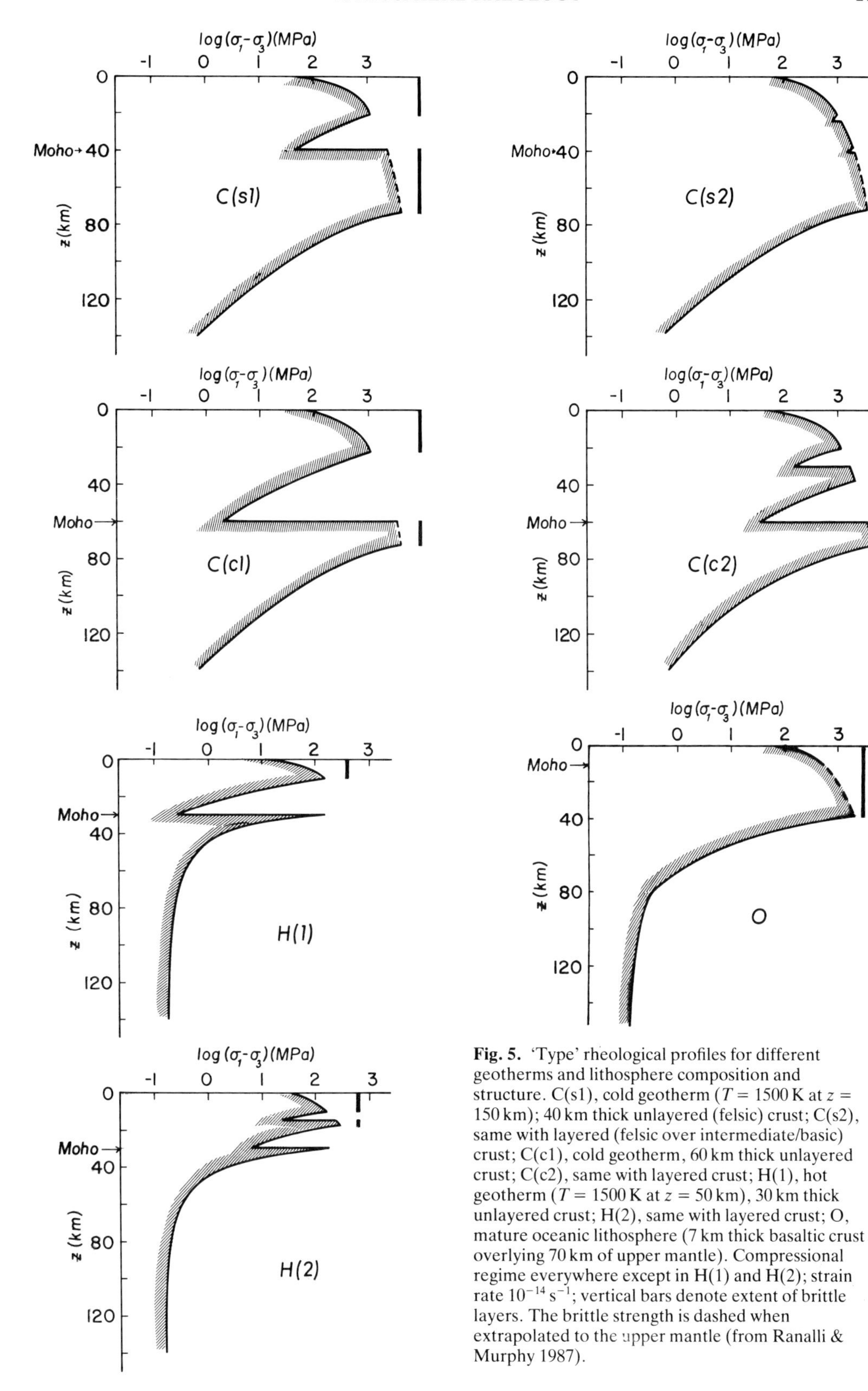

Fig. 5. 'Type' rheological profiles for different geotherms and lithosphere composition and structure. C(s1), cold geotherm ($T = 1500$ K at $z = 150$ km); 40 km thick unlayered (felsic) crust; C(s2), same with layered (felsic over intermediate/basic) crust; C(c1), cold geotherm, 60 km thick unlayered crust; C(c2), same with layered crust; H(1), hot geotherm ($T = 1500$ K at $z = 50$ km), 30 km thick unlayered crust; H(2), same with layered crust; O, mature oceanic lithosphere (7 km thick basaltic crust overlying 70 km of upper mantle). Compressional regime everywhere except in H(1) and H(2); strain rate 10^{-14} s^{-1}; vertical bars denote extent of brittle layers. The brittle strength is dashed when extrapolated to the upper mantle (from Ranalli & Murphy 1987).

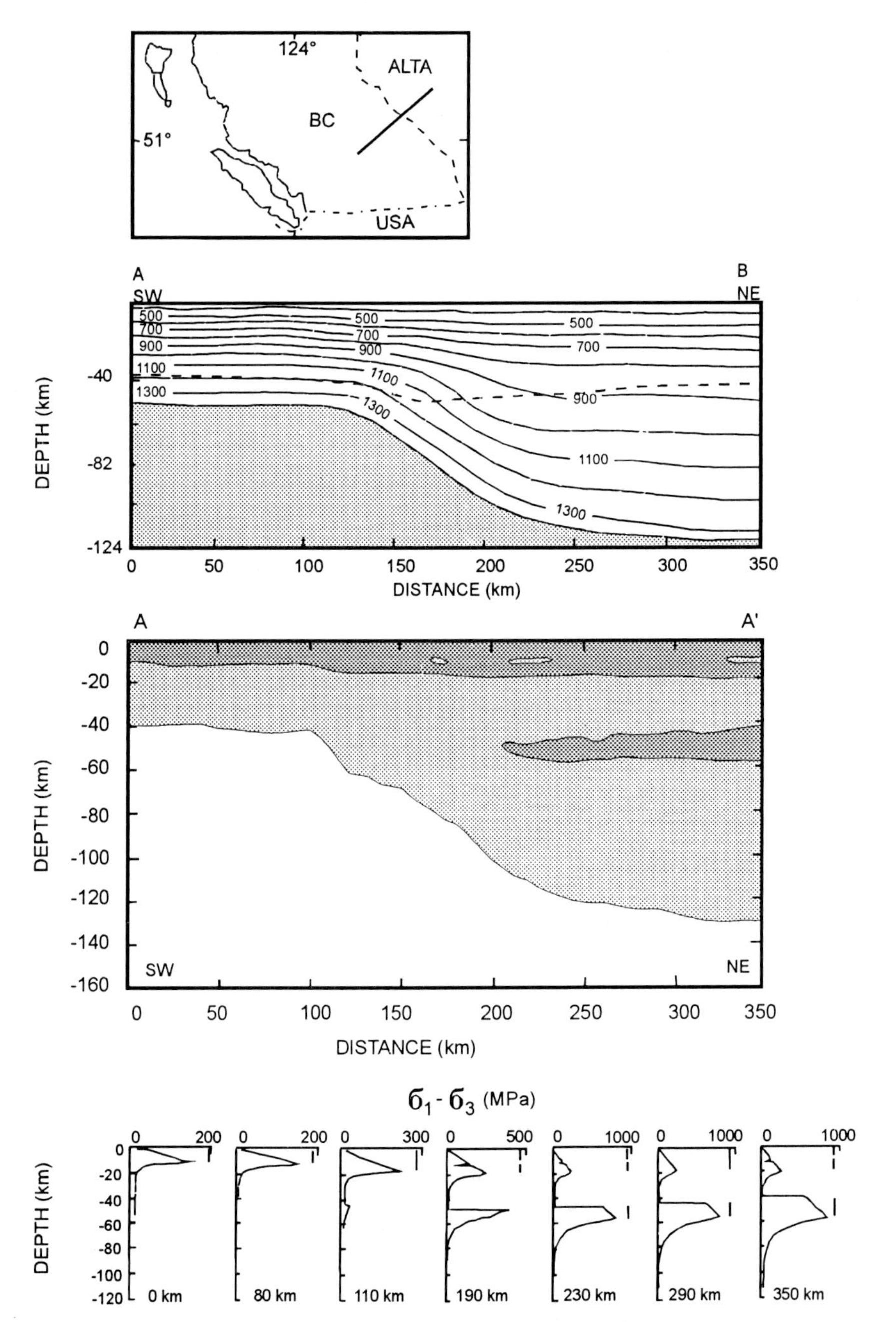

Fig. 6. Estimated temperature and rheology in a cross-section across the southern Canadian Cordillera (area shown in inset). Isotherms in degrees Kelvin. Stipple in the temperature section indicates asthenosphere; dashed line the seismic Moho. In the two-dimensional rheological profile, brittle and ductile regions are denoted by heavy and light stipple, respectively; distances from the SW end of the cross-section are marked in the one-dimensional profiles. Tectonic regime: normal faulting, strain rate 10^{-15} s^{-1} (from Lowe & Ranalli 1993).

ingh *et al.* 1994; Zeyen & Fernandez 1994 and references therein). As previously discussed, there is usually a good correlation between the extent of predicted brittle layers and the depth distribution of seismicity (Sibson 1982; Ranalli 1993; Lamontagne & Ranalli 1996). The often observed laminated structure of the lower crust, determined by seismic reflection profiling (Mooney & Meissner 1992; Clowes 1993), can be interpreted as an indication of horizontal ductile flow. In general, the first-order predictions of rheological studies are confirmed by geophysical observation, although one should be aware of the highly inhomogeneous distribution of deformation in the lower crust mentioned in the previous section (Rutter & Brodie 1992).

Rheological profiles can also be applied to palinspastic reconstructions of mountain belts at different times during their evolution. An application to the Jurassic continental margin and related obducted terranes of the southeastern Canadian Cordillera (Ranalli & Murphy 1987) shows that the large-scale structural characteristics of the system (for instance, sub-horizontal crustal decollements) correlate with the predicted position of the soft ductile layers.

Another approach to the study of lithospheric rheology uses vertical averages of material properties (see e.g., Sonder & England 1986), which can be related to the modelling of continental lithosphere as a thin viscous sheet (England & McKenzie 1982). In this context, rheology is parameterized in terms of Moho temperature, and a detailed knowledge of the geotherm is not necessary. This approach, however, cannot deal by definition with the variation of rheology with depth.

On the basis of rheological profiles, *total lithospheric strength* can be defined as

$$\Sigma = \int_S^H (\sigma_1 - \sigma_3)(z)\mathrm{d}z \qquad (13)$$

where S is the surface and H the depth of the lower boundary of the rheological lithosphere, defined for instance as the level where $\sigma_1 - \sigma_3$ reaches a low value (say, 1–10 MPa) and below which there are no further discontinuities in strength. The strength defined by equation (13) varies inversely with geothermal gradient. For a given geotherm, it is lower in regions with thick continental crust, since crustal materials are softer than mantle materials. The total strength of continental lithosphere usually falls between 10^{12} and 10^{13} N m^{-1} as orders of magnitude (these values, for instance, apply respectively to the thinned and to the Precambrian lithosphere of Fig. 6). The total strength of oceanic lithosphere varies between approximately the same limits (for a review see Ranalli 1991). For typical lithospheric thicknesses, these values correspond to vertically averaged horizontal stresses of the order of 10–100 MPa. These strength estimates match well the values obtained from stability analysis of lateral mass anomalies (see e.g. Molnar & Lyon-Caen 1988). Tectonic forces per unit length of plate boundary are usually estimated to be in the range (1–5) $\times$ 10^{12} N m^{-1}, with 10^{13} N m^{-1} as a probable upper limit (see e.g. Bott 1990; Ranalli 1995). Consequently, young oceanic lithosphere and continental lithosphere with a relatively hot geotherm ($q_0 \gtrsim 80$ mW m^{-2}) are liable to fail in extension under the action of plate boundary forces.

Consequences of rheological stratification of the lithosphere

The variation of lithosphere rheology with depth has many consequences for tectonic deformation. As an example, we examine in this section three types of processes, which play an important role in the dynamic evolution of mountain belts. These are: stresses caused by lateral mass heterogeneities and related gravitational collapse; tectonic inversion consequent upon lithospheric root detachment; and relaxation of Moho topography by lower crustal ductile flow. Physically, all three processes involve flow in the lower ductile crust. However, their separate analysis allows identification of different factors (stress, temperature, time) which – besides the magnitude and extent of the mass anomalies – affect the tectonic outcome.

(a) Lateral mass anomalies and gravitational collapse

Lithospheric plates are subject not only to plate boundary forces and mantle drag, but also to forces arising 'locally' from lateral mass anomalies, either by lateral variations in density or by variations in topography (of the free surface, of the Moho, or of the lower boundary of the lithosphere). Thus, orogenic belts are associated with tensional stresses related to the existence of topography and crustal roots; on the other hand, the presence of a lithospheric root tends to generate compressive stresses (see Ranalli 1995 for a discussion). Naturally, in many cases these 'local' stresses are less than far-field (plate boundary) tectonic stresses, and less than the strength of the lithosphere. Under certain conditions, however, they can be of relative importance and cause significant tectonic deformation.

There are different ways to express lateral mass anomalies. One measure is the density mo-

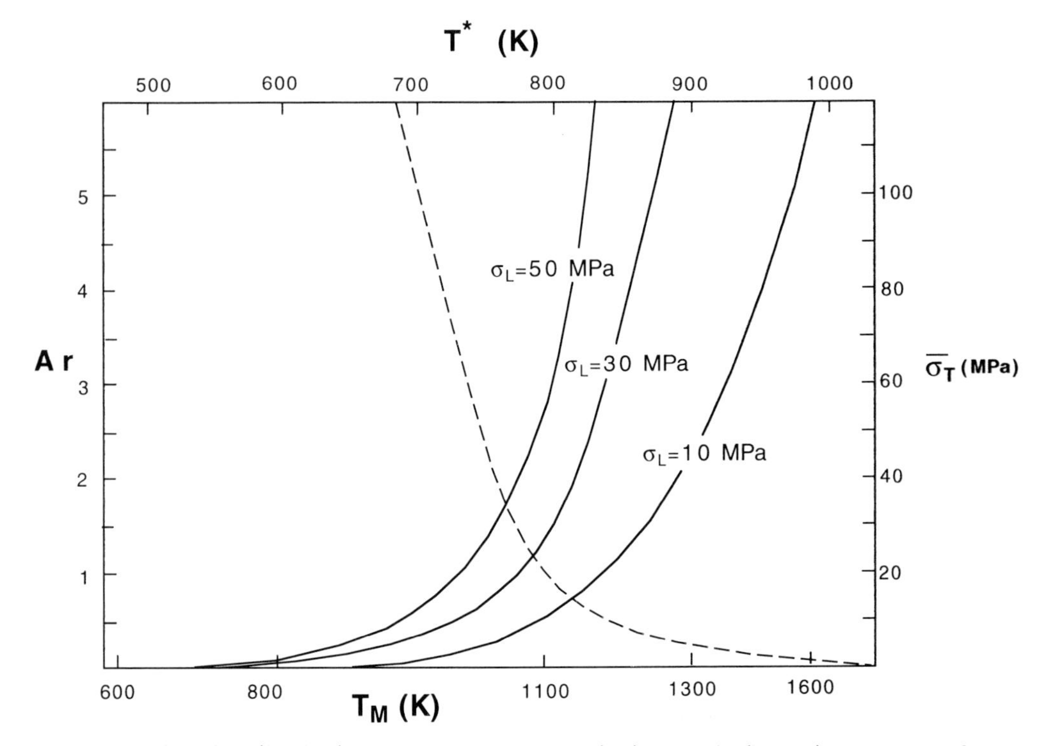

Fig. 7. Argand numbers (full lines) as a function of mid-crustal (top) and Moho (bottom) temperatures, for different values of 'local' stresses σ_L. The tectonic stress for deforming the lower crust is denoted by $\bar{\sigma}_T$ (dashed lines). See text for details.

ment, related to the integral over depth of density anomalies times depth of occurrence (Fleitout & Froidevaux 1982); another is the gravitational potential energy per unit area, which is larger in thickened regions (Molnar & Lyon-Caen 1988). A useful measure of the relative importance of local and far-field stresses is the *Argand number* (England & McKenzie 1982), defined as the ratio between the stress generated by mass inhomogeneities and the stress required to deform the lithosphere at tectonic strain rates (we use 'stress' for simplicity, to be interpreted as deviatoric stress or, in simple stress systems, as maximum stress difference). Where $Ar \to 0$ (low local stress and/or strong lithosphere), the tectonic style is not affected by lithospheric mass anomalies; as Ar increases (high local stress and/or weak lithosphere), local mass anomalies play a role in the tectonic evolution of the region. Not only mass anomalies, therefore, but also the thermal state of the lithosphere are factors in determining whether local stresses affect the tectonic evolution.

Values of the Argand number for different local stresses and lithospheric temperatures are plotted in Fig. 7. The lower crust is assumed to deform by power-law creep, and the tectonic stress necessary to deform it at a rate of 10^{-15} s^{-1} is calculated according to equation (12) with $A = 10^6$ GPa^{-3} s^{-1}, E = 240 kJ mol^{-1}, and n = 3 (typical values for average continental lower crust). Both creep strength and Argand numbers are given, the latter for different values of the local stress, in terms of temperature in the mid-crust (which has been used in calculations) and corresponding reference temperature at the Moho T_M (assuming that the temperature increase in the lower crust is 2/3 of the temperature increase in the upper crust). Results show that, if the lithosphere remains cool after a compressional episode ($T_M \leqslant 600$°C), local stresses, even if rather high, do not cause tectonic inversion and gravitational collapse. If, however, Moho temperatures reach values higher than 600–700°C, gravitational collapse is likely to occur (assuming that the brittle upper crust is dragged along by the flow of the lower crust). These simple order-of-magnitude estimates should be compared with more sophisticated models of stresses in mountain belts (see e.g. Zhou & Sandiford 1992), which lead to similar results.

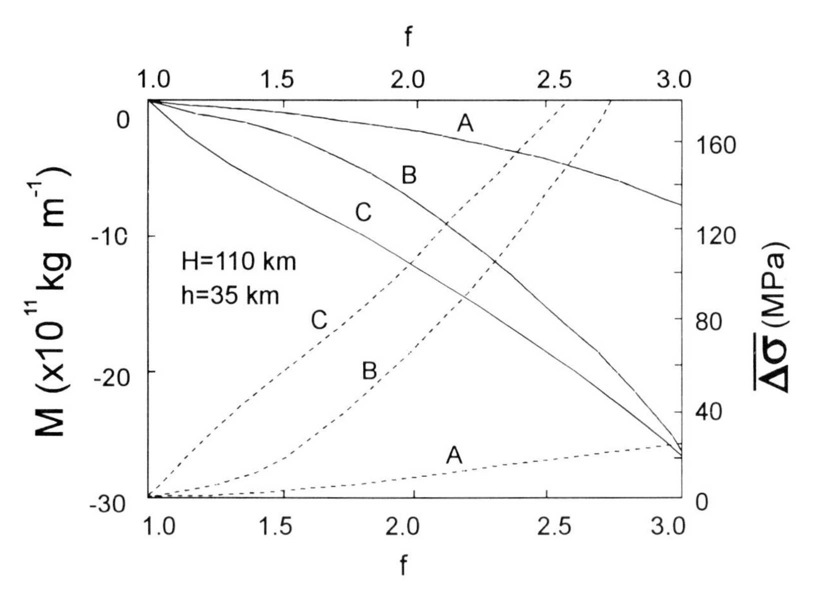

Fig. 8. Density moment M (full lines) and resulting average extensional horizontal stress $\overline{\Delta\sigma}$ (dashed lines) *vs.* thickening factor f for thickened lithosphere (A), lithospheric root detachment at the base of the standard lithosphere (B), and at the base of the thickened crust (C) (from Ranalli *et al.* 1989).

(b) Detachment of lithospheric root and tectonic inversion

The previous considerations neglect the role of the sub-crustal lithosphere. If the lithosphere as a whole, and not only the crust, is thickened, the positive mass anomaly caused by the lithospheric root, since it is cooler than the surrounding mantle, approximately counterbalances the negative mass anomaly generated by the crust and the Moho deflection. However, a thickened upper thermal boundary layer in mantle convection is gravitationally unstable, and is delaminated and entrained in the convective circulation at most within a few tens of million years after its formation (Houseman *et al.* 1981).

This lithospheric decollement has important consequences for the tectonic evolution of mountain belts, and can account for inversion from compression to extension if far-field tectonic stresses are favourable. Figure 8 shows the results of an order-of-magnitude estimate of the effect of root detachment on the evolution of the core zone (Omineca Belt) of the southeastern Canadian Cordillera, which underwent inversion about 59 ± 1 Ma ago (Ranalli *et al.* 1989 and references therein). It is assumed that the crust and the lithosphere, with initial thicknesses of 35 and 110 km, respectively, were uniformly thickened. For different values of the thickening factor f, the figure gives values of the density moment and of the resulting 'local' stress. Three cases are considered: (A) thickened lithosphere; (B) lithosphere after detachment of the root at depth H (pre-shortening thickness of lithosphere); (C) lithosphere delamination at the Moho. For realistic thickening factors $f \leq 2.0$, the thickened lithosphere is in approximate local equilibrium. After delamination, however, (vertically averaged) horizontal extensional stresses of up to 50–100 MPa are generated. (This is an upper limit estimate because the structure is assumed to be of infinite lateral extent.) Consequently, Tertiary extension of the Omineca Belt in the southeastern Cordillera, and possibly in other metamorphic core complexes, can be accounted for in terms of delamination of the lithosphere. Changes in far-field stresses can play an accommodating role, but thickening with consequent increase in potential energy, followed by root detachment, seems to be required to account for inversion in at least some tectonic environments (see Mareschal 1994 for a more rigorous analysis).

Processes related to potential energy, thickening, and delamination are considered to have been active in several mountain belts (for instance, the European Variscides, Rey *et al.* 1991; the Himalayas and Tibet, Molnar *et al.* 1993). Delamination may also be related to magmatic activity (through the heating of the lower crust which is in contact with asthenospheric material after root detachment). Other phenomena such as the rapid exhumation of high pressure/low temperature metamorphic rocks (Platt 1993) and the occurrence of relative motion between the crust and the subcrustal lithosphere with décollement at the Moho (Willett *et al.* 1993) are related to the softness of the lower crust, which under suitable circumstances allows decoupling between crust and upper mantle.

(c) Relaxation of Moho topography

Isostatic equilibrium does not imply hydrostatic equilibrium: disturbances to the interface of two fluids of different densities in the field of gravity tend to disappear, and the interface to become an equipotential surface, if the creep strength of the fluids is sufficiently low. The possibility of lateral extrusion of the ductile lower crust from beneath regions of high topography is well recognized (see e.g. Bird 1991), and the required lower crustal viscosity has been estimated to be in the order of 10^{18}–10^{20} Pa s for long-wavelength flow (500 km or more; Kruse *et al.* 1991).

Lower crustal flow results in relaxation of Moho topography, and, together with erosion and isostatic rebound, can affect both the depth of the Moho and the uplift/exhumination histories of metamorphic terranes in orogenic belts. The Moho deflection consequent upon crustal thickening can be considered as the boundary between two fluids (the crust and the upper mantle) with densities and viscosities ρ_1, η_1, and ρ_2, η_2, respectively. To gain physical insight into this problem, an analytical solution obtained by Ramberg (1968) can be used. Given two incompressible semi-infinite fluids in plane strain, with interface of initial sinusoidal shape with amplitude z_o' and wavelength λ_0, the relaxation at time t is

$$z_0(t) = z_0'\exp(kt) \qquad (14a)$$

$$k = \frac{\lambda_0 g(\rho_1 - \rho_2)}{4\pi(\eta_1 + \eta_2)} \qquad (14b)$$

(note that $k<0$ since $\rho_1<\rho_2$). Therefore the relaxation rate depends on the effective viscosities of both the lower crust and the uppermost mantle. Since the viscosities depend strongly on temperature, whether the Moho relaxes with time after a shortening (orogenic) episode depends on the post-orogenic thermal history of the lithosphere.

Equations (14a) and (14b) are valid for Newtonian fluids (diffusion creep). Nevertheless, since linear and nonlinear creep viscosities down to the mantle transition zone are within an order of magnitude of each other for reasonable values of strain rate and grain size (see Ranalli 1995 for a detailed discussion), the use of effective nonlinear viscosities should give a first-order idea of the relaxation time as a function of temperature. Using the definition $\eta = \sigma/\dot{\epsilon}$, effective viscosity for power-law creep can be expressed from equation (10) as a function of stress, strain rate, or viscous dissipation ($\Delta = \sigma\dot{\epsilon}$) as

$$\eta(\sigma) = A^{-1}\sigma^{1-n}\exp(E/RT) \qquad (15a)$$

$$\eta(\dot{\epsilon}) = A^{-1/n}\dot{\epsilon}^{(1-n)/n}\exp(E/nRT) \qquad (15b)$$

$$\eta(\Delta) = A^{-2/(n+1)}\Delta^{(1-n)(1+n)}\exp[2E/(n+1)RT]. \qquad (15c)$$

Substituting equations (15a)–(15c), with material parameters appropriate for lower crustal (plagioclase-rich) rocks and for upper mantle (peridotitic) rocks, into equations (14a) and (14b) with ρ_1–$\rho_2 = -500$ kg m^{-3}, one obtains the relaxation of the Moho as a function of viscosities and therefore temperature. Figure 9 shows relaxation curves (at constant stress, strain rate, or viscous dissipation), as functions of time for different Moho temperatures, and a wavelength of 1000 km. Given the assumptions and approximations involved, the results are only indicative. For $T_M \lesssim 500$°C (corresponding to surface heat flow $q_0 \lesssim 60$ mW m^{-2}; see Fig. 1) relaxation does not occur over geologic time. As temperature increases ($T_M \simeq 700$°C, i.e. $q_0 \simeq 70$ mW m^{-2}), significant relaxation takes times of the order of 100 Ma (which is also the time that the Moho must be maintained at that temperature). For $T_M \gtrsim 900$°C ($q_o \gtrsim 80$ mW m^{-2}), the characteristic time for relaxation is 10 Ma or less. An intense lower crustal heating episode, therefore, even if relatively short, is sufficient to cause significant relaxation of Moho topography. Lower crustal effective viscosities at $T = 900$°C are in the 10^{18}–10^{20} Pa s range (as also concluded by Kruse *et al.* 1991), and uppermost mantle viscosities are of the order of 10^{22} Pa s. Lateral extrusion of the lower crust in regions of previous crustal thickening is certainly a factor in the late- and post-orogenic evolution of mountain belts if the geotherm is sufficiently high.

As stated at the beginning of this section, the three processes described above are only a small sample of the tectonic consequences of rheological stratification. This stratification affects in a general way the style of deformation both in extensional (e.g. Cloetingh *et al.* 1994) and compressional (e.g. Davy & Cobbold 1991) environments. The reader is referred to these works for further discussions and references.

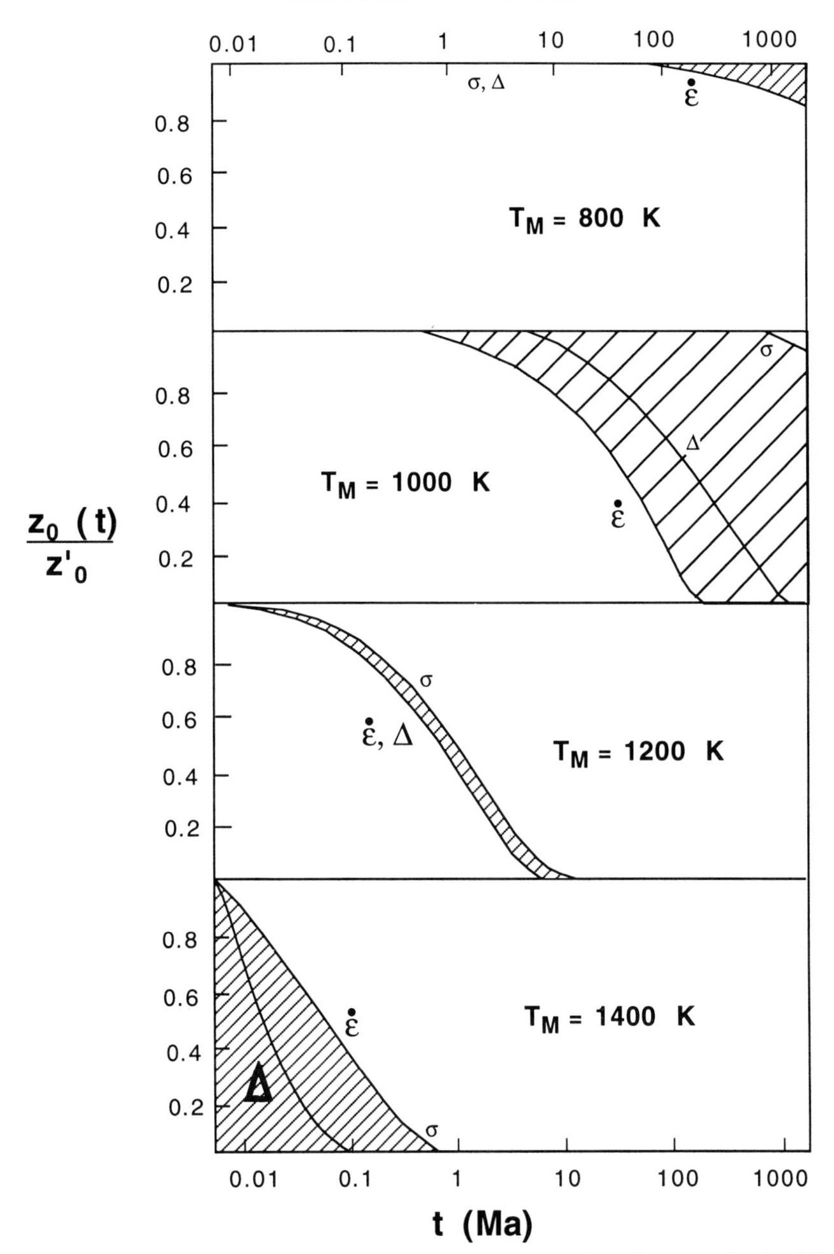

Fig. 9. Relaxation of Moho topography as a function of time after cessation of compression for different values of Moho temperature. Estimates based on constant stress, constant strain rate, and constant viscous dissipation are denoted by σ, $\dot{\epsilon}$, and Δ, respectively ($\sigma = 10\,\mathrm{MPa}$, $\dot{\epsilon} = 10^{-15}\,\mathrm{s}^{-1}$, $\Delta = 10^{-8}\,\mathrm{J\,m^{-3}\,s^{-1}}$).

The Archaean lithosphere: oceanic flake tectonics?

The question of whether the rheology of the lithosphere has changed over geological time is relevant to the problem of the persistence of plate tectonics throughout the history of the Earth. The problem, from the geophysical viewpoint, is poorly constrained. As an example, we discuss in this section a possible significant change.

The mantle in the Archaean was a few hundred degrees hotter than at present (see Durrheim & Mooney 1994 for a discussion of the

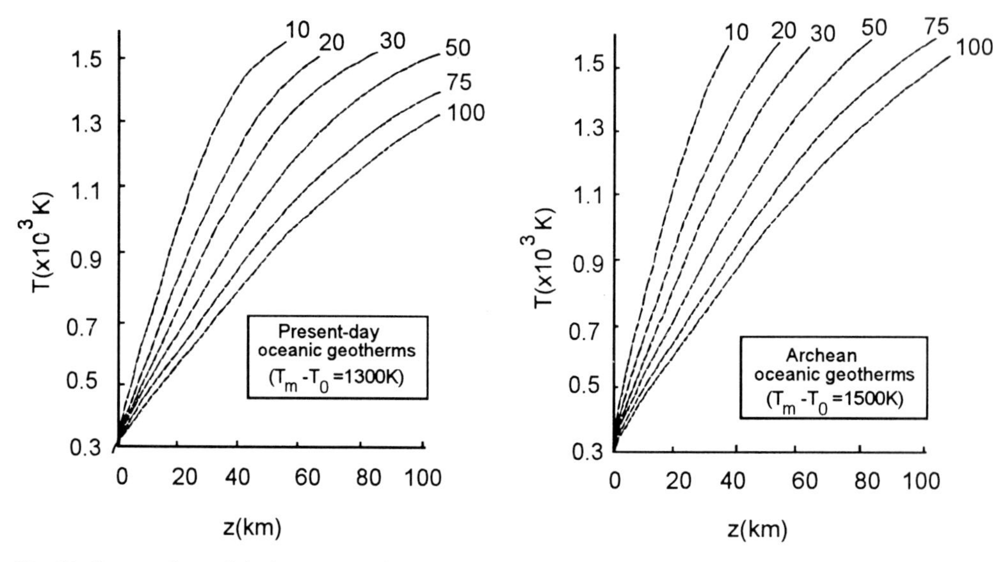

Fig. 10. Present-day and Archaean oceanic geotherms as a function of age of oceanic lithosphere (Ma) assuming the Archaean asthenosphere to have been 200 K hotter than today (from Hoffman & Ranalli 1988).

Precambrian lithosphere). A minimum estimate of asthenospheric temperature, based on the eruption temperature of komatiitic lavas, is 1500°C (Bickle 1986), as compared to the present value of 1300°C. On the other hand, pressure-temperature conditions in the continental Archaean lithosphere were not very different from today (England & Bickle 1984). This has lead to the idea that the extra heat was dissipated by increased oceanic spreading activity (total ridge length times average spreading rate), i.e. by the faster motion (compared to the present) of smaller plates (see e.g. Nisbet & Fowler 1983).

A higher-temperature mantle resulted in a higher degree of partial melting and consequently a thicker oceanic crust, probably in the 25–50 km range (Bickle 1986). At first sight, the presence of this thick basaltic crust, combined with a younger average age of oceanic lithosphere, should have made a part (and probably most) of Archaean oceanic lithosphere incapable of negative-buoyancy subduction. However, a straightforward application of oceanic geotherms (equation 6) to infer the strength profile of Archaean oceanic lithosphere, shows that subduction was made possible by delamination along the Moho (Hoffman & Ranalli 1988).

Oceanic geotherms for the present ($T_m - T_0 = 1300$ K) and for the Archaean ($T_m - T_0 = 1500$ K) are shown in Fig. 10. Present-day and Archaean strength envelopes are shown in Fig. 11 (oceanic crust is modelled with the rheology of diabase, upper mantle with the rheology of peridotite). The main difference is the thickness of the crust (7 km at present, 25 km in the Archaean). As a result, the present crust is completely in the brittle field, and consequently the lithosphere is not layered (see also Fig. 5), while Archaean lower crust was ductile, and the lithosphere was therefore rheologically layered. For $t \leqslant 50$ Ma, the lower crust was very soft. Hoffman & Ranalli (1988) hypothesized that delamination of the oceanic lithosphere would have been easy along this layer, and that in this way the buoyancy problem of Archaean subduction may be overcome: the low-density oceanic crust was accreted or obducted onto the overriding plate (i.e., it formed 'flakes'), and the remaining delaminated lithosphere sank due to its negative buoyancy.

It is therefore possible that the rheological behaviour of the Archaean lithosphere was the factor that allowed, at least in some cases, negative-buoyancy subduction to occur, and consequently the operation of plate tectonics as we know it today. This, together with other factors (for instance, the average viscosity of the mantle was lower because of the higher temperature, resulting in more vigorous convection), points to the conclusion that, while the rate of tectonic processes has probably changed over time, the type of processes has remained the same throughout most of the Earth's history.

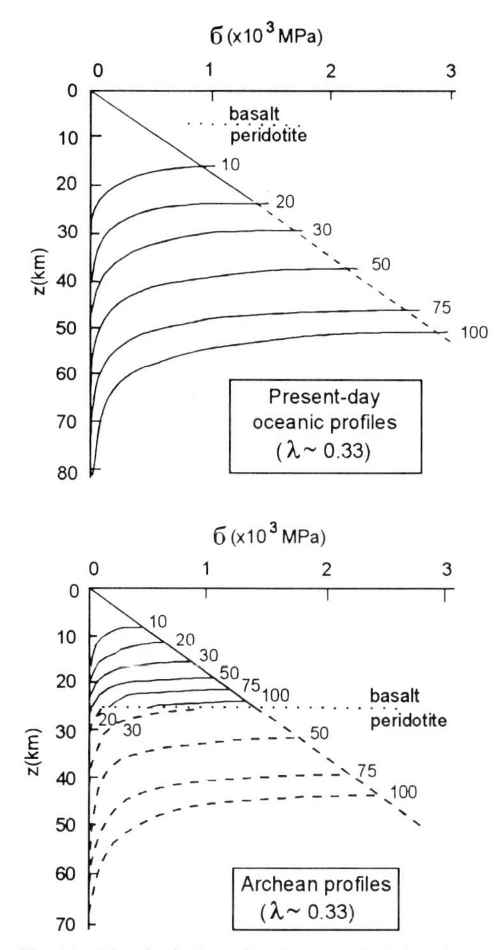

Fig. 11. Rheological profiles for oceanic lithosphere at present and in the Archaean, for compressive regime, strain rate 10^{-14} s^{-1}, and pore fluid factor as shown. See discussion in the text (from Hoffman & Ranalli 1988).

This work has been financed by a Research Grant from the Natural Sciences and Engineering Research Council of Canada. Thanks are due to Jean-Pierre Burg for his kind invitation to lecture at ETH, Zürich, to Philippe Davy and Dave Olgaard for helpful reviews, and to Mary Ford for her friendly support.

References

Bickle, M. J. 1986. Implications of melting for stabilization of the lithosphere and heat loss in the Archaean. *Earth and Planetary Science Letters,* **80**, 314–324.

Bird, P. 1991. Lateral extrusion of lower crust from under high topography, in the isostatic limit. *Journal of Geophysical Research,* **96**, 10275–10286.

Bott, M. H. P. 1990. Stress distribution and plate boundary force associated with collision mountain ranges. *Tectonophysics,* **182**, 193–209.

Byerlee, J. D. 1967. Frictional characteristics of granite under high confining pressure. *Journal of Geophysical Research,* **72**, 3639–3648.

Carslaw, H. S. & Jaeger, J. C. 1959. *Conduction of heat in solids*, 2nd ed. Clarendon Press, Oxford.

Chapman, D. S. & Furlong, K. P. 1992. Thermal state of the continental lower crust. *In*: Fountain, D. M., Arculus, R. & Kay, R. W. (eds) *Continental lower crust.* Elsevier, Amsterdam, 179–199.

Chen, W. P. & Molnar, P. 1983. Focal depths of intracontinental and intraplate earthquakes and their implications for the thermal and mechanical properties of the lithosphere. *Journal of Geophysical Research,* **88**, 4183–4214.

Cloetingh, S., Sassi, W. & Task Force Team 1994. The origin of sedimentary basins: a status report from the task force of the International Lithosphere Program. *Marine and Petroleum Geology,* **11**, 659–683.

Clowes, R. M. 1993. Variations in continental crustal structure in Canada from lithosphere seismic reflection and other data. *Tectonophysics,* **219**, 1–27.

Davy, P. & Cobbold, P. R. 1991. Experiments on shortening of a four-layer model of the continental lithosphere. *Tectonophysics,* **188**, 1–25.

Durrheim, R. J. & Mooney, W. D. 1994. Evolution of the Precambrian lithosphere: seismological and geochemical constraints. *Journal of Geophysical Research,* **99**, 15359–15374.

England, P. & Bickle, M. 1984. Continental thermal and tectonic regimes during the Archaean. *Journal of Geology,* **92**, 353–367.

—— & McKenzie, D. 1982. A thin viscous sheet model for continental deformation. *Geophysical Journal of the Royal Astronomical Society,* **70**, 295–321.

Fadaie, K. & Ranalli, G. 1990. Rheology of the lithosphere in the East African Rift System. *Geophysical Journal International,* **102**, 445–453.

Fleitout, L. & Froidevaux, C. 1982. Tectonics and topography for a lithosphere containing density heterogeneities. *Tectonics,* **1**, 21–56.

Fountain, D. M., Arculus, R. & Kay, R. W. (eds) 1992. *Continental lower crust.* Elsevier, Amsterdam.

Frost, H. J. & Ashby, M. F. 1982. *Deformation mechanism maps – The plasticity and creep of metals and ceramics.* Pergamon Press, Oxford.

Handy, M. R. 1994. Flow laws for rocks containing two non-linear viscous phases: a phenomenological approach. *Journal of Structural Geology,* **16**, 287–301.

Hirth, G. & Tullis, J. 1989. The effects of pressure and porosity on the micromechanics of the brittle-ductile transition in quartzite. *Journal of Geophysical Research,* **94**, 17825–17838.

Hoffman, P. F. & Ranalli, G. 1988. Archaean oceanic flake tectonics. *Geophysical Research Letters,* **15**, 1077–1080.

HOUSEMAN, G. A., MCKENZIE, D. P. & MOLNAR, P. 1981. Convective instability of a thickened boundary layer and its relevance for the thermal evolution of continental convergent belts. *Journal of Geophysical Research,* **86**, 6115–6132.

HYNDMAN, R. D. & SHEARER, P. M. 1989. Water in the lower continental crust: modelling magnetotelluric and seismic reflection results. *Geophysical Journal International,* **98**, 343–365.

JAEGER, J. C. & COOK, N. G. W. 1979. *Fundamentals of rock mechanics*, 3rd ed. Chapman & Hall, London.

JI, S. & ZHAO, P. 1993. Flow laws of multiphase rocks calculated from experimental data on the constituent phases. *Earth and Planetary Science Letters,* **117**, 181–187.

KIRBY, S. H. 1983. Rheology of the lithosphere. *Reviews of Geophysics and Space Physics,* **21**, 1458–1487.

—— & KRONENBERG, A. K. 1987. Rheology of the lithosphere: selected topics. *Reviews of Geophysics,* **25**, 1219–1244.

KNIPE, R. J. & RUTTER, E. H. (eds) 1990. *Deformation Mechanisms, Rheology and Tectonics*. Geological Society, London, Special Publications, **54**.

KRUSE, S., MCNUTT, M., PHIPPS-MORGAN, J. & ROYDEN, L. 1991. Lithospheric extension near Lake Mead, Nevada: a model for ductile flow in the lower crust. *Journal of Geophysical Research,* **96**, 4435–4456.

KUSZNIR, N. J. 1991. The distribution of stress with depth in the lithosphere: thermo-rheological and geodynamic constraints. *Philosophical Transactions of the Royal Society of London,* **A337**, 95–110.

LAMONTAGNE, M. & RANALLI, G. 1996. Thermal and rheological constraints on the earthquake depth distribution in the Charlevoix, Canada, intraplate seismic zone. *Tectonophysics*, **257**, 55–69.

LOWE, C. & RANALLI, G. 1993. Density, temperature, and rheological models for the southeastern Canadian Cordillera: implications for its geodynamic evolution. *Canadian Journal of Earth Sciences,* **30**, 77–93.

MARESCHAL, J. C. 1994. Thermal regime and post-orogenic extension in collision belts. *Tectonophysics,* **238**, 471–484.

MOLNAR, P. & LYON-CAEN, H. 1988. Some simple physical aspects of the support, structure, and evolution of mountain belts. *In*: CLARK, S. P. JR. (ed.) *Processes in Continental Lithospheric Deformation*. Geological Society of America, Special Papers, **218**, 179–207.

——, ENGLAND, P. & MARTINOD, J. 1993. Mantle dynamics, uplift of the Tibetan Plateau, and the Indian monsoon. *Reviews of Geophysics,* **31**, 357–396.

MOONEY, W. D. & MEISSNER, R. 1992. Multi-genetic origin of crustal reflectivity: a review of seismic reflection profiling of the continental lower crust and Moho. *In*: FOUNTAIN, D. M., ARCULUS, R. & KAY, R. W. (eds) *Continental lower crust.* Elsevier, Amsterdam, 45–79.

NISBET, E. G. & FOWLER, C. M. R. 1983. Model for Archaean plate tectonics. *Geology,* **11**, 376–379.

PLATT, J. P. 1993. Exhumation of high-pressure rocks: a review of concept and processes. *Terra Nova,* **5**, 119–133.

POIRIER, J. P. 1985. *Creep of crystals – High temperature deformation processes in metals, ceramics and minerals.* University Press, Cambridge.

RAMBERG, H. 1968. Fluid dynamics of layered systems in the field of gravity, a theoretical basis for certain global structures and isostatic adjustment. *Physics of the Earth and Planetary Interiors,* **1**, 63–87.

RANALLI, G. 1991. Regional variations in lithosphere rheology from heat flow observations. *In*: CERMAK, V. & RYBACH, L. (eds) *Terrestrial heat flow and the lithosphere structure*. Springer-Verlag, Berlin, 1–22.

—— (ed.) 1993. Heat flow, rock mechanics, and seismicity. *Tectonophysics,* **217**, 1–115.

—— 1995. *Rheology of the Earth*, 2nd ed. Chapman & Hall, London.

—— & MURPHY, D. C. 1987. Rheological stratification of the lithosphere. *Tectonophysics,* **132**, 281–295.

—— & YIN, Z. M. 1990. Critical stress difference and orientation of faults in rocks with strength anisotropies: the two-dimensional case. *Journal of Structural Geology,* **12**, 1067–1071.

——, BROWN, R. L. & BOSDACHIN, R. 1989. A geodynamic model for extension in the Shuswap core complex, southeastern Canadian Cordillera. *Canadian Journal of Earth Sciences,* **26**, 1647–1653.

REY, P., BURG, J. P. & CARON, J. M. 1991. Middle and late Carboniferous extension in the Variscan belt: structural and petrological evidence from the Vosges massif (eastern France). *Geodinamica Acta,* **5**, 17–36.

RUTTER, E. H. & BRODIE, K. H. 1992. Rheology of the lower crust. *In*: FOUNTAIN, D. M., ARCULUS, R. & KAY, R. W. (eds) *Continental lower crust.* Elsevier, Amsterdam, 201–267.

SCHOLZ, C. H. 1990. *The mechanics of earthquakes and faulting.* University Press, Cambridge.

SHIMADA, M. 1993. Lithosphere strength inferred from fracture strength of rocks at high confining pressures and temperatures. *Tectonophysics,* **217**, 55–64.

—— & CHO, A. 1990. Two types of brittle fracture of silicate rocks under confining pressure and their implications in the Earth's crust. *Tectonophysics,* **175**, 221–235.

SIBSON, R. H. 1974. Frictional constraints on thrust, wrench and normal faults. *Nature,* **249**, 542–544.

—— 1982. Fault zone models, heat flow, and the depth distribution of earthquakes in the continental crust of the United States. *Bulletin of the Seismological Society of America,* **72**, 151–163.

SONDER, L. T. & ENGLAND, P. 1986. Vertical averages of rheology of the continental lithosphere: relation to thin sheet parameters. *Earth and Planetary Science Letters,* **77**, 81–90.

STEIN, C. A. 1995. Heat flow of the Earth. *In*: AHRENS, T. J. (ed.) *Global Earth Physics: A handbook of physical constants*. American Geophysical Union, Washington, 144–158.

TULLIS, T. E., HOROWITZ, F. G. & TULLIS, J. 1991. Flow laws of polyphase aggregates from end-members flow laws. *Journal of Geophysical Research,* **96**, 8081–8096.

TURCOTTE, D. L. & SCHUBERT, G. 1982. *Geodynamics – Applications of continuum physics to geological problems.* Wiley, New York.

WILKS, K. R. & CARTER, N. L. 1990. Rheology of some continental lower crustal rocks. *Tectonophysics,* **182**, 57–77.

WILLETT, S. D., BEAUMONT, C. & FULLSACK, P. 1993. Mechanical model for the tectonics of doubly vergent compressional orogens. *Geology,* **21**, 371–374.

YIN, Z. M. & RANALLI, G. 1992. Critical stress difference, fault orientation and slip direction in anisotropic rocks under non-Andersonian stress systems. *Journal of Structural Geology,* **14**, 237–244.

ZEYEN, H. & FERNANDEZ, M. 1994. Integrated lithospheric modelling combining thermal, gravity, and local isostasy analysis: application to the NE Spanish geotransect. *Journal of Geophysical Research,* **99**, 18089–18102.

ZHOU, S. & SANDIFORD, M. 1992. On the stability of isostatically compensated mountain belts. *Journal of Geophysical Research,* **97**, 14207–14221.

Rheology of planetary lithospheres: a review from impact cratering mechanics

P. G. THOMAS, P. ALLEMAND & N. MANGOLD

Laboratoire des sciences de la Terre (URA 726 ENS Lyon et UCB Lyon I), Ecole Normale Supérieure de Lyon, 46 allée d'Italie, 69364 Lyon cedex 07, France

Abstract: Impact craters are the only universal drills to probe planetary lithospheres. The relationship of the lithospheric thickness/crater diameter ratio and the morphology of the outer parts of the crater are controlled by the rheology of the target. These relationships indicate the following: (1) The lithosphere is thicker on silicated bodies than on icy bodies of the same size. (2) The lithosphere is thicker on small planetary bodies than on larger ones. (3) There is a secular lithospheric thickening and a planetary cooling for silicated and icy bodies. (4) Because of the thickness and the continuity of the lithosphere of one-plate planets, local mechanical disturbances may be transmitted over a long distance and may affect the entire planet. (5) Thermal properties of the lithospheres of satellites are incompatible with the chondritic abundance of radioactive nucleids. This indicates that the origin of satellites is more complex than the accretion–collision of chondritic/icy planetesimal swarms. (6) There is no direct relation between plate tectonics and lithospheric thickness. Some planets have (or have had) the same lithospheric thickness as the Earth, but have not developed plate tectonics. This indicates that plate tectonic development is controlled by parameters other than lithospheric thickness.

The term 'lithosphere' was introduced at the beginning of this century, and has been widely used by geologists and geophysicists since the development of the plate tectonic theory: the lithosphere of the Earth is divided into eight large undeformable rigid plates which are moving with respect to one another, and with respect to a deep, weak layer called the asthenosphere. Before the acceptance of this dynamic meaning, the notion of lithosphere was not used much by geologists. After the plate tectonic theory was accepted, the lithosphere was more precisely defined in several ways in order to distinguish different types of lithospheres: seismic, thermal or mechanical lithospheres.

With the exception of Venus and possibly Europa and Enceladus, there is no strong evidence of large-scale horizontal motions on planetary surfaces. On Venus, the Magellan space probe revealed unquestionable horizontal motions, but only of little importance, without spreading centres or subduction zones. Thus, the Earth seems to be the only planetary body with active or past plate tectonics. The dynamic concept of lithosphere is thus difficult to apply to planets or satellites. Horizontal displacements associated with some tectonic activity exist on some planets and satellites. But these displacements are generally less than a few hundred metres for each individual structure, which are generally less than 10 km in width and thus probably affect only the upper part of the shallow layers. High resolution topographic and gravimetric data exist only for Venus, and partially for Mars and the Moon. Seismic and thermal data exist only for the Moon. Therefore, it is very difficult to define the lithosphere on the planets, to study their rheology and to compare these few results to the lithosphere of the Earth.

Fortunately impact craters and related phenomena that affect most of the planets are ideal sources of information that can be used to understand the rheology of planetary lithospheres. According to their diameter, craters affect lithospheres to different depths and are thus natural drills. These natural and universal drills allow us to compare different planets and satellites under similar mechanical conditions. Impacts can be dated using cratering time scales. They thus reveal the rheology of a lithosphere at different times, which allows us to identify an evolution of the properties of lithospheres. In this paper, we present a brief reminder of impact cratering mechanics, then we describe the rheology and stress state of the lithosphere of Mercury, the Moon and Mars, as revealed by

From Burg, J.-P. & Ford, M. (eds), 1997, *Orogeny Through Time*,
Geological Society Special Publication No. 121, pp. 39–62.

impact craters. For the Moon and Mars, these lithospheric properties are compared with results obtained by flexural data. On the Moon, the results are compared to the Apollo seismic and thermal data. Data on two Jovian satellites, Ganymede and Callisto are presented to show the similarities between silicate and icy lithospheres, and to determine the thermal state of these bodies during the first billion years. The cases of the lithosphere of Venus and of the shallow structure of the Martian lithosphere are briefly examined without the use of impact craters.

Origin and morphology of impact craters

Impact cratering has been recognized as a field of study for only a few decades. The first key study was by G. K. Gilbert at the end of the nineteenth century (1893), but important geological studies truly began with the discovery of coesite and stishovite near two Arizonan and German craters (Chao *et al.* 1960), and were intensified with planetary exploration. Since then, references are numerous. Three books allow further reading and provide comprehensive reference lists: *Impact and Cratering explosions*, (Roddy *et al.* 1977), *Multi-ring basins* (Lunar and Planetary Institute 1980), and *Impact Cratering, a Geological Process* (Melosh 1989).

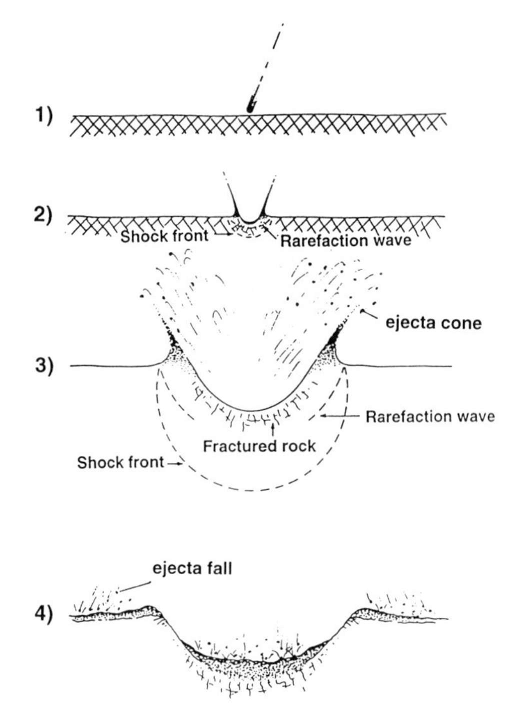

Fig. 1. Schematic diagrams representing a sequence of four events during the formation of a simple crater. (Diagrams adapted from illustrations of Shoemaker 1960; Gault *et al.* 1968; Dence 1968.)

Simple and complex craters

Many planets and satellites exhibit a surface older than 3.5 Ga. These old surfaces have been impacted by several large meteorites, which were abundant during the first billion years of the solar system. These large meteorites produced impact craters larger than 100 km in radius, and deeper than 30 km. These large impact craters represent natural 'drills' through the planetary lithospheres.

When a meteorite hits the ground, a shock wave is produced that transfers the kinetic energy of the meteorite into the ground. The shock wave advances through the ground and a rarefaction wave forms behind. Material compressed and crushed by the passage of the shock front is set in motion by the rarefaction wave. An ejecta cone is formed as melted and shocked rocks are ejected from the expanding cavity. Material behind the rarefaction wave is thrown out along ballistic trajectories. As the shock and the following rarefaction waves grow in the ground, the energy of compression and ejection by surface unit decreases. The excavation stops when the energy of the rarefaction wave is insufficient to eject target material (Gault *et al.* 1968). Most of the ejected material falls to the surface around the crater, forming an ejecta blanket. Some of the ejecta fall back into the crater to form a breccia lens within the crater. The ejected material contains an amount of molten material that increases with the size of the impact. Such craters are named simple craters (Fig. 1). Their bowl-like shape is identical whatever the planet, its gravity, or the nature of the crust. Their final depth is about one fifth of their radius. The depth of the cavity without the interior ejecta (the transient cavity) is about one third of the radius. The ejected blocks of target material induce secondary impact craters (the secondaries) in the ejecta blanket. The distribution of ejecta and secondaries is strongly dependent on gravity (Gault *et al.* 1975).

On the Earth and the Moon, craters wider than 3 km in radius exhibit central uplift and a flat floor. This morphology is produced by the motion of the target during the terminal stage of crater formation, and would be due to the slumping of the cavity walls and decompression and rebound of rocks below the crater floor (Figs 2 & 3). This motion is 'geologically' instantaneous, as sug-

Fig. 2. The Copernicus crater, on the Moon, example of a complex crater 96 km in diameter. Note the central peak, and the secondaries (arrow) which are mainly located 190 km from the central peak (4 copernicus radii). (NASA photo Lunar Orbiter 4.)

gested by the interior ejecta which are generally not deformed by the uplift. These craters are named complex craters. Inside terrestrial craters greater than 15 km in radius, the central uplift becomes a central ring structure with alternating high and low topographic features. The largest impact craters show more than one ring structure and are named ring-basins (Fig. 3). The change in critical radius of crater type is strongly dependent on the gravity. The radius transition of simple to complex crater is 3 km on the Earth, and 8 km on the Moon. The radius transition of complex to ring crater is 15 km on the Earth and 100 km on the Moon (Gault *et al.* 1975).

The multi-ring basins

On the Moon, there are 29 basins larger than 300 km in radius. The Orientale basin, 450 km in radius, is very well preserved and not embayed or buried by later events (Fig. 4). This basin has four rings, R1 (140 km in radius), R2 (220 km), R3 (320 km) and R4 (450 km). The R4 ring is somewhat angular in plan, with sections that are not concentric with the inner rings. A comparison of the Orientale Basin to a classical complex crater such as Copernicus (Fig. 2), was undertaken by several authors (e.g. Head 1974, 1977; Moore *et al.* 1974; McCauley 1977; Scott *et al.*

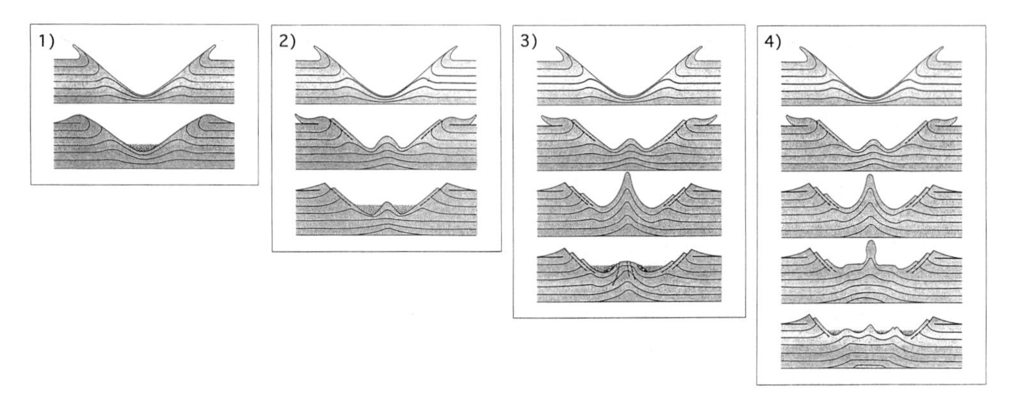

Fig. 3. Schematic cross sections showing succesive stages in formation of impact craters in four different size ranges. Each column corresponds to one diameter of characteristic morphology (lunar value): 1, 10 km in diameter, 2, 30 km in diameter, 3, 80 km in diameter, 4, 180 km in diameter. (From Thomas 1986, adapted from illustrations of Murray 1980.)

1977; McCauley *et al.* 1981). Around fresh complex craters, some parameters are identical, whatever the size of the craters (Fig. 5). (1) Morphological studies show differences among the ejecta, which depend on their position; for example, interior ejecta are different from outer ejecta near the crater rim. (2) The edge of the cavity exhibits characteristic faulted and tilted blocks. (3) The ratio between the cavity radius and the largest secondary crater radius is about 1:15 (Shoemaker 1966). (4) The maximum density of the secondary craters is located about 4 radii from the center, and the most distant secondary is at about 8 radii (Shoemaker 1966). The use of these morphological features and geometric proportions indicate that the Orientale cavity would have a radius of 320 km. The limit of the impact cavity would be the R3 ring. This interpretation implies that the most external ring (R4) was produced outside the transcient impact cavity. Some theories were proposed to explain R4 as an external ring, but the main subject of debate concerned its nature (external ring or transient cavity limit ?). The Martian exploration, which revealed one basin with possible outer rings (Hodges 1980), and the Voyager mission which explored the Jovian satellites proved the reality of outer rings (Smith *et al.* 1979): the Gilgamesh Basin on Ganymede is a seven-ring basin (Fig. 6) and the Valhalla Basin on Callisto (about 300 km in radius) is surrounded by 15 outer rings, which extend more than 1500 km from the crater itself (Fig. 7). This supports the Orientale R4 ring as an external structure.

Three kinds of theories have been proposed to explain external rings. The first one was proposed for the single outer ring of Orientale: the outside ring (R4) would be the result of a mega-terrace which slumped inward (Head 1974). If this mechanism is possible for one single outer ring relatively near the limit of the impact excavation, it cannot be accepted for multiple outer rings which extend more than five crater radii from the transient cavity. The Tsunami model is the second hypothesis, proposed for the Moon (Van Dorn 1968), and later for the Galilean satellites (Smith *et al.* 1979). If the asthenosphere is very fluid, tsunami-like oscillations of the transient cavity would propagate in the manner of tsunami waves. The resulting strains should show a pattern of radial and mainly concentric fractures within the thin lithosphere. The third hypothesis, first proposed by Melosh & McKinnon (1978), was improved by these authors after the Voyager results (McKinnon & Melosh 1980; Melosh 1982; Fig. 8). The lithosphere is supposed to be a thin elastic/brittle plate, and the asthenosphere is modelled as a viscous fluid. If the excavation is deeper than the lithospheric thickness and penetrates the asthenosphere, the asthenosphere, driven by gravity, flows as a viscous fluid towards the crater. The flow exerts a shear stress on the base of the lithosphere and creates a state of radial tension. Analytical solutions of the static stress field around relaxing craters have been found by Melosh & McKinnon (1978), and McKinnon and Melosh (1980) using both elastic and plastic rheologies for the upper part of the deforming lithosphere. In these models the radial asthenospheric flow is considered as a stress boundary condition acting at the base of the overlying brittle plate. These

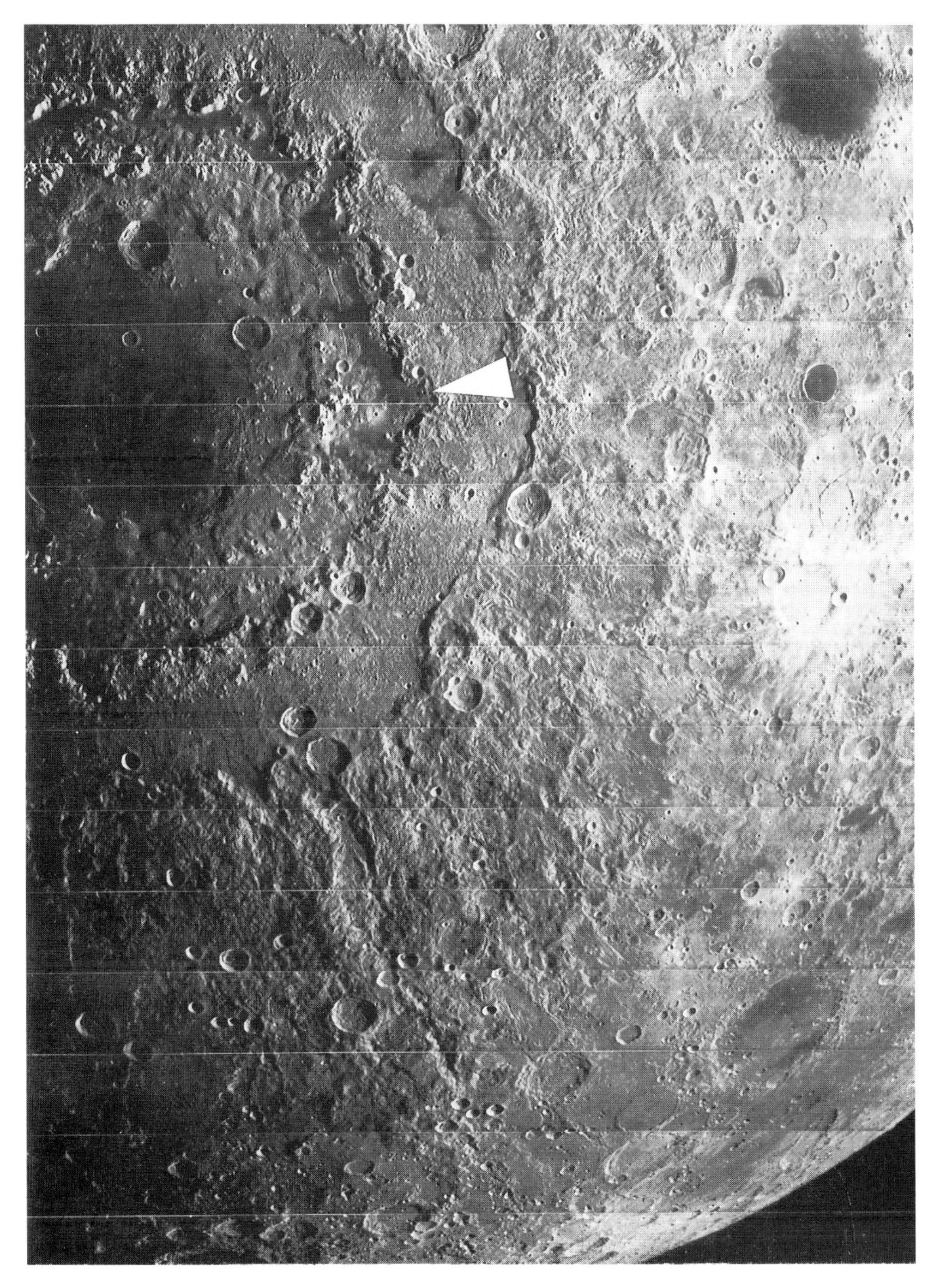

Fig. 4, The Orientale Basin, on the Moon, 900 km in diameter. This basin is a four-ring basin. The limit of the transient cavity is the R3 ring (arrow). (NASA Photo Lunar Orbiter 4.)

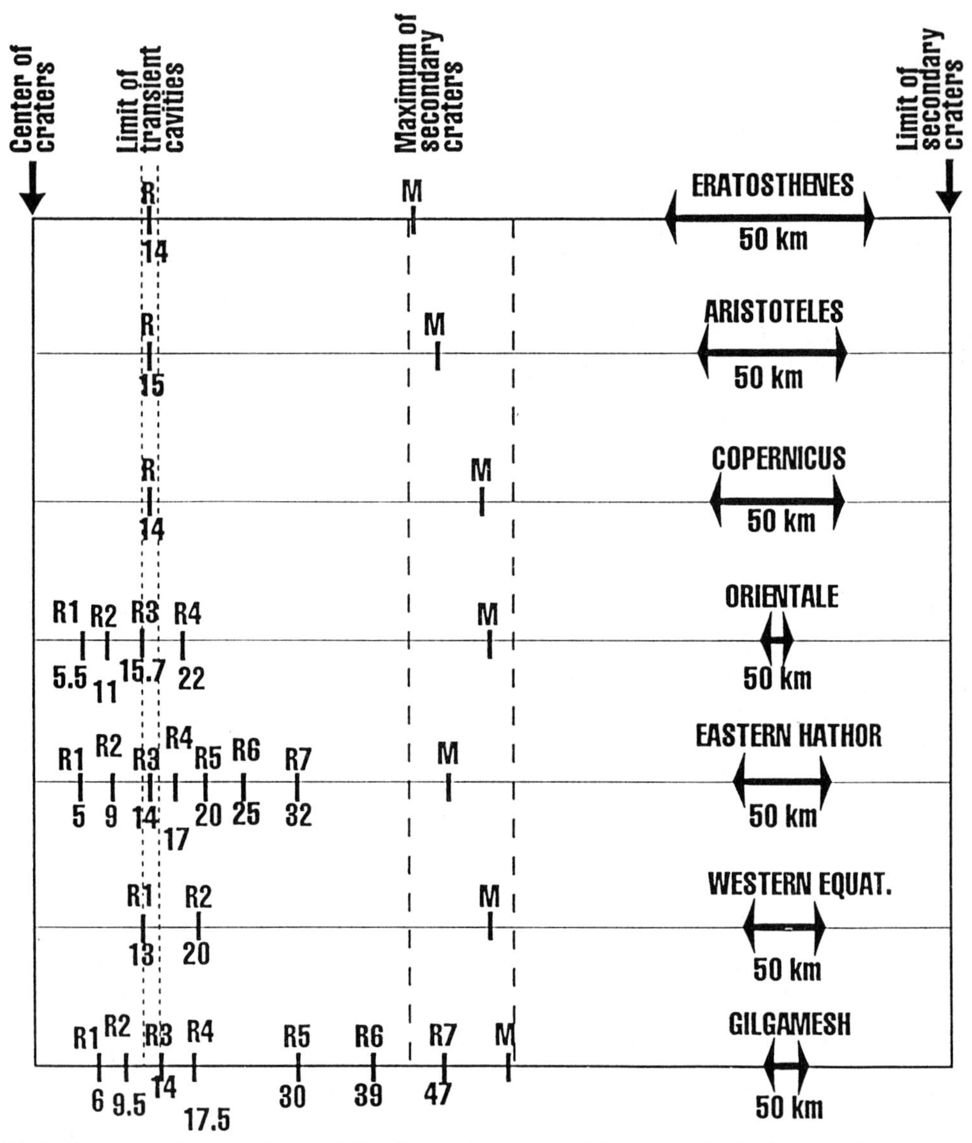

Fig. 5. Comparison between characteristics of the main craters and the secondaries for Lunar and Ganymedean basins. All basins are reported here with the same size. Scales are shown by arrows under the name of craters. The centres of each basin are on the left, the end of each swarm of secondaries are at the right. Above each line that represents the basins are reported the different rings (Rx), and the position of the maximum density of secondaries (M). The number under each ring represents the main crater to largest secondary radii ratio. This shows the position of the transient cavity limits, R3 for Orientale (Fig. 4) and Gilgamesh (Fig. 6). (From Thomas *et al.* 1986.)

calculations show that the stress field around the cavity is compatible with the observed structures: the radial tension induces strike-slip faults and mainly concentric normal faults which represent the external rings.

If the lithosphere is thin relative to the depth of the cavity, the lithosphere rifts in multiple concentric grabens or tilted blocks. For thicker lithospheres, only one or very few irregular, steep inward-facing normal faults form. If the lithosphere is largely thicker than the transient cavity depth, there is no ring formation outside the transient cavity, but a central inner peak or inner ring may exist.

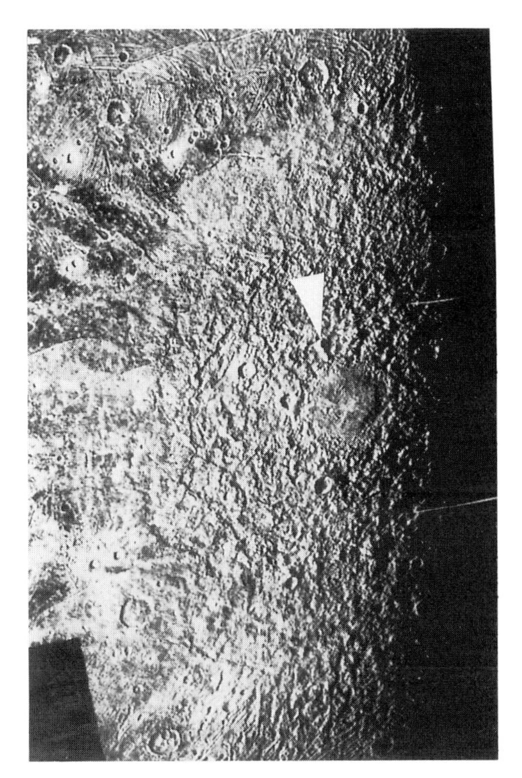

Fig. 6. The Gilgamesh basin on Ganymede, a seven-ring basin. The limit of the transient cavity is the R3 ring (arrow), 240 km in diameter. (NASA Photo Voyager 2.)

In order to understand the strain evolution around a crater and the effect of coupling between asthenospheric flow and brittle deformation, we have undertaken analogical simulations (Fig. 9) using the technique developed by Faugère & Brun (1984), Davy & Cobbold (1988) and Vendeville *et al.* (1987). The upper part of the lithosphere is assumed to follow a Coulomb behaviour so that brittle strength increases with depth. Because temperature increases with depth, the material becomes progressively ductile at depth. The brittle upper part is modelled using quartz sand which obeys Coulomb criteria with an internal friction angle of 30° and negligible cohesion. The ductile layer is modelled using silicone putty with a viscosity of around 10^4 Pa s. The model is constructed in a 1 m^2 box. A circular hole is slowly, and without shock, made in the center of the sand-silicone double layer. The pressure gradient produced by the hole in the ductile level induces a radial flux of the silicone, which simulates the asthenospheric flow around the crater. The parameters studied are: the respective thickness of brittle and ductile layers and the diameter of the hole with respect to the total thickness of the model.

For an experiment with 2 cm of silicone, 2 cm of sand, a crater diameter of 24 cm and a crater depth of 3 cm, the first structures that appear at the surface are strike-slip faults organized in a spiral around the crater and extending to one radius from the border of the cavity. The number of faults decreases with the distance from the border of the hole. During the experiment, the innermost strike-slip faults progressively acquire a reverse component. Conversely the external strike-slip fault segments become progressively normal, locally bounding grabens. They connect together, and finally look like concentric structures. For craters with larger radii, one outer ring of purely normal faults develops in addition to the previously described structures. When the brittle layer is thicker, the number of faults decreases and the displacement on each individual fault increases. These experiments show that the conceptual model for multi-ring basins origin is realistic. However, it seems mechanically more successful in explaining the external purely extensional structures and the far strike-slip faults which have evolved into normal faults, than the inner, oblique strike-slip faults that are more conspicuous in models than around natural craters.

Melosh & McKinnon's (1978) model is now largely accepted, and may be a universal way to determine the lithospheric thickness during the time of large impacts (before 3.5 Ga). It should be noted that the definition of lithosphere and asthenosphere is here mechanical, and depends upon very high stresses and relatively short time scales. The lithosphere so determined is named the 'impact lithosphere'. It is not the same as conventional specifications of the thermal or mechanical lithosphere of the Earth which involve low stress and long geological times.

The lithospheres of terrestrial planets

The Moon and Mars

The Moon and Mars each exhibit one large well preserved basin: Orientale on the Moon, and Argyre on Mars. These two basins are approximately contemporary (3.8–3.9 Ga) and may be used to compare the lithospheres of two planetary bodies at the same time.

The Orientale basin, described in the previous section is the most famous example of a multi-ring basin; it is interpreted as a crater 320 km in radius, with one outer ring, 450 km in radius. The angular shape of the outer ring may

Fig. 7. The Valhalla basin on Callisto, the most complex multi-ring basin in the solar system. (NASA Photo Voyager 1). The patch of high albedo material in the center of the basin corresponds to the transient cavity, 600 km in diameter.

be interpreted as the result of the coalescence of strike-slip fault fragments evolving into normal fault rings as described in the previous section. The Melosh & McKinnon model (1978) implies that the lithospheric thickness is the same or slightly thinner than the transient cavity depth, i.e around, or slightly less than, $320/3 \approx 100$ km.

On Mars, the Argyre basin is eroded and subdued, but detailed morphologic study allows us to identify the transient cavity boundary with a radius of about 300 km. This boundary is surrounded by discontinuous, inward-facing scarps at a distance of 150 km, 275 km, 450 km and 550 km respectively from the basin rim (Hodges 1980; Thomas & Masson 1984). The Melosh & McKinnon model implies that the lithospheric thickness is much thinner than the cavity depth, i.e. less than $300/3 = 100$ km (Thomas & Masson 1984).

The comparison of Orientale and Argyre is important, because the two basins are identical in size ($R \approx 300$ km) and time of formation (3.8–3.9 Ga). The only changing parameters are the planetary size and the lithospheric thickness. The smaller body, the Moon, whose radius is 1738 km, had a thick lithosphere (about 100 km) while Mars, 3393 km in radius had a lithosphere thinner than 100 km. The 'impact lithosphere' lower limit may be interpreted as a thermal boundary. The deeper level of this boundary on the Moon compared to Mars for the same time indicates that the Moon was cooler. This temperature difference is consistant with the volcanic and tectonic histories of these two bodies: a small planetary body interior is cooler than a larger one. This is related with the ratio of the production of heat, which increases with the planetary volume (cube of the radius), to the loss of heat, which increases with the planetary surface (square of the radius).

Mars and the Moon allow us to compare the 'impact lithosphere' with one of the classic notions of terrestrial lithosphere: the elastic thickness of the lithosphere. The elastic bending theory assumes that if an elastic plate resting on a viscous fluid is loaded, the plate is deflected.

1)

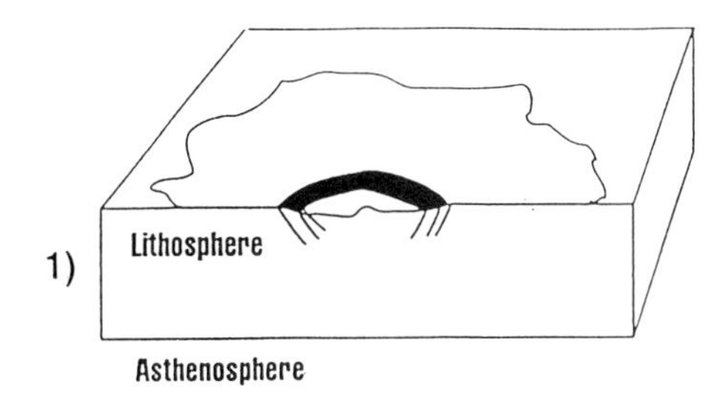

2)

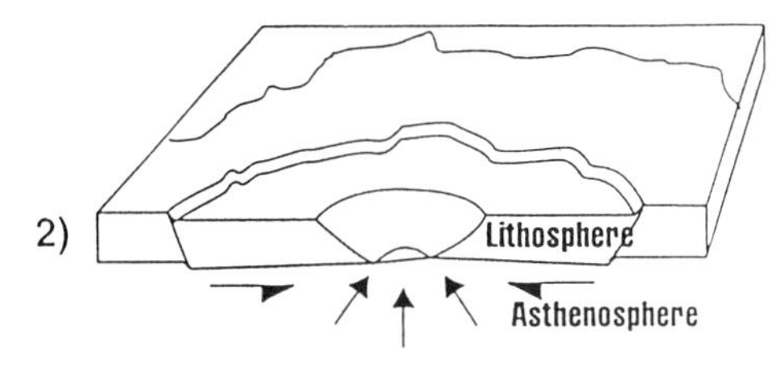

3)

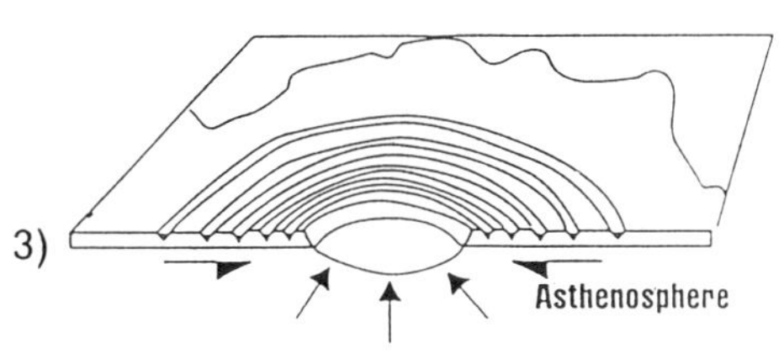

Fig. 8. Ring formation as a function of lithosphere thickness. Details of rim deformation, inner ring or peak development and ejecta are omitted in this schematic treatment. (1) Complex crater; (2) is equivalent to Orientale (Fig. 4) ; (3) is equivalent to Valhalla (Fig. 7). (Diagrams adapted from illustrations of Melosh & McKinnon 1978; McKinnon & Melosh 1980.)

This deflection is larger than the width of the load and is laterally continuous with an outer bulge whose elevation is more than ten times less than the depth of the maximum deflection. The distance from the centre of the deflection to the top of the peripheral bulge is diagnostic for the elastic plate thickness and increases when this thickness increases. If the fibre stress generated in the plate by its torsion exceeds the yield stress of the plate, grabens develop at some distance from the centre of the load before or near the top of the bulge and compressive structures develop inside the load. The distance from the centre of the load to the graben field is characteristic of the elastic plate thickness (Turcotte & Schubert 1982). On the Earth, it has been shown that the base of the elastic oceanic lithosphere roughly corresponds to the 400°C isotherm. Combining the concept of strain envelopes with the mechanics of flexure, more realistic models can be made in which the elastic thickness decreases with the intensity of the load. By connecting the strain envelopes with the thermal state, it is possible to compute the thermal state assuming a realistic strain rate (Carter & Tsenn 1987).

On the Moon, an important volcanic event took place from 3.8 to 3.5 Ga. Several kilometres of lava filled pre-existing topographical depressions of old large impact craters. This mare material is affected by sinuous wrinkle ridges (Fig. 10). These ridges are positive relief, less than 5 km in width, less than 500 m in height and greater than 300 km in length. Such structures, based on the Apollo Lunar Sounding Experiment conducted during the Apollo 17 mission, have been interpreted as compressive structures involving folds and faults (Maxwell 1978). The external border of basins are occupied by concentric grabens less than 10 km in width (Fig. 10). The elastic plate theory explains this association of structures well. Wrinkle ridges result from compression in the basin generated by differential subsidence of the basin floor, and grabens are due to the extensional stress associated with the peripherical bulge. The characteristic distance from the centre of the basin to the graben field indicates an elastic lithosphere of about 50–60 km (Solomon & Head 1979).

This flexural thickness is determined for mare volcanism times, only 1 to 3.10^8 years after the Orientale impact. Neglecting the effect of secular thickening, the 'impact lithosphere' is twice as thick as the elastic lithosphere. Taking into account the secular thickening, the thickness of the 'impact lithosphere' is greater than twice the elastic lithosphere.

On Mars, two broad topographical bulges (2000 and 6000 km in diameter, 10 km in height) exhibit enormous shield volcanoes (20 km in height). These large volcanoes are younger evidence of Martian endogenic activity ($10^9 \pm 5.10^8$ years). The origin of these bulges has been intensely debated for 20 years, and may be related, at least partially, to a deep thermal anomaly. Some of the volcanoes are located on the top of the bulge, above the thermal anomaly, where the lithosphere is thought to be abnormally thin. Others, such as Olympus Mons, are located on the limit of the bulge, where the lithosphere is supposed to be of normal thickness. These enormous volcanoes have induced peripheral lithospheric flexure, as evidenced by graben fields. The corresponding elastic lithospheric thickness was about 20 km on the top of the bulge, and about 120 km at the

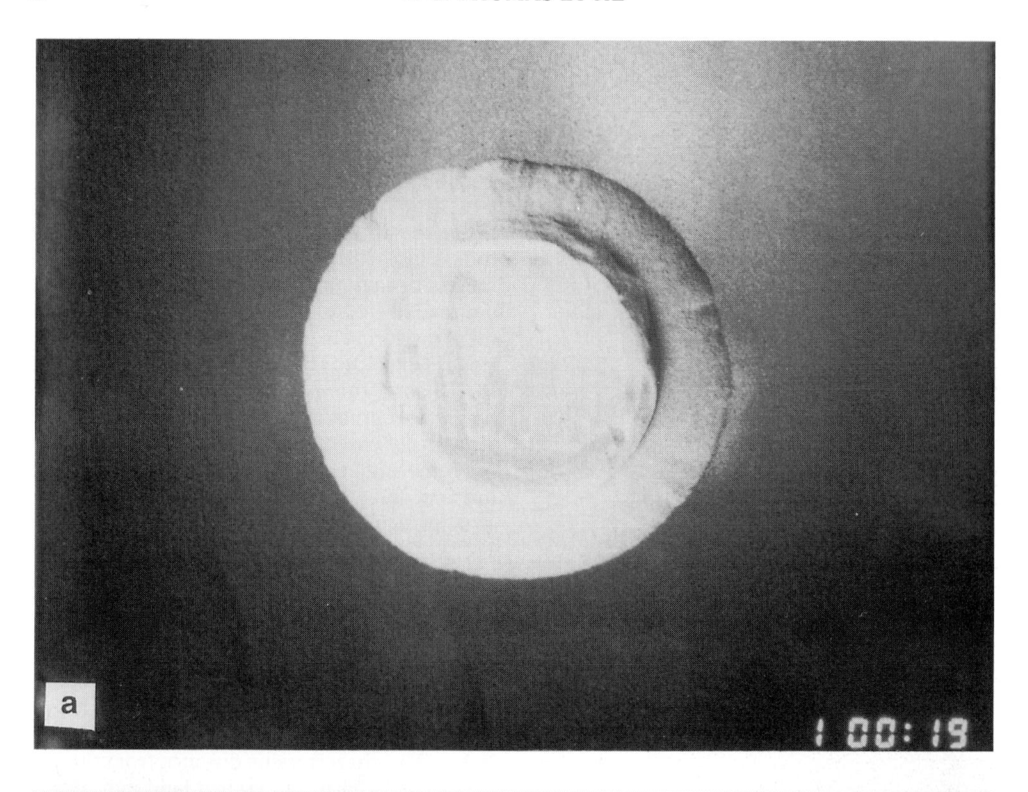
a
1 00:19

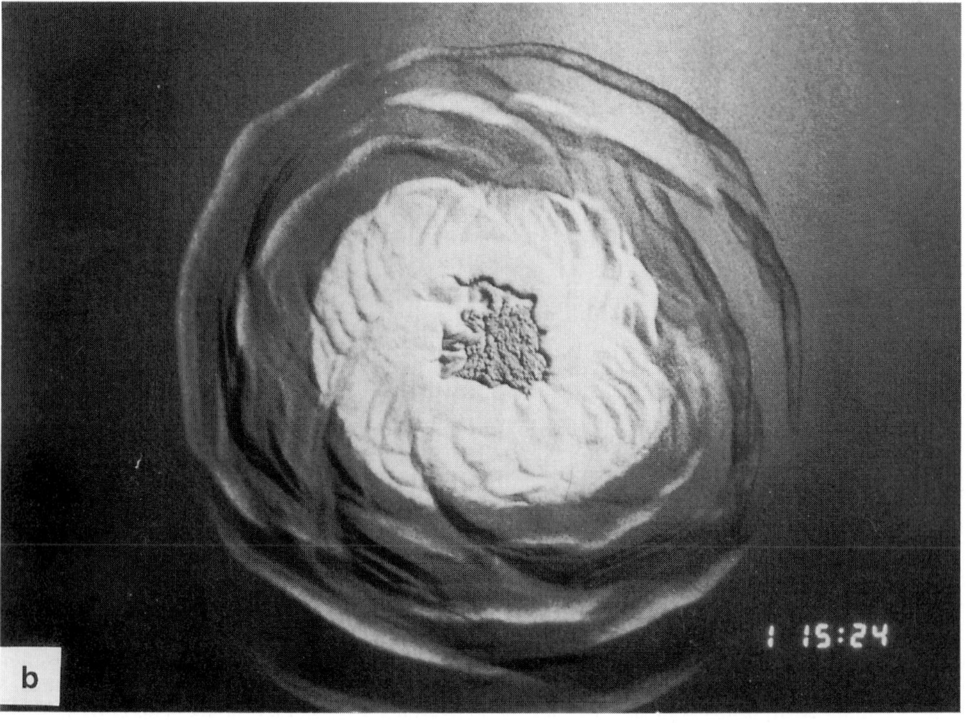
b
1 15:24

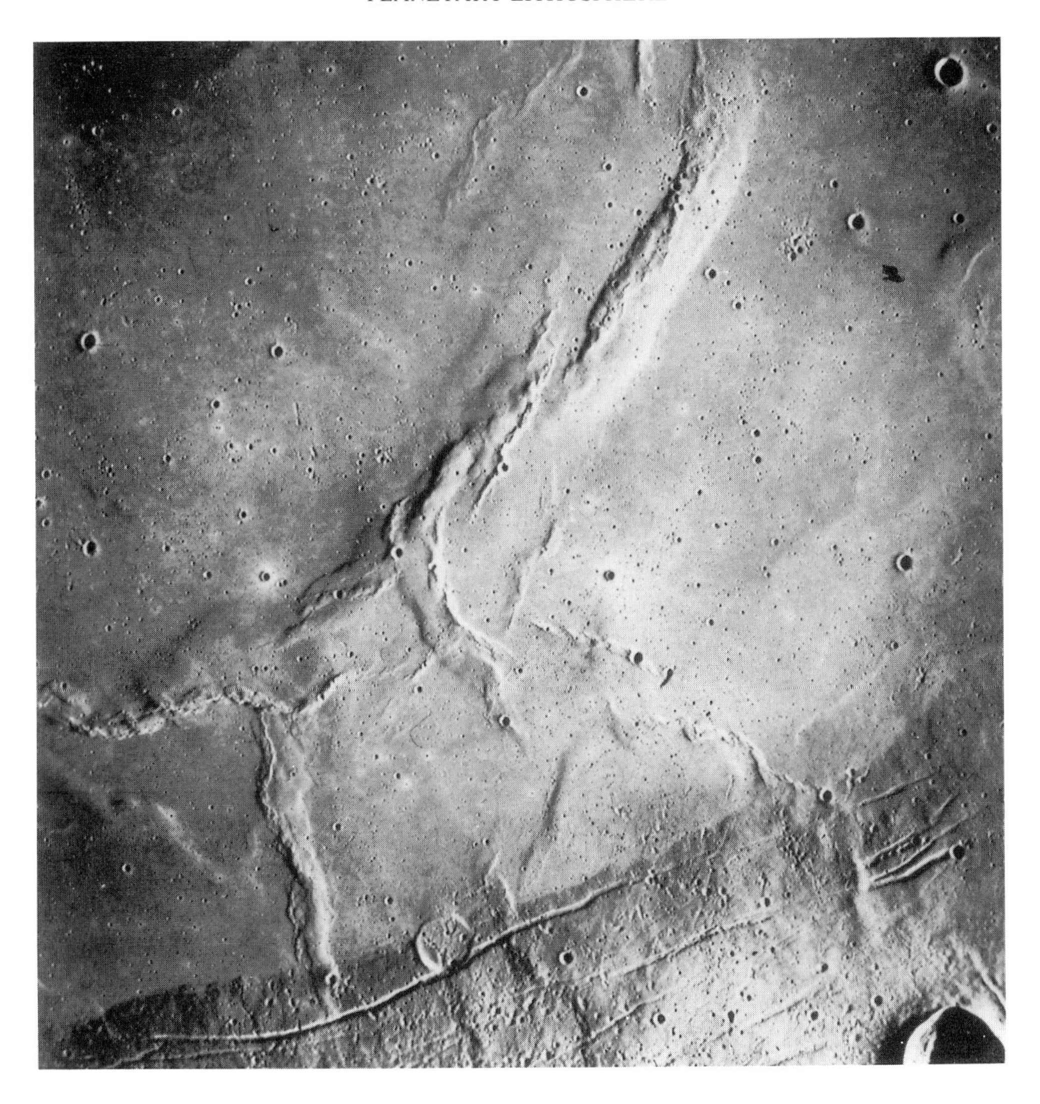

Fig. 10. Southern limit of Mare Serenitatis. The north of this area corresponds to the filling of Mare Serenitatis, and the south corresponds to its borders. The mare material is affected by compressive ridges, and the borders are affected by grabens. The spatial distribution of these extensional and compressional zones allow us to estimate the thickness of the 'elastic lithosphere'. Picture is 160 km wide. (NASA Photo Apollo 17.)

limit of the bulge (Solomon & Head 1982, 1990). Three billion years earlier, the Argyre impact lithospheric thickness was much less than 100 km, which corresponds to a theoretical elastic thickness much less than 50 km if we apply the lunar proportion. This indicates a secular lithospheric thickening from less than 50 km to about 120 km during the last 3 Ga years of Martian history.

The Moon is the only extraterrestrial body

Fig. 9. Analogue simulation of a multi-ring basin (see text for details). (**a**) Image of the simulation just after (one minute) the crater development. The crater was made slowly, without shock. The limit of the cavity is illustrated by the limit of white and dark sand. The crater is 24 cm in diameter. (**b**) Image of the simulation at the end of the relaxation movements, twenty hours after the beginning of the experiment. Note the development of outside ring structures.

where seismic and thermal measurements have been undertaken. Five seismometers have been successfully placed on the Moon during the Apollo missions. Compared with the Earth, the Moon is seismically very quiet. The total energy released by Moonquakes is 12. 10^{10} J a^{-1} compared to the 10^{18} J a^{-1} of the Earth (Goins *et al.* 1981). The Moonquakes fall into two groups: deep events occurring between 800 and 1000 km in depth and shallow events occurring at depths less than 100 km. The periodicity of the quakes is the same as the revolution/rotation period, which indicates a tidal influence. Seismic velocity models have been computed (Toksöz *et al.* 1973; Dainty *et al.* 1973). They show that the Moon's crust is approximately 60–100 km thick. The wave velocities are compatible with the anorthositic to gabbroic composition found on the surface of the continents. Below the crust, the P wave velocity increases to 8 km s^{-1} and the S wave velocity to 4.3 km s^{-1} for V_s, which is compatible with a peridotitic composition of the Moon's mantle. From 1000 to 1300 km depth, a strong decrease in velocity of both types of waves is registered, comparable with the terrestrial Low Velocity Zone, probably due to temperature being near the melting temperature. Thus, in the Moon the terrestrial scheme of a lithosphere resting on a low viscosity mantle is applicable. The lithosphere of the Moon, however, is 1000 km thick, which indicates a very cold planetary body. The regular occurrence of seismic events at 800 km depth probably induced by tidal stress indicates that at these short time scales, the lithosphere of the Moon deforms in a brittle manner.

Two values of heat flux were measured on the Moon, 21 and 16 mW m^{-2} respectively. A mean value of 18 mW m^{-2} has been proposed for the entire body (Langseth *et al.* 1976). If this value is correct, it implies that the abundance of radioactive material in the Moon is twice that of the bulk Earth. Using a global heat flux value of 18 mW m^{-2}, Keihm & Langseth (1977) estimated temperatures of 800–1100°C at 300 km depth. These values seem too high relative to the depth of the seismic low velocity zone which assumes temperatures of 1200°C at 1000 km depth. Further geophysical studies need to be undertaken to obtain a good understanding of the lunar interior.

Mercury

The case of Mercury is different: Mercury does not exhibit unquestionable multi-ring basins. Two possible multi-ring basins are described (Strom *et al.* 1975; McKinnon 1981). One of them, the Tolstoj basin, is very subdued and eroded. Portions of three rings may be identified (Schaber & McCauley 1980; McKinnon 1981). If the 150 km inner ring radius is the cavity's limit (Schaber & McCauley 1980), then the impact lithospheric thickness is ≤50 km. The other basin, the Caloris basin, 600 km in radius is less subdued, and younger than Tolstoj. It is not a true multi-ring basin, because no prominent ring exists outside the basin's limit, except a weak outer scarp which is developed close to the main scarp in the north and southeast (Strom *et al.* 1975; Thomas & Masson 1984). One possible explanation for this lack of a prominent outer ring is the thickness of the lithosphere, which would be thicker than 200 km at the time of the Caloris impact. The difference in lithospheric thickness derived from both impacts would be due to secular thickening (McKinnon 1981). Another explanation is possible: a compressional stress within the Mercurian lithosphere at Caloris time would prevent outer scarps from forming, because it compensated the extensional stress caused by the asthenospheric motions induced by the Caloris cavity (Fleitout & Thomas 1981). The entire surface of Mercury exhibits lobate and arcuate scarps, which have been interpreted as compressive features (Strom *et al.* 1975). The lack of contemporaneous extensional features on Mercury indicates a global compressional stress within the Mercurian lithosphere. This global surface shortening is interpreted as the consequence of (1) the secular cooling of Mercury, and (2) the internal structure of Mercury which is unique in the solar system: a relatively thin silicate mantle (600 km) covering an important iron core (1840 km in radius). The thermal dilatation coefficient of iron is larger than that of silicate. Hence, cooling of such a planet induces differential contraction, the core contraction being more significant than the mantle one, which produces large horizontal stresses. The secular cooling would produce a 2 km decrease in core radius and stresses of nearly 2 kbar on the Mercurian surface (Strom *et al.* 1975; Solomon 1977). The Caloris impact occurred during this compressive state.

The compressive scarps produced by such a mechanism would be randomly oriented, which is not the case in the Mercurian southern hemisphere, the only part of Mercury where stereoscopic coverage is available (O'Donnel 1979). The compressive structures are statistically radial to the Caloris Basin (Thomas *et al.* 1982), which implies that the global compressive stress would be oriented, at least temporarily, by the Caloris cavity. This Caloris effect was computed (Fleitout & Thomas 1982; Thomas *et al.* 1988), assuming: (1) a lithospheric pre-Caloris compressive stress of 2 kbar, (2) that the Caloris interior was a zero stress and strength area, (3) that the crater rim was a free boundary. The result of these calculations (Fig. 11) is a lithospheric motion of the entire lithosphere towards the cavity as around classic basins, but with a different stress field: the compressive stresses oriented radially to Caloris become less intense near the basin, but are still compressive, whereas the compression in the tangential direction is increased. Near the basin rim, the stress field does not induce concentric normal faults as around 'classic' basins, but produces radial reverse faults or oblique strike slip faults, as observed (Thomas *et al.* 1988). Because of the thickness and the continuity of the lithosphere, this stress disturbance may be transmitted over a long distance. The concentric stress is still 100 bars higher than the radial one 90° away from the crater's rim, which explains the preferential orientation of the compressive structures far away from the basin.

Venus

Venus has about the same size and mass as the Earth. Thus, it is important to compare the terrestrial and venusian lithospheres. Unfortunately, unquestionable multi-ring basins do not exist on Earth or Venus. On Earth, double or triple-ring basins, larger than 50 km in diameter, exist but they are so eroded or covered by sediments that the position of their transient cavity limit has sparked intense debates over the last 20 years. On Venus, there is no erosion or sedimentary embayment, and 33 double and triple-ring basins have been imaged by Magellan (Schaber *et al.* 1992). Nevertheless the outer ring exhibits morphological characteristics which look very similar to the cavity limit of simple/complex venusian craters, except for crater Mead, the largest impact crater on Venus (diameter = 280 km). Mead is the only crater whose outer ring may be interpreted from morphological data as a tectonic scarp outside the transient cavity edge (Schaber *et al.* 1992). For all the venusian basins, the presence of the dense atmosphere strongly modifies the morphology and distribution of ejecta and secondaries. So the use of ejecta morphology or characteristic geometric proportions is not appropriate on Venus. Thus, it is impossible to compare the venusian impact lithosphere with that of the other planet's one using the same 'mechanical drill', except for Mead, if it is truly a basin with one single tectonic scarp (280 km in diameter), surrounding an impact cavity 200 km in diameter. With this hypothesis, the transient cavity depth is 30 km, and the impact venusian lithosphere would be equivalent (30 km). If the Mead's outer ring is the transient cavity limit, the lithospheric thickness would be greater than the transient cavity depth, i.e. >45 km. Nevertheless, the Magellan data give us some information about the venusian lithosphere.

The Magellan probe has revealed that Venus is an active planet. Magmatic production has been about $0.5\,km^3\,a^{-1}$ for the last 500 Ma (Phillips *et al.* 1992). On its surface, numerous volcanoes and structural features interpreted as faults or folds show that the lithosphere has been recently deformed. The crater distribution is identical on every part of the planet, indicating that a global resurfacing affected the entire planet 500 million years ago (Schaber *et al.* 1992; Phillips *et al.* 1992) However, these processes are different from those occurring on the Earth, where resurfacing is progressive for oceanic lithosphere, and does not occur for continental crust. Another indication of this difference is the lack of morphologic and topographic signatures of oceanic ridges or subduction zones. The temperature at the surface of Venus is around 450°C, and no water exists either in the atmosphere or in the rocks. Wind is the only, very weak erosional and depositional process.

According to their elevation relative to the mean planetary radius (MPR), terrains of Venus are classified into three domains: the plains which lie below the MPR, the mesolands with an elevation between 0 and 2 km, and the highlands which are above 2 km. Each domain has its own structural and geophysical properties. Plains are affected by extensional fractures and compressive wrinkle ridges, parallel for distances over 1000 km, with a bimodal spacing distribution: 20–30 km and 100–300 km. From these characteristic spacings, Zuber (1987) and Zuber & Parmentier (1990) propose that the rheology of the crust would be governed by properties of diabase, the thickness of the crust would be less than 30 km, and the thermal gradient would be less than $25\,K\,km^{-1}$. Assuming a strong rheological transition between crust and mantle and a linear viscous rheology for the crust, Grimm & Solomon (1988) have estimated from crater depth distribution a crustal thickness around 20 km.

The Coronae are quasi-circular tectono-volcanic features ranging from 100 to 2600 km in diameter. There are characteristic of the mesolands. The tectonic pattern of coronae displays two families of faults: radial fractures that seem to occur early in the development of the coronae and concentric fractures. The coronae are

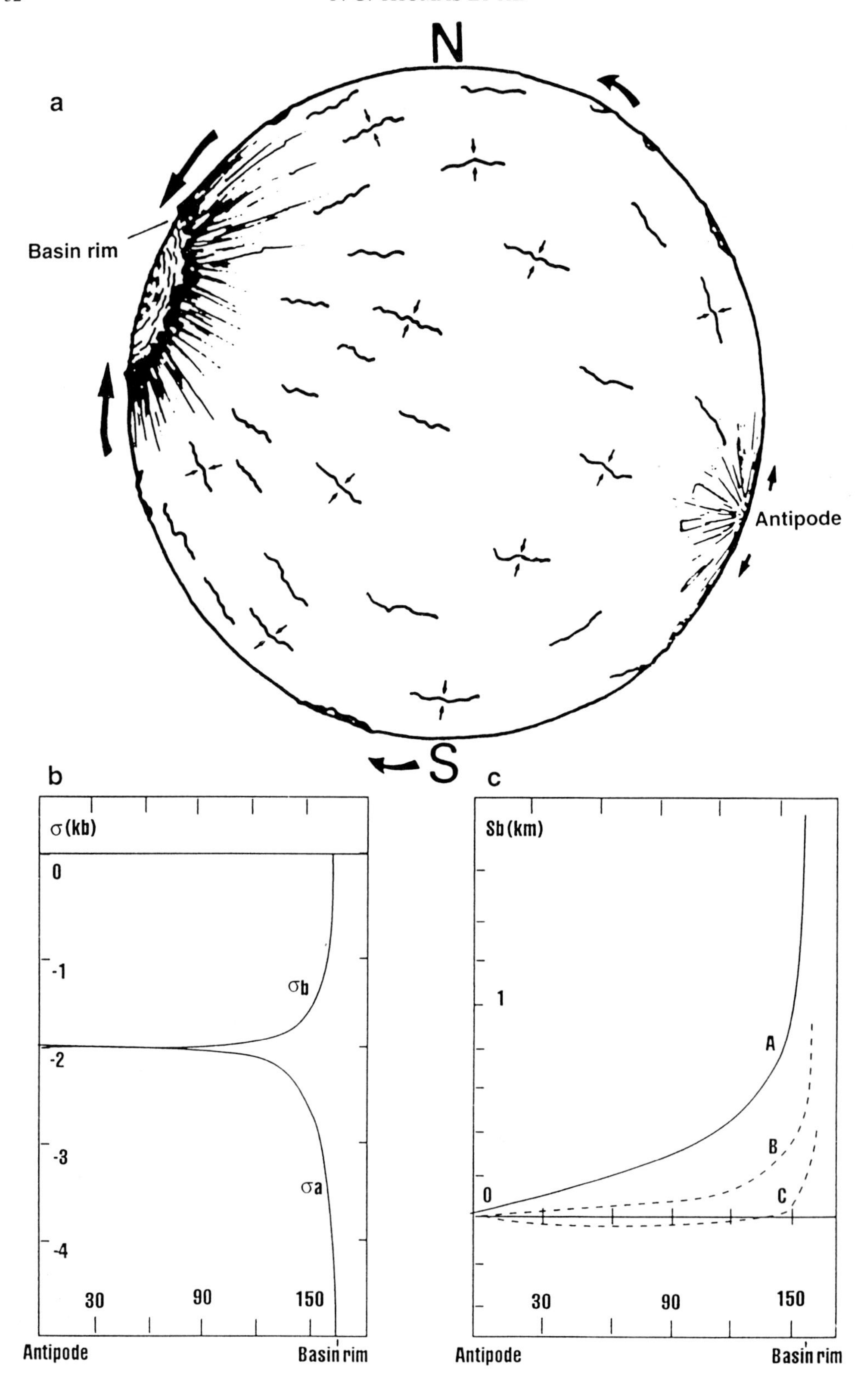
N
a
Basin rim
Antipode
S
b
σ (kb)
0
-1
-2
-3
-4
σb
σa
30
90
150
Antipode
Basin rim
c
Sb (km)
1
0
A
B
C
30
90
150
Antipode
Basin rim

limited by an external annulus whose elevation dominates approximately 1 km above the internal part of the coronae and 1.5 km above the surrounding plains. From the inversion of flexural data, Sandwell & Schubert (1992) found an elastic thickness of 10–60 km, and thermal gradient under coronae ranging from 35 to 77 K km^{-1}. Due to their size, their shape and the thermal gradient of the crust, coronae are interpreted as the interaction of a mantle diapir with the lithosphere. Radial faults would result from the initial doming stage and concentric faults from relaxation of the topography. The elastic thickness found under coronae is thus calculated for unusual areas, and would be thicker on other places of the planet.

The highlands are divisible into volcanic rises, and crustal plateaux including Ishtar Terra, which is the highest region of Venus. Volcanic rises are more than 1000 km in diameter. They are affected by radial grabens and covered by large shield volcanoes. Their apparent depth of compensation, deduced from the gravity field, is generally deeper than 200 km (Smerkar & Phillips 1991). The volcanic rises are generally interpreted as the superficial expression of mantle upwellings. Crustal plateaux and Ishtar Terra, are strongly deformed regions affected by compressional structures similar to an emergant decollement level in front of terrestrial mountain chains. In contrast with the volcanic rises, the apparent depth of compensation of these regions are often less than 50 km. This value supports a crustal thickening process under these areas with a level of compensation located in the ductile lower crust. These regions seem to be the only features in the solar system similar to the collision zones of the lithosphere of the Earth. One problem which still remains is that, despite the high temperature of the crust, there is no evidence of gravitational collapse of the mountain chains of Venus. This implies that the compression is maintained somewhere at the boundary of the system, or that the mountains are made of very strong material.

One of the most striking features of the gravity field of Venus is the strong positive correlation between topography and long wavelength gravity anomalies. This fact, associated with the large values of apparent depth of compensation under volcanic rises, suggests that the volcanic rises are dynamically supported by convection. This implies a strong coupling between lithosphere and convecting mantle, stronger than on Earth (Phillips & Hansen 1994). The model that best explains the venusian mountain chains assumes that these regions are located above stable downwelling zones of mantle flow. This downwelling motion in the mantle would produce shortening of the overlying crust and prevent any extension. The strong coupling between lithosphere and convecting mantle suggests a lack of low viscosity asthenosphere. This is possibly due to the lack of water or volatiles in the interior of Venus, which would, if they were present, lower the viscosity of silicates (Phillips 1986).

The icy lithospheres

The icy satellites

The satellites of Jupiter, Saturn, Uranus and Neptune have a very low surface temperature (<130 K). With the exception of Europa and Io, their density is less than 2000 kg m^{-3}, which indicates a proportion of light component >50%. Because H_2O is the main condensible light component in the nebula, and because the light reflected from these bodies bears the characteristic spectral signature of water ice, these satellites are thought to be icy bodies, that contain less than 50% of heavy components (silicates, carbonaceous component). This study only concerns Ganymede and Callisto, because the other satellites are too small to have a significant gravity, or because their surfaces do not exhibit large impact craters (Europa, Io and Triton), or are unexplored (Titan).

Ganymede (2631 km in radius) and Callisto (2400 km in radius) are the two largest satellites of Jupiter. They are the only known cratered icy bodies comparable in size to the terrestrial planets. Their ice/silicates ratio by mass is

Fig. 11. (**a**) Schematic representation of the mechanism proposed for the preferential orientation of the lobate scarps on Mercury. The exterior arrows show the direction of the displacement of the lithosphere. The interior small arrows indicate the main compressive stress. (**b**) Tangential σ_a and azimuthal stress σ_b as a function of the angular distance δ to the Caloris antipode for a 50 km thick lithosphere with a 2 kbar pre-Caloris stress. (**c**) Azimuthal displacement Sb towards Caloris as function of the angular distance δ to the Caloris antipode. Curve A corresponds to the case of a 50 km thick lithosphere with a 2 kbar pre-Caloris stress. Curve B corresponds to the case of a 100 km thick lithosphere with a 1 kbar pre-Caloris stress. Curve C corresponds to the case of a 250 km thick lithosphere with a 0.5 kbar pre-Caloris stress. The basin rim is at δ = 164°. (Figures from Thomas *et al.* 1988.)

around 1. We do not have direct evidence of the differentiation of Ganymede and Callisto. Some models propose differentiated bodies with a silicate core and an icy mantle (Consolmagno & Lewis 1976; Fanale 1977). Some other models propose undifferentiated bodies of 'dirty' ice, water ice containing small grains of silicates (Parmentier & Head 1979; Schubert *et al.* 1981), while a third kind of model proposes a three layered model with a rocky core, a dirty ice mantle and an icy crust (Schubert *et al.* 1981; Mueller & McKinnon 1988). Whatever the chosen model, the rheology of the shallow layers would be controlled by the physical properties of ice, which is less viscous than the silicates. The detailed morphological studies of topographic features provide a basis for the estimation of viscosity and temperature of the first tens of kilometres of their lithospheres. On the heavily cratered terrains of Ganymede and Callisto, the oldest craters are extremely subdued, lacking any significant topography (Smith *et al.* 1979). Using the cratering time scale, Shoemaker & Wolfe (1982) showed that on Callisto and Ganymede craters older than 4.1 Ga are lacking and craters younger than 4.1 Ga are found in a wide range of degradational stages from featureless to fairly fresh. This distribution of crater morphologies may be explained by a complete relaxation of the older reliefs and a very significant increase in viscosity of the surface layer from 4.1 Ga to the present time. The increase in viscosity can be interpreted in terms of decrease in internal temperature. The large, younger craters exhibit high topography and a deep depression. The existence of such relief some billion years after the impact, implies a very high viscosity of the superficial ice and a very low thermal gradient. Looking for the highest topography of different ages, Passey & Shoemaker (1982) calculated the Ganymedean thermal gradient using the Newtonian behaviour for ice: 3 K km^{-1} at 3.9 Ga, 0.3 K km^{-1} at 3.5 Ga and 0.1 K km^{-1} at 2 Ga.

Ganymede

On Ganymede three well-imaged multi-ring basins are preserved, the Eastern Hathor and Western Hathor Basins (3.7–3.8 Ga) and the Gilgamesh Basin (3.5 Ga, Fig. 6) which are six/seven-ringed basins. There is also one double-ring basin, the Western Equatorial Basin, 3.5 Ga old (Passey & Shoemaker 1982). As on the Moon, the problem is to distinguish inner rings which were formed inside the transient cavity, from outer ones, which were formed outside the transient cavity limit. On Ganymede, some one-ring craters with unambiguous cavity limits allow us to determine the ejecta and secondary crater distribution, and the values of some characteristic ratios (secondary craters to main cavity radii ratio), which have been used on the Moon to determine the transient cavity radius. These values and ratios are identical on Ganymede and on the Moon, which have the same gravity (Passey & Shoemaker 1982). The use of these values applied to Ganymedean multi-ring basins allows us to identify the transient cavity radius, the number of outside rings, and the transient cavity depth using the 1:3 ratio between cavity depth and radius (Thomas *et al.* 1986, Fig. 5). The Hathor transient cavities radii are 30 km, with four continuous outside rings, the Western Equatorial transient cavity radius is 60 km with one single continuous outside ring, and the Gilgamesh transient cavity radius is 120 km with four discontinuous outside rings. The corresponding lithospheric thickness was less than 10 km at the Hathor impacting time, and about 20 km at Western Equatorial–Gilgamesh impacting time (Thomas *et al.* 1986). These values show a lithospheric thickening and a planetary secular cooling which are in agreement with the superficial cooling deduced from the topographic relaxation of older impact craters. These values allow us also to compare the lithospheric thickness of one silicated body, the Moon, and one icy body, Ganymede, which are about identical in mass (10^{23} kg), at approximately the same time (3.9–3.8 Ga): the icy lithosphere was less than 10 km while the silicate one was about 100 km. Despite the difference in external temperature, the icy layers are less viscous than the silicate one.

Callisto

Two well-imaged multi-ring impact structures exist on Callisto, which are named Asgard and Valhalla (Fig. 7), and are respectively 4 and 3.8–3.9 Ga (Passey & Shoemaker 1982). Valhalla is the most complex and the largest crater of Callisto. The diameter of its cavity corresponds to a patch of high albedo material approximately 300 km in radius. From the border of the crater to 1500 km, the surrounding terrains are affected by more than 10 concentric structures. The Asgard basin is the second largest multi-ring basin on Callisto. The centre is occupied by a high albedo patch, 115 km in radius interpreted as the crater cavity. This cavity is surrounded by more than 6 concentric features up to 900 km from the basin centre. The number of outer rings indicates a lithospheric

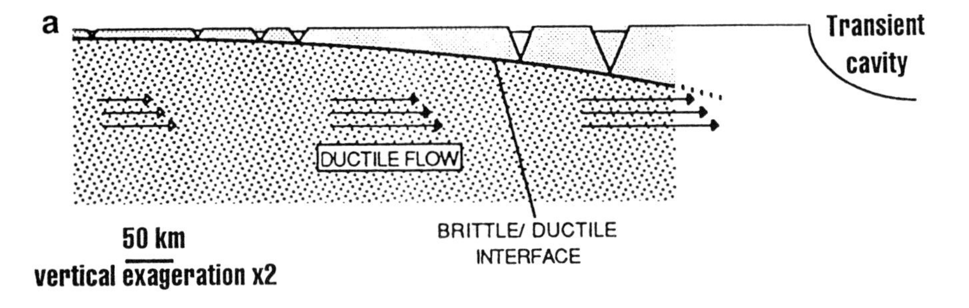

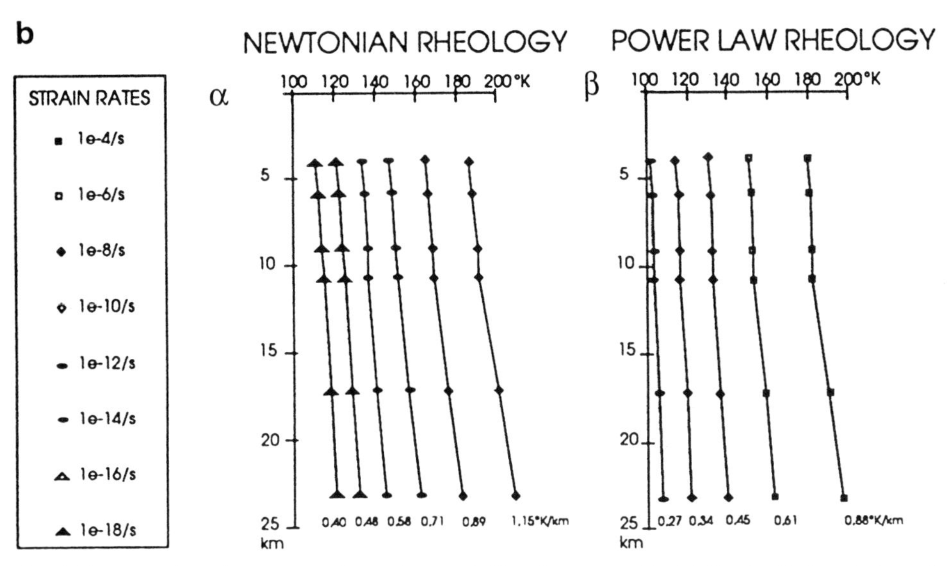

Fig. 12. (**a**) Cross section of the surroundings of Asgard. The displacement inside the ductile layer induces the fractures of the brittle part. The width of the grabens is controlled by the depth of the brittle–ductile interface. If the superficial temperature and the thermal gradient are constant over the entire extended zone, and if there is no chemical layering in the Callistean crust, the depth of the brittle–ductile interface depends only on the strain rate, which is decreasing from the crater. (**b**) Various thermal profiles for Callisto, calculated for different possible strain rates at the crater rim, for Newtonian rheology (α) and power law rheology (β). The order of magnitude of the calculated thermal gradient is independent from the chosen rheological law and strain rates. (Figures from Allemand & Thomas 1991.)

thickness less than 30 km at Valhalla/Asgard impact times.

The concentric structures around Valhalla are sinuous ridges, and mainly concentric outward or inward facing scarps. The concentric structures around Asgard are in some places grouped in pairs. Near the cavity limit, the structures are pairs of ridges. The individual ridges are about 500 m high (Passey & Shoemaker 1982) and the pairs of ridges look like the furrows of Gallileo Regio on Ganymede which are interpreted as relaxed grabens (McKinnon & Melosh 1980). Far from the crater border, structures which are still grouped in pairs are very similar to classic non-relaxed grabens. The width of these grabens decreases from 23 km at 200 km from the cavity limit to 6 km at 800 km from the cavity limit (Allemand & Thomas 1991). It is assumed that the width of grabens depends on the depth of the brittle–ductile transition level where the two normal faults are connected (Golombeck & Banerdt 1986; Allemand & Brun 1991). The depth of the brittle–ductile transition level depends on the superficial temperature and the thermal gradient (which are supposed to be constant all over the studied area, sufficiently far from the cavity), but also depends on the differences in the strain rate around the basin

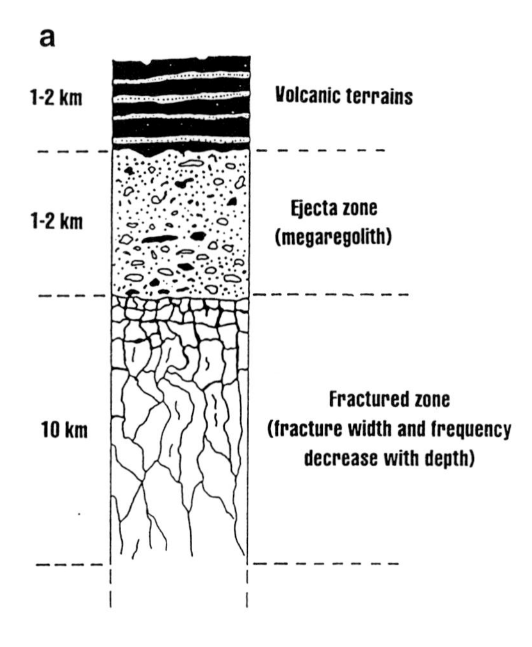

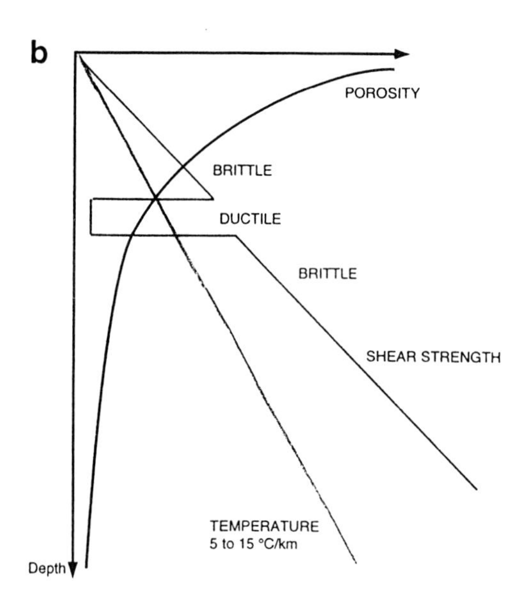

Fig. 13. (**a**) Generally accepted model for the martian regolith. The ejecta zone lies on a fractured basement and is covered with volcanic terrains due to Tharsis volcanoes. (**b**) Porosity, shear strength and temperature profile in the upper crust of Tharsis. The porosity decreases with depth due to the compaction of the megaregolith under its weight. The temperature increases with a gradient ranging from 5 to 15°C km^{-1}. The rheological behaviour of a rock–ice mixture depends on temperature and the proportion of ice versus particles. Ice alone tends to be brittle at low temperatures and ductile at high temperatures. Near the surface, the temperature is low and the mixture is brittle. Near two kilometres depth, the temperature is sufficient for the mixture to become ductile. At larger depths, the porosity becomes too small for the mixture to remain ductile.

during the relaxation of the initial cavity (Fig. 12a). Laws of mass conservation indicate that in cylindrical coordinates the strain rate around the relaxing cavity decreases from the crater centre in an r^{-3} law (Melosh & McKinnon 1978). Using different values of strain rates at the cavity limit, it is possible to compute the temperature at each brittle/ductile transition zone, the depth of which is known by the graben width (Allemand & Thomas 1991). The surface temperature is chosen between 100 and 150K. Calculations show that: (1) the Newtonian flow law for ice gives a strain rate at the crater rim of about $10^{-14} s^{-1}$, whereas a power law gives a strain rate at the crater rim of about $10^{-6} s^{-1}$; (2) the order of magnitude of the calculated thermal gradient is independent of the chosen flow law, and is calculated between 1.5 and 0.5K km^{-1} (Fig. 12b). This calculated thermal gradient is lower than those calculated theoretically with a chondritic abundance of radioactive nucleids in the silicates of Callisto, which assume a thermal gradient ranging from 6.5 to 7.9K km^{-1} (Mueller & McKinnon 1988). This very low thermal gradient, incompatible with chondritic abundance, was already presumed (but not quantified) for Ganymede looking at the high topography of some young impact craters (Passey & Shoemaker 1982).

On the Moon, the radioactive abundance deduced from the present measured thermal gradient is twice the chondritic abundance. On Callisto, the radioactive abundance is seven times lower than the chondritic one. Thus, the chondritic abundance, at least for radioactive elements, seems not to be generally applicable to the whole solar system.

Shallow structure of the icy part of Martian lithosphere

Mars is too small to have ever developed any active plate tectonics and is supposed to be geologically dead now. Nevertheless, Mars presents many tectonic features that have to be explained by other mechanisms and that can inform us about the crustal behaviour. On the one hand there are very wide structures: the Tharsis bulge, one fourth of the Martian surface, and the Elysium bulge. Their origin is not clear at the present time, but is probably related to endogenic processes (e.g. Banerdt *et al.* 1992). There is also Valles Marineris, an important canyon (7 km deep, more than 100 km wide), that is thought by some to be a rift, but its genesis is not well understood. These wide structures are probably rooted in deep martian layers. On the other hand, Tharsis and Elysium areas, and many other places of the martian surface, exhibit grabens and ridges. These structures are less than 10 km wide, and are rooted in the shallow martian layers. Some structural analyses have been done on the grabens which are especially well developed on Tharsis (Tanaka & Davies 1988; Plescia 1991; Thomas & Allemand 1993). These analyses show a global average graben width varying from 3500 m to 2330 m, and an average graben depth of 120 m. Assuming that the grabens are bound by conjugate normal faults connected on a mechanical discontinuity at depth, these parameters imply a 2 km deep discontinuity (Thomas & Allemand 1993). The nature of the ridges is less understood. In Lunae and Solis planum, structural analyses show that ridges are pop-up structures bounded by two reverse faults (Allemand & Thomas 1995). The width of the ridges, which are three times wider than the graben in the same areas, indicates that the two reverse faults are connected on the same mechanical discontinuity as the normal faults, at a depth of about 2 km (Allemand & Thomas 1995). This study shows too that the ridge location is strongly influenced by the occurrence of middle size impact craters (several kilometres in diameter). Indeed, reduction of lithostatic pressure under the crater is probably of central importance to ridge initiation. This confirms that ridges are very shallow structures and shows that impact craters are useful to understand thin-shell tectonic processes.

In some areas, grabens or ridges cross a boundary between two geological units. The graben or ridge width is the same on each side of the geological boundary. This consistency indicates that the depth of the mechanical discontinuity does not depend on superficial geology. The depth of this mechanical discontinuity indicates that it lies inside the mega-regolith. The mega-regolith is the shallow layer that represents the ejecta zone formed by impact processes during the first 500 million years of the planet (Fig. 11). Therefore, the mega-regolith is a clastic porous layer overlying a fractured basement, sometimes disturbed by lava flows and perhaps weathering products with a thickness varying from 1 to 10 km depending on authors (De Hon 1982; Clifford 1993). The mega-regolith was covered by later terrains such as volcanic flows. The thickness of these superficial layers is estimated at approximately 1 km (De Hon 1982). Many authors have shown the possible occurrence of a discontinuity involving the martian regolith (McKinnon & Tanaka 1989; Davis & Golombek 1990; Tanaka *et al.* 1991; Plescia 1991; Thomas & Allemand 1993). Three levels are proposed for the discontinuity: (1) at the regolith–basement interface, (2) inside the regolith involving either a cemented or non-cemented zone or volcanic and interbedded strata, or (3) the top of the megaregolith, which is a lava flow–clastic porous level interface (McKinnon & Tanaka 1989; Davis & Golombek 1990; Tanaka *et al.* 1991; Thomas & Allemand 1993). But these three hypotheses do not explain the consistency of the discontinuity's depth below two different geological units.

The presence of ice in the martian regolith has been well established, based on many morphological features (see e.g. Squyres 1989). For example, many craters show fluidized ejecta indicating the presence of volatiles, probably water, in the martian shallow crust. The presence of an ice-rich layer is very important because of the low viscosity and particularly ductile properties of ice as we can observe in terrestrial glaciers. Furthermore, there are striking correlations between the occurrence of fluidized ejecta craters and the occurrence of ridges, that let us suppose a strong relationship between volatiles and ridges (Thomas & Masson 1984; Chicarro *et al.* 1985). If the 2 km deep decollement is related to an icy layer, the nearly constant depth could be related to thermal constraints in the ice-rich layer that influences the ductility of ice. Because ridges and grabens are common over more than half of the martian surface, this mechanical discontinuity, probably related to low viscosity of a shallow ice-rich layer, seems to exist 2 km below the majority of the martian surface (Fig. 12).

The existence of this decollement level explains why the origin of the Tharsis bulge has sparked intense debates over the last 20 years. Different kinds of models are proposed to explain this immense topographic relief and its tectonic features (Banerdt *et al.* 1982; Solomon & Head 1982; Phillips *et al.* 1990; Tanaka *et al.* 1991). For each model, theoretical lithospheric stress fields were calculated and the related tectonic features were predicted. The discrepancy between theoretical tectonic features and the observed data explains the intense debates.

Table 1 *Evolution of planetary lithospheric thicknesses (in km) versus time*

	R (km)	4 Ga	3.5 Ga	■■■■■■■■	1 Ga	0.5 Ga	Present
Earth	6678						up to 50 (ML) up to 100 (SL) 1000 (SL)
Moon	1732		50 (ML) 100 (IL)				
Mercury	2439	≤50 (IL)					
Mars	3398		>200 (IL) ≪100 (IL)		*c.* 120 (ML)		
Venus	6050						*c. 30 (IL)* 10 to 60 (ML)
Ganymede	2631	<10 (IL)	*c.* 20 (IL)				
Callisto	2400	≪30 (IL)					

The studied planets are the Earth and the planetary bodies with a mass $>10^{22}$ kg. This study does not concern small satellites that have a very low gravity. Europa, Io, Triton and Titan are not studied, because they do not exhibit large impact craters, or are unimaged.
The values given correspond to impact lithosphere (IL), mechanical (or flexural) lithosphere (ML) and seismic lithosphere (SL). It is possible to compare the different kinds of lithosphere using the terrestrial and lunar relations IL ≈ 2ML ≈ SL. The terrestrial values concern oceanic plates.
The italicized values are based on poorly constrained impact data.

A decollement level at 2 km everywhere below the surface of Tharsis explains that the strains of the superficial layer are independent of the bulge-related lithospheric stress and strain fields. Thus, because of this ice-rich layer, it is difficult to constrain the bulge origin and the lithospheric strains by using shallow tectonic features.

Lithospheric thickness, plate tectonic and orogenesis

In 1980, one year after the Voyager mission, but before the Magellan data, the ideas about plate tectonics on the planets were clear: only the Earth, and maybe Venus, had developed plate tectonics. Io would be too hot, and its surface may be interpreted as a 'hot spot field'. The other planetary bodies do not exhibit plate tectonic features, but exhibit an old surface with many impact craters that indicated a lack of resurfacing by plate tectonics. The lack of plate tectonics was attributed to the low internal temperature of these bodies, which induced a single-shell, thick and motionless lithosphere. Fifteen years later, new studies and data have modified these ideas. It seems true that small-sized silicate bodies such as the Moon and Mercury had a very thick lithosphere, even 3.5–4 Ga ago. Nevertheless, same-sized icy bodies had a thin lithosphere at the same time, because the ice is less viscous than silicates. Mars, a medium-sized silicate planet has now a thick lithosphere, but had a thin lithosphere 3.5–4 Ga ago, and the Magellan data indicate that the present Venusian lithosphere is as thin as the terrestrial one. Nevertheless, despite these past thin lithospheres, Mars, Ganymede and Callisto do not exhibit any plate tectonic signature. Venus has a present-day thin lithosphere and exhibits tectonic features that look like terrestrial collision zones. But Venus does not exhibit morphological features interpretable as subduction zones, oceanic ridges or other plate tectonic features. Thus, the Earth seems to be unique in the solar system. The reason for this terrestrial singularity is now being discussed, and some partial and qualitative explanations have already been orally proposed. A high viscosity contrast between a rigid lithosphere and a low viscosity asthenosphere would be necessary for the development of plate tectonics. On the Earth, this viscosity contrast may be due both to the very low superficial temperature and to the presence of water in the mantle, which reduces the asthenosphere's viscosity. On Venus, the 450°C superficial temperature and the lack of water in the asthenosphere lower the viscosity contrast. On Ganymede and Callisto, the very low thermal gradient may induce very low and gradual viscosity gradient. On Mars, nobody knows why this hydrated silicated body would have had a low viscosity gradient. Further work, data and modelling are necessary to understand planetary mantle dynamics.

Conclusions and summary

Because of their abundance on the majority of terrestrial and icy bodies, impact craters and related structures are useful tools to understand the rheology of the lithosphere of low-activity planetary bodies, and also give information on the partition of the chemical elements in the solar nebula. Some general conclusions can be drawn from their study.

(1) There are different ways to define lithosphere, so that we can distinguish between the seismic, mechanical and impact lithospheres. Thickness values of these different lithospheres may be proposed for different planetary bodies at different times (Table 1). The relationships of these different lithosphere types may be derived by using the Earth and the Moon: on the Moon, between 3.8 and 3.5 Ga the impact lithosphere (IL) was twice as thick as the mechanical one (ML), neglecting the secular thickening. On the present Earth, the seismic lithosphere (SL) is twice as thick as the mechanical one (ML). Thus it is possible to compare the different kinds of lithosphere at different times using the relation $IL \approx 2ML \approx SL$.

(2) At a given time, the thickness of the lithospheres of a silicate body is an inverse function of the diameter and mass of the body. This is related with the ratio of the production of heat which increases with the cube of the radius, to the surface of the planet which increases with the square of the radius.

(3) For a given silicated or icy body, there is a secular lithospheric thickening. This is related to secular cooling and natural radioactive decay.

(4) Even if the surface temperature of icy bodies is low (for example, 120K for Ganymede and Callisto), icy lithospheres are five to ten times thinner than silicate lithospheres at a given time for two bodies of the same mass.

(5) Because of the thickness and the continuity of the lithosphere of one-plate planets, local mechanical disturbances may be transmitted over a long distance and may affect the entire planet.

(6) Impact craters are natural thermal probes of the lithosphere. On Callisto, the thermal gradient is estimated to be about 1 K km^{-1} at 3.9 Ga. This implies a radiogenic heat production seven times lower than a chondritic one.

Conversely, on the Moon, the heat flux indicates a radiogenic production twice as high as a chondritic one. Thus, the study of planetary lithospheres gives information on the distribution of radioactive elements in the solar nebula, and indicates that the origin of satellites would be more complex than the accretion–collision of chondritic–icy planetesimal swarms.

(7) There is no direct relation between plate tectonics and lithospheric thickness. The 3.5–4 Ga old Mars, Ganymede and Callisto, and the present Venus have a lithospheric thickness identical or thinner than the present Earth's one. Despite this thickness similarity, the Earth is the only body that has developed plate tectonics.

References

ALLEMAND, P. & BRUN, J. P. 1991. Relation between rift width and rheological layering of the lithosphere. *Tectonophysics*, **188**, 63–69.

—— & THOMAS, P. G. 1991. The thermal gradient of Callisto constrained by Asgard basin: rheological and chemical implications. *Journal of Geophysical Research*, **96**, E4, 20 981–20 988.

—— & —— 1995. Localisation of Martian ridges by impact craters: mechanical and chronological implications. *Journal of Geophysical Research*, **100**, E2, 3251–3262.

BANERDT, W. B., PHILLIPS, R. J., SLEEP, N. H. & SAUNDERS, R. S. 1982. Thick shell tectonics on one-plate planets: applications to Mars. *Journal of Geophysical Research*, **87**, 9723–9733.

——, GOLOMBEK, M. P. & TANAKA, K. L. 1992. Stress & tectonics on Mars. *In*: KIEFFER, H. H., JAKOSKY, B. M., SNYDER, C. W. & MATTHEWS, M. S. (eds) *Mars*. University of Arizona Press, Tucson, 249–297.

CARTER, N. L. & TSENN, N. C. 1987. Flow properties of continental lithosphere. *Tectonophysics*, **132**, 247–263.

CHAO, E. C., SHOEMAKER, E. M. & MADSEN, B. M. 1960. First natural occurrence of coesite. *Science*, **132**, 220–222.

CHICARRO, A. F., SCHULTZ, P. H. & MASSON, P. 1985. Global and regional ridges patterns on Mars. *Icarus*, **63**, 153–174.

CLIFFORD, A. 1993. Model for the Hydrologic and Climatic Behavior of Water on Mars. *Journal of Geophysical Research*, **98**, 10 973–11 016.

CONSOLMAGNO, G. J. & LEWIS, J. S. 1976. Structural and thermal models of icy Galilean satellites. *In*: GEHRELS, T. (ed.) *Jupiter*. University of Arizona Press, Tucson, 1035–1051.

DAINTY, A. M., TOKSÖZ, M. N., SOLOMON, S. C., ANDERSON, K. R. & GOINS, N. R. 1973. Constraints on lunar structure. *Proceedings of the 5th Lunar and Planetary Conference*, **3**, 3091.

DAVIS, P. A. & GOLOMBEK, M. P. 1990. Discontinuities in the Shallow Martian crust at Lunae, Syria, and Sinai Plana. *Journal of Geophysical Research*, **95**, 14 231–14 248.

DAVY, P. & COBBOLD, P. R. 1988. Indentation tectonics in nature and experiment. Part 1: Experiments scaled for gravity. *Bulletin of the Geological Institution of Uppsala*, **14**, 129–141.

DE HON, R. A. 1982. Martian volcanic material: Preliminary thickness estimate in the eastern Tharsis region. *Journal of Geophysical Research*, **87**, 9821–9828.

DENCE, M. R. 1968. Shock zoning at Canadian craters: petrography and structural implications. *In*: FRENCH, B.M. & SHORT, N.M. (eds) *Metamorphism of Natural Material*. Mono Book Corp, Baltimore, 169–184.

FANALE, F., JOHNSON, T. & MATSON, D. 1977. Io's surface and the histories of the Galilean satellites. *In*: BURNS, J. A. (ed.) *Planetary Satellites*. University of Arizona Press, Tucson, 379–405.

FAUGÈRE, G. & BRUN, J. P. 1984. Modelisation expérimentale de la distension continentale. *Comptes Rendus de l'Académie des Sciences, Paris*, **299**, 7, 365–370.

FLEITOUT, L. & THOMAS, P. G. 1982. Far-field tectonics associated with a large impact basin: Applications to Caloris on Mercury and Imbrium on the Moon. *Earth and Planetary Science Letters*, **58**, 104–115.

GAULT, D. E., QUAIDE, W. L. & OBERCHER, V. R. 1968. Impact cratering mechanism and structures. *In*: FRENCH, B.M. & SHORT, N.M (eds) *Shock metamorphism of natural material*. Mono, Baltimore, 87–99.

——, GUEST, J., MURRAY, J. B. DZURIZIN, D. & MALIN, M. C. 1975. Some comparisons of impacts on Mercury and the Moon. *Journal of Geophysical Research*, **80**, 2444–2462.

GILBERT, G. K. 1893. The Moon's face, a study of the origin of its features, *Philosophical Society of Washington Bulletin*, **12**, 241–292.

GOLOMBEK, M. P. & BANERDT, W. B. 1986. Early thermal profile and lithospheric strength of Ganymede from extensional tectonics features. *Icarus*, **68**, 252–265.

GOINS, N. R., DAINTY, A. M. & TOKSÖZ, N. F. 1981. Seismic energy release of the Moon. *Journal of Geophysical Research*, **86**, 378–388.

GRIMM, R. E. & SOLOMON, S. C. 1988. Viscouse relaxation of impact crater relief on Venus: constraints on crustal thickness and thermal gradient. *Journal of Geophysical Research*, **93**, 11 911–11 929.

HEAD, J. W. 1974. Orientale multi-ringed basin interior and implications for the petrogenesis of lunar highland samples. *The Moon*, **11**, 327–356.

—— 1977. Origin of the outer rings in lunar multi-ring basins: Evidence from morphology and ring spacing. *In*: RODDY, D. J., PEPIN, R. O. & MERRIL, R. B. (eds) *Impact and Explosion Cratering*. Pergamon, New York, 567–573.

HODGES, C. A. 1980. *Geologic map of the Argyre quadrangle of Mars*. USGS Miscellaneous Investigations Series Map I–1181 (Mc 26).

KEIHM, S. J. & LANGSETH, M. G. 1977. Lunar thermal regime to 300 km. *Proceedings of the 8th Lunar and Planetary Conference*, 499–514.

LANGSETH, M. G., KEIHM, S. J. & PETERS, K. 1976.

Revised lunar heat flow values. *Proceedings of the 7th Lunar and Planetary Conference*, 3143–3171.
LUNAR AND PLANETARY INSTITUTE 1981. *Multi-Ring Basins*. Proceedings of Lunar and Planetary Science Institute, **12**, Part A.
MCCAULEY, J. F. 1977. Orientale and Caloris. *Physics of the Earth and Planetary Interiors*, **15**, 220–250.
——, GUEST, J. E., SCHABER, G. G., TRASK, N. J. & GREELEY, R. 1981. Stratigraphy of the Caloris basin, Mercury. *Icarus*, **47**, 184–202.
MCKINNON, W. B. 1981. Application of ring tectonic theory to Mercury and other solar system bodies. *In*: SCHULTZ, P. H. & MERRIL, R. B. (eds) *Multi-Ring Basins*. Proceedings of the Lunar and Planetary Science Institute, **12A**, 259–273.
—— & MELOSH, H. J. 1980. Evolution of planetary lithospheres: evidences from multi-ringed structures on Ganymede and Callisto. *Icarus*, **44**, 454–471.
—— & TANAKA, K. L. 1989. The impacted Martian crust: structure, hydrology, and some geologic implications. *Journal of Geophysical Research*, **94**, 17 359–17 370.
MAXWELL, T. A. 1978. Origin of multi-ring basin ridge system: an upper limit to elastic deformation based on finite-element model. *Proceedings of the 9th Lunar and Planetary Conference*, 3541–3559.
MELOSH, H. J. 1982. A simple mechanical model of Valhalla basin, Callisto. *Journal of Geophysical Research*, **87**, 1880–1890.
—— 1989. *Impact Cratering, A Geological Process*. Oxford University Press, New York.
—— & MCKINNON, W. B. 1978. The mechanics of ringed basins formation. *Geophysical Research Letters*, **5**, 985–988.
MOORE, H. J., HODGES, C. A. & SCOTT, D. H. 1974. Multi-ring basins illustrated by Orientale and associated features. *Geochimica et Cosmochimica Acta*, **1**, Supplement 5, 71–100.
MUELLER, S. W. & MCKINNON, W. B. 1988. Three layered models of Ganymede and Callisto: compositions, structures and aspect of evolution. *Icarus*, **76**, 437–464.
MURRAY, J. B. 1980. Oscillating peak model of basin and crater formation. *The Moon and the Planets*, **22**, 269–291.
O'DONNEL, W. P. 1979. *The Surface History of the Planet Mercury*. PhD Thesis, University of London Observatory.
PARMENTIER, E. M. & HEAD, J. W. 1979. Internal processes affecting surfaces of low density satellites: Ganymede and Callisto. *Journal of Geophysical Research*, **84**, 6263–6276.
PASSEY, Q. R. & SHOEMAKER, E. M. 1982. Craters and basins on Ganymede & Callisto: morphological indicators of crustal evolution. *In*: MORISSON, D. (ed.) *The satellites of Jupiter*. University of Arizona Press, Tucson, 379–434.
PHILLIPS, R. J. 1986. A mechanism for tectonic deformation on Venus. *Geophysical Research Letters*, **13**, 1141–1144.
—— & HANSEN, W. L. 1994. Tectonic and magmatic evolution of Venus. *Annual Review of Earth and Planetary Science*, **22**, 597–654.
——, SLEEP, N. H. & BANERDT, W. B. 1990. Permanent uplift in magmatic systems with application to Tharsis region of Mars. *Journal of Geophysical Research*, **95**, 5089–5100.
——, HERRICK, R. R., GRIMM, R. E., RAUBERTAS, R. F., SARKAR, I. C., ARVIDSON, R. E. & IZENBERG, N. 1992. Impact crater distribution on Venus: implication for resurfacing history. *Journal of Geophysical Research*, **97**, 15 923–15 948.
PLESCIA, J. B. 1991. Graben and extension in northern Tharsis. *Journal of Geophysical Research*, **96**, 18 883–18 895.
RODDY, D. J. , PEPIN, R. O. & MERRIL, R. B. (eds) 1977. *Impact and Explosion Cratering*. Pergamon, New York.
SANDWELL, D. T. & SCHUBERT, G. 1992. Flexural ridges, trenches and outer rises around coronae on Venus. *Journal of Geophysical Research*, **97**, 16 069–16 083.
SCOTT, D. H., MCCAULEY, J. F. & WEST, N. M. 1977. *Geological map of the West side of the Moon*. USGS Miscellaneous Investigations Series, Map I–1034.
SCHABER, G. G. & MCCAULEY, J. F. 1980. *Geological map of the Tolstoj quadrangle of Mercury*. USGS *Miscellaneous Investigations Series*, Map I–1199 (H8).
——, STROM, R. G., MOORE, H. J., SODERBLOM, L. A., KIRK, R. L., CHADWICK, D. J., DAWSON, D. D., GADDIS, L. R., BOYCE, J. M. & RUSSEL, J. 1992. Geology and distribution of impact craters on Venus: what are they telling us? *Journal of Geophysical Research*, **97**, 13257–13301.
SHOEMAKER, E. M. 1960. Penetration mechanics of high velocity meteorites, illustrated by Meteor crater, Az. *In*: *Structure of the earth's crust and deformation of Rocks*. International Geological Congress, session 21, part 18, Copenhagen, 418–434.
—— 1966. Preliminary analysis of the fine structure of the lunar surface in mare cognitum. *In*: *The nature of Lunar Surface*. Proceedings of the 1965 IAU/NASA Symposium. Johns Hopkins Press, Baltimore, 23–68.
—— & WOLFE, R. F. 1982. Cratering time scale for the Galilean satellites. *In*: MORRISON, D. (ed.) *The satellites of Jupiter*. University of Arizona Press, Tucson, 277–339.
SCHUBERT, G., STEVENSON, D. J. & ELLSWORTH, K. 1981. Internal structures of the Galilean satellites. *Icarus*, **47**, 46–59.
SMITH, B. A. & THE VOYAGER IMAGING TEAM 1979. The Jupiter system through the eyes of Voyager 1. *Science*, **204**, 951–953.
SMERKAR, S. & PHILLIPS, R. J. 1991. Venusian highlands: geoid to topography ratios and their implications. *Earth and Planetary Science Letters*, **107**, 582–587.
SOLOMON, S. C. 1977. The relationship between crustal tectonics and internal evolution in the Moon and Mercury. *Physics of Earth and Planetary Interiors*, **15**, 135–145.
—— & HEAD, J. W. 1979. Vertical movement in mare basins; relation to mare emplacement, basin

tectonic and lunar thermal history. *Journal of Geophysical Research*, **84**, 1667–1682.
—— & —— 1982. Evolution of the Tharsis province of Mars: the importance of heterogeneous lithospheric thickness and volcanic construction. *Journal of Geophysical Research*, **87**, 9755–9774.
—— & —— 1990. Heterogeneities in the thickness of the elastic lithosphere of Mars: constraints on heat flow and internal dynamics. *Journal of Geophysical Research*, **95**, 11073–11083.
SQUYRES, S. W. 1989. Urey Prize Lecture: Water on Mars. *Icarus*, **79**, 229–288.
STROM, R. G., TRASK, N. J. & GUEST, J. E. 1975. Tectonism and volcanism on Mercury. *Journal of Geophysical Research*, **80**, 2478–2507.
TANAKA, K. L. & DAVIS, P. A. 1988. Tectonic History of Syria Planum province of Mars. *Journal of Geophysical Research*, **93**, 14 893–14 917.
——, GOLOMBEK, M. P. & BANERDT, W. B. 1991. Reconciliation of Stress and Structural Histories of the Tharsis region of Mars. *Journal of Geophysical Research*, **96**, 15 617–15 633.
THOMAS, P. G. 1986. La Lune. *In*: *Le Grand Atlas de l'Astronomie*. Encyclopaedia Universalis, 98–113.
—— & ALLEMAND, P. 1993. Quantitative analysis of the Extensional Tectonics of Tharsis Bulge, Mars: Geodynamic Implications. *Journal of Geophysical Research*, **98**, 13 097–13 108.
—— & MASSON, P. 1984. Geology and tectonics of the Argyre area on Mars; comparison with other basins in the solar system. *Earth, Moon and Planets*, **31**, 25–42.
——, —— & FLEITOUT, L. 1982. Global volcanism and tectonism on Mercury; comparison with the moon. *Earth and Planetary Science Letters*, **58**, 95–103.
——, FORNI, O. & MASSON, P. 1986. Geology of large impact craters on Ganymede: implications on thermal and tectonic histories. *Earth, Moon and Planets*, **34**, 35–53.
——, MASSON, P. & FLEITOUT, L. 1988. Tectonic History of Mercury. *In*: VILAS, F., CHAPMAN, C. R. & MATTHEWS, M. S. (eds) *Mercury*. University of Arizona Press, Tucson, 401–428.
TOKSÖZ, M. N., DAINTY, A. M. & SOLOMON, S. C. 1973. Velocity structure and evolution of the Moon. *Proceedings of the 4th Lunar and Planetary Conference*, 2529–2547.
TURCOTTE, D. & SCHUBERT, G. 1982. *Geodynamic applications of continuum physics to geological problems*. John Wiley, New York.
VAN DORN, W. G. 1968. Tsunami on the Moon. *Nature*, **220**, 1102–1103.
VENDEVILLE, B., COBBOLD, P. R. , DAVY, P., BRUN, J. P. & CHOUKROUNE, P. 1987. Physical models of extensional tectonics at various scales. *In*: COWARD M. P., DEWEY, J. F. & HANCOCK, P. L. (eds) *Continental Extension Tectonics*. Geological Society, London, Special Publications, **28**, 95–107.
ZUBER M. T. 1987. Constraints on the lithospheric structure of Venus from mechanical models and tectonic surface features. *Journal of Geophysical Research*, **92**, E541–E551.
—— & PARMENTIER, E. M. 1990. On the relationship between isostatic elevation and the wavelengths of tectonic surface features on Venus. *Icarus*, **85**, 290–308.

Archaean crustal growth and tectonic processes: a comparison of the Superior Province, Canada and the Dharwar Craton, India

P. CHOUKROUNE[1], J. N. LUDDEN[2], D. CHARDON[1], A. J. CALVERT[3] & H. BOUHALLIER[4]

[1]*Archaean Team, Géosciences-Rennes (UPR CNRS 4661), Campus de Beaulieu, 35042 Rennes Cedex, France*

[2]*Centre de Recherche Pétrographiques et Géochimiques (CRPG-CNRS), 54501, Vandoeuvre-les-Nancy Cedex, France & Departement de Géologie, Université de Montréal, Montréal, CP 6128, Québec, Canada*

[3]*École Polytechnique, CP 6079, Succ. Centre-Ville, Montréal, Québec, H3C 3A7, Canada*

[4]*Deceased 1995*

Abstract: We present a comparison of the processes involved in the tectonic evolution of two Archaean cratons, the Superior Province of Canada and the Dharwar Craton of India. These two cratons exhibit distinct map patterns, the Superior Province being dominated by elongate belts, while the Dharwar Craton is characterized by dome and basin features.

We suggest that certain tectonic processes operating in the Phanerozoic, such as enhanced mantle plume activity, subduction of young (warm) oceanic crust, and faster than usual accretion of crust, may have been the norm during the Archaean. In the Superior Province rapid crustal growth occurred, largely due to horizontal tectonic forces. Models analagous to modern plate tectonics are applicable, but the rates of convergence and accretion exceeded those normal for the present day. Accreted crust was warm and subject to more ductile deformation than in modern accretionary zones. These accreted arc, ocean-floor and ocean-plateau fragments would have been underlain by a thick refractory, buoyant, warm lithospheric root that was rapidly underplated (or imbricated) below the recently accreted terranes. In the Dharwar craton a major thermal event appears to characterize its evolution at 2.5 Ga. Reheating of the lower and middle crust in response to magmatism and metamorphism, resulted in diapirism and growth of crust in a vertical sense.

The southern Superior Province's evolution reflects accretion at the margins of a protocraton, while the Dharwar craton's tectonic environment may reflect plume impact and incipient rifting in the centre of an Archaean craton.

The Archaean geological period represents one third of the history of our planet, its record on Earth starts at about 4.0 Ga, the age of the oldest rocks known, and ends around 2.5 Ga. Paradoxically, despite the fact that much of the growth of continental crust probably occurred at this time, this period of the Earth's history is poorly known. Two fundamental reasons for this are: (i) the complexity of the tectonic and thermal history of Archaean regions, where any traces of the original tectonic or petrological character of the rocks have often been eradicated, and (ii) the difficulty in applying modern analytical methods such as palaeontology and palaeomagnetism to Archaean terranes.

The tendency for most geologists has been to use modern plate tectonic models to explain Archaean tectonics. This approach emphasizes the similarities rather than the differences in tectonic processes. Such a uniformitarian approach involves the general application of plate-tectonic principals as we now know them to the Archaean, despite obvious differences particularly in thermal regimes and rates of crustal growth.

The modern Earth is characterized by a variety of tectonic settings of which normal spreading at oceanic ridges and subduction of cold and thickened oceanic lithosphere represent well defined common end-member processes. Nonetheless, voluminous and short-lived pulses of mafic magmatism also punctuate the plate-tectonic cycle and are inconsistent with what we understand about steady-state convec-

From Burg, J.-P. & Ford, M. (eds), 1997, *Orogeny Through Time*, Geological Society Special Publication No. 121, pp. 63–98.

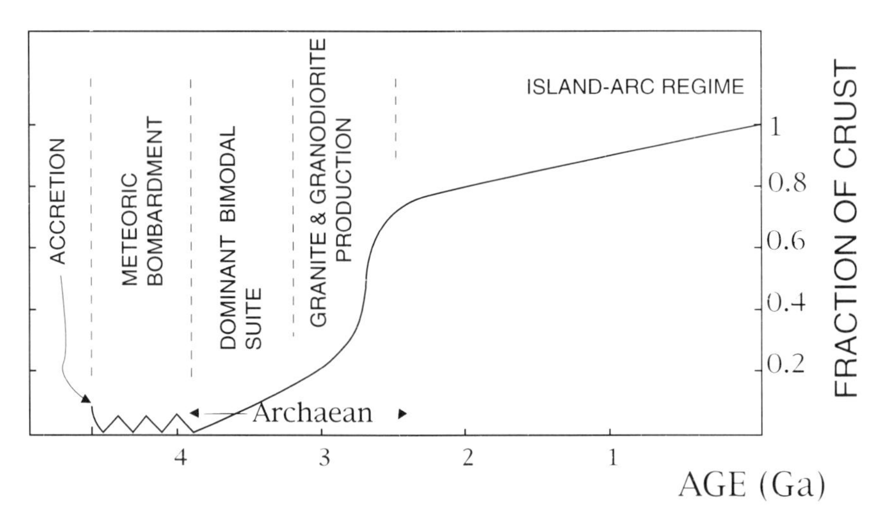

Fig. 1. A generalized model for the growth of the continental crust throughout geologic time based on a selection of crustal growth models.(Reymer & Schubert 1984; Armstrong 1981; Fyfe 1978; Hurley 1968; Hurley & Rand 1969; Veizer *et al.* 1979; Taylor & McLennan 1985).

tion and mass-fluxes in the mantle (Stein & Hofmann 1994). Locally, subduction of young hot oceanic crust results in significant differences in the heat budget and the tectonic characteristics of subduction zones. Large portions of thick oceanic crust (Ontong–Java) are impinging on subduction zones or have already been subducted (Caribbean plateau). We suggest that fundamental geological differences did exist between modern and Archaean times and that some of the more unusual characteristics of the modern Earth, such as those listed above, may be generally comparable to the normal tectonic regime of the Archaean.

A period of crustal growth

Numerous authors have addressed the problem of how the Archaean crust formed (i.e., Arth & Hanson 1975; Moorbath 1975; Arth 1979; Barker 1979; Condie 1981, 1994; Dewey & Windley 1981; Jahn *et al.* 1981; Martin 1986, 1994; Arculus & Ruff 1990; Rapp *et al.* 1991; Winther & Newton 1991). It is clear that the Archaean represented a period during which crust was formed and stabilized as cratons; the oldest crust still present on Earth (Bowring *et al.*. 1989) was stabilized at around 4.0 Ga, and a significant proportion of the present Archaean crust formed between about 3.0 and 2.5 Ga (Fig. 1). The Archaean is a period in the Earth's history during which as much as 70–80% of the present cratons formed. In marked contrast, since about 2.0 Ga, geotectonic processes have maintained an approximately constant rate of crust formation and destruction (Hurley 1968; Hurley & Rand 1969; Veizer *et al.* 1979; McLennan & Taylor 1982; Allègre 1985; Taylor & McLennan 1985).

Proponents of Archaean tectonic models must consider the fact that crust was being created and converted into stable cratons. Thus any reflection on Archaean tectonic processes must take into account the fact that the volume of crust was increasing.

Dominantly bimodal lithologies

Twenty-five cratons and associated supracrustal assemblages have been identified on the surface of the Earth. In addition, recent geophysical and geochemical data indicate that significant volumes of Archaean crust are buried below mid- to late Precambrian orogenic zones (Lucas *et al.* 1993; Martignole & Calvert 1996). The extent of the Archaean cratons is thus considerably larger than their surface exposures indicate. All of the exposed cratons are dominated by two lithological associations: (i) tonalitic–trondhjemitic–granodioritic assemblages (TTG-suites) and (ii) volcanic and sedimentary associations (greenstone belts). The combined associations are generally referred to as granite–greenstone belts or terranes.

Examination of geological compilations of the cratons reveals a consistent geometrical relationship between TTG suites and greenstone belts, with the plutonic suites representing broad domed features and the greenstone belts being 'trapped' between them (Goodwin 1981). These relationships extend over regions on the scale of a 1000 km. Only in large greenstone belts, such as the Abitibi belt of the Superior Province and the Norsman–Wiluna belt of the Yilgarn Craton, do the greenstone belts form relatively large linear bodies.

Tonalite–trondhjemite–granodiorite (TTG) series

This magmatic suite, dominated by variable proportions of tonalite, trondhjemite and granodiorite, represents the major constituent of exposed Archaean continental crust. The composition of these TTG suites remains remarkably constant over the Archaean period. They are commonly quartzo-feldspathic gneisses, rich in plagioclase and containing biotite and hornblende.

The most ancient examples of the Archaean continental crust define variable ages: 3.96 Ga in the Slave Province (Bowring *et al.* 1989), 3.9 Ga in Greenland (Kinny 1986) and in the Antarctic (Black *et al.* 1986), 3.65 Ga in South Africa (Compston & Kröner 1988), 3.45 Ga in western Africa (Potrel 1994), 3.6 Ga in the Siberian craton (Bibikova 1984), 3.36 Ga in India (Beckinsale *et al.* 1982). They are generally considered to be derived from partial melting of basaltic protoliths either in the crust or mantle (Arth & Hanson 1975; Glikson 1979; Jahn *et al.* 1981; Jahn & Zhang 1984; Martin 1986, 1994; Evans & Hanson 1992). However, some authors have proposed that these early felsic remnants of crust may represent the products of fractional crystallisation of a hydrated basaltic magma (Barker 1979; Arth *et al.* 1978; Kramers 1988). Notwithstanding their mode of formation, these magmatic products were formed in an environment with a high heat-flux closely associated with mafic magmatism.

Greenstone belts and komatiites

The greenstone belts comprise volcanic and sedimentary sequences of which the stratigraphy, sedimentary features and the geochemical characteristics of the volcanic rocks are comparable from one craton to another. Greenstone belts are found in all of the Archaean cratons and their age is generally between 3.5 Ga and 2.5 Ga (Condie 1981). They usually comprise basal sequences which are commonly mafic–ultramafic volcanics belonging to the tholeiite–komatiite lineage, which pass towards summits dominated by felsic volcanics, often of calc-

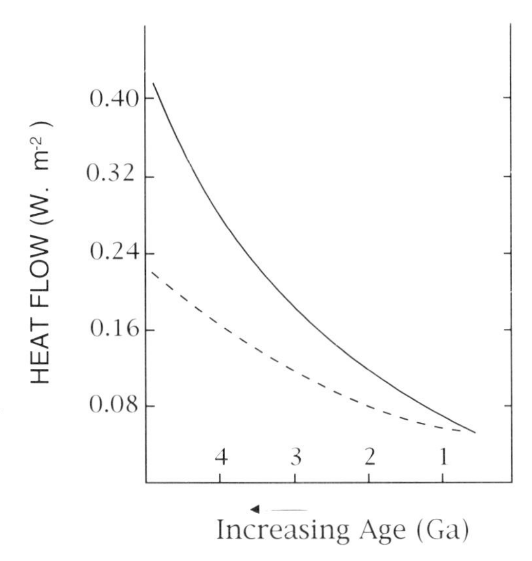

Fig. 2. Variation of average terrestrial heat flow (0.06 W m^{-2}) with time in a model of the Earth with an initial temperature sufficently great to permit convection throughout the mantle. The solid curve is for a model with chondritic abundances of radioactive elements; the dashed line is for a model with a heat-flux equivalent to present values, but with a K/U ratio derived from measurements of crustal rocks (Wasserburg *et al.* 1964). Diagram adapted from Hargraves (1986).

alkaline lineage, and sedimentary rocks, normally turbidites or late sedimentary sequences in basins associated with faulting (Mueller *et al.* 1994). Some of these greenstone assemblages have been compared to modern ophiolitic complexes (de Witt *et al.* 1987*b*).

Komatiitic ultramafic lavas are widespread in the Archaean, but their occurrence after this period is rare and, when present, their compositions indicate generally lower eruption temperatures (Nisbet *et al.* 1993). Practically all of the Archaean cratons contain komatiite lavas (South Africa, Canada, Australia, Finland, India, Brazil, etc.). Although these lavas are ubiquitous to most cratons (Viljoen & Viljoen 1969; Arndt 1994), they commonly constitute less than 10% of the volcanic sequence. With an original primary mineralogy of olivine, pyroxene and glass, they are characterized by an MgO content of at least 18% (Arndt & Nisbet 1982; Nisbet *et al.* 1993). Given their magnesium-rich composition, their low content of incompatible trace elements and their calculated temperatures of eruption (approx. 1600–1700°C), these lavas are considered to reflect both high degrees of partial melting of mantle peridotite and melting at great depths in the hottest portion of an ascending mantle plume (Campbell *et al.* 1989; Griffith & Campbell 1992). Considering an adiabatic rise for these magmas in the mantle (1°C km^{-1}, Nisbet 1982), mantle temperatures would have reached 1700°C at depths of 50 km. A non-adiabatic rise implies even higher temperatures and the possibility that these lavas were superheated, resulting in their unusually non-viscous nature and rapid quench-texture crystallization features (Arndt 1984). These lavas represent liquids that formed from mantle that exceeded ambient temperatures, predicted for secular cooling of the Earth, by as much as 300°C (Richter 1988). Given our present understanding of the Earth, the only regime capable of generating these magmas is within deeply-rooted mantle plumes (Griffith & Campbell 1992; Nisbet & Walker 1982).

Many of the fundamental principles of tectonic models for the Archaean are based on the structural relationships between greenstone belts and TTG series. As will be discussed in the following section, some greenstone belts show a structural complexity resulting from an interplay between sedimentation, magmatism and deformation on time scales of 1–10 Ma. Others show conformable sequences or discordant relationships with their basement.

A unique Archaean thermal structure

As shown in Fig. 2, the Archaean ambient heat-flow was certainly higher (Fyfe 1978; Nisbet *et al.* 1993) than at present. This heat originated from both heat inherited from the earliest phases of the Earth's accretion, and latent sources of heat associated with short-lived radioactive isotopic elements stored in the mantle and core, ^{129}I, ^{26}Al, or long-lived radioisotopes such as ^{40}K, ^{235}U and ^{238}U and ^{232}Th (Wasserburg *et al.* 1964; McKenzie & Weiss 1975; Lambert 1976). While from 4.5 Ga to 2.5 Ga the thermal structure of the Earth may have changed, the ambient heat flow in the Archaean must have been different to that of the present-day Earth.

The dissipation of internal heat in the Earth, by transforming thermal energy into kinetic energy, today results in the different convection processes at ridge axes, hot-spots, marginal basins etc. While the Archaean Earth may have simply been characterized by a more active convection regime, more hot-spots, thicker oceanic crust etc. (Sleep 1979), some authors have proposed quite different convection regimes for the early Earth (Richter 1985, 1988). These models require that the heat was dissipated rapidly by the convection process, but that the convection regime may not have been as ordered as in the modern Earth. Thus the modes of convection may have been quite distinct in the Archaean Earth (Fyfe 1974, 1978; McKenzie & Weiss 1975; Ridley & Kramers 1990).

The result of these models is a profusion of proposals for Archaean convection regimes, from chaotic mantle convection (Campbell & Jarvis 1984) to convection in small unstable cells on which rode oceanic plates. Today, the accretion of oceanic crust, its cooling, thickening and subduction can account for dissipation of 85% of the internal energy of the Earth. Calculations by Davies (1993) demonstrate that plate-tectonic processes alone were unable to dissipate the internal heat of the early Earth (Vlaar *et al.* 1994) and that a different mantle convection mechanism must have operated in the Archaean.

Several authors suggest that the Archaean Earth was characterized by many more mantle plumes than at present (Lambert 1976 ; Reymer & Schubert 1987; Campbell *et al.* 1989), or that the evolution of the planet has been punctuated by periods of enhanced plume activity (Stein & Hofmann 1994). These plumes may have provided juvenile crust from which the continental crust was derived by differentiation. In particular, Kröner (1991), Kröner & Layers (1992) and Malloe (1982) have proposed that Iceland may be an analogue of Archaean crust formation; Iceland, sited atop a mantle plume and a zone of active oceanic extension, is the site of a local production of TTGs. Storey *et al.* (1991) propose that oceanic plateaux are the sites of komatiite formation, and Desrochers *et al.* (1993) and Kimura *et al.* (1993) conclude that many greenstone belts may contain a significant proportion of accreted oceanic plateaux. These plateaux have undergone syn-formational crustal differentiation, as in Iceland, and/or later differentiation, due to passage over a hot-spot, or subduction below the plateau (e.g. Ontong-Java, Tarduno *et al.* 1991).

The absence of HP–LT metamorphism in the Archaean crust

An important characteristic of the Archaean crust is the absence of high-pressure metamorphic minerals (Windley & Bridgewater 1971; Saggerson & Owen 1971; Saggerson & Turner 1972). Lambert (1976), suggests that the absence of blueschist and eclogite facies rocks reflects the absence of tectonic processes during the Archaean which can form these rocks. In fact, the maximum exposed depths in Archaean terranes rarely exceed 30 km and are associated with temperatures of 650–900°C (e.g. Chinner & Sweatman 1968). Glaucophane, a HP–LT mineral, is predominantly found in the Phanerozoic and appears most common in the Mesozoic (Ernst 1972). To our knowledge the oldest blueschist facies rocks have been identified in China and are of approx. 1.8 Ga (Liou *et al.* 1988). The rarity of these metamorphic rocks, although perhaps a result of their poor preservation in the geological record, probably reflects a different subduction environment in the Archaean (Kröner 1981*a, b*).

Geodynamic models in the Archaean

Two modern environments exist in which continental crust is produced: (i) in subduction zones (e.g. island arcs and Andean-type arcs) and (ii) oceanic and continental intra-plate environments associated with mantle plumes (e.g. Iceland and/or the Ontong–Java plateau on oceanic crust; Kerguelen, on oceanic and continental crust; Karoo, on continental crust). Despite the fact that continental crustal production is at present lower than in the Archaean (e.g. Taylor & McLennan 1985), these two geodynamic environments represent the two principal candidates for the production of the Archaean continental crust. While both environments involve the addition and differentiation of crust from below, the constraints on lithosphere dynamics are very different: the former requires global lateral displacement of lithospheric entities with crust being produced in convergent zones, the latter involves transport of material into the crust by vertical processes.

Active plate margins have long been recognised as sites of crustal production (Dewey & Horsfield 1970; Oxburgh & Turcotte 1970). Dehydration of the subducted oceanic plate may induce fusion of the mantle wedge or of the subducting plate (Ringwood 1974) producing linear regions of juvenile and recycled crust parallel to the subduction zones. This environment is often proposed for the formation of juvenile crust and, in large part, can account for the general andesitic composition of Archaean continental crust (Talbot 1973). The model relies on the assumption that oceanic crust was generated in ridge systems, which were more extensive in the Archaean, associated with rapid spreading, and which dissipated the Archaean heat (Burke & Kidd 1978; Bickle 1978; Dewey & Windley 1981; Windley 1984; Hargraves 1986). Although definitive Archaean ophiolitic remnants have not yet been located in greenstone belts (Bickle 1994), this fact cannot be used to argue against subduction processes, as one cannot discount the fact that important quantities of juvenile crust were accreted, reworked and preserved at this time.

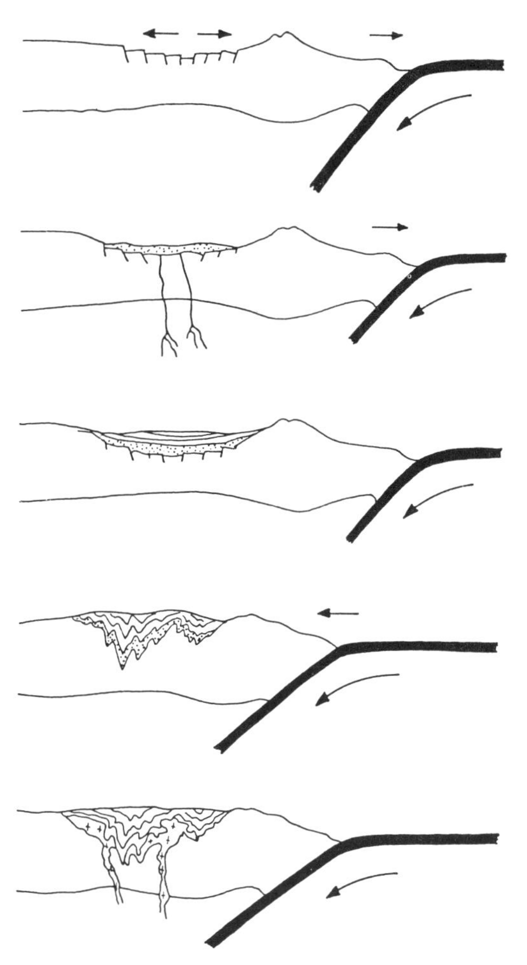

Fig. 3. Suggested development of an Archaean greenstone belt according to the 'Rocas Verdes' model. In back-arc extension magma production (dotted pattern) exceeds that required for simple extension. The sediment sequence (shown as overlying the volcanics) is mixed volcanogenic and continental clastic and may include calc-alkaline andesitic and felsic lavas from the adjacent volcanic arc. Later migration of the arc towards the continent produces deformation and the synclinal form of greenstone belt. Andean-type tonalitic to granitic plutons (crossed pattern, derived from the mantle by a two- or three-stage process; cf. Ringwood 1974) may be syn- to post-tectonic, with compositions dependent upon the depth of melting of the subducted oceanic crust. Adapted from Tarney *et al.* (1976).

A number of authors suggest that Archaean TTG suites result from fusion of hydrated

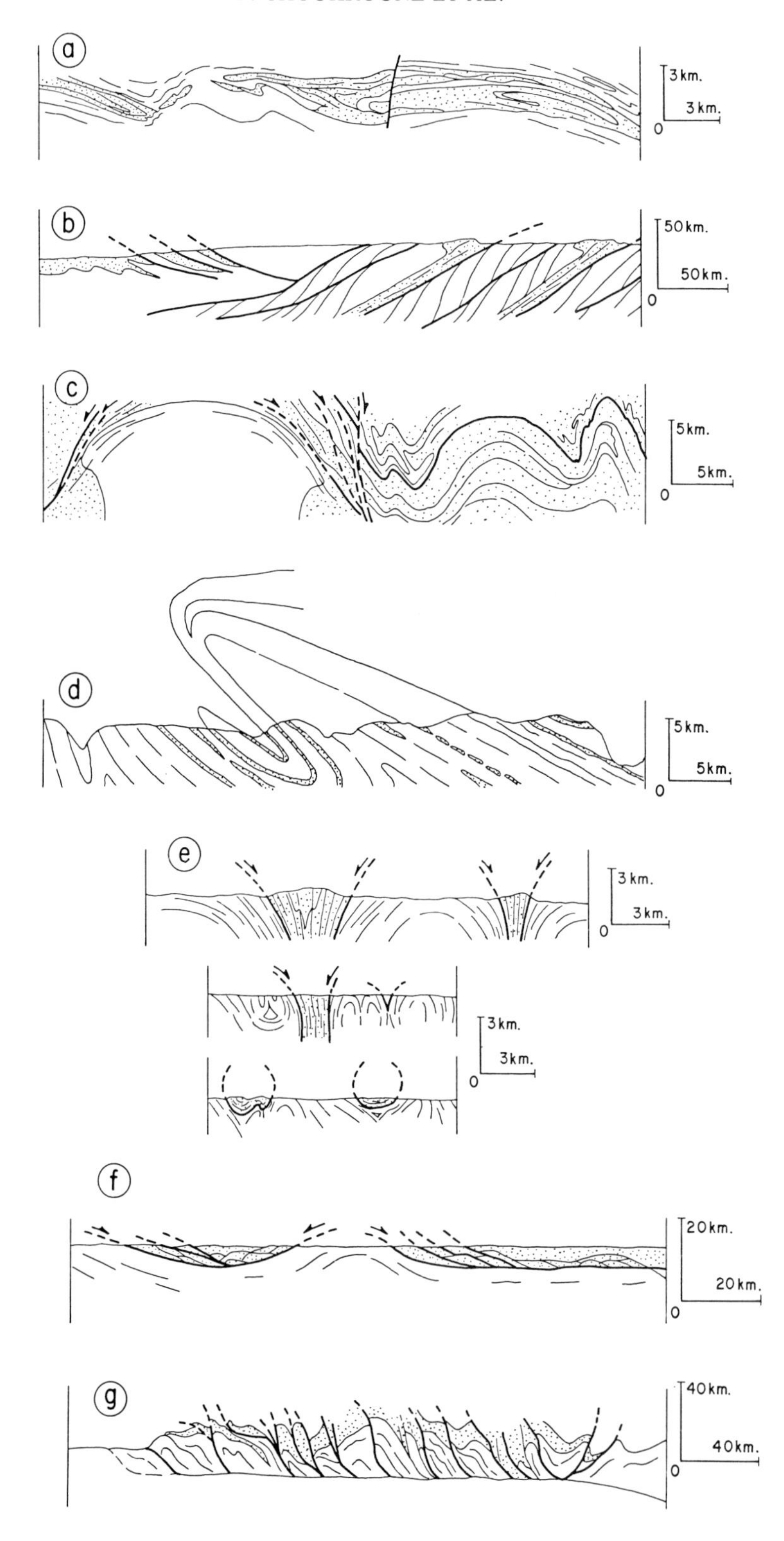
a
3km.
3km.
0
b
50km.
50km.
0
c
5km.
5km.
0
d
5km.
5km.
0
e
3km.
3km.
0
3km.
3km.
0
f
20km.
20km.
0
g
40km.
40km.
0

basaltic sources. To explain their characteristics, specifically the fact that their source material requires residual garnet, the conditions in which these suites formed must have exceeded pressures of 12 kbar and T of 1000–1100°C (Rapp *et al.* 1991). Martin (1986) suggests that subduction in a regime of high thermal gradient could explain these observations. In fact, the coupling of a high geothermal gradient with the subduction of young hot oceanic crust is the current situation in the Rocas Verdas region in Chile which facilitates TTG production by melting of subducted oceanic crust (Fig. 3; Tarney *et al.* 1976). The consequences of this model for Archaean tectonics are important as the geodynamic environment requires that the plates were organised in a similar fashion to that of modern plate tectonics.

Despite numerous studies, the origin and tectonic significance of Archaean greenstone belts remains controversial. For some workers the greenstones represent the remains of a primitive crust which once covered the entire globe (e.g. Fryer *et al.* 1979). Windley (1973) considers that the greenstones developed in rift-zones produced between divergent continental plates, while some workers consider that the greenstone belts represent marginal basins associated with subduction zones (e.g. Groves *et al.*. 1978; Tarney & Windley 1981; Drury 1983). Others (e.g. Taira *et al.* 1992; Kimura *et al.* 1993) consider the greenstone belts to represent complex terranes involving remnants of volcanic arcs, oceanic plateaux, microcontinents, etc.

A radically different vision is based on geochemical constraints of mafic rock composition in greenstone belts (Arndt 1994). These tholeiitic rocks with low Al, high Fe, Ni and Cr and depleted incompatible trace elements are comparable to modern flood basalt sequences and could be related to a hot-spot environment (Tarduno *et al.* 1991). The recognition of the large volume of volcanics produced in large igneous provinces (Coffin & Eldholm 1994) on both the modern ocean crust and in continental trappe sequences, and the geochemical similarities of these sequences to Archaean tholeiites (Arndt 1994), provide evidence that magmas related to mantle plumes may have been important in building Archaean crust (Lambert 1981; Hill *et al.* 1992; Peucat *et al.* 1993*b*). Major plume events may contribute on a cyclic basis to the formation of new crust (Stein & Hofmann 1994).

An overview of structural characteristics of Archaean cratons

Archaean nappe sequences

In recent years many authors have suggested that the structural evolution of the Archaean cratons records collisional tectonics, generally involving accreted oceanic lithologies (arcs, plateaux, etc.; Spray 1985; Ludden *et al.* 1986; Hoffman 1989; Taira *et al.* 1992). Other authors envisage collision between different continental domains (e.g. Burke *et al.* 1976; Light 1982; Shackleton 1986; de Wit *et al.* 1992). The small-scale structures in some of these areas define regions of imbricated terranes which resemble those of modern orogens. A particular example would be the collision between the Kapvaal and Zimbabwe cratons, where the Limpopo belt would resemble the roots of a

Fig. 4. Interpretative cross-sections through the Archaean crust illustrating the main types of tectonic models proposed in granite–gneiss–greenstone terranes. Supracrustal rocks and greenstones are shaded, gneissic rocks are white. (**a**) The Selukwe greenstone belt, Zimbabwe craton (modified after Stowe 1984). The structure of the belt is interpreted as a nappe within gneissic terranes. (**b**) The granite–greenstone terranes of the Dharwar craton, India (modified after Drury *et al.* 1984). This section illustrates late Archaean crustal shortening and thickening of this craton by thrusting before the development of major transcurrent shear belts. Supracrustals and greenstone units may represent tectonic slices and tectonized 'ophiolitic' relicts of crust from a marginal basin. (**c**) The Tavani area, Hearn province, Keewatin, Northwest Territories, Canada (modified after Ralser & Park 1992). The structure of this area is interpreted as refolded supracrustal nappes. (**d**) The Fiskenaesset region, Southwest Greenland (modified after Myers 1984). Granitoids, anorthosites and amphibolites are folded into large recumbent nappe-like folds and refolded by two sets of folds with axial surface normal to each other. (**e**) Granite–greenstone terranes of the Dharwar craton, India (modified after Bouhallier *et al.* 1993; Bouhallier 1995). At different structural levels, the granite–greenstone dome-and-basin pattern is interpreted in terms of sagging of dense supracrustals into their migmatitic gneissic basement. (**f**) The Leonora area, Yilgarn craton, Australia (modified after Williams & Currie 1993). The structure is interpreted as a result of extensional tectonics, with the development of a metamorphic core complex. (**g**) Granite–greenstone terranes of the Yilgarn craton, Australia (modified after Hammond & Nisbet 1992). This section illustrates the style of large-scale thrust imbrication for a shortening event which postdates early extension.

Himalayan-type mountain chain (Wilks 1988; Treloar *et al.* 1992). Figure 4 shows various tectonic models for Archaean terranes which are discussed below.

Low-angle structures involving large tracts of Archaean volcanic and sedimentary terranes have been described by Bickle *et al.* (1980), de Wit (1982), de Wit *et al.* (1987*a,b*) and Camiré & Burg (1993) for many Archaean terranes. Inside greenstone belts stratigraphic repetitions correspond to tectonic stacks and nappes. For example, isoclinal folds and mylonitic shear zones have been described in the Selukwe greenstone belt in Zimbabwe (Fig. 4a), where, due the large amount of flattening of the fabrics, these structures have been interpreted as nappes resulting from gravitational spreading (Stowe 1984). The evidence for low angle thrusting in the Archaean is derived largely from geochronological relationships. For example, in Greenland (Bridgewater *et al.* 1974) thrust-intercalated amphibolites, metapelites and quartzites are implicated in regional deformation (Myers 1976, 1984; Chadwick & Nutman 1979). Other arguments for thrust tectonics in the Archaean are based on low-angle isoclinal folds (Coward *et al.* 1976; Bickle *et al.* 1980). Thrusting following regional shortening has been proposed for the Yilgarn Block in Australia, where the folding is recognised due to stratigraphic repetitions and isoclinal folds in the greenstones (Fig. 4 f, g; Swager & Griffin 1990).

At the shallowest crustal levels, olistostromes, inverted sedimentary features and juxtaposed sediments have been observed in Barberton, South Africa (de Wit 1982; Jackson *et al.* 1987). For these reasons, Heubeck & Lowe (1994) have proposed that the Barberton greenstones have been deformed by thrust tectonics since the earliest evolution of the basin. Furthermore, they cite evidence for refolded thrusts, which include basement gneisses.

Basal decollements have been invoked at the interface of contrasting lithologies, for the Barberton (de Wit 1982), and Selukwe greenstone belts in Zimbabwe (Fig 4a; Stowe 1984). The geological environment in which these low-angle thrusts have developed is compared to obduction of oceanic crust. Crustal thickening due to stacking of crustal slices has been invoked to explain the exposure of the Limpopo granulite terranes by isostatic readjustment between the Kaapval and Zimbabwe Archaean cratons (Coward & Fairhead 1980 ; de Wit *et al.* 1992 ; Van Reenen *et al.* 1992). The granulite domains of the Archaean Dharwar craton have also been explained in this way (Fig. 4b). However, in this case the thickening is deduced from the geometries of greenstone belts which are considered to have been thrust to the north (Fig. 4b; Drury, *et al.* 1984).

Dome and basin structures

Dome and basin structures are common in Archaean cratons and characterize high-grade (Windley & Bridgewater 1971) as well as low-grade metamorphic terranes (McGregor 1951). The dimension and the shapes of the domes are variable, from a few hundred kilometres to several tens of kilometres in diameter. In general, the structural interpretation of these domains implies that they resulted from fold interference (Snowden 1984; Drury *et al.* 1984; Fyson 1984; Myers & Watkins 1985). Such interpretations have been proposed to explain the regional distribution of greenstone belts by Snowden & Bickle (1976). Fyson (1984) and Myers & Watkins (1985) indicate that two phases of regional orthogonal shortening are required. The development of nappes or refolded isoclinal folds during late regional shortening is also proposed as an explanation for local complexities in the regional map patterns. Often the structural arguments by which the gneiss domes in granite–greenstone terranes are defined are neither supported by the regional deformation field, nor do they satisfy the simple tests defined by Brun (1983*a*) for their origin.

Diapiric domes. For some authors many dome and basin structures are considered to be the result of the development of gravitational instabilities (Brun *et al.* 1981; Collins 1989; Ramsay 1989; Bouhallier *et al.* 1993; Jelsma *et al.* 1993; Choukroune *et al.* 1995).

Rayleigh–Taylor-type instabilities, known since the end of the last century (Rayleigh 1883), are related to local perturbations of multiple layers of different density. Wegmann (1935) and Eskola (1949) proposed the origin of gneissic domes by diapirism, and McGregor (1951) used this idea to explain the Archaean craton of Rhodesia. He suggested that local overloading due to basic volcanic rocks overlying a granitic basement resulted in simultaneous deformation of the cover and basement. This process, later termed sagduction, has been applied to several Archaean cratons (Anhaeusseur *et al.* 1969; Anhaeusseur 1973; Glikson 1972; Drury 1977). This process has also been applied to the Superior Province (Schwerdtner *et al.* 1979; Goodwin & Smith 1980), but was later questioned (see section on Superior Province in this paper).

Gorman *et al.* (1978), based on experiments

performed by Ramberg (1967, 1971, 1973), propose a dynamic model involving different stages of sagduction of a greenstone belt. In this model thrust structures, which result from the development and ballooning of the diapir, surround the zones of subsidence. Following a similar scenario, West & Mareshal (1979) and Mareshal & West (1980) produce a series of thermo-mechanical models which indicate that it is possible to reproduce diapiric structures when a layer of granitic crust is covered with a thick layer of basaltic lavas. If the crust is heated from below, the development of mechanical instabilites and diapirism are enhanced.

Strike-slip structures. The presence of large strike-slip features on the Archaean cratons are generally related to late orogenic, intracratonic processes. These structures are often closely spaced and interfere with, and obscure, earlier deformation structures. They are commonly responsible for the juxtaposition of terranes of different metamorphic grade and sometimes are proposed as the cause of interference structures (Drury *et al.* 1984).

Early transpression has been invoked as providing the framework for the late history of greenstone belts (Swager & Griffin 1990; Chadwick *et al.* 1989). For example, Platt (1980) constructed a strike-slip evolutionary system model for Australian greenstone belts. He suggested that they had developed as intra-continental transpressional basins in which dome and basin structures and 'en échelon' folds developed contemporaneously.

The Superior Province of Canada is crosscut by numerous E–W shear zones (Park 1981) which have been interpreted as post-collisional extensional features (Percival & Williams 1989; Williams 1990) and as strike-slip faults controlling the accretion of terranes in a zone of oblique convergence (Ludden *et al.* 1986).

Post-orogenic extension. Post-orogenic extension can constitute an important stage in continental collision (Dewey 1988). This phenomenon is related to thermal relaxation in zones of continental thickening and is associated with the development of metamorphic core complexes which are often described in Phanerozoic orogens (Brun & Van Den Driessche 1994). Despite the fact that metamorphic evidence for extensional juxtaposition of high-grade and lower-grade terranes is often lacking in the Archaean, models invoking post-orogenic extension have been proposed by Williams & Currie, (1993), Sawyer & Barnes (1994), and Kusky (1993) for terranes in Australia and Canada (Fig. 4f).

Pre-orogenic extension. Models involving post-orogenic extension (Williams & Currie 1993) or pre-orogenic extension (Hammond & Nisbet 1992) have been proposed for the Yilgarn Block in Australia. The latter model was defined from the following structural features: (i) the presence of low-angle mylonitic shear-zones between the greenstone belts and basement gneisses; (ii) the geometry of lineations and structural features indicating a uniform transport direction; (iii) the presence of steep metamorphic gradients within the greenstone belts. Based on similar criteria Passchier (1994) proposed that the first deformation episode in the Kalgoorlie region was associated with early extension.

Summary

The majority of the scientific community working in Archaean tectonic environments considers that the Archaean continental crust was formed in an environment comparable to that of modern island-arcs. Many structural geologists contend that, by comparison with modern terranes, Archaean terranes contain structural elements related to global dynamic processes involving horizontal displacements of the lithosphere. Thus, Archaean continental collisions should preserve many of the characteristics of modern orogenic zones. However, because of the enhanced thermal regime and rapid growth of continental crust at this time, Archaean tectonic models should perhaps also involve structural characteristics specific to this period.

In the following sections we present geological data for two regions of Archaean crust which display significant differences in regional structural trends: (i) the Dharwar craton is dominated by dome and basin tectonics related to diapiric movements in the mid to lower crust. The entire craton appears to have been coaxially strained due to E–W shortening; (ii) the Superior Province displays linear subprovinces, which increase in age away from a protocraton and appear to be consistent with a model involving growth of crust by lateral accretion and only limited reworking of the lower and mid crust. We suggest that both regions may reflect end-member processes in the formation of Archaean continental crust.

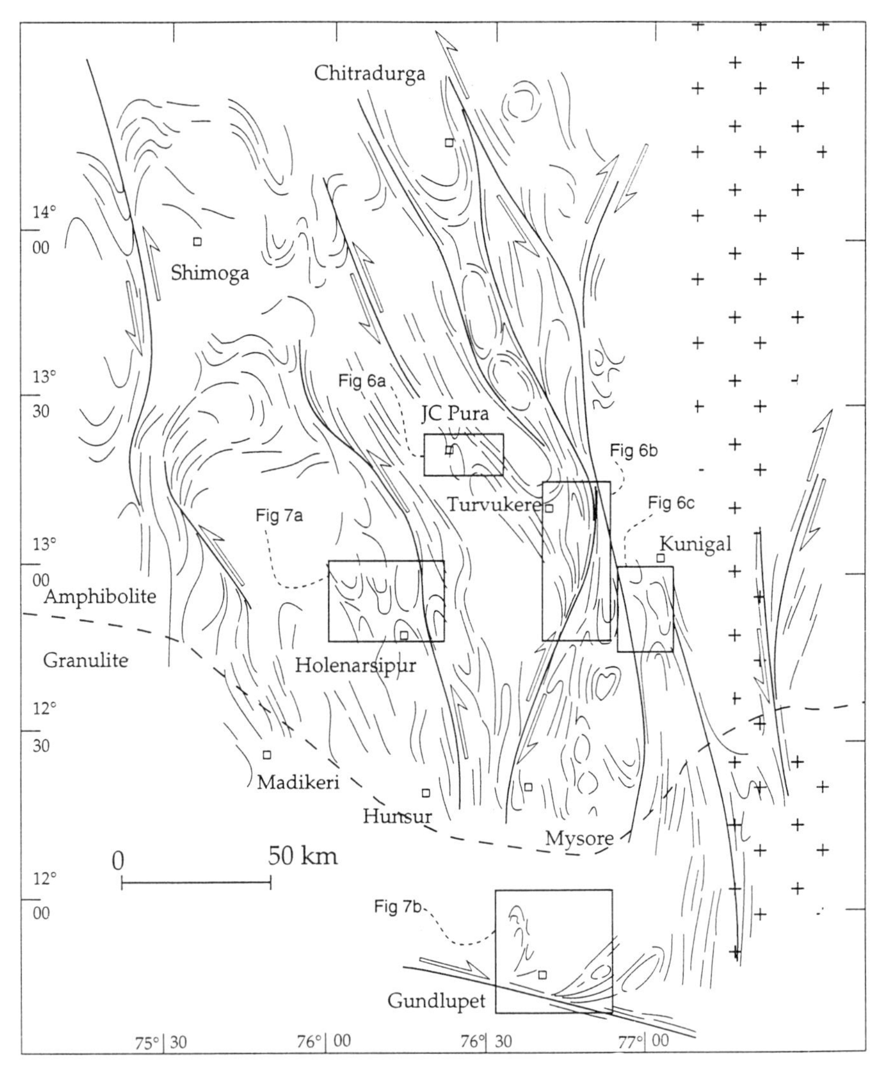

Fig. 5. Structural map of the western part of the Dharwar craton, Karnataka, South India; modified after Bouhallier (1995). Crosses indicate the Closepet granite. Arrows indicate major shear zones, curvilinear lines indicate direction of foliation, the boxes correspond to the location of detailed maps shown in Figs 6 and 7.

Dome and basin granite–greenstone patterns of the Dharwar craton: specific structural constraints and mechanisms

As already discussed, dome and basin structures are a fundamental characteristic of many TTG–greenstone terranes. Their structural interpretation is the subject of a debate between authors who favour models comparable to those of the present crust (fold interference, extensional domes, etc.) and those who consider that extensive crustal diapirism may have been characteristic of the Archaean period. In this section we will summarize the results obtained from structural studies in the Dharwar craton of India (Bouhallier 1995; Chardon 1996). Tectonic models for the formation of these structures have been tested by comparing field data with analogue models. The results indicate that at least in the Indian craton, diapirism did play an important part in the evolution of the crust and that significant proportions of the crust may have been formed during vertical accretion by magmatic underplating, melting of the lower crust and the development of gravitational instabilities.

Granite–greenstone pattern in the Indian Shield

Geological background. In the Dharwar craton (Swami Nath & Ramakrishnan 1981; Ramakrishnan & Viswanatha 1983; Naqvi & Rodgers 1983) both types of classic Archaean terranes are exposed: i.e., a succession of low- to high-grade volcanic and sedimentary rocks making up linear and curved belts of 'greenstone', 'supracrustals' or 'schist', surrounding more extensive areas of high-grade gneiss of low-K tonalite, trondhjemite and granodiorite (TTG series, here referred to as 'granite terrain'). The ages of the latter vary between 3.35 Ga and 2.5 Ga (Fig. 5; Crawford 1969; Peucat *et al.* 1989, 1993*b*).

On the craton scale, LANDSAT satellite images reveal a N–S-trending fabric which results partly from a late Archaean transcurrent ductile shearing episode (Fig. 5; Drury & Holt 1980; Chadwick *et al.* 1989). The distribution of the regional transcurrent shear zones appears nearly symmetric. This suggests that the whole craton has been coaxially strained during a near E–W regional shortening (Choukroune *et al.* 1987). This episode is thought to have been coeval with the emplacement of the large Closepet batholith (Jayananda & Mahabaleswar 1990) which is dated at 2.5 Ga (Crawford 1969; Friend & Nutman 1991; Jayananda *et al.* 1995).

Another important characteristic of the Dharwar craton is a transition from a low- to medium-grade granite–greenstone terrain in the north to a high-grade granulitic terrain in the south. The paleopressures in gneissic and mafic rocks increase from about 3 kbar in the north and in central Karnataka to 8–9 kbar in the Sargur area in the south (Fig. 5; Harris & Jayaram 1981; Raase *et al.* 1986). Most authors (Pichamuthu 1962; Raith *et al.* 1982; Raase *et al.* 1986) have attributed this feature to post-metamorphic tilting of the craton; the southernmost part of the Closepet batholith would then form the deepest structural level in a heterogeneous granitic complex which was affected by the granulite-facies overprint (Janardhan *et al.* 1979*a*, *b*, 1982; Stähle *et al.* 1987) at around 2.5 Ga (Grew & Manton 1984; Peucat *et al.* 1989, 1993*a*).

In the northern part of the craton the presence of two generations of greenstone assemblages has been known for some time (Chadwick *et al.* 1978, 1981; Viswanatha *et al.* 1982; Ramakrishnan & Viswanatha 1983, 1987); a younger series (the Dharwar volcanics) which is clearly discordant on the gneissic basement and on older greenstones (the Sargur volcanic series). In this region the youngest series is metamorphosed to greenschist facies.

In the central medium-grade terrain of the Dharwar craton (Raith *et al.* 1982) the distinction between the two series is less evident. Nevertheless, this part of the craton contains one of the oldest 'greenstone belts' in India (Hussain & Naqvi 1983). The surrounding gneisses, which yield Rb–Sr and Pb–Pb ages between 3.35 Ga (Beckinsale *et al.* 1980, 1982) and 3.305 Ga (Taylor *et al.* 1988), are intruded by 3.1–3.0 Ga trondhjemitic plutons (Beckinsale *et al.* 1982; Taylor *et al.* 1984; Meen *et al.* 1992; Bhaskar Rao *et al.* 1992). The trondhjemites are also seen to intrude supracrustal rocks (Chadwick *et al.* 1978; Bouhallier *et al.* 1993). Recent geochronological studies on supracrustal rhyolites yield an age of 3.3 Ga (Peucat *et al.* 1995).

In the southernmost area, greenstones and TTG sequences display a metamorphic paragenesis that is indicative of upper amphibolite to transitional hornblende-granulite facies metamorphism (Fig. 5; Janardhan *et al.* 1979*a*). Structures in this area are comparable with those recognized in the neighbouring area of Sargur (Chadwick *et al.* 1978; Viswanatha & Ramakrishnan 1975). Regional dome and basin patterns (Janardhan *et al.* 1979*b*) are deformed by a major dextral transcurrent shear zone (Drury & Holt 1980). *P–T* estimates for the surrounding gneissic and mafic areas north of the shear zone and east of the Gundlupet area have been estimated at about 700–750°C/8 kbar (Janardhan *et al.* 1982; Raith *et al.* 1983; Raase *et al.* 1986). As in the central portion of the craton, the principal metamorphic and deformational events appear to be related to the emplacement of the Closepet granite at 2.5 Ga (Bouhallier 1995).

Maps of foliation trajectories in the Dharwar craton indicate extensive dome and basin structures (Figs 6 & 7). Supracrustal rocks coincide with synforms in the basement which define elliptical antiforms and foliation planes that are parallel to the contact with the supracrustal rocks. The domes become more elliptical towards the western contact of the Closepet granite and also towards the higher-grade metamorphic terranes to the south. In both cases these regions correspond to areas of intense transcurrent deformation and the shape of these gneissic domes reflects the intensity of the bulk horizontal shortening at 2.5 Ga.

Before outlining the principal differences in deformation across the region, we define the deformation characteristics of five key regions.

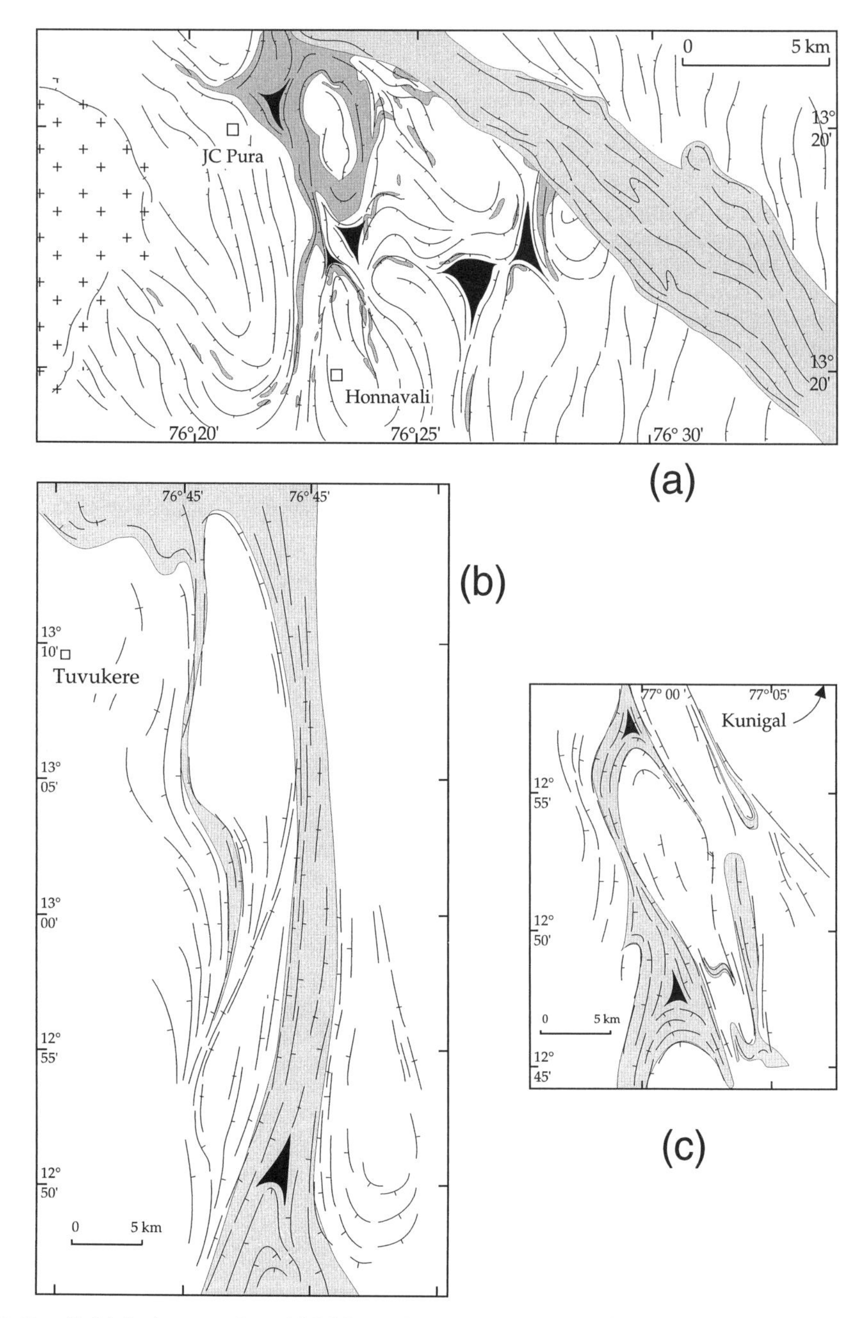

Fig. 6. Detailed foliation maps from (**a**) JC Pura where two superimposed greenstone series were recognized (the youngest is in light grey) (Chardon 1996), (**b**) Tuvukere and (**c**) Kunigal areas (modified after Bouhallier 1995). Greenstones and supracrustal rocks are in grey and gneissic rocks (TTG) in white. Triple points are in black. Areas are located on Fig. 5. Light grey shading indicates supracrustal rocks.

These regions provide a section of the crust from north to south, i.e., from lower grade to higher grade metamorphic conditions. The regions are shown in Fig. 5 and are: JC Pura; Kunigal, Tuvukere, Holenarsipur and Gunlupet (the most easterly region). The characteristics of the different strain patterns are shown in Figs 6 and 7.

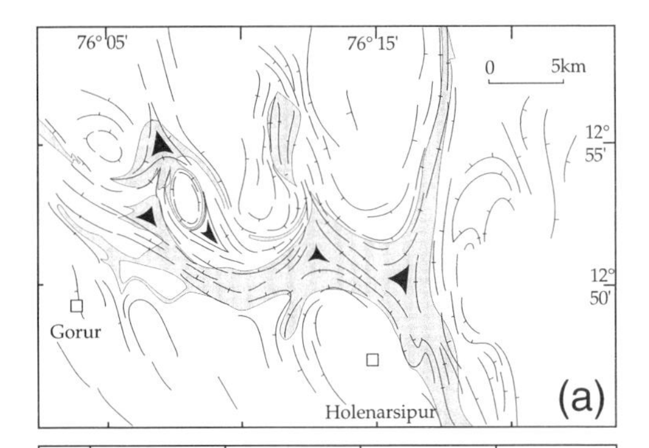

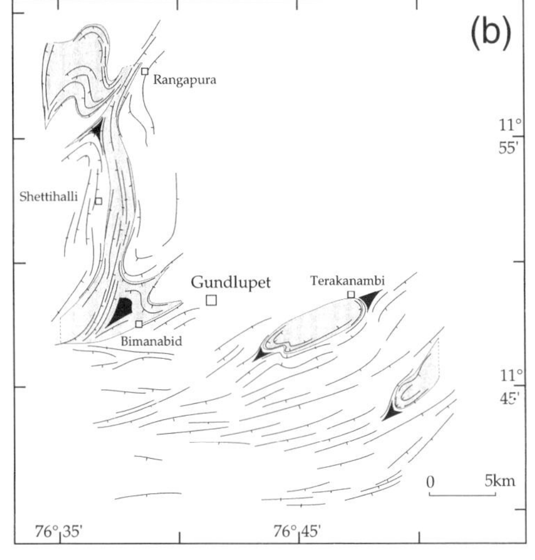

Fig. 7. Detailed foliation maps from (**a**) Holenarsipur (modified after Bouhallier *et al.* 1993) and (**b**) Gunlupet areas (modified after Bouhallier 1995). Same legend as Fig. 6, located on Fig. 5.

Characteristics of the strain field. These characteristics are described by Bouhallier *et al.* (1993, 1995) and Chardon (1996) and are summarized below.

(1) The supracrustal rocks and the gneissic basement define an increase in strain towards their contacts. In most cases these contacts are steeply dipping. Exposures of basement gneiss with sub-horizontal foliations planes are mostly restricted to the central parts of the elliptical gneissic antiforms. Locally, dome-in-dome structures are observed.

(2) Foliation triple points are probably the most noteworthy features and are defined by foliation trajectories. They occur as follows: (a) between domes where triangular arrangements of foliations define vertical or horizontal triple junctions whose dip depends on the observed level of erosion; these areas occur in supracrustal rocks; (b) at the terminations of elliptical synformal or domal closures, where they are generally horizontal (Fig. 7b, c 6b). These features are present in both supracrustal rocks and the gneissic basement; (c) in the interiors of the crests of the domal structures where they are gently dipping. They run parallel to the long axis of the dome when it is elliptical in shape.

(3) Far from synformal and domal structures, the plunge of stretching lineations L1 (which are assumed to represent the direction of the maximum stretching axis of the strain ellipsoid) is generally high in the supracrustal rocks. Near synformal or domal closures L1 lineations are evident in both supracrustal rocks and basement and is generally moderately plunging. They lie downdip on the foliation and parallel to the supracrustal-basement interface. L1 trajectories converge towards supracrustal foliation triple points where they are vertical. Within these triple junctions, stretching lineations are very well developed and parallel to the interface between supracrustals and the gneissic basement.

(4) Fabrics occurring in the rocks have the following characteristics: (a) poorly developed fabrics are located within the central parts of domes; (b) planar fabrics occur on dome limbs, where the foliation is vertical, and especially at the interfaces between supracrustals and the gneissic basement; (c) planar-linear fabrics occur around foliation triple junctions; (d) horizontal or vertical linear fabrics are located at the horizontal or vertical triple points respectively.

(5) Non-coaxial strain regimes are defined using observable criteria and well defined kinematic indicators, are generally restricted to localities near the interface between the gneissic basement and the supracrustal rocks. They indicate a systematic downward displacement of the supracrustal rocks relative to the basement. (Bouhallier *et al.* 1993). Strain regimes can be characterized some distance away from the central axes of the vertical foliation triple points in greenstone belts.

Of particular note is the unconformity between the Kibanahalli belt (Dharwar-type series) and the underlying oldest assemblages (Fig. 6a, the JC Pura belt, Sargur-type series). The younger belt is barely deformed and relatively unmetamorphosed and, as such, provides a test of the behaviour of the volcanic assemblages relative to their basement. The detailed structural characteristics of this site are as follows.

(a) Evidence for two deformation episodes, D1 and D2.

(b) D1 is defined as a dome and basin

structure that affects the oldest belt (J.C. Pura). The NW–SE deformation field (D2) affects only the younger greenstone belt (Kibanahalli) and produces a complex syncline which is discordant on the D1 structure.

(c) The 'unconformity' is tectonized and defines a 'decollement' on both sides of the syncline: down-dip displacements on both limbs of the syncline define a clear convergence of the greenstones towards the central axis of the syncline. This is only compatible with a tectonic model involving the progressive sinking of the greenstones into the granite-gneiss substratum which was previously deformed by D1.

(6) In the Holenarsipur area, metamorphic assemblages record a prograde variation in pressure from 3 to 7 kbar at a given point situated within a vertical triple junction, during the same tectonic event. This indicates that as much as 10 kilometers vertical displacement of the supracrustal rocks relative to the gneissic basement occurred.

Interpretation and discussion

Diapiric models. Many experiments have been performed to define the shape of domes during the evolution of gravitational instabilities (initiation, amplification, locking) and their periodicity according to parameters such as density, viscosity and relative thickness ratios of the buoyant and overburden layers (Ramberg 1981, 1971, 1973; Berner *et al.* 1972; Woidt 1978; Talbot 1977; Talbot *et al.* 1991). Furthermore, some analogue models lead to a quantitative estimate of strain throughout diapiric structures (Dixon 1975; Dixon & Summers 1983; Cruden 1988; Schmeling *et al.* 1988; Guglielmo 1993). All of the models indicate that the location of areas of highest strain is mainly determined by the geometry of the overburden–source (cover–basement) interface through the course of the movement of a diapir.

These experiments suggest a high variability in the distribution of strain regimes through the dome and basin structures. Non-coaxial strain regimes occur along the limbs of domes. The shear strain, which is always greatest at the interface, increases during amplification of the dome. The sense of shear shows a downward displacement of the cover with respect to the basement, with an opposite shear sense on both sides of a given dome. The result is a high variability in the distribution of fabric types through the structures and across the dome crests. In all the models, the different types of strain appear to be controlled by the dome and basin geometry. All of these model characteristics have been observed in the Dharwar craton. However, some of the structural features observed in the studied areas are inconsistent with experimental models of diapirs. This is mainly due to the horizontal displacements that have occurred along regional transcurrent shear zones. In the Holenarsipur area (Fig. 7a), the easternmost branch of the supracrustal belt is a north–south-trending linear and sinistral shear zone. In the Gundlupet area (Fig. 7b), evidence of horizontal shearing along vertical zones is also observed. In the Kunigal area (Fig. 6c), shear zones appear on a smaller scale and are dextral for strikes of 20° E, and sinistral for strikes of 140° E and cut the crust into lenses, the shape of which is almost constant.

For simple diapiric evolution, the limbs of a dome should display pitches of L1 close to 90° irrespective of foliation dip attitude. However, this is not the case in the study areas where some domains are controlled by horizontal shearing. A simple explanation in terms of interfering strain fields can be given. In the JC Pura area (Fig. 6a), the greenstone–gneiss boundary displays only downdip lineation. The diapiric process (i.e. vertical shearing) is dominant. In the Holenarsipur area (Fig. 7a), a N–S vertical shear zone bounds the eastern part of the belt. The stretching direction in the N–S foliation planes near this linear domain is horizontal. However, high L1 pitches resulting from simple diapiric evolution are observable where the effect of horizontal shear is minimal. In the Kunigal (Fig. 6a), and Tuvukere areas (Fig. 6b), the intense horizontal ductile shearing is dominant and the vertical motion is poorly preserved. The ellipticity of the domes is maximized.

Triple junctions. Horizontal triple junctions which are not directly predicted by experimental models may also be the result of interfering strain fields. Some of these junctions are situated at the terminations of the domes and sinking basins, while others are internal (Figs 6 & 7).

The significance of triple junctions has been studied by Brun 1983*a*, *b* and Brun & Pons 1981. At the terminations of rising elliptical domes, or of sinking elliptical basins, horizontal triple junctions can result from simultaneous horizontal shortening due to regional deformation and vertical shortening due to vertical motion of the diapir or the sagducted bodies. This is the case for triple points situated at the termination of supracrustal basins south of the Gundlupet area (Fig. 7b).

Horizontal triple junctions can also develop within domes or basins. Two situations of superimposed structures have been observed near the top of the domes or near the bottom of sagducted basins. The first case is illustrated by the top of a dome situated northwest of Gundlupet (Fig. 7b). Here, a vertical foliation is affected by folds with horizontal axial planes indicating vertical shortening. The resulting bulk strain is constrictive, with a sub-horizontal principal stretching direction.

Another type of situation is illustrated by the central part of the supracrustal basins south of Terakanambi (Gunlupet area, Fig. 7b). In this case, the fabric is planar–linear with S1 horizontal and horizontally shortened. The resulting lineation is horizontal and parallel to the axes of crenulation.

In order to understand the differences between these two situations, we have to explain why a vertical diapiric foliation can only be vertically shortened, while a horizontal diapiric foliation can be horizontally shortened. In the first case, the area of constriction can be a simple consequence of diapiric motion. A given point in the domain of vertical flattening planes migrate into the domain of horizontal flattening planes when the diapir moves upward. In the second case, the folding of a horizontal foliation can only be the result of superimposed horizontal regional shortening.

Lastly, we consider the vertical triple points which are present mainly in the Holenarsipur area. Models consider the strain field as due to a single diapiric body. In the field, however, diapirs are seen to interfere. It has been pointed out that the final stages of diapir emplacement are often characterized by an increase in horizontal diameter termed 'ballooning' (Holder 1979; Ramsay 1989). The interference between various shortening directions linked to the horizontal spreading of two, three or four domes generate triple junctions in which the finite strain type is constrictional (Brun & Pons 1981). The vertical triple junctions observed in greenstones situated between domes result from such a process.

To conclude, the internal tectonic style of the Archaean continental crust of the Dharwar craton appears to result from the interfering strain fields produced by Raleigh–Taylor diapiric structures (body forces) and horizontal regional shortening (boundary forces). However, the regional shortening has not erased the basic indicators of diapirism.

Diapirism in the Archaean. Since diapiric features are observed at different levels of the Archaean crust (from greenschist to granulite facies) and over very wide areas, it is possible to conclude that body forces were operating on a large scale at various periods in the evolution of a young continental crust. The observed patterns characterizing the supracrustal basins in the hornblende-granulite zone of the craton can only be interpreted in terms of sinking, drop-like blobs of supracrustal material, whereas the spoke-like patterns of linear 'greenstone belts' in the amphibolitic zone represent subsiding troughs of the overburden (supracrustal cover) at a higher structural level. The difference in the shape of regional structures between the amphibolite and the hornblende-granulite zones is simply due to the fact that the field observations concern different horizontal sections of the same structures.

These deformation conditions have been defined from pressures of 3 kbar to near 7–8 kbar and affect at least two thirds of the crust. In plan form they cover the entire craton and are clearly related to a horizontal flattening expressed by a series of conjugate strike-slip faults in amphibolite- and granulite-grade crust. The volumes of material involved in this deformation event are enormous and on a scale that is not observed in modern orogeny. We suggest that crustal reheating and diapirism on the scale described for the Dharwar craton is a distinctly Archaean phenomenon.

Although the limits of the craton are not observed in the study region, these forces seem to have acted on the interior of a mature craton. As discussed below, this deformation style is very different to that of the Superior Province which was formed by rapid accretion of crust at the margins of a craton.

The southern Superior Province: rapid accretion of crust due to horizontal tectonic forces

Evidence and models for lateral accretion of the Superior Province

The Superior Province, the largest of the Archaean cratons, provides important clues to understanding the mechanisms of crustal growth in the Late Archaean. The extensive high-precision U/Pb geochronological data available for the province (e.g. Corfu & Davis 1991), the numerous geological compilations (Card 1990; Thurston *et al.* 1991; Williams *et al.* 1991), coupled with a relatively large data-base for Nd-isotopes (Stern *et al.* 1994; Shirey & Hanson 1989; Bedard & Ludden in press) enable a comprehensive reconstruction of tectonic accretion.

The progressive increase in ages of volcanic and plutonic rocks from south to north has been used as a convincing argument for the accretion

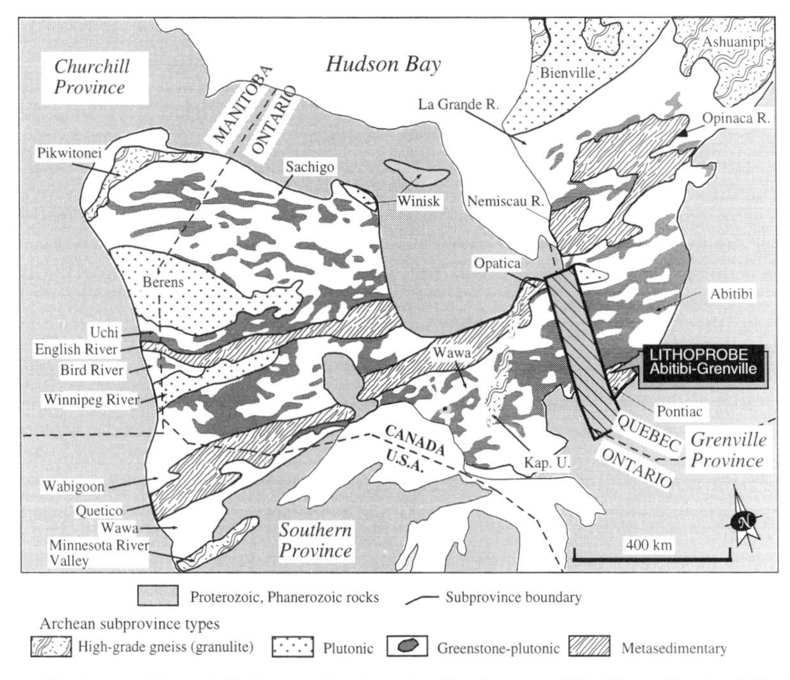

Fig. 8. Generalised map of the subdivisions of the Superior Province, modified from Card and Czcielski (1986). The northwestern Sachigo Block may be correlated below the Phanerozoic cover and southern Hudson Bay to the Minto Block, which lies north of Bienville and Ashuanipi (see text and Percival *et al.* 1994). These two blocks may form a 3.0 Ga protocontinent against which west-northwest directed subduction resulted in thermal reworking of the protocontinent (at about 2730–2700 Ma) and the formation of plutonic belts such as the Barens River, Minto and Utsalik. The Southern Superior Province represents juvenile accreted material amalgamated and accreted to the protocraton between 2750 and 2680 Ma.

The Kapuskasing uplift (Kap.U) represents a 25 km section of the accreted crust in the Abitibi-Wawa region which was uplifted in the Paleoproterozoic.

of lithological assemblages against a protocraton in the northwestern Superior Province (indicated as the Sachigo Block, Fig.8). This block comprises *c*.3.0 Ga tonalitic gneisses and relicts of sedimentary rocks (Thurston & Chivers 1990). Recent mapping in the northeastern Superior Province (the Minto Block, Percival *et al.* 1992, 1994) has pinpointed assemblages of a similar age and lithology. Percival *et al.* (1994), suggest a correlation between the two protocratonic regions below Palaeozoic cover and Hudson Bay. They further suggest that WNW-directed subduction underneath the protocraton resulted in thermal reworking and construction of an Andean-type arc; the igneous rocks of the arc being represented in the Barens river plutonic belt, the Bienville and Lake Minto and Utsalik plutonic terranes (both located in the northeastern Superior Province), all of which have ages of 2730–2710 Ma (Stern *et al.* 1994; Corfu & Davis 1991). Furthermore, the Minto Block suites have relatively evolved ϵ_{Nd} values of +1 to zero reflecting the presence of an older component in their petrogenesis (Stern *et al.* 1994). In contrast, the Wabigoon, Wawa and Abitibi are dominated by volcano-plutonic assemblages of 2730 Ma and younger, and all have juvenile ϵ_{Nd} values of +2 to +4. Several structural models for these regions (Williams 1990; Hubert *et al.* 1992*a*; Sawyer & Benn 1993) indicate a regime involving WNW-directed lateral accretion accompanied, and followed by, sinistral displacements. Percival *et al.* (1994), propose that this tectonic history may be related to oblique subduction-related collision with the southern Superior Province. This changes to orthogonal collision to the east of the Minto Block. In this model, the linear metasedimentary belts, which differentiate the Superior Province relative to most other Archaean cratons, would represent accretionary prism assemblages caught between the accreting oceanic arcs, plateaux and older microcontinents.

While an accretionary model involving some

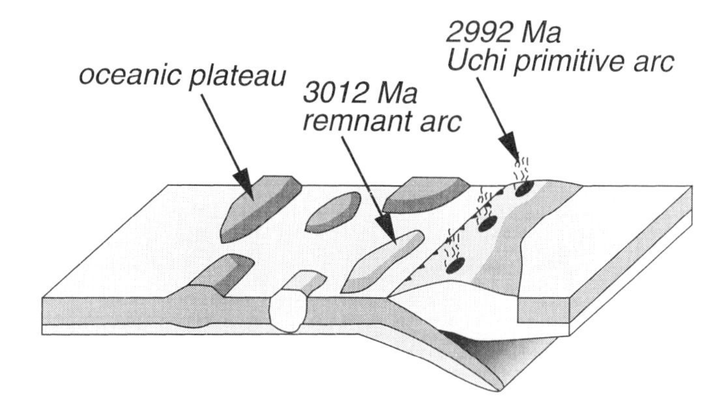

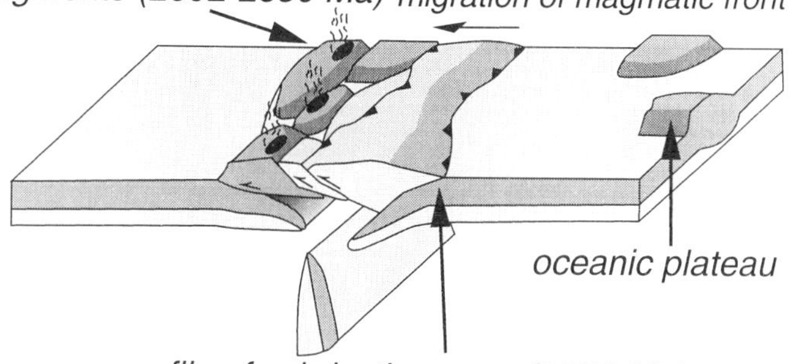

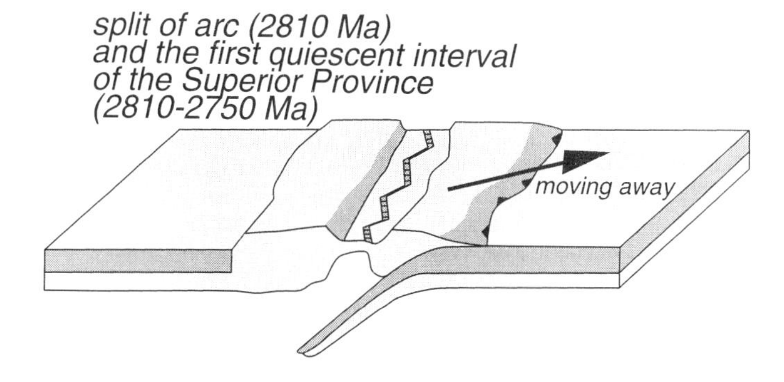

Fig. 9. Examples of tectonic models for accretion of the Superior craton (Kimura *et al.* 1993). These diagrams illustrate the following: (i) the role of accreted inactive 'exotic' material in the migration of the locus of subduction; (ii) the formation of duplexes by lateral accretion of immature oceanic material; (iii) the superposition of the products of younger igneous events (autochtonous extrusive sequences and associated intrusive rocks) on previously accreted volcano-plutonic terranes.

form of plate convergence for the Superior Province is attractive, particularly in the light of the elongate metasedimentary and metavolcanic belts, the mechanisms of accretion are the subject of debate. Suggested accretionary models from Kimura *et al.* (1993) are shown in Fig. 9. The most widely proposed models (Percival & Williams 1989; Williams 1990; Hoffman 1989, 1991; Thurston & Chivers 1990; Percival *et al.* 1994) suggest collisions involving paired arcs and associated sedimentary prisms. However, paired arcs with the same or opposite subduction polarity would not define a progression in ages, and sediments preserved in the accretionary prism cannot be younger than the youngest igneous rocks in the paired arc. In contrast, the accretion of oceanic or continental fragments to construct greenstone belts as described by Kimura *et al.* (1993) and by Hoffman (1991) dictates that both the timing and the cessation of arc magmatism show an oceanward migration (Fig. 9). Choking of the subduction zone would result in an oceanward migration of the trench and renewed arc volcanism and plutonism in previously accreted sedimentary and mafic material (Hoffman 1991; Taira *et al.* 1992). An oblique collision involving a prograding arc-accretionary complex (i.e. Jackson *et al.* 1994) would enhance the process of accretion of exotic fragments. The possibility that oceanic crust may have been thicker due to the higher thermal regime (Bickle 1978; Richter 1985), or involve numerous oceanic plateaux fragments (Desrochers *et al.* 1993) would result in a continuous process of choking the subduction zone and accretion of juvenile crust.

Petrological arguments for a juvenile character (light-REE depleted, high ϵ_{Nd}) for many of the volcanic sequences are convincing, as is the presence of light-REE-enriched volcanics and plutonics, zircon inheritance, and lower ϵ_{Nd} in regions where crustal reworking associated with Andean-type subduction is proposed. Nonetheless, the rate of magma production and accretion of juvenile crust in the Superior Province was high relative to modern subduction regimes. In the model for the assembly of the Superior Province proposed by Percival *et al.* (1994), a 400–500 km wide Andean arc and accretion of an additional band of 300–500 km of juvenile crust occurred over a period of 50–60 Ma. In the Abitibi region, where accretion ages and U/Pb formation ages are relatively well constrained, tectonic accretion occurred less than 30 Ma after formation of the volcano-plutonic sequences (Hubert *et al.* 1992*a*; Ludden *et al.* 1995). In modern subduction regimes the average age of subducted oceanic crust is >60 Ma old and most of the oceanic plateaus, that may eventually contribute to a period of crustal growth (e.g. Stein & Hofmann 1994) are >100 Ma old. Only in ophiolites, or immature arc/back-arc systems, are the volcano-plutonic assemblages tectonically emplaced soon after their formation.

A period of activity occurred between 2.75–2.7 Ga in all of the Archaean cratons. The ages of the main volcanic events during this time are surprisingly similar in all cratons, including the Superior Province. The Earth appears to have either, (i) undergone a major period of igneous activity involving production of crust from the mantle in response to one (or a series) of unusual thermal events (e.g. Stein & Hofmann 1994), and/or, (ii) due to a particular combination of tectonic circumstances preserved more crust than usual of this time interval. As discussed in the introduction, even in a regime of high heat-flux, a convincing petrological case can be argued for the origin of komatiite assemblages from a mantle plume. Komatiites occur throughout the western Superior Province (Thurston & Chivers 1990) and are particularly abundant in the southern Abitibi belt (Arndt 1984; Barnes 1983). Desrochers *et al.* (1993) and Kimura *et al.* (1993) have argued that many of the mafic–ultramafic assemblages of the southern Superior Province are more consistent with an origin as oceanic plateaux rather than a primitive arc sequence. Nonetheless, the amalgamation and accretion of the various components of the Superior Province occurred in a relatively well defined compressional–transpressional regime at about 2.7 Ga. This (these) episode(s) involved young warm juvenile crust in the southern Superior Province and extensively remelted crust in the northern Superior Province. Melting of older tonalite in the cores of regions such as the Opatica plutonic belt (Fig. 8; Sawyer & Benn 1993; Bedard & Ludden 1996), and in the metasedimentary belts (Rive *et al.* 1990) to generate granite and peraluminous granite, indicate collision and burial to at least upper amphibolite grade conditions.

The causes of a period of enhanced crustal production and accretion in the southern Superior Province, and on all of the Archaean cratons, remain enigmatic. Did a surge in plume activity result in the formation of large accretionary assemblages and/or enhance back-arc basin formation? When were the cratonic roots added to the cratons, and what role did they play in protecting the cratons from erosion, both from below through magmatism and from the craton margins through subduction? What lessons can be learned from the Earth at 1.8–2.0 Ga and in the Cretaceous, where similar surges in

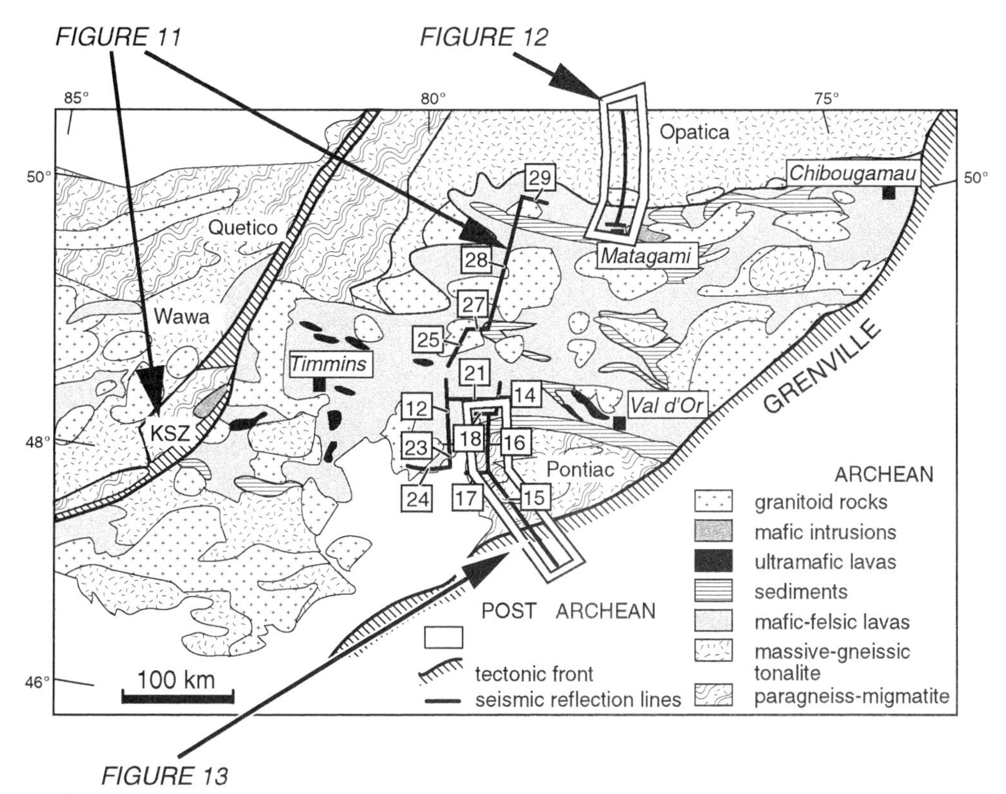

Fig. 10. A simplified geological map of the southwestern Superior Province showing the locations of the sections given in Figures 11, 12 and 13. This figure presents in simplified form the fundamental geological relationships of the region. The Kapuskasing section (high grade rocks exposed to the west of the Abitibi belt) has been described in detail by Percival & West (1994), and other papers in the same volume. The general subdivision of the Abitibi belt from Ludden *et al.* (1986) appears to conform to different volcanic assemblages in the accretionary collage (dominantly komatiite–tholeiite lineages of <2715 Ma in the south, and Fe-tholeiites calc-alkaline assemblages of 2730–2720 Ma in the north). The Opatica Plutonic Belt is dominated by gneissic tonalite or 2850–2720 Ma (Sawyer & Benn 1993; Davis *et al.* 1995). The Pontiac and the Quetico paragneiises have zircon populations as young as 2685 Ma and are intruded by peraluminous granites as young as 2640 Ma (Rive *et al.* 1990; Davis *et al.* 1995; Mortensen & Card 1994). Crustal-scale faults (not shown) transect the region and were active until late in the evolution of the crust.

plume activity probably occurred, and from the lithospheric structure of oceanic plateaux, such as the Ontong–Java, which, assuming 10–20% melting, should also preserve roots of 4–5 times the crustal thickness. In the following section we focus our attention on the deep structure of the southern Superior Province along the LITHOPROBE transect shown in Fig. 8, in an attempt to address some of the questions of late Archaean crustal accretion and stabilization.

The deep structure of the Superior Province

As part of the LITHOPROBE project a series of studies along geological and geophysical research corridors were carried out in the southeastern Superior Province (Fig. 8). The two Archaean regions studied in detail are the Kapuskasing Uplift (KU), (Percival & West 1994) and the Abitibi greenstone belt (Ludden *et al.* 1993; Hubert *et al.* 1992*a*; Calvert *et al.* 1995). The Kapuskasing Uplift, is a mid-Proterozoic structure which exposes a late Archaean section from granulite-grade mafic–amphibolite to greenschist-grade supracrustal rocks, with the deepest sections representing late Archaean crustal levels equivalent to pressures of about 8 kbar (approx. 25 km depth), (Percival & West 1994). This geological section through the Archaean craton is approximately 100 km west of the LITHOPROBE research corridor in the Abitibi greenstone belt. 250 km of seismic reflection were completed across the northern boundary of the Abitibi greenstone belt and the

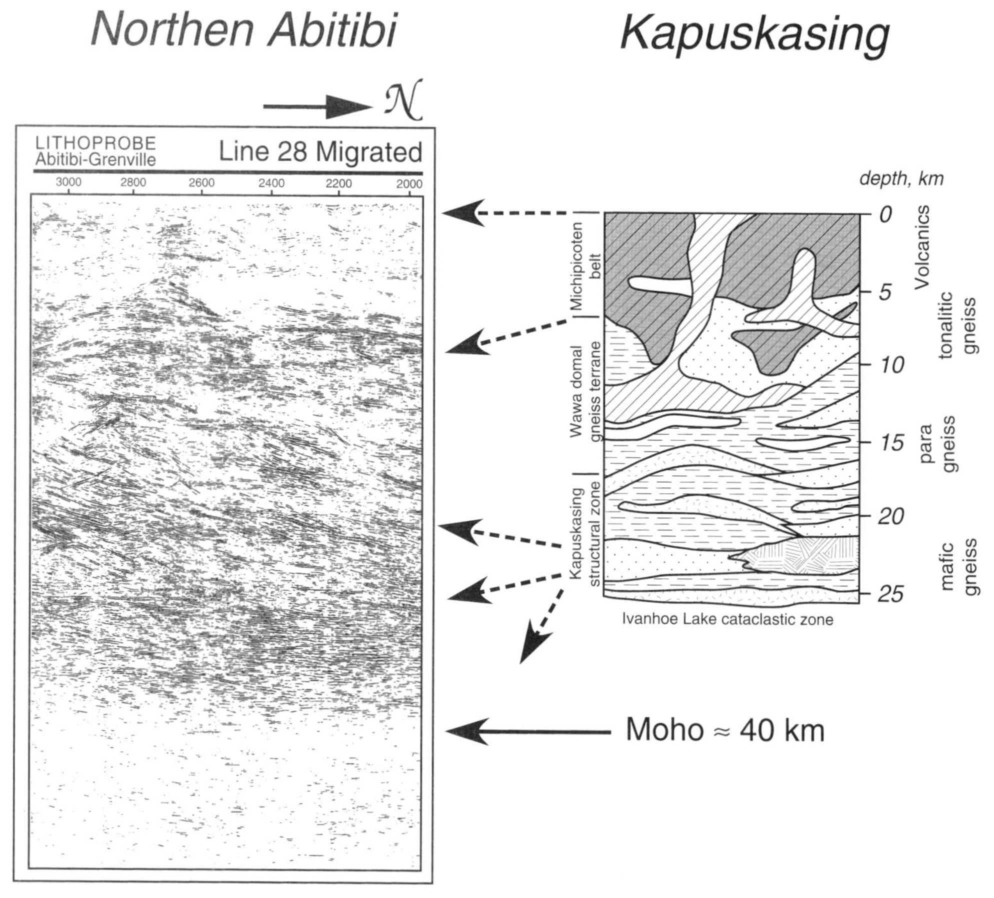

Fig. 11. A segment of LITHOPROBE line 28 (Northern Abitibi), compared with a geological section constructed for Archaean crust from the Kapuskasing Uplift (Percival & West 1994), showing a gross subdivision of Archaean crust into three crustal domains (Ludden *et al.* 1993).

Opatica Plutonic Belt (OPB), an elongate belt of TTG gneisses that may represent the orogenic core to the northern Abitibi belt (Sawyer & Benn 1993; Lacroix & Sawyer, 1995; Calvert *et al.* 1995).

In the following section we summarise the LITHOPROBE results from the following: (i) a comparison of a seismic profile through the northern Abitibi greenstone belt with the lithologies exposed in the Kapuskasing Uplift (KU); (ii) a seismic section across the northern limit of the Abitibi belt and into the Opatica Plutonic Belt; (iii) a composite seismic section across the Pontiac Subprovince; (iv) implications of the seismic data for deep crustal and mantle structure. The locations of the study areas are given in Fig. 10.

A section through late Archaean crust. Seismic reflection results from LITHOPROBE (Line 28) in the northern Abitibi belt are shown in Fig. 11 in relation to a lithological summary of crustal exposures recognized in the KU (Ludden *et al.* 1993). A broad three-fold division of crust can be observed: the lower crust is reflective and exhibits flat layered structures; the mid crust is characterized by well defined low-angle reflectors that show a general northeasterly dip, which is evident in both the Abitibi and the Pontiac Subprovinces (Hubert *et al.* 1992*a*); the upper 3–5 km of the upper crust are generally poorly reflective, especially in regions dominated by plutons. These broad divisions correspond to the lithologies in the KU where the upper crust, comprises volcanics and sediments, the mid-upper crust is dominated by the Wawa domal gneisses and the mid-lower crust contains layered amphibolitic gneiss. An important aspect of the interpretations of the crustal structure is the fact that the lower–mid-crustal section exposed in the Kapuskasing zone comprises a significant proportion of paragneiss (Leclair *et al.* 1995; Percival & West 1995). Given the correlation of the KU with the Abitibi greenstone belt, it is probable that similar paragneiss sequences lie underneath the Abitibi greenstones.

As evident from Fig. 11, the seismic images indicate generally shallow-dipping structures for the late Archaean crust. This result is inconsistent with the vertical structures which are characteristic of many of the granite-greenstone belts on the Earth (see introduction). Steep-dipping thrust faults have been identified in northern Abitibi supracrustal sequences (Bellefleur *et al.* 1995; Lacroix & Sawyer 1995); these faults root into a basal décollement which is traced to the limits of the OPB and the Quetico paragneisses (Fig. 11). The dips observed in the supracrustal assemblages of the Abitibi belt are steepened along the shear zones that define the various supracrustal blocks (Hubert *et al.* 1984). These anastomosing strike-slip shear zones transect the accretionary assemblages on the scale of the entire southern Superior Province.

The northern Abitibi belt comprises a 4–8 km thick carapace of volcanic–plutonic and sedimentary rocks that overlies plutonic and tonalitic gneisses and plutons of the OPB as well as part of the Quetico metasedimentary terrane (Sawyer & Benn 1993; Hubert *et al.* 1992*a*; Bellefleur *et al.* 1995; Lacroix & Sawyer 1995). The OPB represents a relatively homogeneous terrane north of the Abitibi greenstone belt that extends under the greenstone assemblages of northern Abitibi.

The northern Abitibi volcanic and plutonic lithologies are geochemically primitive and were probably formed in an oceanic basin and not on mature or thickened crust (Ludden *et al.* 1986; Vervoort *et al.* 1994). They bear no geochemical relationship to the underlying plutonic assemblages (Bedard & Ludden 1996). Thus, given the geometrical relationships obtained from the seismic interpretations, the supracrustal package is interpreted as being allochthonous relative to both the OPB and the paragneisses located between the KU and the Abitibi belt (Fig. 12).

A suture at a greenstone–granite boundary. Deciphering the relationships at the boundaries of the granite-gneiss terranes and the greenstone belts is critical to understanding the formation of Archaean crust. Part of the LITHOPROBE project was carried out across the boundary between the OPB and the Northern Abitibi greenstone belt (Fig. 10).

The OPB is around 500 km long and 200 km in width. The tonalitic gneiss and granitoid rocks of the belt are highly deformed with moderately dipping fabrics that contrast sharply with many of the subvertical structures mapped in the Abitibi belt. The OPB formed over a period of about 125 Ma, from before 2825 Ma to 2702 Ma and contains plutonic rocks which are significantly older than those in the Abitibi belt to the south (Davis *et al.* 1995). A northward increase in metamorphic grade to upper amphibolite facies, in addition to the structural evidence for pervasive crustal-scale WSW-vergent high temperature, ductile shearing and a subsequent a SSE-vergent thrusting event led Sawyer & Benn (1993) to propose that the OPB was the deeply eroded core of an Archaean orogen. This orogen was associated with a southward-propagating foreland fold and thrust belt. In their model the Opatica gneisses were thrust beneath the Northern Volcanic Zone of the Abitibi belt.

The seismic reflection survey was designed to test these relationships, and the results are shown in Fig. 12 and discussed by Calvert *et al.* (1995). The superb seismic image is characterized by high reflectivity to deep levels within the crust, and, in particular, by a zone of reflectivity that extends about 40 km into the mantle. The boundary between the Abitibi belt and the OPB is characterized by a marked change in reflective character in the upper crust. The greenstone lithologies, as noted for the central Abitibi, are generally unreflective, with the exception of the Bell River layered igneous complex (shot-point 700) which displays well defined southerly dips. The upper crust in the OPB is highly reflective. The shallow V-shaped fabric in the core of the OPB corresponds to well defined shear zones in the upper crust related to D2 deformation features; the strain pattern may relate to underthrusting which is pinned in the lower crust and mantle at the the Abitibi–OPB boundary (Calvert *et al.* 1995). The lower crust is highly reflective, the Moho is very well defined and both are truncated by the reflector which penetrates into the mantle. South of the OPB deep crustal reflectivity is patchy, the Moho less well defined and the gross crustal characteristics are similar to those imaged in northern Abitibi (Fig. 12).

These data are interpreted in the lower portion of Fig. 12. As for the northern Abitibi section the greenstones are thin slices (6–10 km). In this case they overlie the OPB tonaltic gneisses. Lower Abitibi crust is clearly imaged as underthrusting the OPB gneisses. The name Abitibi is applied to the entire crust, but it is clear that greenstones tectonically overlie the gneisses. The pre-collision (pre-2693 Ma)

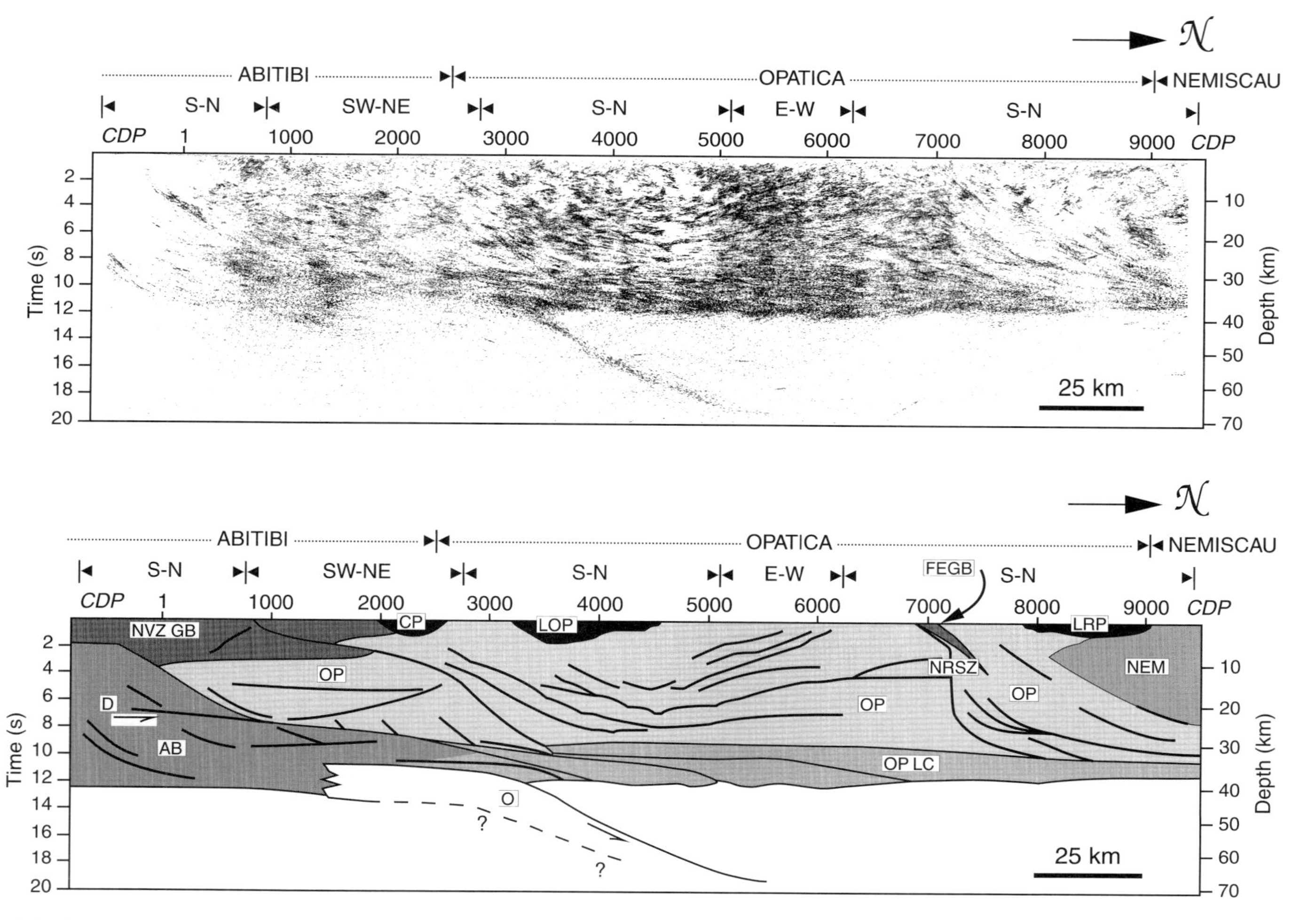

Fig. 12. Seismic reflection results and a geological interpretation of the limit between the northern Abitibi and the Opatica Plutonic belt, showing a suture extending from lower crust into the mantle, and interpreted as resulting from a collision between the OPB and the Abitibi accretionary collage (Calvert *et al.* 1995).

Opatica crust appears to have possessed a reflective lower unit and to have been separated from the Abitibi belt by a north-dipping subduction zone, possibly associated with its formation. At an early stage of the collision, the greenstone belt, which contrasts markedly with the Opatica plutons in having the geochemical characteristics of a primitive oceanic succession (Ludden *et al.* 1986; Vervoort *et al.* 1994), was emplaced. At around the same time the Abitibi margin was thrust beneath Opatica, displacing its lower crust to the north, and creating the shallowly north-dipping lower crustal decollement observed in the seismic data. The final stage of the collision produced the Opatica orogen (Sawyer & Benn 1993).

Both the seismic reflection images and geological mapping (Benn *et al.* 1992; Sawyer & Benn 1993; Calvert *et al.* 1995; Senechal *et al.* 1996) support the contention that the contact between the low-grade Abitibi and high-grade Opatica subprovinces is associated with an Archaean suture zone. The seismic reflection data show that shortening was accommodated differently above and below a shallowly north-dipping decollement, and that the rapid exhumation of mid-crustal rocks in the Opatica orogen was linked to subduction of the Abitibi lower plate into the mantle as implied more generally by geodynamical models (Beaumont & Quinlan 1994).

In this model, and as discussed below, the Abitibi belt would represent an accretionary collage of oceanic material that was trapped between the OPB and the granite–gneiss belt in the southern Pontiac subprovince. The greenstones are thus rafts of oceanic material (crust, plateaux and arcs) that have a distinct petrological and geo-tectonic association relative to the OPB and many of the plutons that later intruded this sequence. Of particular note is the relative age of formation of these exotic slabs of crust and their age of accretion and incorporation into crust; in most cases primitive crust is rarely 20–30 Ma older than the age of accretion. This crust was therefore much hotter than accreted oceanic fragments in most modern subduction zones where subducting oceanic crust is 100 Ma or older. However, rapidly evolving arc and back-arc basins are characteristic of modern plate boundary zones, and such environments are probably those in which accretion of immature crust and subsequent crustal growth can occur.

Imbrication in paragneiss sequences. The Pontiac Subprovince at the southern limit of the Abitibi belt is dominated by metawackes and several suites of crystalline rocks such as tonalite–granodiorite gneiss and plutons of granodioritic, monzonitic and granitic composition (Rive *et al.* 1990). Minor komatiite and Fe-rich tholeiites with dacites constitute a minor component of the metasediments-granite assemblage. From north to south, the metamorphic grade increases across the belt and the highest grades are indicated by the appearance of biotite, garnet, staurolite, kyanite and sillimanite (Jolly 1978; Dimroth *et al.* 1983; Camiré & Burg 1993). The youngest age of detrital zircons (approx. 2686 Ma) is younger than all of the metavolcanic assemblages of the Abitibi subprovince, but correlative with post-accretion calc-alkaline plutons in the Abitibi belt (Davis 1992; Mortensen & Card 1994).

A composite seismic section across the belt is shown in Fig. 13. The northern part of line 16A crosses the Pontiac paragneisses. These are imaged clearly as a series of shallow north dipping imbricates in the upper crust. From field mapping (Camiré & Burg 1993; Hubert *et al.* 1992*b*) this subprovince comprises imbricated thick duplexes involving sheets of gneissic and granitic rocks, ultramafic-mafic lavas and sedimentary rocks. The dominant fabric in the middle crust is nearly horizontal except near the northern boundary of the belt with the Abitibi belt where it dips progressively, from south to north, from 10° to 60° N (Dimroth *et al.* 1983; Camiré & Burg 1993; Hubert *et al.* 1992*b*). The F1 folds are SW-verging and trend NW-SE. Sheath and rootless folds are common (Benn *et al.* 1992). D2 deformation has resulted in thin-skinned northward thrusting of Pontiac metasediments over gneiss (Camiré & Burg 1993; Benn *et al.* 1992, 1994). NE–SW-trending sinistral shear zones represent another set of late ubiquitous structures. The lower-grade Baby and Belleterre volcanic rocks have been tectonically emplaced on top of higher-grade gneisses, granites and metasediments in the western part of the Pontiac Subprovince. Dimroth *et al.* (1982) interpreted these contacts to be unconformable, whereas Rive *et al.* (1990) and Hubert *et al.* (1991*b*) have postulated a fault; these relationships are perhaps best related to an extensional detachment (Sawyer & Barnes 1994).

The seismic images of the Pontiac subprovince (Fig. 13) define the strong reflectivity of the crust. Three well defined superimposed packages each with a different seismic character are observed on the profiles of lines 16A, 16 and 15 (Fig. 13). The lower package is characterized by a layer-parallel fabric that is homogeneous throughout the Pontiac. These parallel layers delineate broad

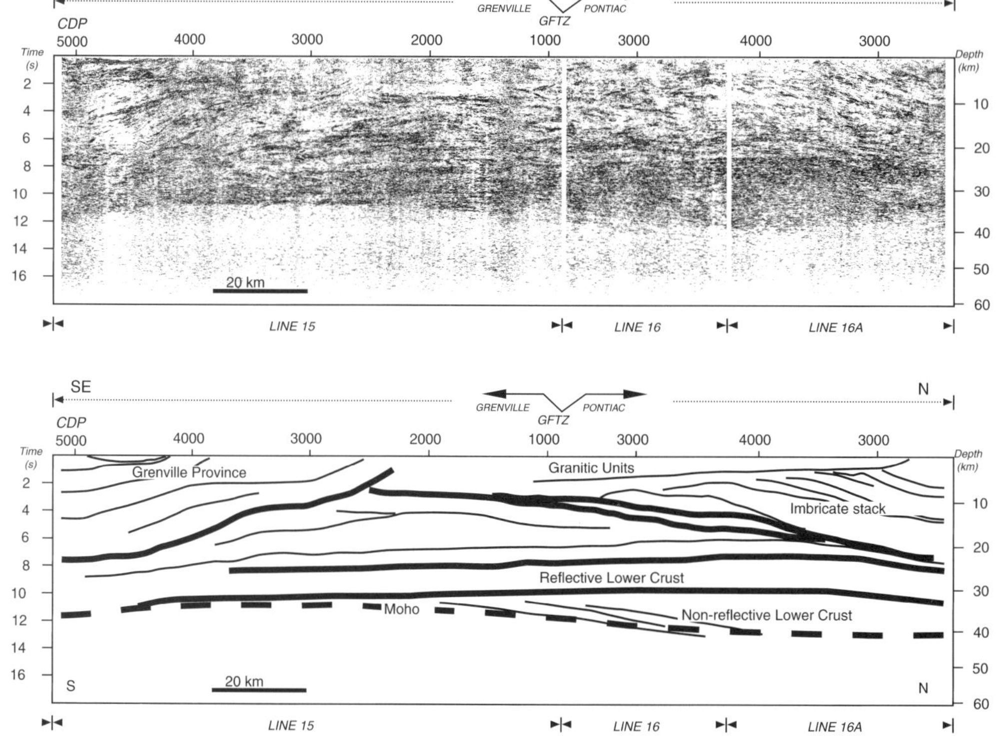

Fig. 13. A composite seismic reflection section across the Pontiac subprovince, demonstrating the southwest verging imbrication of the Pontiac sediments relative to Pontiac basement whch extends below the Grenville Province.

antiformal flexures and define two lower crustal units of different reflectivity. The middle crustal package exhibits numerous truncations and convergence between sets of strong and weak seismic reflectors. The weak reflectors form extensive lenses of homogeneous and isotropic material that are interpreted here to be sheet-like granite bodies that are commonly associated with subhorizontal domains of orthogneisses, and are intercalated with the metasediments and lavas of the Pontiac Group; these are defined as an imbricate stack on Fig. 13. The spatial arrangement and geometry of the packages is reminiscent of the imbricate slices found in duplex structures and antiformal stacks associated with accretionary prisms and in the cores of Phanerozoic orogenic belts. The upper crust is non-reflective in the granitic core of the Pontiac Subprovince. The Grenville Front transects the Pontiac Subprovince and its relationships with the Archaean rocks are discussed in Kellet *et al.* (1994).

Nature of the lower crust and upper mantle. The craton has not been extensively reworked by younger orogenies, thus the characteristics of the lower crust are considered to be Archaean. The base of the crust on seismic profiles (Moho) is defined by a decrease in the high reflectivity of the lower crust. Reflections cease at 12–13 s and define a crustal thickness of approximately 40–45 km. Minor topography on the Moho of 1–2 s is observed, with crust being thickest in southern Abitibi and thinnest adjacent to the Grenville Front in the Pontiac Subprovince (Grandjean *et al.* 1995). The Moho is relatively well defined in northern Abitibi and is extremely well defined where the Archaean crust is at its thinnest (approx. 35 km) in the southern Pontiac Subprovince adjacent to the Grenville Front (Fig. 13; Kellett *et al.* 1995). These crustal thicknesses are comparable to those of other Precambrian cratons (Durrheim & Mooney 1991) and not substantially different from mature crust in Phanerozoic regions. Bulk crustal compositions must therefore have been similar in the late Archaean and Phanerozoic, and protolith compositions and crustal differentiation processes would have been comparable. The thinned crust near the Grenville Front

(Fig. 13) is the only part of the region that has been extended. This late extension was either in response to loading along the Grenville Front, or possibly due to Palaeozoic extension associated with the St Lawrence–Ottawa graben (Percival & West 1994).

A reflective lower crust is characteristic of the entire region. This reflectivity appears to truncate the reflectivity in the mid-crust and also the extension to depth of E–W faults shown on Fig. 11. Based on the differences in seismic character of adjacent terranes (Green *et al.* 1990; Calvert *et al.* 1995) and electrical responses (Zhang *et al.* 1995) these crustal-scale structures can be traced down to depths of about 20 km (Senechal *et al.* 1996). These faults are also traceable into mid-crust exposures in the KU where they displace Archaean granulite blocks (Leclair *et al.* 1995). The vertical shear zones were active until late in the evolution of the southern Superior Province. They provided the locus for eruption of late autochthonous volcanic sequences (Desrochers *et al.* 1993) and were the site of deposition of sedimentary sequences in local pull-apart basins (Mueller *et al.* 1994). They were also the locus of major fluid pathways associated with carbonate-rich solutions and gold mineralisation (Kerrich 1989; Kerrich & Wyman 1990). Mareshal *et al.* (1995) have observed electrical anisotropy in the mantle below the Abitibi belt which correlates with the orientation of these shear zones and is Archaean in age as it cannot be traced into the Grenville Province. The anisotropy is interpreted to be related to graphite deposited in Archaean mantle and associated with the post-accretionary degassing of the subcrustal lithosphere.

If, as the arguments cited above indicate, these vertical shear zones did at one time penetrate to the mantle, their extension through the lower crust has since been obliterated as the lower crust reflectors can commonly be traced across the vertical traces of the faults. This lower crustal signature must therefore post-date the major movements on these faults and may result from lower crust extension and/or magmatic intraplating. Recent U/Pb geochronological results on growth of zircon in granulites of the KU (Krogh 1993), indicate thermal requilibration and possibly magma injection into the lower crust until as much as 100 Ma after the emplacement of the youngest igneous bodies in the upper crust.

Towards a synthesis

The basic premise of this paper is that the Archaean was dominated by a higher thermal regime than that of present and that this should be evident in specific tectonic styles. The present-day Earth is characterized by a steady-state evolution which involves a plate tectonic process dominated by the formation and destruction of oceanic crust with a mean age of about 60 Ma. Nevertheless, the modern plate tectonic cycle is punctuated by major thermal events, such as the Cretaceous superplume (Larson 1991). In areas such as the North American Cordillera and the southern Andes, subduction of young, hot oceanic crust also occurs. As discussed in the introduction, many authors suggest that both plume tectonics and the subduction of young oceanic crust were commonplace in the Archaean. Furthermore, geochronological data from mineral inclusions in diamonds (Richardson *et al.* 1984) indicate that thick roots of highly refractory mantle formed below the cratons in the Archaean. These roots now act as stable anchors to the cratons, and they have both protected the cratons during collisional events, deflecting the influence of thermal plumes away from the lower crust towards the reworked margins of the cratons (Cox 1989). Thinner and warmer lithospheric roots in the Archaean may have been easily eroded, resulting in very different heat budgets in the crust during both orogenic and anorogenic events.

We emphasize the following points.

(a) The Dharwar craton is characterized by the presence of diapiric structures recording vertical displacements which were shortened horizontally by a transpressional event.

(b) The dominant structural characteristics of the southern Superior Province indicate lateral accretion of crust. Thus, in the late Archaean, this region was dominated by horizontal relative displacements.

(c) The TTG terranes in the Dharwar appear as large rounded masses without any clear zonality in age, and which extend over a region of 1000 km by 1000 km.

(d) The Superior craton shows a relatively well-defined outward decrease in age compatible with progressive accretion. Tonalite-dominated terranes such as the OPB appear to be made of relatively homogeneous elongated linear tonalite bodies for which an origin as a plutonic arc appears appropriate.

(e) The northern Abitibi–OPB boundary shows geophysical and geological evidence for its having formed as an accretionary collage between an accreting greenstone terrane and a plutonic 'arc'.

(f) The greenstone assemblages appear as thin slices resting tectonically on underthrust

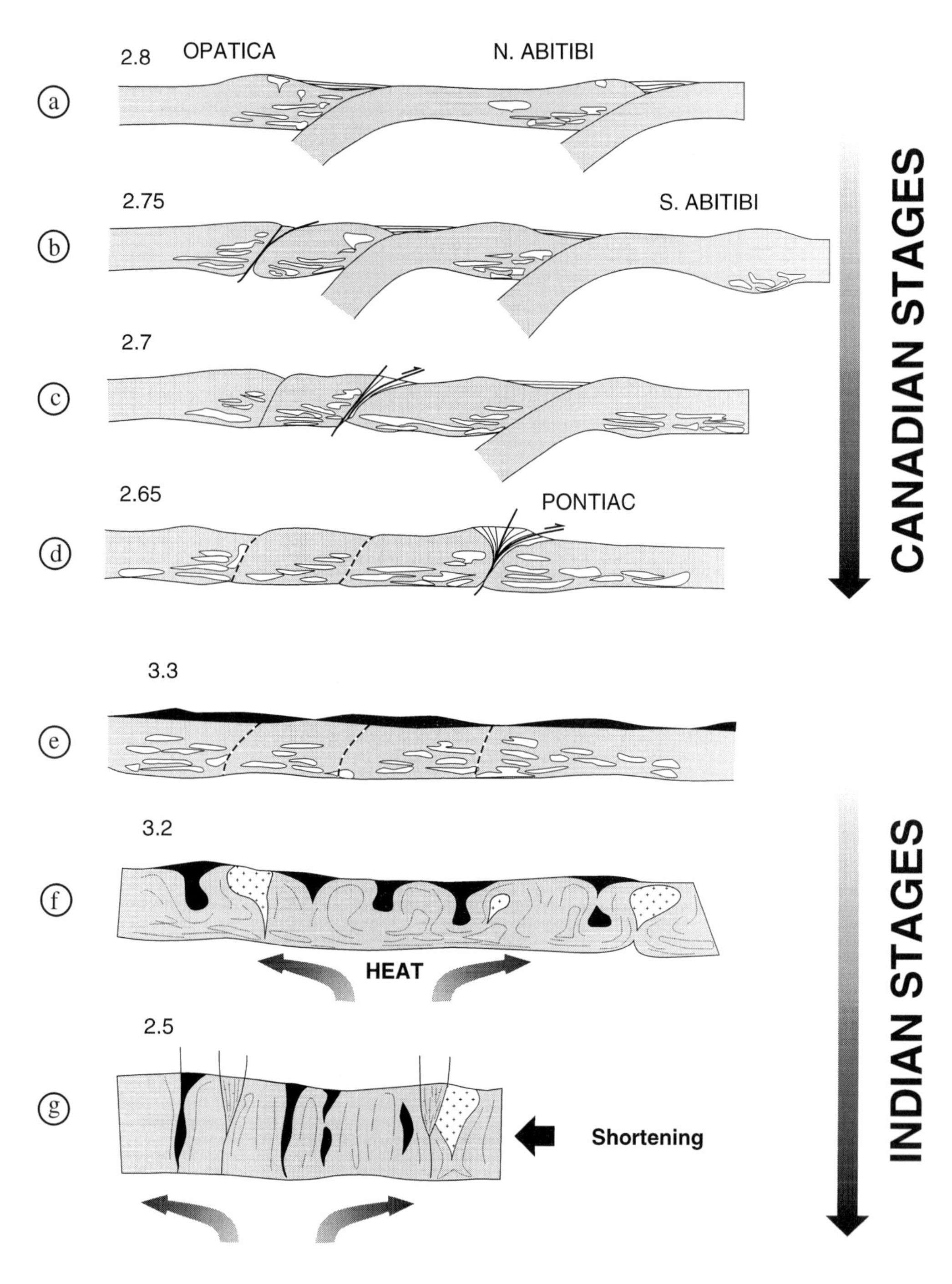

Fig. 14. Cartoon showing the stages of evolution of the Superior Province and the Dharwar craton. In these models the Superior craton was formed by lateral accretion relative to a protocraton (Opatica), while the Dharwar craton was reworked by a thermal plume which impacted on a stable protocraton. Grey indicates primitive accreted crust; white, intraplated plutons; white layered, sediments; black, volcanics; crosses, late plutons.

deeper crustal lithologies of plutonic (OPB), volcanic and sedimentary belts (i.e., Quetico and Pontiac); the latter must also include the original basement to the overthrust assemblages.

(g) Inverted and low angle contacts have been defined by surface mapping and by geophysics in the Abitibi region (i.e., the Pontiac imbricated sediments and the OPB tonalite thrust sheets).

(h) Low-angle thrust relationships cannot be demonstrated in the Dharwar craton.

(i) The material being accreted in the southwestern Superior Province was immature in terms of crustal residence and mantle source.

(j) Dharwar lithologies show long crustal residence times and represent a differentiated stage of crustal evolution.

To summarize, both cratons represent a net addition of juvenile crustal material. The Dharwar crust appears to have been fed from the bottom due to plume impact (underplated and intrusive material), while the Superior Province was accreted laterally and later annealed by intraplating events. Are these two situations incompatible, and can we reconcile the different processes? We suggest that the southern Superior Province represents an example of rapid lateral accretion of Archaean crust occurring at the limits of an Archaean protocraton. Dharwar represents a protocraton for which the initial material (no longer observable) is probably comparable to that observed in the Superior Province, but which was cratonised by underplating, reheating and reworking. These two examples thus represent different stages of the evolution of ancient crust which are illustrated in Fig. 14.

The Abitibi belt represents an accretionary complex caught between the colliding Opatica belt and a plutonic belt in the southern Pontiac subprovince now largely part of the Grenville province (Fig. 14a, b; Ludden *et al.* 1993; Martignole & Calvert 1996). Subduction below these warm accreted terranes resulted in TTG plutonism, which peaked in the Abitibi belt at about 2695 Ma (Fig. 14c). The final collision occurred at about 2685 Ma and the entire southern Superior Province then underwent sinistral transpression (Fig. 14d; Percival *et al.* 1994). The Opatica belt provided a buttress against which the greenstones were accreted. Stacking to produce crust of at least mid-amphibolite grade (20 km) occurred in the Opatica belt and in the Pontiac Subprovince (Fig. 14d). Sheets of peraluminous granite intruded into the sediments of the Pontiac and Quetico Subprovinces and the pink-granodiorites in the Opatica belt (Bedard & Ludden in press); all provide evidence for post-collisional melting of crust. However, crustal melting was localised and relates to tectonic stacking and/or intraplating of magmas in the lower crust. Mafic amphibolites in the exposed section of the Kapuskasing structural zone give U/Pb ages indicating late ductile flow and magma intrusion (Krogh 1993). The highly reflective lower crust in the seismic images may record these late intraplating events. From the teleseismic and electrical responses of the mantle, it is evident that the lithosphere below the Superior Province was formed by the end of the late Archaean (Silver & Kaneshima 1993; Mareschal *et al.* 1995). Constraining the time of formation of this mantle lithosphere in the sequence of tectonic events shown in Fig. 14, would provide important constraints on craton stabilization. The extensive ultramafic volcanism at around 2.72 Ga in southern Abitibi may have been responsible for forming a buoyant, thick, refractory lithosphere, which during subduction may have been underplated beneath immature plutonic belts such as the OPB and the Pontiac subprovince. The suture imaged in the mantle may be evidence for this process.

We further suggest that the different tectonic relationships in the Dharwar craton reflect a thermal event that reworked an earlier-formed protocraton which itself may have had an earlier accretionary history such as that for the Superior craton. The extensive diapirism may reflect the impact and incubation of a thermal plume below the craton which also produced a layer of dense mafic lavas on the pre-existing craton (Fig. 14e). Transfer of heat and magma into the lower and middle crust resulted in partial melting. Gravitational instabilities drove a diapiric regime which was simultaneous with shortening in a regional transpressional event (Fig. 14g). In order to transport the heat into the crust, either no mantle lithosphere existed, or the lithosphere was winnowed away by the plume itself.

Conclusions

It is probable that certain tectonic processes operating in the Phanerozoic, such as enhanced mantle plume activity, subduction of young (warm) oceanic crust, and faster than usual accretion of crust, may have been the norm during the Archaean. Evidence from the Superior Province indicates that rapid crustal growth occurred, largely due to horizontal tectonic forces. Models analagous to modern plate tectonics are applicable, but the rates of convergence and accretion exceeded those normal for the present day. Accreted crust was warm and subject to more ductile deformation than in modern accretionary zones. These accreted arc, ocean-floor and ocean plateau fragments would have been underlain by a thick refractory, buoyant, warm lithospheric root that was rapidly underplated (or imbricated) below the recently accreted terranes.

The southern Superior Province evolution reflects accretion at the margins of a protocraton; perhaps the widespread preservation of magmas as old as 2.7 Ga on Earth and their rapid accretion reflects the impact of a plume (or series of plumes) in an oceanic basin. Nonetheless, as was the case for the event during the Jurassic–Cretaceous, tectonic convergence dominated over 'plume tectonics'. In the Dharwar craton a major thermal event appears to characterise its evolution at 2.5 Ga. Reheating of the lower and middle crust in response to magmatism and metamorphism, resulted in diapirism and growth of crust in a vertical sense. This tectonic environment may reflect plume impact and incipient rifting in the centre of an Archaean craton.

Lithoprobe contribution No. 756.

References

ALLÈGRE, C. J. A. 1985. The evolving Earth system. *Terra Cognita,* **5**, 5–14.

ANHAEUSSER, C. R. 1973. The evolution of the early Precambrian crust of Southern Africa. *Philosophical Transactions of the Royal Society, London,* **273**, 359–388.

——, MASON, R. & VILJOEN, R. P. 1969. A reappraisal of some aspects of Precambrian shield geology. *Geological Society of America Bulletin,* **80**, 2175–2200.

ARMSTRONG, R. L. 1981. Radiogenic isotopes: the case for crustal recycling on a near steady-state no continental growth Earth. *Philosophical Transactions of the Roy*^*' Society, London,* **A301**, 443–472.

ARCULUS, R. J. & RUFF, L. J. 1990. Genesis of the continental crust: evidence from island arcs, granulites and exospheric processes. *In*: VIELZEUF, D. & VIDAL, P. (eds) *Granulites and crustal evolution*. Kluwer Academic, Dordrecht, 7–23.

ARNDT, N. T. 1994. Archaean komatiites. *In*: CONDIE, K. C. (ed.) *Archaean crustal evolution*. Elsevier, Amsterdam, 11–44.

—— 1984. Magma mixing in komatiitic lavas from Munro township, Ontario. *In*: KRÖNER, A., HANSON, G. N. & GOODWIN, A. M. (eds) *Archaean geochemistry*. Springer-Verlag, Berlin, 99–115.

—— & NISBET E. G. (ed.) 1982. *Komatiites*. Allen & Unwin, London.

ARTH, J. G. 1979. Some trace-elements in trondhjemites, their implication to magma genesis and paleotectonic setting. *In*: BARKER, F. (ed.) *Trondhjemites, Dacites and Related Rocks*. Elsevier, Amsterdam, 123–132.

—— & HANSON, G. N. 1975. Geochemistry and origin of the early Precambrian crust of northeastern Minnesota. *Geochimica et Cosmochimica Acta*, 325–362.

——, BARKER, F., PETERMAN Z. E. & FRIEDMAN I. 1978. Geochemistry of Gabbro-diorite-tonalite-trondhjemite suite of southwest Finland and its implication for the origin of tonalite and trondhjemite magmas. *Journal of Petrology,* **19**, 289–316.

BARKER, F. 1979. Trondhjemites: definition, environment and hypotheses of origin. *In*: BARKER, F. (ed.) *Trondhjemites, Dacites and Related Rocks*. Elsevier, Amsterdam, 1–12.

BARNES, S. J. 1983. A comparative study of olivine and clinopyroxene spinifex flows from Alexo, Abitibi greenstone belt, Canada. *Contributions to Mineralogy and Petrology,* **83**: 293–308.

BEAUMONT, C. & QUINLAN, G. 1994. A geodynamic framework for interpreting crustal-scale seismic reflectivity patterns in compressional orogens. *Geophysical Journal International,* **116**, 754–783.

BECKINSALE, R. D., DRURY, S. A. & HOLT, R. W. 1980. 3.360 Myr old gneisses from the south Indian craton. *Nature,* **283**, 469–470.

——, REEVES-SMITH, G., HOLT, R. W. & THOMPSON, B. 1982. Rb-Sr and Pb-Pb isochron ages and REE data for Archaean gneisses and granites, Karnataka state, South India. *In*: ASHWAL L. D. (ed.) *Indo-US Workshop on the Precambrian of South India*. National Geophysical Research Institute, Hyderabad, 35–36.

BEDARD, L.-P. & LUDDEN, J. N. 1997. Nd-isotope evolution of Archaean plutonic rocks in the Opatica, Abitibi and Pontiac subprovinces (Québec, Canada), *Canadian Journal of Earth Sciences*. In press.

BELLEFLEUR, G., BARNES, A., CALVERT, A., HUBERT, C. & MARESCHAL, M. 1995. Seismic reflection constraints from Lithoprobe line 29 on the upper crustal structure of the northern Abitibi greenstone belt. *Canadian Journal of Earth Sciences,* **32**, 128–134.

BENN, K., SAWYER, E. W. & BOUCHEZ, J.-L. 1992. Orogen parallel and transverse shearing in the Opatica belt, Quebec: implications for the structure of the Abitibi subprovince. *Canadian Journal of Earth Sciences,* **29**, 2429–2444.

——, MILES, W., GHASSEMI, M. R. & GILET, J. 1994. Crustal structure and kinematic framework of the north-western Pontiac subprovince, Quebec: an integrated structural and geophysical study. *Canadian Journal of Earth Sciences,* **31**, 271–281.

BERNER, H., RAMBERG, H. & STEPHANSSON, O. 1972. Diapirism in theory and experiment. *Tectonophysics,* **15**, 197–218.

BHASKAR RAO, Y. J., SIVARAMAN, T. V., PANTULU, G. V. C., GOPALAN, K. & NAQVI, S. M. 1992. Rb-Sr ages of late Archaean metavolcanics and granites, Dharwar craton, South India and evidence for early Proterozoic thermotectonic event(s). *Precambrian Research,* **59**, 145–170.

BIBIKOVA, E. V. 1984. The most ancient rocks in the USSR territory by U-Pb data on accessory zircons. *In*: KRÖNER, A. *et al.* (eds) *Archaean geochemistry*, Springer-Verlag, Berlin, 235–250.

BICKLE, M. J. 1978. Heat loss from the Earth: a constraint on Archaean tectonics from the relation between geothermal gradients and the rate of plate production. *Earth and Planetary Science Letters,* **40**, 301–315.

—— 1994. Archaean greenstone belts are not oceanic crust. *Journal of Geology,* **102**, 121–138.

——, BETTENAY, L. F., BOULTER, C. A. & GROVES, D. I. 1980. Horizontal tectonic interaction of an Archaean gneiss belt and greenstones, Pilbara block,Western Australia. *Geology,* **8**, 525–529.

BLACK, L. P., WILLIAMS, I. S. & COMPSTON, W. 1986. Four zircon ages from one rock : the history of a 3930 Ma old granulite from Mount Sones, Enderby Land, Antarctica. *Contributions to Mineralogy and Petrology,* **94**, 427–437.

BOUHALLIER, H. 1995. *Evolution structurale et metamorphique de la croûte continentale archéenne (Craton de Dharwar, Inde du Sud)*. Memoires de Géosciences-Rennes.

——, CHOUKROUNE, P. & BALLÈVRE, M. 1993. Diapirism, bulk homogeneous shortening and transcurrent shearing in the Archaean Dharwar craton: the Holenarsipur area, southern India. *Precambrian Research,* **63**, 43–58.

——, CHARDON, D. & CHOUKROUNE, P. 1995. Strain patterns in Archaean dome and basin structures: the Dharwar craton (Karnakata,South India). *Earth and Planetary Science Letters,* **135**, 57–75.

BOWRING, S. A., WILLIAMS, I. S. & COMPSTON, W. 1989. 3.96 Ga gneiss from the Slave Province Northwest Territories, Canada. *Geology,* **17**, 971–975.

BRIDGWATER, D., MCGREGOR, V. R. & MYERS, J. S. 1974. A horizontal tectonic regime in the Archaean of Greenland and its implications for early crustal thickening. *Precambrian Research,* **1**, 179–197.

BRUN, J.-P. 1983*a*. L'origine des dômes gneissiques: modèles et tests. *Bulletin de la Société Géologique de France, (7)* **25**, 219–228.

—— 1983*b*. Isotropic points and lines in strain fields. *Journal of Structural Geology,* **5**, 321–327.

—— & PONS, J. 1981. Patterns of interference between granite diapirism and regional deformation (abstract). *Journal of Stuctural Geology,* **3**, 93.

—— & VAN DEN DRIESSCHE, J. 1994. Extensional gneiss domes and detachment fault systems: structure and kinematics. *Bulletin de la Société Geologiques de France,* **165**, 519–530.

——, GAPAIS, D. & LE THEOFF, B. 1981. The mantled gneiss domes of Kuopuo (Finland) : interfering diapirs. *Tectonophysics,* **74**, 283–304.

BURKE, K. & KIDD, W. S. F. 1978. Were Archaean geothermal gradients much steeper than those of today? *Nature,* **272**, 240–241.

——-, DEWEY, J. F. & KIDD, W. S. F. 1976. Dominance of horizontal movements, arc and microcontinental collisions during the later permobile regime. *In*: WINDLEY B. F. (ed.) *The Early History of the Earth*. Wiley, London, 113–129.

CALVERT, A. J., SAWYER, E. W., DAVIS, W. J. & LUDDEN, J. N. 1995. Archaean subduction inferred from a mantle suture in the Superior Province. *Nature,* **375**, 670–674

CAMIRÉ, G. E. & BURG J. P. 1993. Late Archaean thrusting in the northwestern Pontiac Subprovince, Canadian Shield. *Precambrian Research,* **61**, 51–66.

CAMPBELL, I. H. & JARVIS, G. T. 1984. Mantle convection and early crustal evolution. *Precambrian Research,* **26**, 15–56.

——, GRIFFITHS, R. W. & HILL, R. I. 1989. Melting in an Archaean mantle plume: head it's basalts, tail it's komatiites. *Nature,* **339**, 697–699.

CARD, K. D. 1990. A review of the Superior Province of the Canadian shield, a product of Archaean accretion. *Precambrian Research,* **48**, 99–156.

—— & CIESIELSKI, A. 1986. DNAG#1 Subdivisions of the Superior Province of the Canadian shield. *Geoscience Canada,* **13**, 5–13.

CHADWICK, B. & NUTMAN, A. P. 1979. Archaean structural evolution in the northwest of the Buksefjorden region,southern West Greenland. *Precambrian Research,* **9**, 199–226.

——, RAMAKRISHNAN, M., VISWANATHA, M. N. & SRINIVASA MURTHY, V. 1978. Structural studies in the Archaean Sargur and Dharwar supracrustal rocks of the Karnataka craton. *Journal of the Geological Society of India,* **19**, 531–549.

——, ——, & —— 1981. Structural and metamorphic relations between Sargur and Dharwar supracrustal rocks and Peninsular gneiss in Central Karnataka. *Journal of the Geological Society of India,* **22**, 557–569.

——, ——, VASUDEV, V. N. & VISWANATHA, M. N. 1989. Facies distributions and structures of a Dharwar volcano. sedimentary basin : evidence for late Archaean transpression in southern India? *Journal of the Geological Society, London,* **146**, 825–834.

CHARDON, D. 1996. *Les déformations continentales archéennes: exemples naturels et modélisation thermomécanique*. PhD. Thesis. Univ. Rennes.

CHINNER, G. A. & SWEATMAN, T. R. 1968. A former association of enstatite and kyanite. *Mineralogical Magazine,* **36**, 1052–1060.

CHOUKROUNE, P., GAPAIS, D. & MERLE, O. 1987. Shear criteria and structural symmetry. *Journal of Structural Geology,* **9**, 525–530.

——, BOUHALLIER, H. & ARNDT, N. T. 1995. Soft lithosphere during periods of Archaean crustal growth or crustal reworking. *In*: COWARD, M. P. & RIESS, A. C. (eds) *Early Precambrian Processes*. Geological Society, London, Special Publications, **95**, 67–86.

COFFIN, M. F. & ELDHOLM, O. 1994. Large igneous provinces: crustal structure, dimensions and external consequences. *Reviews in Geophysics,* **32**, 1–36.

COLLINS, W. J. 1989. Polydiapirism of the Archaean

MOUNT EDGAR BATHOLITH, PILBARA BLOCK, WESTERN AUSTRALIA. *Precambrian Research,* **43**, 41–62.

COMPSTON, W. & KRÖNER, A. 1988. Multiple zircon growth within early Archaean tonalitic gneiss from the Ancient Gneiss Complex, Swaziland. *Earth and Planetary Science Letters,* **87**, 13–28.

CONDIE, K. C. (ed.) 1981. *Archaean Greenstone Belts.* Elsevier, Amsterdam.

—— (ed.) 1994. *Archaean Crustal Evolution.* Elsevier, Amsterdam.

CORFU, F. & DAVIS, D. W. 1991. A U-Pb geochronological framework for the western Superior Province, Ontario. *In*: *Geology of Ontario.* Ontario Geological Survey, Special Volume **4**, part 1,

COWARD, M. P. & FAIRHEAD, J. D. 1980. Gravity and structural evidence for the deep structure of the Limpopo belt, southern Africa. *Tectonophysics,* **68**, 31–43.

——, LINTERN, B. C. & WRIGHT, L. I. 1976. The pre-cleavage deformation of the sediments and the gneisses of the northern Limpopo belt. *In*: WINDLEY B. F. (ed.) *The Early History of the Earth.* Wiley, London, 323–330.

COX, K. G. 1989. The role of mantle plumes in the development of continental drainage patterns. *Nature,* **342**, 873–877.

CRAWFORD, A. R. 1969. Reconnaissance Rb-Sr dating of the Precambrian rocks of Southern Peninsular India. *Journal of the Geological Society of India,* **10**, 117–166.

CRUDEN, A. R. 1988. Deformation around a rising diapir modelled by creeping flow past a sphere. *Tectonics,* **7**, 1091–1101.

DAVIES, G. 1993. Conjectures on the thermal and tectonic evolution of the Earth. *Lithos,* **30**, 281–289.

DAVIS, D. 1992. Age constraints on the deposition and the provenance of Archaean detrital sediments in the southern Abitibi and Pontiac subprovinces from U-Pb analyses of detrital zircons. *In*: *Lithoprobe report.* University of British Columbia, **25**, 147–150.

——, MACHADO, N, GARIÈPY, C., SAWYER, E. W. & BENN, K. 1995. U-Pb geochronology of the Opatica tonalite-gneiss belt and its relationship to the Abitibi greenstone belt, Superior Province, Quebec. *Canadian Journal of Earth Sciences,* **32**, 113–127.

DESROCHERS, J.-P., HUBERT, C., LUDDEN, J. & PILOTE, P. 1993. Accretion of Archaean oceanic plateau fragments in the Abitibi Greenstone Belt, Canada. *Geology,* **21**, 451–454.

DE WIT, M. J. 1982. Gliding and overthrust nappe tectonics in the Barberton greenstone belt. *Journal of Structural Geology,* **4**, 117–136.

——, ARMSTRONG, R. A., HART, R. J. & WILSON, A. H. 1987*a*. Felsic igneous rocks within the 3.5 Ga Barberton greenstone belt: high crustal level equivalents of the surrounding tonalite-trondhjemite terrain, emplaced during thrusting. *Tectonics,* **6**, 529–549.

——, HART, R. A. & HART, R. J. 1987*b*. The Jamestown ophiolite complex, Barberton mountain belt: a section through 3.5 Ga oceanic crust. *Journal of African Earth Sciences,* **6**, 681–730.

——, ROERING, C., HART, R. J., ARMSTRONG, R. A., DE RONDE, C. E. J., GREEN, R. W. E., TREDOUX, M., PEBERTY, E. & HART, R. A. 1992. Formation of an Archaean continent. *Nature,* **357**, 553–562.

DEWEY, J. F. 1988. Extensional collapse of orogens. *Tectonics,* **7**, 1123–1139.

—— & HORSFIELD, B. 1970. Plate tectonics, orogeny and continental growth. *Nature,* **225**, 521–525.

—— & WINDLEY, B. F. 1981. Growth and differentiation of the continental crust. *Philosophical Transactions of the Royal Society, London,* **A301**, 189–206.

DIMROTH, E., IMREH, L., ROCHELEAU, M. & GOULET, N. 1983. Evolution of the south-central part of Archaean Abitibi belt. Part I: stratigraphy and paleogeographical model. *Canadian Journal of Earth Sciences,* **19**, 1729–1758.

DIXON, J. M. 1975. Finite strain and progressive deformation in models of diapiric structures. *Tectonophysics,* **28**, 89–124.

—— & SUMMERS, J. M. 1983. Patterns of total and incremental strain in subsiding troughs: experimental centrifuged models of inter-diapir synclines. *Canadian Journal of Earth Sciences,* **20**, 1843–1861.

DRURY, S. A. 1977. Structures induced by granite diapirs in the Archaean greenstones belt at Yellowknife, Canada : implications for Archaean geotectonics. *Journal of Geology,* **85**, 345–358.

—— 1983. A regional tectonic study of the Archaean Chitradurga greenstone belt, Karnataka. *Journal of the Geological Society of India,* **24**, 167–184.

—— & HOLT, R. W. 1980. The tectonic framework of the South Indian craton: a reconnaissance involving LANDSAT imagery. *Tectonophysics,* **65**, 111–115.

——, HARRIS, N. B., HOLT, R. W., REEVES-SMITH, G. J. & WIGHTMAN, R. T. 1984. Precambrian tectonics and crustal evolution in South India. *Journal of Geology,* **92**, 3–20.

DURRHEIM, R. J. & MOONEY, W. D. 1991. Archaean and Proterozoic crustal evolution: evidence from crustal seismology. *Geology,* **19**, 606–609.

ERNST, W. G. 1972. Occurence and mineralogic evolution of blueschist belts with time. *American Journal of Science,* **272**, 657–668.

ESKOLA, P. E. 1949. The problem of mantled gneiss domes. *Quarterly Journal of the Geological Society of London,* **104**, 461–476.

EVANS, O. C. & HANSON, G. H. 1992. Most late Archaean tonalites, trondhjemites and granodiorites (TTG) in the SW Superior Province were derived from mantle melts, not by melting of basalts. *EOS, Transactions of the American Geophysical Union,* **22D–3**, 330.

FRIEND, C. R. L. & NUTMAN, A. P. 1991. SHRIMP U-Pb geochronology of the Closepet granite and Peninsular gneisses, Karnataka, South of India. *Journal of the Geological Society of India,* **38**, 357–368.

FRYER, B. J., FYFE, W. S. & KERRICH, R. 1979.

Archaean volcanogenic oceans. *Chemical Geology,* **24**, 25–35.
FYFE, W. S. 1974. Archaean tectonics. *Nature,* **249**, 338.
—— 1978. Evolution of the Earth crust: modern plate tectonics to ancient hot spot tectonics? *Chemical Geology,* **23**, 89–114.
FYSON, W. K. 1984. Fold and cleavage patterns in Archaean metasediments of the Yellowknife supracrustal domain, Slave province, Canada. *In*: KRÖNER A. & GREILING R. (eds) *Precambrian Tectonics Illustrated*. E. Schweizerbartsche Verlags, Stuttgart, 281–293.
GLIKSON, A. Y. 1972. Early Precambrian evidence of a primitive ocean crust and island nuclei of sodic granite. *Geological Society of America Bulletin,* **83**, 3323–3344.
—— 1979. Early Precambrian tonalite-trondhjemite sialic nuclei. *Earth Science Reviews,* **15**, 1–73.
GOODWIN, A. M. 1981. Archaean plates and greenstone belts. *In*: KRÖNER A. (ed.) *Precambrian Plate Tectonics*. Elsevier, Amsterdam, 105–135.
—— 1991. *Precambrian Geology*. Academic Press.
—— & SMITH, I. E. M. 1980. Chemical discontinuities in Archaean metavolcanic terrains and the development of Archaean crust. *Precambrian Research,* **10**, 301–311.
GORMAN, B. E., PEARCE, T. H. & BIRKETTE, T. C. 1978. On the structure of Archaean greenstone belts. *Precambrian Research,* **6**, 23–41.
GRANDJEAN, G., WU, H., WHITE, D., MARESCHAL, M. & HUBERT, C. 1995. Crustal velocity models for the Archaean Abitibi greenstone belt from seismic refraction data. *Canadian Journal of Earth Sciences,* **32**, 149–166.
GREW, E. S. & MANTON, W. I. 1984. Age of allanite from Kabbaldurga quarry, Karnataka. *Journal of the Geological Society of India,* **25**, 193–195.
GREEN, A. G., MILKEREIT, B., MAYRAND, L. J. & LUDDEN, J. N. 1990. Deep structure of an Archaean greenstone terrane. *Nature,* **344**, 327–330.
GRIFFITH, R. W. & CAMPBELL, I. H. 1992. On the dynamics of long-lived plume conduits in the convecting mantle. *Earth and Planetary Science Letters,* **103**, 214–227.
GROVES, D. I., ARCHIBALD, N. J., BETTENAY, L. F. & BINNS, R. A. 1978. Greenstone belts as ancient marginal basins or ensialic rift zones. *Nature,* **273**, 460–461.
GUGLIELMO, G. 1993. Interference between pluton expansion and non-coaxial tectonic deformation: three dimensional computer model and field implications. *Journal of Structural Geology,* **15**, 593–608.
HAMMOND, E. C. & NISBET, B. W. 1992. Towards a structural and tectonic framework for the central Norseman-Wiluna greenstone belt, Western Australia. *In*: GLOVER J. E. & HO, S. E. (eds) *The Archaean: Terrains, Processes and Metallogeny*, The Geology Key Centre & University extension, the University of Western Australia, Perth, 39–50.
HARGRAVES, R. B. 1981. Precambrian tectonic style: a liberal uniformitarian interpretation. *In*: KRÖNER, A. (ed.) *Precambrian Plate Tectonics*. Elsevier, Amsterdam, 21–56.
—— 1986. Faster spreading or greater ridge length in the Archaean? *Geology,* **14**, 750–752.
HARRIS, N. B. W. & JAYARAM, S. 1981. Metamorphism of cordierite gneisses from the Bangalore region of the Indian Archaean. *Lithos,* **15**, 89–98.
HEUBECK, C. & LOWE, D. R. 1994. Late syndepositional deformation and detachment tectonics in the Barberton greenstone belt, South Africa. *Tectonics,* **13**, 1514–1536.
HILL, R. I., CAMPBELL, I. H., DAVIES, G. F. & GRIFFITH, R. W. 1992. Mantle plumes and continental tectonics. *Science,* **256**, 186–193.
HOFFMAN, P. F. 1989. Precambrian geology and tectonic history of North America. *In*: BALLY, A. W. & PALMER, A. R. (eds) *The Geology of North America; an Overview*. Geological Society of America, Boulder, 447–512.
—— 1991. On accretion of granite-greenstone terranes. *In*: ROBERT, F., SHEAHAN, P. A. & GREEN, S. B. (eds) *Nuna conference on Greenstone gold and crustal evolution*. Geological Association of Canada, Mineral Deposits Division, 32–45.
HOLDER, M. T. 1979. An emplacement mechanism for post-tectonic granites and its implications for their geochemical features. *In*: ATHERTON M. P. & TARNEY J. (eds) *Origin of Granite Batholiths – Geochemical Evidence*, Shiva Pub., Orpington, 116–128.
HUBERT, C., TRUDEL, P. & GELINAS, L. 1984. Archean wrench fault tectonics and structural evolution of the Blake River Group, Abitibi belt, Quebec. *Canadian Journal of Earth Sciences,* **21**, 1024–1032.
——, SAWYER, E., BARNES, A., DAIGNEAULT, R., LACROIX, S., LUDDEN, J. N., MILKEREIT, B. & RIVE, M. 1992*a*. Geological interpretations of seismic lines in the northern and central Abitibi greenstone belt: evidence for regional thrust imbrication and crustal-scale Archaean wrench fault systems. *Lithoprobe Project, Reports* **25**, University of British Columbia, 33–36.
——, LUDDEN, J. N., BARNES, A., BENN, K., MILKEREIT, B., SAWYER E. & RIVE, M. 1992*b*. Interpretation of seismic lines in the Pontiac subprovince: evidence for regional thrust imbrication and the allocthonous nature of the Baby and Belleterre metavolcanic belts. *Lithoprobe Project, Reports* **25**, University of British Columbia, 19–21.
HURLEY, P. M. 1968. Absolute abundance and distribution of Rb, K and Sr in the Earth. *Geochemica et Cosmochemica Acta,* **32**, 273.
—— & RAND, J. R. 1969. Pre-drift continental nuclei. *Science,* **164**, 1229.
HUSSAIN, S. M. & NAQVI, S. M. 1983. Geological, geophysical, and Geochemical studies over the Holenarsipur schist belt, Dharwar craton, India. *In*: NAQVI, S. M. & ROGERS, J. J. W. (eds) *Precambrian of South India*. Geological Society of India Memoirs, **4**, 473–495.
JACKSON, M. P. A., ERIKSSON, K. A. & HARRIS, C.

W. 1987. Early Archaean foredeep sedimentation related to crustal shortening: a reinterpretation of the Barberton sequence, Southern Africa. *Tectonophysics,* **136**, 197–221.

JACKSON, S. L., FYON, J. A. & CORFU, F. 1994. Review of Archaean supracrustal assemblages of the southern Abitibi greenstone belt in Ontario, Canada: products of microplate interaction within a large-scale plate-tectonic setting. *Precambrian Research,* **65**, 183–205.

JAHN, B. M. & ZHANG, Z. Q. 1984. Archaean granulite gneisses from eastern Hebei Province, China: rare Earth geochemistry and tectonic implications. *Contributions to Mineralogy and Petrology,* **85**, 224–243.

——, GLIKSON, A. Y., PEUCAT, J. J. & HICKMAN, H. A. 1981. REE geochemistry and isotopic data of Archaean silicic volcanics and granitoids from the Pilbara Block, Western Australia: implications for the early crustal evolution. *Geochimica et Cosmochimica Acta,* **45**, 1633–1652.

JANARDHAN, A. S., NEWTON, R. C. & SMITH, J. V. 1979*a*. Ancient crustal metamorphism at low pH2O : charnockite formation at Kabbaldurga, south India. *Nature,* **278**, 511–514.

——, RAMACHANDRA, H. M. & RAVINDRA KUMAR, G. R. 1979*b*. Structural history of Sargur supracrustals and associated gneisses, southwest of Mysore, Karnataka. *Journal of the Geological Society of India,* **20**, 61–72.

——, NEWTON, R. C. & HANSEN, E. C. 1982. The transformation of amphibolite facies gneiss to charnockite in Southern Karnataka and Northern Tamil Nadu, India. *Contributions to Mineralogy and Petrology,* **79**, 130–149.

JAYANANDA, M. & MAHABALESWAR, B. 1990. Relationship between shear zones and igneous activity: the Closepet Granite of Southern India. *Proceedings of the Indian Academy of Sciences (Earth and Planetary Sciences),* **100**, 31–36.

——, MARTIN, H., PEUCAT, J. J. & MAHABALESWAR, B. 1995. Late Archaean crust-mantle interactions: geochemistry of LREE-enriched mantle derived magmas. Example of the Closepet batholith, South India. *Contributions to Mineralogy and Petrology,* **119**, 314–329.

JELSMA, H. A., VAN DER BEEK, P. A. & VINYU, M. L. 1993. Tectonic evolution of the Bindura-Shamva greenstone belt (northern Zimbabwe): progressive deformation around diapiric batholiths. *Journal of Structural Geology,* **15**, 163–176.

JOLLY, W. T. 1978. Metamorphic history of the Archaean Abitibi belt. *In*: FRASER J. A. & HEYWOOD, W. W. (eds) *Metamorphism in the Canadian Shield*. Geological Survey of Canada, Papers, **78–10**, 63–78.

KELLETT, R. L., BARNES, A. E. & RIVE, M. 1994. The deep structure of the Grenville Front: a new perspective from western Quebec. *Canadian Journal of Earth Sciences,* **31**, 282–292.

KERRICH, R. 1989. Source processes for Archaean Au-Ag vein deposits: evidence from lithophile element systematics of the Hollinger-McIntyre and Buffalo-Ankerite deposits, Timmins. *Canadian Journal of Earth Sciences,* **26**, 55–78

—— & WYMAN, D. 1990. Geodynamic setting of mesothermal gold deposits: an association with accretionary tectonic regimes. *Geology,* **18**, 882–885.

KIMURA, G., LUDDEN, J. N., DESROCHERS, J. P. & HORI, R. 1993. A model of ocean-crust accretion for the Superior Province, Canada. *Lithos,* **30**, 337–355.

KINNY, P. 1986. 3820 Ma zircons from a tonalitic gneiss in the Godthab district of southern West Greenland. *Earth and Planetary Science Letters,* **79**, 337–347.

KRAMERS, J. D. 1988. An open-system fractional crystallisation model for very early continental crust formation. *Precambrian Research,* **38**, 281–295.

KROGH, T. E. 1993, High precision U-Pb ages for granulite metamorphism and deformation in the Archaean Kapuskasing zone, Ontario: implications for structure and development of the lower crust. *Earth and Planetary Science Letters,* **119**, 1–18.

KRÖNER, A. 1981*a*. Precambrian plate tectonics. *In*: KRÖNER A. (ed.) *Precambrian plate tectonics*. Elsevier, Amsterdam, 57–90.

—— 1981*b*. Precambrian crustal evolution and continental drift. *Geologisches Rundschau,* **70**, 412–428.

—— 1991. Tectonic evolution in the Archaean and Proterozoic. *Tectonophysics,* **197**, 393–410.

—— & LAYERS, P. W. 1992. Crust formation and plate motion in the Early Archaean. *Science,* **256**, 1405–1411.

KUSKY, T. M. 1993. Collapse of Archaean orogens and the generation of late- to postkinematic granitoids. *Geology,* **21**, 925–928.

LACROIX, S. & SAWYER, E. W. 1995. An Archaean fold-thrust belt in the northwestern Abitibi greenstone belt: structural and seismic evidence. *Canadian Journal of Earth Sciences,* **32**, 97–112.

LAMBERT, R. S. J. 1976. Archaean thermal regimes, crustal and upper mantle temperatures, and a progressive evolutionary model for the Earth. *In*: WINDLEY B. F. (ed.) *The Early History of the Earth*. Wiley, London, 363–373.

—— 1981. Earth tectonics and thermal history: review and a hot-spot model for the Archaean. *In*: KRÖNER, A. (ed.) *Precambrian Plate Tectonics*, Elsevier, Amsterdam, 453–467.

LARSON, R. L. 1991. Latest pulse of the Earth: evidence from a mid-Cretaceous superplume, *Geology,* **19**, 547–550.

LECLAIR, A. D., PERCIVAL, J. A. GREEN, A. G., WU, H. & WEST, G. F. 1995. Seismic reflection profiles across the central Kapuskasing uplift. *Canadian Journal of Earth Sciences,* **31**, 1027–1041.

LIGHT, M. P. R. 1982. The Limpopo mobile belt: a result of continental collision. *Tectonics,* **1**, 325–342.

LIOU, J. G., MARUYAMA, S., WANG, X., GRAHAM, S., XIAO, S., FENG, Y., LIANG, Y., ZHO, M. & TANG,

Y. 1988. Geological evidence for a major Proterozoic coherent blueschist terrane in Aksu, Xinjiang, China. *EOS, Transactions of the American Geophysical Union,* **69**, 1513.

LUCAS, S. B., GREEN, A. G., HAJNAL, Z., WHITE, D., LEWRY, J., ASHTON, K., WEBER, W. & CLOWES, R. 1993. Deep seismic profile across a Proterozoic collision zone: surprises at depth. *Nature,* **365**, 339–342.

LUDDEN, J. N., HUBERT, C. & GARIEPY, C. 1986. The tectonic evolution of the Abitibi belt, Canada. *Geological Magazine,* **123**, 153–166.

——, —— BARNES, A., MILKEREIT, B. & SAWYER, E. 1993. A three dimensional perspective on the evolution of Archaean crust: LITHOPROBE seismic reflection images in the southwestern Superior Province. *Lithos,* **30**, 357–372.

——, MARESCHAL, J. C. & CALVERT, A. J. 1995. Accretion of late-Archaean crust. *Terra Abstracts, EUG 8,* **7**, 100

MCGREGOR, A. M. 1951. Some milestones in the Precambrian of Southern Rhodesia. *Transactions of the Geological Society of South Africa,* **54**, 27–71.

MCKENZIE, D. & WEISS, N. 1975. Speculations on the thermal and tectonic history of the Earth. *Geophysical Journal of the Royal Astronomical Society,* **42**, 131–174.

MCLENNAN, S. M. & TAYLOR, S. R. 1982. Geochemical constraints on the growth of the continental crust. *Journal of Geology,* **90**, 342–361.

MALLOE, S. 1982. Petrogenesis of Archaean tonalites. *Geologisches Rundschau,* **71**, 328–346.

MARESHAL, J.-C. & WEST, G. F. 1980. A model for Archaean tectonism. Part 2. Numerical models of vertical tectonism in greenstone belts. *Canadian Journal of Earth Sciences,* **17**, 60–71.

MARESCHAL, M., KELLETT, R., KURTZ, R., LUDDEN, J. N., JI, S. & BAILEY, R. C. 1995. Archaean cratonic roots, mantle shear zones and deep electrical anisotropy. *Nature,* **375**, 134–137.

MARTIGNOLE, J. & CALVERT, A. J. 1996. Crustal-scale shortening and extension across the Grenville province of western Quebec. *Tectonics,* **15**, 376–386.

MARTIN, H. 1986. Effects of a steeper geothermal gradient on geochemistry of subduction-zone magmas. *Geology,* **14**, 753–756.

—— 1994. The Archaean grey gneisses and the genesis of continental crust. *In*: CONDIE, K. C. (ed.) *Archaean Crustal Evolution*, Elsevier, Amsterdam, 205–259.

MEEN, J. K., ROGERS, J. J. & FULLAGAR, P. D. 1992. Lead isotopic composition of the Western Dharwar craton, southern India: evidence for distinct middle Archaean terranes in a late Archaean craton. *Geochimica et Cosmochimica Acta,* **56**, 2455–2470.

MOORBATH, S. 1975. Evolution of Precambrian crust from strontium isotopic evidence. *Nature,* **254**, 395–398.

MORTENSEN, J. K. & CARD, K. D. 1995. U-Pb age constraints for the magmatic and tectonic evolution of the Pontiac subprovince, Quebec. *Canadian Journal of Earth Sciences,* **30**, 1970–1980.

MUELLER, W., DONALDSON, J. A. & DOUCET, P. 1994. Volcanic and tectono-plutonic influences on sedimentation in the Archaean Kirkland basin, Abitibi greenstone belt, Canada. *Precambrian Research,* **68**, 201–230.

MYERS, J. S. 1976. Granitoid sheets, thrusting, and Archaean crustal thickening in West Greenland. *Geology,* **5**, 265–268.

—— 1984. Archaean tectonics in the Fiskenaesset region of southwest Greenland. *In*: KRÖNER, A. & GREILING, R. (eds) *Archaean Tectonics Illustrated*. E. Schweizerbart'sche Verlags, Stuttgart, 95–112.

—— & WATKINS, K. P. 1985. Origin of granite-greenstone patterns, Yilgarn block, Western Australia. *Geology,* **13**, 778–780.

NAQVI, S. M. & RODGERS J. J. P. (eds) 1983. *Precambrian of South India*. Geological Society of India, Memoirs, **4**.

NISBET, E. G. 1982. The tectonic setting and petrogenesis of komatiites. *In*: ARNDT, N. T. & NISBET, E. G. (eds) *Komatiites*. Allen & Unwin, London, 501–520.

—— & WALKER, D. 1982. Komatiites and the structure of the Archaean mantle. *Earth and Planetary Science Letters,* **60**, 103–113.

——, CHEADLE, M. J., ARNDT, N. T. & BICKLE, M. J. 1993. Constraining the potential temperature of the Archaean mantle: a review of the evidence from komatiites. *Lithos,* **30**, 291–307.

OXBURGH, E. E. & TURCOTTE, D. L. 1970. Thermal structure of island arcs. *Geological Society of America Bulletin,* **81**, 1665–1688.

PARK, R. G. 1981. Shear-zone deformation and bulk strain in granite-greenstone terrain of the western Ontario province, Canada. *Precambrian Research,* **14**, 31–47.

PASSCHIER, C. W. 1994. Structural geology across a proposed Archaean terrane boundary in the eastern Yilgarn craton, Western Australia. *Precambrian Research,* **68**, 43–64.

PERCIVAL, J. A. & WEST, G. F. 1994. The Kapuskasing uplift: a geological and geophysical synthesis. *Canadian Journal of Earth Sciences,* **31**, 1256–1286.

—— & WILLIAMS, H. R. 1989. The Quetico accretionary complex, Superior Province, Canada. *Geology,* **17**, 23–25.

——, MORTENSEN, J. K., STERN, R. A., CARD, K. D. & BÉGIN, N. J. 1992. Giant granulite terranes of the northeastern Superior Province: the Ashuanipi complex, and Minto block. *Canadian Journal of Earth Sciences,* **29**, 2287–2308.

——, STERN, R. A., SKULSKI, T. CARD, K. D., MORTENSEN, J. K. & BÉGIN, N. J. 1994. Minto block, Superior Province: missing link in deciphering assembly of the craton at 2.7 Ga. *Geology,* **22**, 839–842.

PEUCAT, J.-J., VIDAL P., BERNARD-GRIFFITHS J. & CONDIE, K. C. 1989. Sr, Nd and Pb isotopic systematics in the Archaean low- to high- grade transition zone of southern India : syn-accretion vs. post-accretion granulites. *Journal of Geology,* **97**, 537–550.

——, GRUAU, G., MARTIN., H., AUVRAY, B., FOURCADE, S., CHOUKROUNE, P., BOUHALLIER H. & JAYANANDA, M. 1993*a*. A 2.5 Ga mega-plume in South India? *Terra Nova,* **5**, 321.

——, MAHABALESWAR, B. & JAYANANDA, M. 1993*b*. Age of younger tonalitic magmatism and granulitic metamorphism in the South Indian transition zone (Krishnagiri area); comparison with older Peninsular gneisses from the Gorur-Hassan area. *Journal of Metamorphic Geology,* **11**, 879–888.

——, BOUHALLIER, H., FANNING, C. M. & JAYANANDA, M. 1995. Age of the Holenarsipur greenstone belt, relationships with the surrounding gneisses (Karnataka,South India). *Journal of Geology,* **6**, 701–710.

—— 1962. Some observations on the structures, metamorphism and geological evolution of Peninsular India. *Journal of the Geological Society of India,* **13**, 106–118.

PLATT, J. P. 1980. Archaean greenstone belts: a structural test of tectonic hypotheses. *Tectonophysics,* **65**, 127–150.

POTREL, A. 1994. *Evolution tectonométamorphique d'un segment de croûte continentale archéenne; exemple de l'Amsaga (Mauritanie).* Memoires de Géosciences-Rennes.

RAASE, P., RAITH, M., ACKERMAND, D. & LAL, R. K. 1986. Progessive metamorphism of mafic rocks from greenschist to granulite facies in the Dharwar craton of South India. *Journal of Geology,* **94**, 261–282.

RAITH, M., RAASE, P. & ACKERMAND, D. 1982. The Archaean craton of southern India: metamorphic evolution and P-T conditions. *Geologishe Rundschau,* **71**, 280–290.

——, —— & —— 1983. Regional geothermobarometry in the granulite facies terrane of South India. *Transactions of the Royal Society Edinburgh, Earth Sciences,* **73**, 221–244.

RALSER, S. & PARK. A. F. 1992. Tectonic evolution of the Archaean rocks of the Tavani Area, Keewatin, N.W.T., Canada. *In*: GLOVER, J. E. & HO, S. E. *The Archaean: Terrains, Processes and Metallogeny.* Proceedings, Third International Archaean Symposium, Perth 1990, 71–76.

RAMAKRISHNAN, M. & VISWANATHA, M. N. 1987. Angular unconformity, structural unity argument and Sargur-Dharwar relations in Bababuban basin. *Journal of the Geological Society of India,* **29**, 471–482.

—— & —— 1983. Crustal evolution in central Karnataka: a review of present data and models. *In*: NAQVI S. M. & ROGERS, J. J. W. (eds) *Precambrian of South India,* Geological Society of India, Memoirs, Bangalore, 96–109.

RAMBERG, H. 1971. Model studies in relation to intrusion of plutonic bodies. *In*: NEWALL, G. & RAST, N. (eds) *Mechanism of igneous intrusion.* Geological Journal Special Issue, **2**, 261–286.

—— 1973. Model studies of gravity-controlled tectonics by the centrifuge technique. *In*: DE JONG, K. A. & SCHOLTEN, R. (eds) *Gravity and Tectonics.* Wiley, New York, 49–66.

—— 1981. *Gravity, deformation and the Earth crust.* Academic Press, London.

RAMSAY, J. G. 1989. Emplacement kinematics of a granite diapir: the Chindamora batholith, Zimbabwe. *Journal of Structural Geology,* **11**, 191–209.

RAPP, R. P., WATSON, E. B. & MILLER, C. F. 1991. Partial melting of amphibolite/eclogite and the origin of Archaean trondhjemites and tonalites. *Precambrian Research,* **51**, 1–25.

RAYLEIGH, L. 1883. Investigation of the character of the equilibrium of an incompressible heavy fluid of variable density. *Proceedings of the London Mathematical Society,* **14**, 170–177.

REYMER, A. P. S. & SCHUBERT, G. 1984. Phanerozoic and Precambrian crustal growth. *In*: KRÖNER, A. (ed.) *Proterozoic Crustal Evolution.* American Geophysical Union, Geodynamic Series, **17**, 1–10.

RICHARDSON, S. H., GUERNEY, J. J., ERLANK, A. J. & HARRIS, J. W. 1984. The origin of diamonds in old enriched mantle. *Nature,* **310**, 198–202.

RICHTER, F. M. 1985. Models for the Archaean thermal regime. *Earth and Planetary Science Letters,* **73**, 350–360.

RICHTER, F. M. 1988. A major change in the thermal state of the Earth at the Archaean-Proterozoic boundary: consequences for the nature and preservation of continental lithosphere. *Journal of Petrology,* **30**, 39–52.

RIDLEY, J. R. & KRAMERS, J. D. 1990. The evolution and tectonic consequences of a tonalitic magma layer within Archaean continents. *Canadian Journal of Earth Sciences,* **27**, 219–228.

RINGWOOD, A. E. 1974. The petrological evolution of island-arc system. *Journal of the Geological Society, London.* **130**, 183–204.

RIVE, M., PINTSON, H. & LUDDEN, J. N. 1990. Characteristics of late Archaean plutonic rocks from the Abitibi and Pontiac subprovinces, Superior Province, Canada. *In*: RIVE M. ET AL. (eds) *The Northwestern Québec Polymetallic belt.* Canadian Institute of Mining and Metallurgy, Special Volumes, **43**, 65–76.

SAGGERSON, E. P. & OWEN, L. M. 1971. Metamorphism as a guide to depth of the top of the mantle in southern Africa. *Proceedings of the 24th International Geological Congress, Montréal,* **1**, 153–161.

—— & TURNER, L. M. 1972. Some evidence for the evolution of regional metamorphism in Africa. *Proceedings of the 24th International Geological Congress, Montréal,* **1**, 153–161.

SAWYER, E. W. & BARNES, S. J. 1994. Thrusting, magmatic intraplating and metamorphic core complex development in the Archaean Belleterre-Angliers greenstone belt, Superior Province, Quebec, Canada. *Canadian Journal of Earth Sciences,* **68**, 183–200.

—— & BENN, K. 1993. Structure of the high-grade Opatica belt and adjacent low-grade Abitibi subprovince and Archaean mountain front. *Journal of Structural Geology,* **15**, 1443–1458.

SCHMELING, H., CRUDEN, A. R. & MARQUART, G. 1988. Finite deformation in and around a fluid sphere moving through a viscous medium: implications for diapiric ascent. *Tectonophysics,* **149**, 17–34.

SCHWERDTNER, W. M., STONE, D., OSADETZ, K., MORGAN, J. & STOTT, G. M. 1979. Granitoïd complexes and the Archaean tectonic record in the southern part of northwestern Ontario. *Canadian Journal of Earth Sciences,* **16**, 1965–1977.

SENECHAL, G., MARESCHAL, M., HUBERT, C., CALVERT, A., GRANDJEAN, G. & LUDDEN, J. N. 1996. Integrated geophysical interpretation of crustal structures in the northern Abitibi Belt: constraints from seismic amplitude analysis. *Canadian Journal of Earth Sciences*, in press.

SHACKLETON, R. M. 1986. Precambrian collision tectonics in Africa. *In*: COWARD, M. P. & RIES, A. C. (eds) *Collision Tectonics*. Geological Society of London, 329–349.

SHIREY, S. B. & HANSON, G. N. 1986. Mantle heterogeneity and crust recycling in Archaean granite-greenstone belts: evidence from Nd-isotopes and trace elements in the Rainy Lake area, Superior Province, Ontario, Canada. *Geochimica et Cosmochimica Acta,* **50**, 2631–2651.

SILVER, P. G. & KANESHIMA, S. 1993. Constraints on mantle anisotropy beneath Precambrian north America from a transportable teleseismic experiment. *Geophysical Research Letters,* **20**, 1127–1130

SLEEP, N. H. 1979. Thermal history and degassing of the Earth: some simple calculations, *Geology,* **87**, 671–687.

SNOWDEN, P. A. 1984. Non-diapiric batholiths in the north of the Zimbabwe shield. *In*: KRÖNER A. & GREILING, R. (eds) *Precambrian Tectonics Illustrated*. E. Schweizerbart'sche Verlagsbuchhandlung, Stuttgart, 135–145.

SNOWDEN, P. A. & BICKLE, M. J. 1976. The Cinamora Batholith: diapiric intrusion or interference fold? *Journal of the Geological Society, London,* **132**, 131–137.

SPRAY, J. G. 1985. Dynamothermal transition zone between Archaean greenstone and granitoid gneiss at Lake Dundas, western Australia. *Journal of Structural Geology,* **7**, 187–203.

STÄHLE, H. J., RAITH, M., HOERNES, S. & DELFS, A. 1987. Element mobility during incipient granulite formation at Kabbaldurga, southern India. *Journal of Petrology,* **28**, 803–834.

STEIN, M. & HOFMANN, A. W. 1994. Mantle plumes and episodic crustal growth. *Nature,* **372**, 63–68.

STERN, R. A., PERCIVAL, J. A. & MORTENSEN, J. K. 1994. Geochemical evolution of the Minto block: a 2.7 Ga continental magmatic arc built on the Superior proto-craton. *Precambrian Research,* **65**, 115–153.

STOREY, M., MAHONEY, J. J., KROENKE, L. W. & SAUNDERS, A. D. 1991. Are oceanic plateaus sites of komatiite formation? *Geology,* **19**, 376–379.

STOWE, C. W. 1984. The early Archaean Selukwe nappe, Zimbabwe. *In*: KRÖNER, A. & GREILING, R. (eds) *Precambrian Tectonics Illustrated*. E. Schweizerbart'sche Verlags, Stuttgart, 41–56,

SWAGER, C. & GRIFFIN, T. J. 1990. An early thrust duplex in the Kalgoorlie-Kambalda greenstone belt, Eastern Goldfields Province, western Australia. *Precambrian Research,* **48**, 63–73.

SWAMI NATH, J. & RAMAKRISHNAN, M. 1981. *Early supracrustals of Southern Karnataka: Present classification and correlation*. Memoirs of the Geological Survey of India, **112**.

TAIRA, A., PICKERING, K. T., WINDLEY, B. F. & SOH, W. 1992. Accretion of Japanese island arcs and implications for the origin of Archaean greenstone belts. *Tectonics,* **11**, 1224–1244.

TALBOT, C. J. 1973. A plate tectonic model for the Archaean crust. *Philosophical Transactions of the Royal Society, London,* **A273**, 413–427.

—— 1977. Inclined and asymmetric upward-moving gravity structures. *Tectonophysics,* **42**, 159–181.

——, RONNLUND, J. P., SCHMELING, H., KOYI, H. & JACKSON, M. P. A. 1991. Diapiric spoke patterns. Tectonophysics, 188, 187–201.

TARDUNO, J. A., SLITER, W. V, KROENKE, L. W., LECKIE, M., MAYER, R., MAHONEY, J. J., MUSGRAVE, R., STOREY, M. & WINTERER, E. L. 1991. Rapid formation of the Ontong-Java plateau by Aptian mantle plume volcanism. *Science,* **254**, 399–403.

TARNEY, J. & WINDLEY, B. F. 1981. Marginal basins through geological time. *Philosophical Transactions of the Royal Society, London,* **A300**, 263–285.

——, DALZIEL, W. D. & DEWIT, M. J. 1976. Marginal basin Rocas Verdes complex from South Chile : a model for Archaean greenstone belt formation. *In*: WINDLEY, B. F. (ed.) *The Early History of the Earth*. Wiley, London, 131–146.

TAYLOR, P. N., CHADWICK, B. MOORBATH, S. RAMAKRISHNAN, M. & VISWANATHA, M. N. 1984. Petrography, chemistry and isotopic ages of Peninsular gneiss, Dharwar acid volcanic rocks and the Chitradurga granite with special reference to the late Archaean evolution of the Karnataka craton, Southern India. *Precambrian Research,* **23**, 349–375.

——, ——, FRIEND, C. R. L., RAMAKRISHNAN, M. & VISWANATHA, M. N. 1988. New age data on the geological evolution of Southern India. *In*: ASHWAL, L. D. (ed.) *Indo-US Workshop on the deep continental crust of South India*. National Geophysical Research Institute, Hyderabad, 181–183.

TAYLOR, S. R. & MCLENNAN, S. M. 1985. *The Continental Crust: its Composition and Evolution*. Blackwell, Oxford.

THURSTON, P. C. & CHIVERS, K. M. 1990. Secular variation in greenstone development. *Precambrian Research,* **46**, 21–58.

——, OSMANI, I. A. & STONE, D. 1991. Northwestern Superior Province: Review and terrane analysis. *In: Geology of Ontario*. Ontario Geological Survey, Special Volume **4**, part 1, 81–142.

TRELOAR, P. J., COWARD, M. P. & HARRIS, N. B. W. 1992. Himalayan–Tibetan analogies for the evolution of the Zimbabwe craton and Limpopo Belt. *Precambrian Research,* **55**, 571–587.

VAN REENEN, D. D., ROERING C., ASHWAL, L. D. & DE WIT, M. J. (eds) 1992. *The Archaean Limpopo Granulite Belt: Tectonics and Deep Crustal Processes*. Elsevier, Amsterdam.

VERVOORT, J. D., WHITE, W. M. & THORPE, R. I. 1994. Nd and Pb isotope ratios of the Abitibi greenstone belt: new evidence for very early differentiation of the Earth. *Earth and Planetary Science Letters,* **128**, 215–229.

VEIZER, J., HOEFS, J., RIDLER, R. H., JENSEN, L. S. & LOWE, D. R. 1979. Geochemistry of Precambrian carbonates: I. Archaean hydrothermal systems. *geochimica et Cosmochimica Acta,* **53**, 845–857.

VILJOEN, M. J. & VILJOEN, R. P. 1969. Archaean vulcanicity and continental evolution in the Barberton region, Transvaal. *In*: CLIFFORD, T. N. & GASS, I. (eds) *African Magmatism and Tectonics*. Oliver & Boyd, Edinburgh, 27–39.

VISWANATHA, M. N. & RAMAKRISHNAN, M. 1975. The pre-Dharwar supracrustal rocks of the Sargur schist complex in Southern Karnataka and their tectono-metamorphic significance. *Industrial Minerals,* **16**, 48–65.

——, Ramakrishnan, M. & SWANI NATH, J. 1982. Angular unconformity between Sargur and Dharwar in Sigegudda, Karnataka craton, south India. *Journal of the Geological Society of India,* **23**, 85–89.

VLAAR, N. J., VAN KEKEN, P. E. & VAN DEN BERG, A. P. 1994. Cooling of the Earth in the Archaean: consequences of pressure-release melting in a hotter mantle. *Earth and Planetary Science Letters,* **121**, 1–18.

WASSERBURG, G. J., MC DONALD, G. L. F., HOYLE, F. & FLOWER, W. A. 1964. Relative contributions of uranium, thorium and potassium to heat production in the Earth. *Science,* **143**, 465–467.

WEGMANN, C. E. 1935. Zur deutung der Migmatite. *Geologishe Rundschau,* **26**, 306–350.

WEST, G. F. & MARESCHAL, J.-C. 1979. A model for Archaean tectonism. Part I. The thermal conditions. *Canadian Journal of Earth Sciences,* **16**, 1942–1950.

WILKS, M. E. 1988. The Himalayas: a modern analogue for Archaean crustal evolution. *Earth and Planetary Science Letters,* **87**, 127–136.

WILLIAMS, H. R. 1990. Subprovince accretion tectonics in the south-central Superior Province. *Canadian Journal of Earth Sciences,* **27**, 570–581.

——, STOTT, G. M., HEATHER, K. B., MUIR, T. L. & SAGE, R. P. 1991, Wawa Subprovince. *In*: *Geology of Ontario*. Ontario Geological Survey, Special Volume, **4**, part 1, 485–539.

WILLIAMS, P. R. & CURRIE, K. L. 1993. Character and regional implications of the sheared Archaean granite-greenstone contact near Leonora, Western Australia. *Precambrian Research,* **62**, 343–365.

WINDLEY, B. F. 1973. Crustal development in the Precambrian. *Philosophical Transactions of the Royal Society, London,* **A273**, 321–341.

—— 1984. *The Evolving Continents*, Wiley, New York.

—— 1993. Uniformitarism today: plate tectonics is the key to the past. *Journal of the Geological Society, London,* **150**, 7–19.

—— & BRIDGWATER, D. 1971. The evolution of Archaean low- and high-grade terrains. *Geological Society of Australia, Special Publications,* **3**, 33–46.

WINTHER, T. K. & NEWTON, R. C. 1991. Experimental melting of a hydrous low-K tholeiite: evidence of the origin of Archaean cratons. *Bulletin of the Geological Society of Denmark,* **39**, 213–228.

WOIDT, W. 1978. Finite element calculations applied to salt dome analysis. *Tectonophysics,* **50**, 369–386.

ZHANG, P., CHOUTEAU, M., MARESCHAL, M., KURTZ, R. & HUBERT, C. 1995. High frequency magnetotelluric investigation of crustal structure in north-central Abitibi, Quebec, Canada. *Geophysical Journal International,* **120**, 406–418.

Geodynamic evolution of the Proterozoic Mount Isa terrain

M. G. O'DEA, G. S. LISTER, T. MACCREADY, P. G. BETTS, N. H. S. OLIVER[1], K. S. POUND, W. HUANG & R. K. VALENTA[2]

Australian Geodynamics Cooperative Research Centre, VIEPS, Department of Earth Sciences, Monash University, Clayton, Vic 3168, Australia

[1]*School of Applied Geology, Curtin University, GPO Box U 1987, Perth, WA 6001, Australia*

[2]*MIM Exploration Pty Ltd, GPO Box 1042, Brisbane, Qld 4001, Australia*

Abstract: The Proterozoic Mount Isa terrain records the effects of four periods of intraplate tectonism. The *c.* 1870 Ma Barramundi Orogeny was characterized by a massive felsic magmatic event, and global correlations suggest a physical link between Australia and Laurentia at this time. Thereafter, the terrain underwent an extensional history spanning 200 Ma involving repeated episodes of rifting, post-rift subsidence and associated depositional and magmatic phases. This protracted rifting history resulted in a cumulative stratigraphic thickness of up to 25 km above attenuated continental crust. Rifting was interrupted prior to the formation of ocean crust by the compressional Isan Orogeny (1590–1500 Ma). The Isan Orogeny was synchronous with low-pressure high-temperature metamorphism and widespread metasomatism. In the waning stages of shortening, the Mount Isa terrain evolved into a wrench system characterized by an extensive network of strike-slip faults. The current level of exposure in this terrain provides spectacular examples of superimposed rifts, basin inversion, and wrench geometries developed at middle to upper crustal levels.

The Mount Isa terrain (Figs 1 & 2) provides one of the world's best records of Proterozoic orogenesis. It displays superb examples of intracontinental rift development, basin inversion, regional low-pressure high-temperature metamorphism, wrench faulting and extensive metasomatism. The processes that characterized the evolution of the Mount Isa terrain are shared by many Proterozoic belts throughout Australia and bear similarities to those of the same age worldwide. In this paper we discuss the geodynamic processes responsible for the development of sedimentological, structural and metamorphic associations within the Mount Isa terrain and present new interpretations for its tectonic evolution.

The Mount Isa terrain of northwest Queensland comprises lower to middle Proterozoic sediments, bimodal volcanic rocks and plutons. Exposed rocks crop out over an area of approximately 50 000 km^2, and the geophysical expression of the Mount Isa terrain, beneath surrounding upper Proterozoic and Phanerozoic sedimentary basins, extends over 160 000 km^2 (Figs 1 & 2). This terrain has been deformed and metamorphosed in a zone of long-lived tectonic activity between 1900 and 1500 Ma.

Beginning around 1900 Ma, much of northern Australia underwent a widespread extensional event leading to the development of intraplate sedimentary basins in which voluminous bimodal volcanic rocks and rift-sag sequences accumulated (Etheridge *et al.* 1987; Etheridge & Wall 1994). These rocks were then deformed and metamorphosed during the ensialic Barramundi Orogeny at approximately 1870 Ma (Etheridge *et al.* 1987; Etheridge & Wall 1994). Barramundi age rocks form the crystalline basement of the Mount Isa terrain and are found throughout northern Australia (Plumb *et al.* 1980). This interval of extension, sedimentation and volcanism, followed by compressional orogenesis, can be correlated globally with the Trans-Hudson of Canada (Hoffman 1988) and the Sveco-Fennian in the Baltic Shield (Gower 1985; Gaal & Gorbatschev 1987).

Following the Barramundi Orogeny, the Mount Isa terrain underwent a long and complex history of intermittent rifting and deposition between *c.* 1800 and 1600 Ma (Carter *et al.* 1961;

From Burg, J.-P. & Ford, M. (eds), 1997, *Orogeny Through Time*, Geological Society Special Publication No. 121, pp. 99–122.

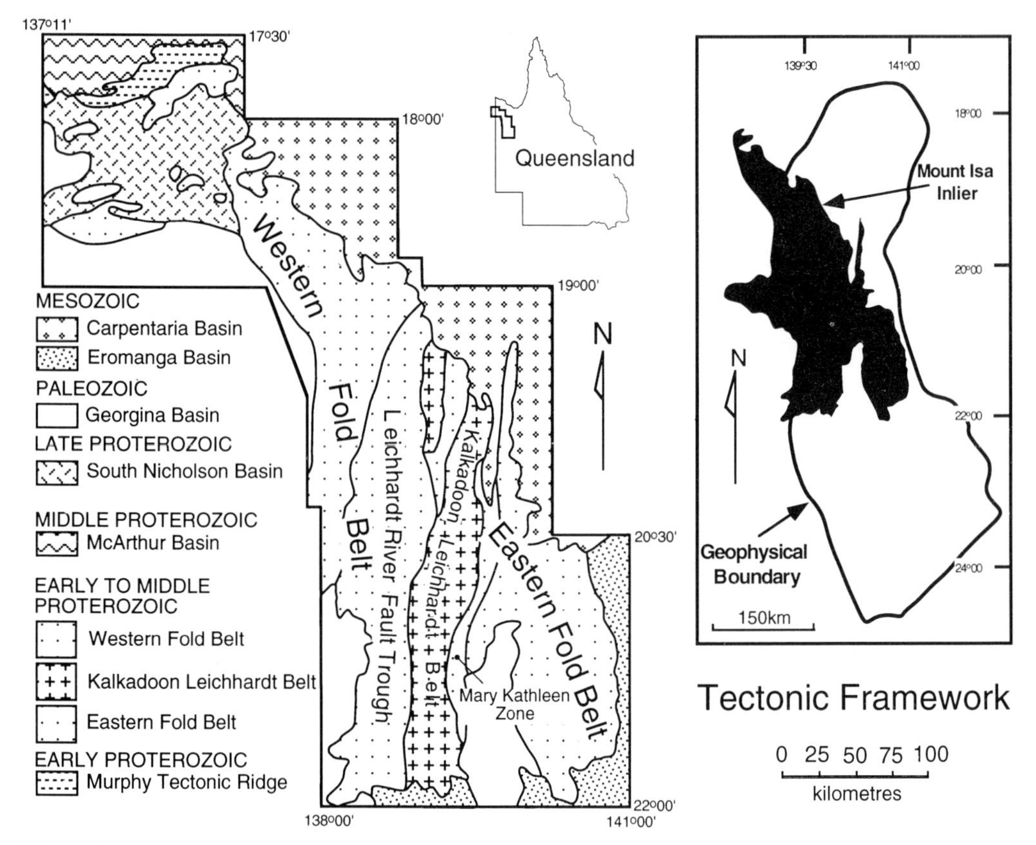

Fig. 1. Simplified tectonic map of the Proterozoic Mount Isa Inlier showing its structural belts and surrounding Phanerozoic basins (modified from Blake 1987). Inset map shows the position and extent of the Mount Isa Inlier along with its geophysical domain boundaries (modified from Wellman 1992*a, b*). Together, the Mount Isa Inlier and its geophysical continuation define the Mount Isa terrain.

Smith 1969; Plumb *et al.* 1980; Derrick 1982; Blake 1987; Page 1988; Etheridge & Wall 1994). Igneous and sedimentary rocks of this period were formed or deposited during at least four episodes of superimposed intracontinental rifting and associated post-rift subsidence. This extensional history was interrupted by compression, basin inversion and regional wrenching during the Isan Orogeny (*c.* 1590–1500 Ma).

The Isan Orogeny involved shortening of a complex intraplate rift system and was not accompanied by subduction. The geodynamic processes that characterised the evolution of the Mount Isa terrain were distinct from those that characterize continental margin orogens (e.g. magmatic arcs, allochthonous terranes and ophiolites do not occur). The deformational and metamorphic history of the Isan Orogeny was extremely heterogeneous, partitioning strain into spectacular zones of up to 80% shortening. Areas of lower strain preserve primary stratigraphic features that provide valuable insight into pre-shortening palaeogeography. Metamorphic assemblages range from sub-greenschist facies to upper amphibolite facies and are associated with one of the most regionally extensive areas of metasomatism known globally.

The Mount Isa terrain consists of three dominant, north-south trending structural belts that display distinctive geophysical trends (Wellman 1992a) and are separated by regionally extensive transcurrent fault zones (Blake & Stewart 1992) (Figs 1 & 2). The Kalkadoon–Leichhardt Block forms the central belt of the Mount Isa terrain, exposing basement rocks deformed and metamorphosed during the Barramundi Orogeny. The Kalkadoon–Leichhardt Block separates an extremely thick succession of basin-fill sedimentary and volcanic rocks with associated plutonic rocks in the Western Fold

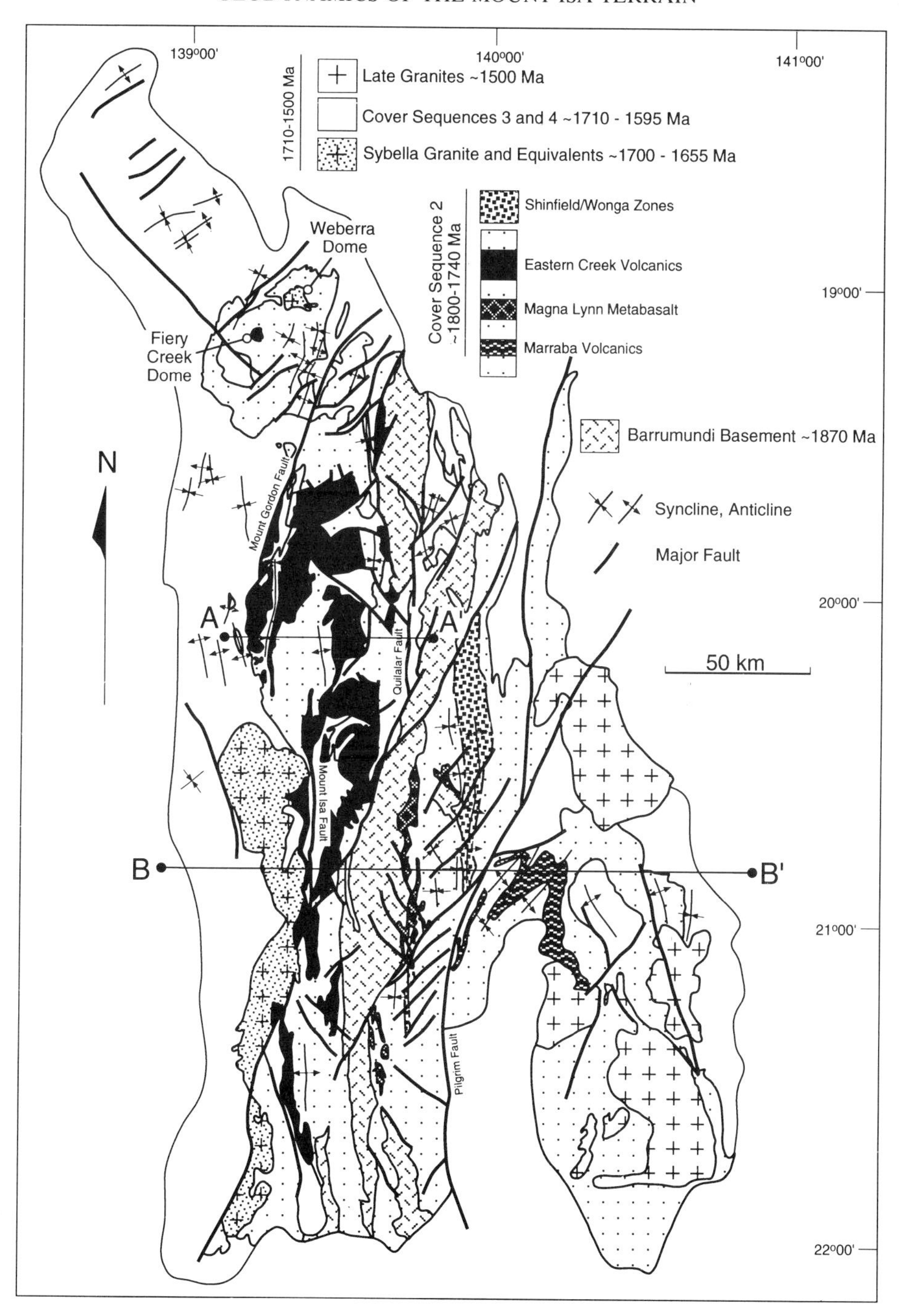

Fig. 2. Regional geological map of the Mount Isa Inlier (modified from Blake 1987). Lines A–A′ and B–B′ refer to corresponding cross sections depicted in Fig. 11.

Belt (which includes the Leichhardt River Fault Trough) from a moderately thick but more intensely deformed and metamorphosed succession in the Eastern Fold Belt.

The Mount Isa terrain is an orogenic belt whose underlying geometrical template was established prior to crustal shortening, and it offers an opportunity to study the superposition

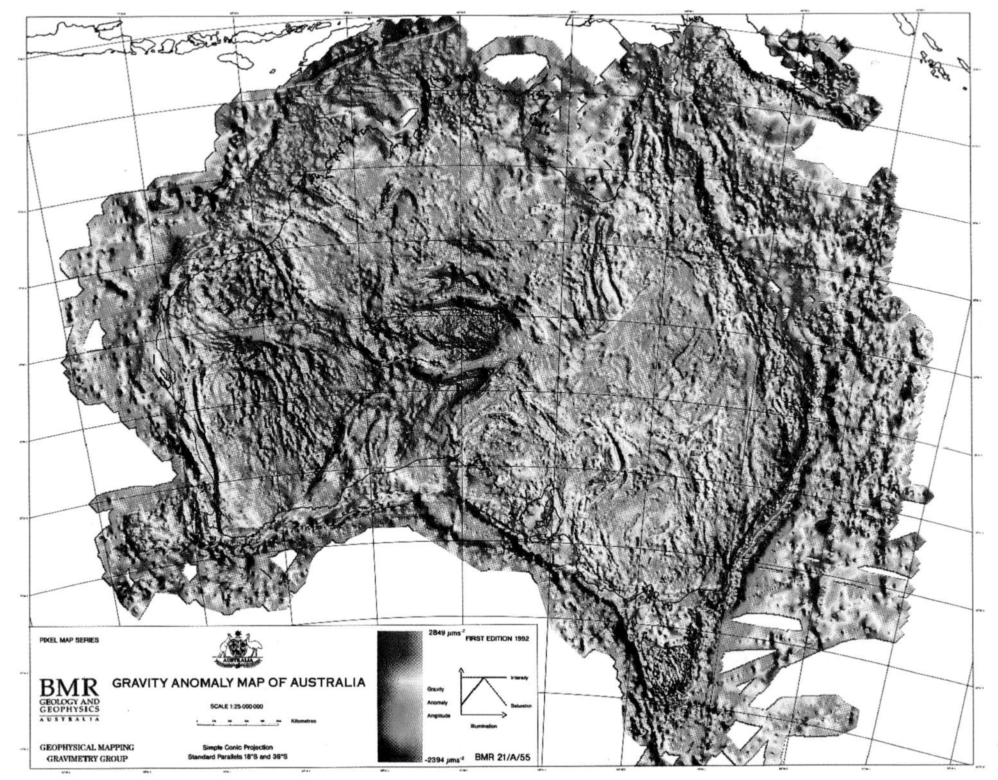

Fig. 3. Gravity Anomaly Map of Australia (Morse *et al.* 1992). Reproduced, with permission, from the Australian Geological Survey Organisation.

of extensional and compressional structures within a now spectacularly exposed stratigraphic sequence. Over the past 25 years many researchers working in the Mount Isa region have ignored the structural implications of the rifting history in their interpretations of local and regional geometries (e.g. Bell 1983; Derrick 1982; Blake 1987). Consequently, all complex field relations were interpreted to have originated during, rather than prior to, the compressional Isan Orogeny. This paper highlights the rifting history and incorporates this history in the geodynamic evolution of the Mount Isa terrain.

The Mount Isa terrain in its Australian context

Numerous studies have outlined correlations of stratigraphy and events between the various Proterozoic areas of Australia, including the Mount Isa terrain (e.g. Plumb 1990; Myers *et al.* 1994; Etheridge & Wall 1994). Interpretations of regional geophysical images indicate that the development of the Mount Isa terrain is linked with other crustal-scale structures throughout northern and central Australia (Figs 3 & 4).

Early to mid-Proterozoic rifting, volcanism and sedimentation formed the rocks within the Mount Isa terrain, Georgetown and Coen Inliers, as well as the McArthur Basin (Fig. 4) on the present eastern portion of the North Australian craton (Plumb *et al.* 1980; Etheridge & Wall 1994). A link between the northern boundary of the Mount Isa terrain and the NW–SE-trending Urapunga Tectonic Ridge may be the remanence of an ancient transfer system that relayed early extension in the Mount Isa terrain to that in the Batten trough in the McArthur Basin to the northwest (Plumb *et al.* 1980; Plumb & Wellman 1987). Sediments in the Batten Trough correlate in age and lithology with high level cover rocks of the Mount Isa terrain.

The Mount Isa terrain is geophysically distinct from the other named areas, because of its magnetic and gravity anomalies (Fig. 3). The stronger geophysical signature of the Mount Isa terrain is largely due to the juxtaposition of rocks with contrasting geophysical properties. Dense

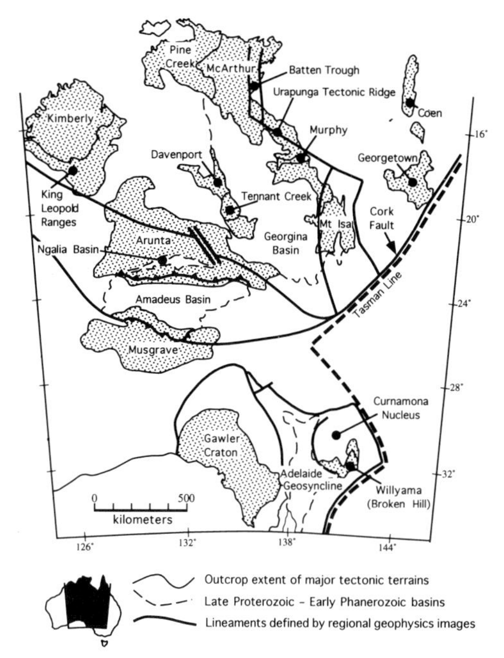

Fig. 4. A geological interpretation of a portion of the gravity map of Australia (see Fig. 3), showing the Proterozoic Mount Isa terrain in its regional context. The interpretation is based on Coney *et al.* (1990), Freeman *et al.* (1990), Murray *et al.* (1989), Parker (1990*a, b*), Plumb & Wellman (1987), Plumb *et al.* (1980), Wellman (1992*a, b*).

and high magnetic susceptibility mafic volcanic rocks (and mafic intrusions at depth) are often in contact with less dense and lower magnetic susceptibility granitoids and basement rocks. These associations may be explained by more advanced rifting and associated basaltic volcanism and/or more underplating of mafic material within the Mount Isa terrain than in other contemporaneous adjacent terrains.

All of the northeastern Australian inliers show the effects of subsequent compressional orogenesis between approximately 1600 and 1400 Ma (Plumb *et al.* 1981). On a terrain-wide scale compressional deformation is favoured by crust that has been thermally weakened by a preceding period of rifting (Ziegler 1982). Being thinned, thermally immature and therefore, weak, the crust underlying the Mount Isa terrain is interpreted to have localized deformation preferentially over the more stable cratonic areas to the west.

Similarities between the Mount Isa terrain and Proterozoic blocks in southern Australia suggest a physical link between northern and southern Australia throughout the middle Proterozoic. The regional geophysical responses of the Mount Isa terrain and the Willyama Block in the Broken Hill region are very similar (see Fig. 3). The timing, style and orientation of deformation features produced during the Isan Orogeny are comparable to those found in the Willyama Complex (Olarian Orogeny) to the south of Mount Isa (Marjoribanks *et al.* 1980). Elsewhere in Australia, however, Isan Orogeny age deformation events are not widely recognized, suggesting that the Mount Isa terrain and Willyama Complex represent the locus of a compressional event. With this correlation it is difficult to accept the hypothesis that the Southern Australian Craton did not form a suture zone with the Northern Australian Craton until 1300–1100 Ma when the Musgrave, Rudall, Albany and Fraser complexes underwent deformation at high metamorphic grades (e.g. Myers *et al.* 1994).

One of the most striking points about the Australian context of the Mount Isa Inlier is the remarkable correlation in time between the sediment-hosted base metal deposits of the McArthur Basin (HYC), the Mount Isa terrain (Mt Isa, Hilton, Lady Loretta, Century, Cannington) and the Broken Hill area (Broken Hill, Pinnacles). This observation implies continent-scale similarities in factors such as source rocks and regions, and tectonically controlled depositional and hydrodynamic regimes.

At present there is little documented evidence for geological processes active in the Mount Isa terrain between 1500 and 500 Ma, yet one of the largest collisional orogenies in Earth history occurred at *c.* 1100 Ma. In Laurentia, for instance, the Grenville Orogeny represents a major collision between proto-Laurentia and Amazonia(?) initiated at *c.* 1200 Ma (Hoffman 1989, 1991). This collision was broadly coeval with major orogenic events in Baltica, South America, Africa and East Gondwanaland (P. Hoffman pers. comm.). The effects of this extensive 1200–1000 Ma orogenic system may have affected the Mount Isa terrain in ways that have not yet been recognized.

The Barramundi Orogeny

Evidence for the Barramundi Orogeny exists throughout many Proterozoic inliers of Australia, such as the Halls Creek (Plumb 1990), Pine Creek (Needham & De Ross 1990), and Tennant Creek (Le Messurier *et al.* 1990) blocks. In these areas, the Barramundi Orogeny was preceded by an episode of rifting and sag-phase sedimentation, followed shortly thereafter by shortening, voluminous igneous activity (Wyborn 1988) and low-pressure meta-

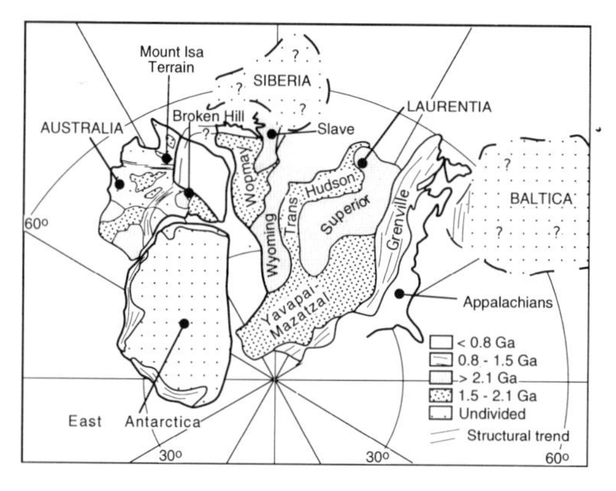

Fig. 5. Plate reconstruction of a portion of the Rodinian supercontinent showing the relative positions of Australia, Antarctica, Laurentia and Baltica as proposed by Dalziel (1991, 1992), Moores (1991) and Hoffman (1991). Parts of this supercontinent may have been assembled as early as *c.* 2000 Ma based on global correlations of the Barramundi and Trans-Hudson Orogenies. Reproduced, with permission, from the Annual Review of Earth and Planetary Sciences Volume 20, © 1992, by Annual Reviews Inc.

morphism (Etheridge *et al.* 1987). The Leichhardt Volcanics and the Kalkadoon Batholith of the Mount Isa terrain (see Fig. 2) are part of an Australia wide 1870–1840 Ma felsic magmatic event possibly related to the Barramundi Orogeny (Wyborn 1988).

The Barramundi Orogeny occurred at around the same time as the Trans-Hudson Orogeny in North America (Laurentia) (Hoffman & Bowring 1984; Hoffman 1988) and the Sveco-Fennian Orogeny in the Baltic Shield (Gower 1985; Gaal & Gorbatschev 1987). This early Proterozoic history shared by Australia, Laurentia and Baltica indicates that this orogenic event was either of global significance or that these terrains (or parts thereof) occupied approximately contiguous positions throughout the early Proterozoic. Alternatively, as with tectonically active regions today, such as the Himalayas, the Andes, the Cascades, New Zealand and Papua New Guinea, it is possible that distinct and widely separated orogenic belts in the Proterozoic developed at the same time.

The 'Southwest US–East Antarctic' (or SWEAT) hypothesis (Moores 1991; Dalziel 1991; Hoffman 1991) proposed the existence of a Proterozoic supercontinent called Rodinia (Powell & Li 1994), which restored the western margin of the North American craton with the eastern margin of Australia–Antarctica (Fig. 5). Evidence for this reconstruction was centred on stratigraphic correlations (Bell & Jefferson 1987), isotopic mapping (Borg & DePoalo 1994) and the use of the 1200–1000 Ma Grenville Front as a plate-scale piercing point (Moores 1991; Dalziel 1991; Hoffman 1991). There is uncertainty regarding the time of assembly of this supercontinent but recent attention has focussed mainly on the Neoproterozoic (e.g. Powell & Li 1994; Myers *et al.* 1994).

Paleomagnetic studies by Idnurm & Giddings (1995) revealed that eastern Australia and western Laurentia share a common apparent polar wander path for the interval between 1700 and 1600 Ma. This suggests that both western Laurentia and eastern Australia were 'fellow travellers' (Hoffman 1991) during the Mid-Proterozoic. It may be possible that parts of this supercontinent (i.e. northern Australia and western Canada) were fused by as early as *c.* 2000 Ma (cf. Hoffman 1991), based on the similarities between the Barramundi Orogeny of Australia and the Trans-Hudson Orogeny of Laurentia. Other blocks, however, that record coeval Early Proterozoic orogenies (e.g. Laurentia and Baltica) are known to have broken apart before being reassembled at 1100 Ma (Hoffman 1991).

It is interesting to note that the geochemical trends of volcanic rocks of the Barramundi Orogen and the Wopmay Orogen (Trans-Hudson equivalents) display remarkable uniformity (compare Hoffman & Bowring 1984; Etheridge *et al.* 1987). Interpretations of their tectonic significance, however, differ radically, since geochemical characteristics alone are insufficient to distinguish tectonic setting. Regardless of its global significance, Barramundi age orogenic events are recorded on many continents.

Intracontinental rifting in the Mount Isa terrain

Following the Barramundi Orogeny, the Mount Isa terrain (between *c.* 1800 Ma and 1600 Ma) underwent a protracted period of intracontinental rifting characterized by voluminous bimodal magmatism and predominantly clastic sedimentation (Carter *et al.* 1961; Smith 1969; Glikson *et al.* 1976; Derrick 1982; Blake 1987; Page 1988; Page *et al.* 1994). Igneous and sedimentary rocks of this period were formed or deposited during at least four episodes of superimposed rifting and associated post-rift subsidence, resulting in the development of a complex extensional rift system comparable in dimensions to those of modern rift systems such as the Baikal Rift and the East African Rift. Evidence for this exten-

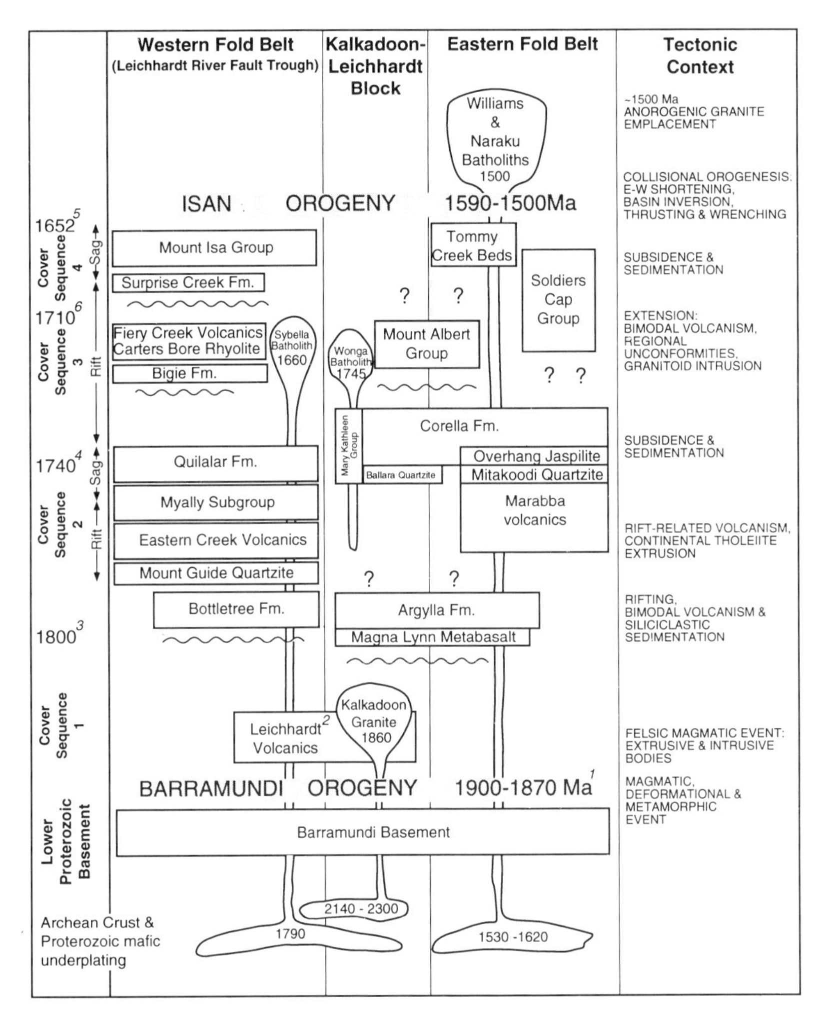

Fig. 6. Time-space diagram for the Mount Isa terrain showing names, ages and correlation of stratigraphic units and intrusive bodies across the Western Fold Belt (Leichhardt River Fault Trough), Kalkadoon–Leichhardt Block and Eastern Fold Belt (modified from Wyborn *et al.* 1988). Regionally significant unconformities are depicted as wavy lines and orogenic events are shown in capital letters and stipple. For more detail regarding stratigraphic units see Blake & Stewart (1992). 1, Barramundi age and older basement includes Yaringa Metamorphics (Page & Williams 1988), and Kurbayia Migmatite (Blake & Page 1988). 2, see Wyborn (1988) and Page (1988). 3, see Page (1983b). 4, see Page (1983*a*). 5, see Page *et al.* (1994). 6, Page (unpublished data).

sion is contained within the geometries of stratigraphic packages along with their lithological and depositional associations. Figure 6 presents a tectonostratigraphic framework for the Mount Isa terrain summarizing the names and ages of stratigraphic units along with their inferred origins in terms of rift and sag cycles.

The rocks of the Mount Isa terrain were divided by Blake (1987) into three cover sequences. This tripartite stratigraphic subdivision has become entrenched and widely used in the geological literature (e.g. see Stewart & Blake 1992). In this paper, however, the rocks of the Mount Isa terrain are divided into four cover sequences with the stratigraphic makeup of Cover Sequences 1 and 2 of Blake (1987) remaining unchanged. The former Cover Sequence 3 of Blake (1987) is here subdivided into Cover Sequences 3 and 4. It is important to note that, despite the usage of the term Cover Sequence, strata of the Mount Isa terrain display significant lateral thickness and facies variations and do not define a layer-cake stratigraphy.

Rocks of Cover Sequence 1 comprise the felsic Leichhardt Volcanics and comagmatic Kalkadoon and Ewen Batholiths (1870–1840 Ma, Blake *et al.* 1990), the extrusion and emplacement of which occurred during or shortly after the Barramundi Orogeny (Blake & Stewart 1992). These Barramundi age rocks form the crystalline basement of the Mount Isa terrain and much of northern Australia (Plumb *et al.* 1980).

The deformed and metamorphosed rocks now exposed within the Eastern and Western Fold Belts belong to the upper three cover sequences. Cover Sequences 2, 3 and 4 are three stacked sequences related to separate basin-forming events and are separated by two pronounced angular unconformities (the Bigie Unconformity and the Surprise Creek Unconformity). Deposition of these three cover sequences spanned approximately 200 Ma (between 1790 Ma and 1595 Ma) and was associated with repeated episodes of rifting and post-rift subsidence, without intervening periods of regional shortening.

Deposition and extrusion of Cover Sequence 2

Deposition of Cover Sequence 2 was associated with a multi-phase episode of continent-wide extension, which produced a linked network of basins across northern and central Australia (Etheridge & Wall 1994). In the Eastern Fold Belt of the Mount Isa terrain, rift-related bimodal volcanism commenced at approximately 1780 Ma and produced 1–5 km of felsic volcanic rocks (e.g. the Argylla Formation, Blake & Stewart 1992) and continental flood basalts (e.g. the Magna Lynn Metabasalt). Coeval sedimentation and volcanism in the Western Fold Belt was associated with the development of an aereally extensive north–south-trending rift basin, termed the Leichhardt Rift, which is inferred to have developed during a period of regional east-west extension. Half-grabens of the Leichhardt Rift are estimated to have been up to 60 km wide and controlled by the development of widely spaced, crustal-scale border faults of alternating dip direction. The deformed remains of the Leichhardt Rift occupy the Leichhardt River Fault Trough (Derrick 1982).

Deposition within the Leichhardt Rift commenced with a basal rift-sag sequence deposited directly on crystalline basement (e.g. the Bottletree Formation and the Mount Guide Quartzite, Eriksson *et al.* 1993; Blake 1987). The Mount Guide Quartzite comprises a basal alluvial syn-rift sequence consisting of conglomerate and lithic to feldspathic sandstone of braided alluvial origin interbedded with stromatolitic dolomite of lacustrine origin (Eriksson *et al.* 1993). The basal Mount Guide Quartzite is overlain by sag-phase sands and orthoquartzite (Derrick 1982).

Following this rift-sag cycle, up to 6 km of continental flood basalts accumulated in the Leichhardt Rift (i.e. the Eastern Creek Volcanics). The geodynamic significance of these flood basalts is unclear. There is no doubt that they were associated with continental rifting, but whether or not they were a direct consequence of this process is arguable. It is possible that their extrusion was localized and facilitated by the rising of hot mantle plumes which caused lithospheric uplift and an increase in gravitational potential energy (Houseman & Hegarty 1987). In this model, flood basalts are extruded when upwelling asthenosphere intersects the solidus. Calculations by Klein *et al.* (1988) and Lister & Etheridge (1989) showed that for such a voluminous melt fraction to separate from the mantle, the upwelling asthenosphere would have to be abnormally hot (>1500°C). These calculations imply that the initiation of volcanism and rifting in the Leichhardt Rift was probably driven by mantle overturn or the arrival of a mantle plume.

Following the basaltic volcanism, the Mount Isa terrain experienced a period of predominantly clastic sedimentation (with minor basaltic volcanism) under coastal and shallow marine shelf conditions, resulting in the development of a southward tapering clastic prism termed the Myally Subgroup, which onlaps underlying strata. Widespread thickness variations in this sequence across east–west-striking faults attest to syn-sedimentary basement block faulting (Smith 1969; Derrick 1982). This wedge has the characteristic morphology and sedimentology of a syn-rift sequence, which is inferred to have been deposited during a period of north–south-extension. By inference, the palaeogeography of the Leichhardt River Fault Trough at this time was characterized by en echelon fault blocks with shallow northerly tilts (Fig. 7a).

Deposition of the Myally Subgroup was followed by a period of thermal subsidence associated with the deposition of a region-wide transgressive-regressive quartzite-carbonate package termed the Quilalar Formation. This formation represents a continental shelf and shoreline sequence deposited within a north–south-trending marginal platform or central trough (Jackson *et al.* 1990). This formation has been correlated with the Corella Formation in

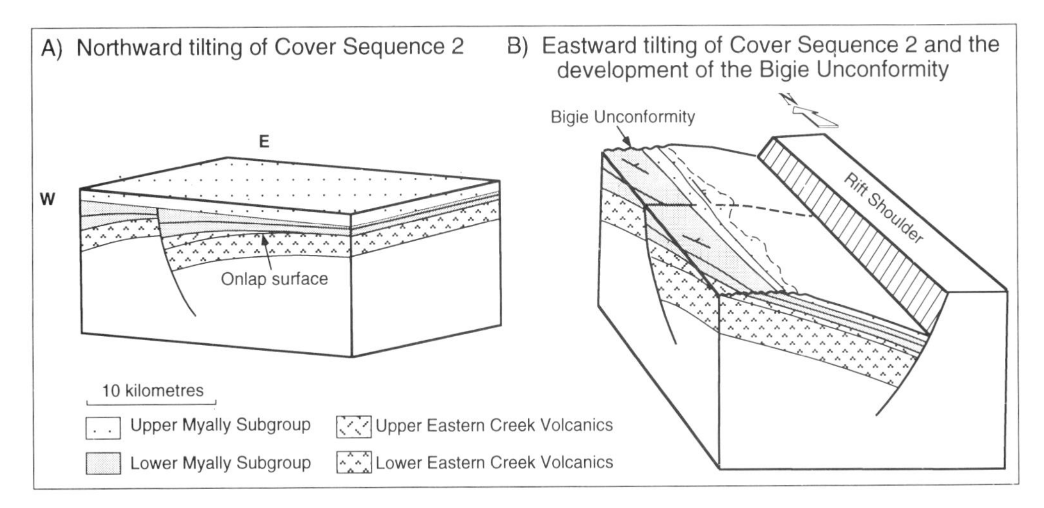

Fig. 7. Figure illustrating the geometry of regional rift-related tilting events inferred for the Western Fold Belt. (**a**) Syn-depositional northward tilting of lower Cover Sequence 2 rocks resulting in a southward tapering onlap sequence. This onlap sequence consists of strata ranging from the upper Eastern Creek Volcanics to the middle Myally Subgroup. (**b**) The geometry of the Western Fold Belt during the development of the Bigie Unconformity. Approximate easterly (and southeasterly) tilting and erosional bevelling were superimposed on previously north-tilted blocks of Cover Sequence 2. This period of tilting is postulated to have initiated at about 1740 Ma during a period of terrane-wide extension related to the development of the Wonga–Shinfield Zone in the Eastern Fold Belt. A subsequent period of approximate easterly (to southeasterly) tilting in the Western Fold belt occurred during the development of the Surprise Creek Unconformity (see Fig. 8).

the Eastern Fold Belt (Fig. 6) implying that its lateral depositional extent transgressed the central Kalkadoon–Leichhardt Block (Derrick *et al.* 1980).

Erosional unconformities and stacked basins

Deposition of the Quilalar Formation was interrupted during a subsequent phase of regional east–west to northwest–southeast extension, which impacted strongly on the Mount Isa terrain. This phase of extension was characterised by a period of rift-related tilting, footwall uplift and erosion producing an array of deeply eroded tilt blocks, the crests of which were truncated by the angular Bigie Unconformity (Fig. 7b).

The Bigie Unconformity was overlain at approximately 1710 Ma by Cover Sequence 3, comprising syn-rift strata of the Bigie Formation and 1709 Ma Fiery Creek Volcanics (Page & Sweet in press). The Bigie Formation is a rift-related, subaerial sequence consisting of texturally and compositionally immature conglomerate and sandstone deposited in an alluvial braid plain environment. The Fiery Creek Volcanics are bimodal and largely subaerial, typical of those developed in continental rifts (e.g. Burke & Kidd 1980).

Following the extrusion of the Fiery Creek Volcanics, fault blocks within the Mount Isa terrain underwent another period of rift-related tilting, footwall uplift and erosion resulting in the development of the Surprise Creek Unconformity and the complete bevelling of Cover Sequence 2 and Cover Sequence 3 rocks.

The Surprise Creek Unconformity was overlain by Cover Sequence 4 comprising strata of the Surprise Creek Formation and the 1652 ± 7 Ma Mount Isa Group (Page *et al.* 1994). Sediments of Cover Sequence 4 stratigraphically above the Mount Isa Group (i.e. the upper McNamara Group) are exposed in the far northwest of the Mount Isa terrain and have been dated at 1595 ± 6 Ma (Page *et al.* 1994).

Deposition of Cover Sequence 4 was preceded and/or accompanied by a period of epizonal granite emplacement between 1700 Ma (Weberra Granite) and 1655 Ma ± 5 Ma (Sybella Granite, Connors & Page 1995). The intrusion of these granites resulted in the development of large domal culminations such as the Weberra Dome and Fiery Creek Dome in the northwestern Mount Isa terrain (see Fig. 2).

The Surprise Creek Formation was deposited unconformably and paraconformably on the Fiery Creek Volcanics, the Bigie Formation, and on most of Cover Sequence 2 rocks. It consists of a westward thinning package of channel-fill conglomerate, sandstone and thinly-bedded sandstone and siltstone deposited in a braided stream, alluvial fan to shallow marine environment (Nijman *et al.* 1992*a, b*; Blake *et al.* 1990). The abundance of coarse detrital mica in the Surprise Creek Formation suggests that clastic material may have been derived from uplifted crystalline basement rocks (Plumb *et al.* 1980) possibly exposed during a period of tilting and uplift inferred to be associated with the development of the Surprise Creek Unconformity.

The overlying Mount Isa Group consists predominantly of carbonates and siltstones interpreted by Neudert & Russell (1981) to have been deposited in a shallow marine environment with intermittent hypersaline to emergent conditions. These sediments host several world class Pb–Zn–Ag deposits. The contact between the Surprise Creek Formation and the Mount Isa Group is complex reflecting the transition from a largely alluvial braid plain and marginal marine depositional environment to that of a quiescent shallow marine environment. This transition, together with the vast lateral extent of the Mount Isa Group (and correlative McNamara Group), strongly suggests that these sediments were deposited during a period of regional post-rift subsidence. Following deposition of rift-related Fiery Creek Volcanics, Bigie Formation and Surprise Creek Formation, the inferred onset of thermal subsidence restored uplifted basement to sea level leading to a transition from subaerial to shallow marine deposition (see Houseman & Hegarty 1987; Keen 1988).

Reconstructed stratal geometries of fault blocks within the Western Fold Belt are strongly asymmetric in cross-section, exhibiting a pronounced rotational thickening of Cover Sequences 3 and 4 towards the east and a tapering of these units towards the west. Both the Bigie Unconformity and the Surprise Creek Unconformity cut down-section to the west in fault blocks suggesting that differential uplift and rotation were responsible for their development. Both unconformities are interpreted to mark pronounced erosion surfaces at the crests of approximately east- to southeast-tilted fault blocks developed during regional east–west to northwest–southeast extension. The degree of angularity with respect to underlying rocks, across both unconformities, decreases towards the eastern edges of fault blocks where all strata of Cover Sequences 2, 3 and 4 are paraconformable. Thus, strata of Cover Sequences 3 and 4 are interpreted to have been deposited in the hangingwalls of rotating tilt blocks, the width of which were between 15 and 40 km (see Fig. 8).

Following the deposition of Cover Sequence 4, the Mount Isa terrain underwent a final episode of north–south extension during which time previously formed east–west-striking faults were reactivated as south-block-down normal faults. Evidence for this extension event is manifested throughout the Western Fold Belt in the form of east–west-trending synclines (cored by Mount Isa Group rocks) developed to the south of east–west-striking faults (Fig. 9). The development of these synclines has long been considered to reflect a period of south-directed thrusting that predated regional east-west shortening (Bell 1983). However, these synclines consistently occur in the hanging walls of south-dipping normal faults and record low strains (Dunnet 1976; Stewart 1992; O'Dea & Lister 1995). Furthermore, bedding to the north of these east–west-striking faults did not develop corresponding east–west-striking anticlines. The east–west-trending synclines throughout the Western Fold Belt are interpreted here to have developed in the hanging walls of rotational normal faults (Hamblin 1965; Nijman *et al.* 1992*a*). Alternatively, these east–west synclines may have developed in response to differential compaction as post-rift sediments of the Mount Isa Group filled remnant accommodation space within pre-existing half grabens. While all of these synclines have been accentuated and refolded during regional east–west shortening, they are interpreted to have originated prior to the Isan Orogeny.

The Bigie Unconformity and implications for extension in the Eastern Fold Belt

A regionally significant extensional interval in the evolution of the Mount Isa terrain occurred between the deposition of Cover Sequence 2 and Cover Sequence 3. Evidence for continental extension during this interval is manifested across the entire region and can be correlated between the Eastern and the Western Fold Belts. In the Eastern Fold Belt, the dominant structure produced during extension occurs in the Wonga–Shinfield Belt. Although this belt has been folded during the Isan Orogeny, its reconstructed geometry (Fig. 10) consists of a shallowly west-dipping, mid-upper crustal ductile deformation zone within rocks of Cover Sequence 2 (Holcombe *et al.* 1991*a*). The Wonga–Shinfield Belt was intruded by syntectonic granitic plutons and dolerite dykes which are thought to have thermally weakened the crust resulting in ductile fabrics overprinting

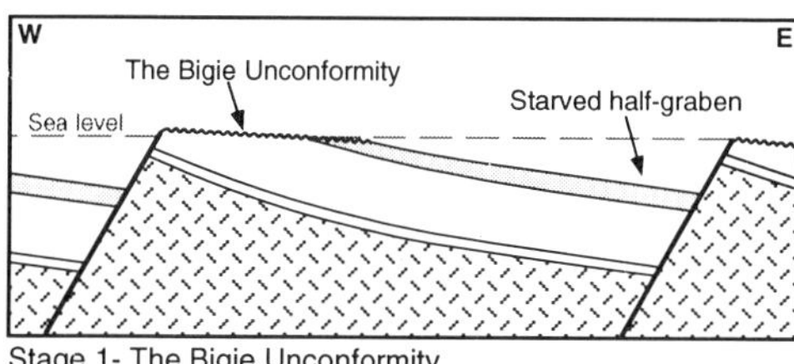

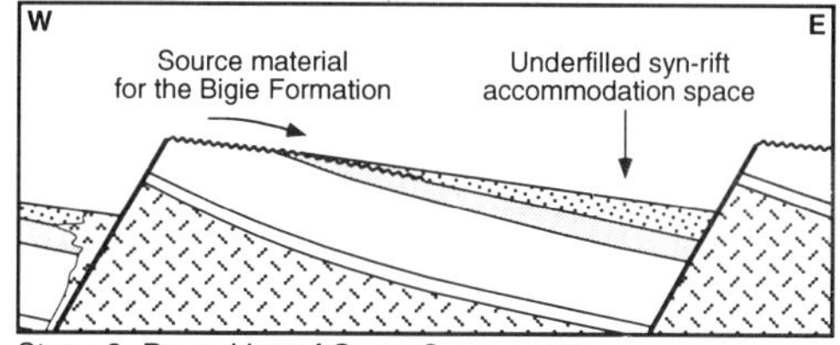

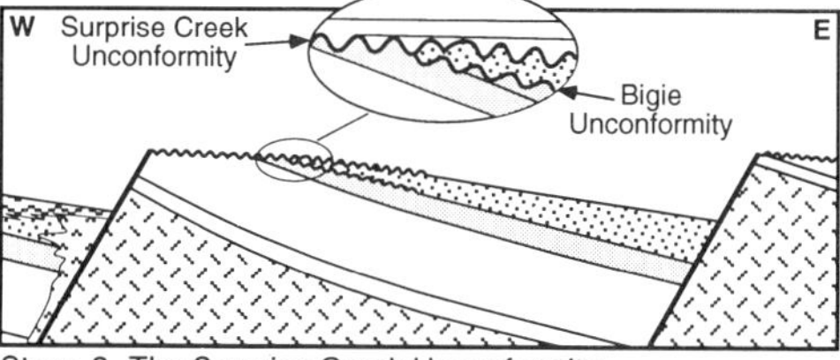

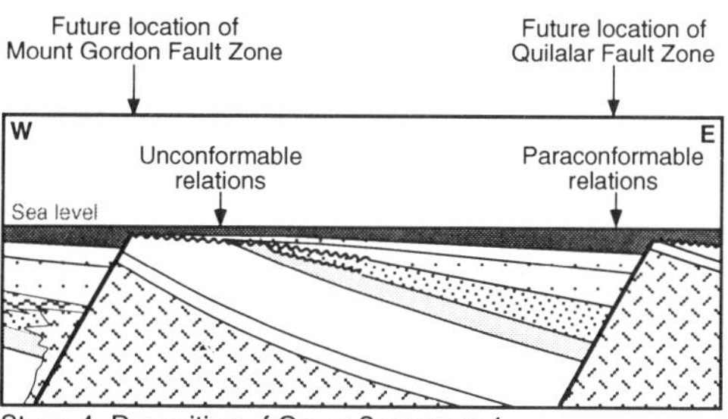

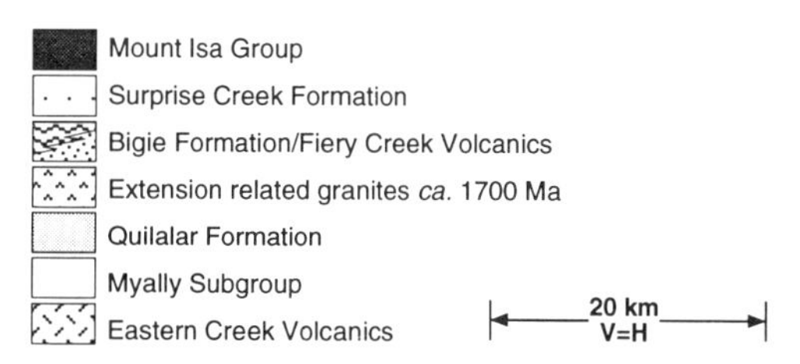

Fig. 8. Schematic E–W cross-sections illustrating the geometrical and temporal evolution of unconformity-bounded sequences within the Western Fold Belt. Stage 1: The Bigie Unconformity. This unconformity developed during rift-related tilting and erosion between 1737 Ma and 1710 Ma. Stage 2: Deposition of Cover Sequence 3. Note that rocks of the Cover Sequence 3 are unconformable in the west and paraconformable in the east. Small volcanic edifices responsible for the extrusion of the Fiery Creek Volcanics are interpreted to have developed along rift faults and been comagmatic with granitoid intrusions (e.g. 1700 Ma Weberra Granite). Stage 3: The Surprise Creek Unconformity. Footwall uplift and fault controlled subsidence caused pre-rift Cover Sequence 2 and the tapering edge of syn-rift Cover Sequence 3 to be elevated, eroded and capped by the Surprise Creek Unconformity. Stage 4: Deposition of the Cover Sequence 4. Strata ranging from the Surprise Creek Formation to the Warrina Park Quartzite (basal Mount Isa Group) were deposited during episodic rifting. The submergence of eroded footwall crests led to the deposition of Cover Sequence 4 on Cover Sequences 2 and 3 across the Surprise Creek Unconformity. Note the rotational thickening of Cover Sequences 3 and 4 towards the east. The stratigraphic geometry of the underlying Eastern Creek Volcanics are not considered in this figure.

slightly earlier brittle extension features (Passchier 1986; Passchier & Williams 1989).

This ductile deformation zone is structurally overlain by an imbricate set of south-dipping listric normal faults. Locally these faults are intruded by dolerite, and merge into the Wonga–Shinfield zone at depth. Other evidence for extension during this period comes from the abundant syn-kinematic intrusives dated at 1758 ± 8, 1742 ± 13 and 1729 ± 5 Ma (Pearson *et al.* 1992) and attenuation of metamorphic isotherms around the lower plate–upper plate transition (Oliver *et al.* 1991). Fluid flow associated with both early development of this extensional regime and the emplacement of the syntectonic Wonga granites has resulted in a spectacular metamorphic-metasomatic zone of wholesale scapolitization, which is largely the

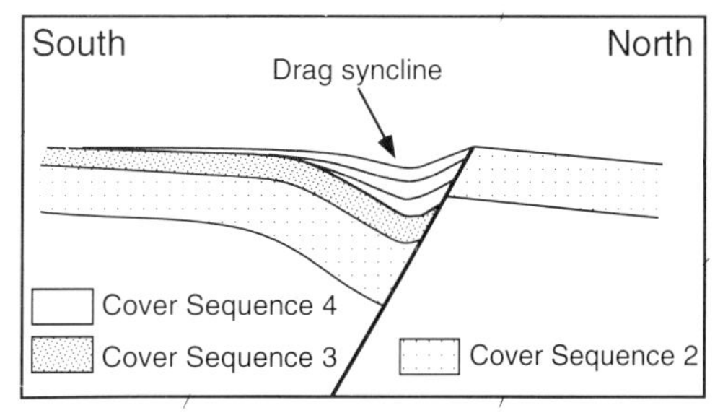

Fig. 9. Extension-related syncline developed in rocks of Cover Sequence 4. These synclines commonly occur in the hangingwalls of south-dipping normal faults throughout the Western Fold Belt, suggesting that folding occurred during faulting. These structures were previously interpreted by Bell (1983) to be first generation (F1) ramp synclines developed during a period of proposed south-directed thrusting.

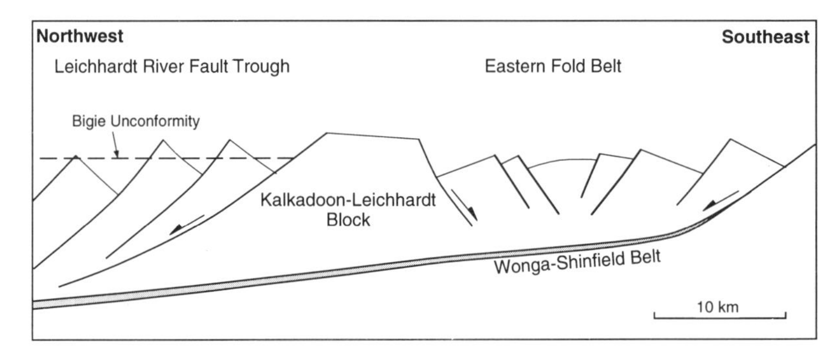

Fig. 10. Schematic diagram showing a postulated reconstructed geometry of the Wonga–Shinfield Zone and its upper crustal manifestations in the Western Fold Belt. Note the development of the Bigie Unconformity above easterly to southeasterly tilted fault blocks.

result of mobilization of sodium from the evaporite-dominated sag-phase sediments of Cover Sequence 2 (Oliver *et al.* 1993).

Evidence for a phase of crustal extension is also manifested in the Western Fold Belt with the development of the Bigie Unconformity during approximate east–west to northwest–southeast extension (Figs 8 & 10). This erosional surface developed after the deposition of the Quilalar Formation and prior to the extrusion of the Fiery Creek Volcanics and is thus, constrained between 1737 and 1710 Ma. Strata deposited during this extensional interval are not preserved within the Western Fold Belt suggesting that associated half-grabens were starved of syn-rift material. Alternatively, basin fill material deposited during this interval may have been removed prior to the deposition of the Bigie Formation. Restoration of the Bigie Unconformity in several localities throughout the Western Fold Belt indicates that the underlying rocks of Cover Sequence 2 had shallow to moderate palaeotilts towards the east and southeast. These tilts are consistent with both the orientations expected from the superimposition of eastward tilting on previously north-tilted blocks (see Fig. 7) and with the inferred extension directions in the Eastern Fold Belt.

The strongest line of evidence for correlating the Bigie Unconformity and the Wonga–Shinfield Belt is their similarity in age. Both of these features developed between the deposition of Cover Sequence 2 and Cover Sequence 3. Reconstructing the geometry of an extensional system that extends from the Eastern Fold Belt, through the Kalkadoon Leichhardt Block and into the Western Fold Belt, is problematic because the extension directions for the two regions are not entirely consistent. Restoration of the Bigie Unconformity suggests eastward tilting during east–west extension. Kinematics in the Wonga–Shinfield Belt indicate a general north–south extension direction. This disparity in extension direction makes a single detachment system difficult to construct. However, a detachment system linking the two belts can be constructed assuming regional northwest–southeast extension that was influenced by pre-existing rift structures and modified by later east–west shortening. The tilt blocks preserved in the Western Fold Belt may have developed in the upper plate of an asymmetric detachment system (Fig. 10).

Widespread rifting, deposition and magmatism within the Mount Isa terrain were interrupted at approximately 1590 Ma, prior to the formation of oceanic crust, when the region was deformed and metamorphosed during the compressional Isan Orogeny. In many respects, the complex extensional architecture established during the preceding 200 Ma rifting history formed the geometrical template upon which Isan compressional structures were superimposed. Understanding the interaction between extensional fault architecture, stratigraphic geometry and regional shortening is critical to interpreting the structural evolution of the Mount Isa terrain as a whole.

The Isan Orogeny

Between approximately 1590 and 1500 Ma the Mount Isa terrain underwent a period of compressional orogenesis, termed the Isan Orogeny (Blake & Stewart 1992). This orogeny involved a variable degree of north–south and east–west shortening and sub-vertical extension accompanied by crustal thickening (Fig. 11).

Historically, the Isan Orogeny was thought to have occurred between approximately 1620 Ma

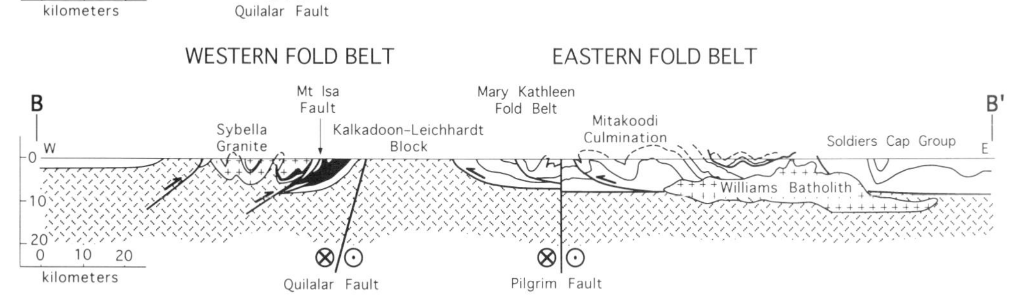

Fig. 11. Regional E–W vertical cross sections through the Mount Isa terrain illustrating the fold–fault relations constrained from surface geology, seismic reflection and gravity. Cross-sections are based on map data by Blake (1991*a, b*), Connors *et al.* (1991), Holcombe *et al.* (1991*b*), Loosveld (1991), Reinhardt (1991), Stewart (1991), and field data collected by T. MacCready. See Fig. 2 for locations of cross section lines.

and 1500 Ma, although the interpretations of individual authors differed considerably (compare Page 1983*b*, Page & Bell 1986, Blake *et al.* 1990, Blake & Stewart 1992, Eriksson *et al.* 1993). In this paper, the Isan Orogeny is interpreted to have occurred between approximately 1590 and 1500 Ma. This interval is constrained by the age of the youngest deformed sediments exposed in the northwest of the Mount Isa terrain (1595 ± 6 Ma for the upper McNamara Group, Page *et al.* 1994), and by the age of the mildly deformed Williams and Naraku plutonic suite (1500 Ma, Blake 1987, Nisbet *et al.* 1983). It is important to note, however, that the Isan Orogeny may have occurred during a considerably shorter interval than 90 Ma and its apparent lengthy duration may simply reflect poor geochronological resolution.

In the high grade parts of the terrain, there is widespread development of an S1 schistosity sub-parallel to bedding. Related regional folds have not been identified making the tectonic significance of this fabric unclear. The S1 fabrics are defined by metamorphic micas (and other metamorphic minerals such as andalusite and sillimanite in the Eastern Fold Belt) and provide evidence for an early period of low-pressure high-temperature metamorphism (M1) prior to the onset of a regional deformation D1. While the tectonic processes that characterized the initial stages of the Isan Orogeny are poorly understood, it is possible that both M1 and D1 were related to a period of extensional tectonism immediately prior to the Isan Orogeny.

In the Eastern Fold Belt, subsequent (D2) deformation produced east-west trending upright folds with well-developed axial planar S2 fabrics. A second low-pressure high-temperature metamorphic event (M2) took place (characterised by the growth of andalusite and sillimanite) after S2 fabrics developed, but deformation continued. M2 involved a sequence of prograde reactions culminating in a metamorphic peak that post-dated S2 and predated S3 fabric development. North–south-trending upright folds with S3 axial planar fabrics formed subsequent to M2. In some localities east–west-trending S2 fabrics were reoriented during D3 into north–south trends.

In the Western Fold Belt, west of the Mount Isa Fault, steeply dipping S2 fabrics had developed by the time low-pressure high-temperature metamorphism reached its peak (Connors *et al.* 1992). The growth of peak metamorphic assemblages statically overprint S2 fabrics, implying either that the duration of blastic mineral growth was short, or that a significant hiatus occurred between the D2 and the D3 deformation. Deformation continued after the peak of metamorphism, but instead of being characterised by pervasive cleavage development strain became more localised and major ductile shear zones evolved. The most significant D3 shear zone is the steeply west-dipping Mount Isa

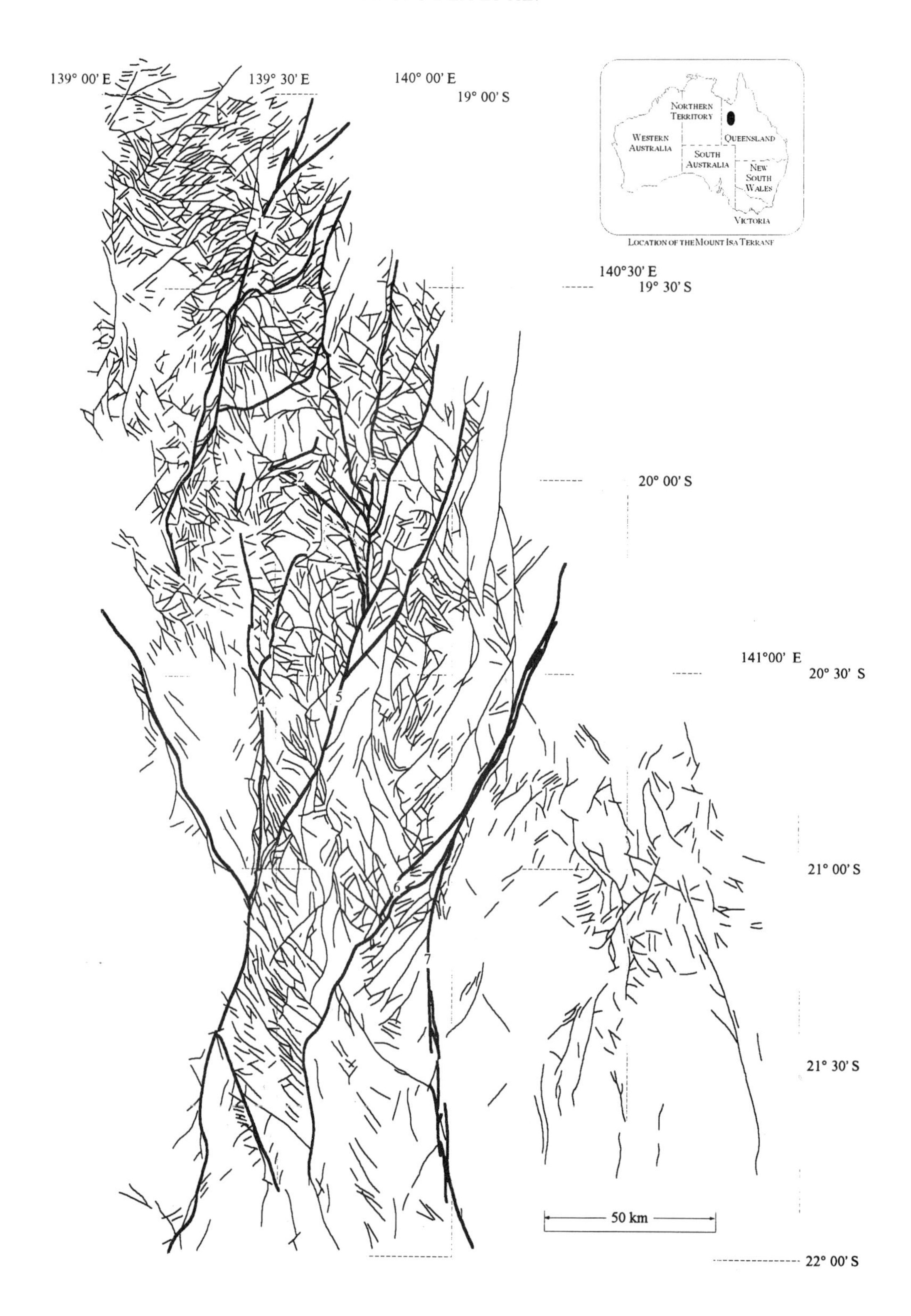
139° 00' E
139° 30' E
140° 00' E
19° 00' S
140°30' E
19° 30' S
20° 00' S
141°00' E
20° 30' S
21° 00' S
21° 30' S
22° 00' S
50 km
1
2
3
4
5
6
7
Northern Territory
Western Australia
Queensland
South Australia
New South Wales
Victoria
Location of the Mount Isa Terrane

Shear Zone (and the subsequently developed Mount Isa Fault, Figs 2 and 12), which juxtaposes amphibolite facies rocks to the west with greenschist facies slates, siltstones and carbonates of the Mount Isa Group to the east. Connors *et al.* (1992) argued that D2, D3, D4 and D5 fabrics developed during successive phases of a progressive regional deformation as the Mount Isa Shear Zone exhumed the rocks west of the Mount Isa Mine.

There is a broad similarity in the tectonothermal histories of the Eastern, Central and Western Fold Belts; however, the intensity of fabric development varies significantly. All three areas display the early layer-parallel S1 fabric. This fabric is best developed in the Eastern Fold Belt where metamorphic zircons have been dated at 1584 ± 17 Ma (Page 1993), possibly constraining the age of M1. The main phase of the Isan Orogeny produced regional S2 and S3 fabrics separated by the M2 metamorphic event. In the area west of the Mount Isa Fault, Connors & Page (1995) dated zircons in pegmatites associated with the peak of M2 at 1537 ± 7 Ma. The orientation of S2 fabrics developed throughout the Mount Isa terrain are extremely variable with orientations ranging from east-west to north–south. S3 fabrics have a more consistent north–south orientation.

In the Western Fold Belt, the structural features that characterize the main phase of shortening include open to tight upright north–south-trending folds, high angle reverse faults, ductile shear zones and local breccias, and the steeply-dipping penetrative S2 and S3 foliations (Fig. 11). North–south-trending folds are often bounded by reverse faults, which are either parallel to the fold axial plane traces or cut through their axial planes at a small angle. In contrast, the Eastern Fold Belt has more in common with fold and thrust belts (e.g. Rodgers 1990). Cover Sequence rocks are detached from the basement along low-angle west-vergent thrusts such as the Roos Mine Thrust beneath the Mitakoodi Culmination (Fig. 11) (Huang 1994). Regional east–west compression during the Isan Orogeny accomplished an estimated 40 to 50% shortening.

One of the enigmatic geometrical aspects of the Mount Isa terrain is that the structures developed in high level cover sequences are commonly more complex than those developed in lower units (O'Dea & Lister 1995). Part of the reason for this contrast may be that the structures developed during the shortening of previously extended crust are more variable than those developed in thrust belts which deform a 'layer-cake' stratigraphy (cf. McClay & Buchanan 1992).

Early fabrics developed in the low grade parts of the terrain (central Leichhardt River Fault Trough) have been explained as the result of thrusting and structural inversion (O'Dea & Lister 1995). Hanging-wall strain gradients, and the occurrence of stratigraphically younger rocks in the hanging wall of faults which show evidence for compressional strain, provide evidence for inversion (see Butler 1989). Furthermore, numerous features such as footwall buttressing and inconsistent offsets across faults strongly resemble those documented in highly inverted basins such as the Kechika Trough of British Columbia (McClay *et al.* 1989) and the margin of the French Alps (de Graciansky *et al.* 1989; Gratier & Vialon 1980; Butler 1989).

In the Western Fold Belt, high level cover rocks, such as the Mount Isa Group, are often characterized by zones of macroscopic superposed folding. In contrast, underlying rocks of Cover Sequence 2 do not record these fold interference patterns. Rocks of Cover Sequence 2 display regional-scale north–south-trending folds, which exhibit remarkable continuity along their axial planes (O'Dea & Lister 1995). Structural inversion removes extensional displacement across pre-rift rocks and produces folds and faults in overlying post-rift rocks. This produces two structural levels characterized by different styles and deformational histories (see Hayward & Graham 1989; Williams *et al.* 1989). The fact that fold interference patterns occur only in Cover Sequence 4 supports the hypothesis that these sediments were deposited on a complex pre-existing rift architecture. When subjected to a regional shortening event, parts of this architecture was reactivated and controlled the geometry of developing structures in overlying cover rocks (O'Dea & Lister 1995).

Fig. 12. Figure illustrating the density and distribution of faults within the Mount Isa Inlier. A large proportion of these faults show evidence for strike-slip displacement and resemble a wrench system. The most regionally significant strike-slip faults are depicted by thicker lines and numbered 1 through 7. 1 is the Mount Gordon Fault; 2 is the Lake Julius Fault; 3 is the Quilalar Fault; 4 is the Mount Isa Fault; 5 is the Mount Remarkable Fault; 6 is the Fountain Range Fault; 6 is the Pilgrim Fault. Many of these faults are not referred to in the text.

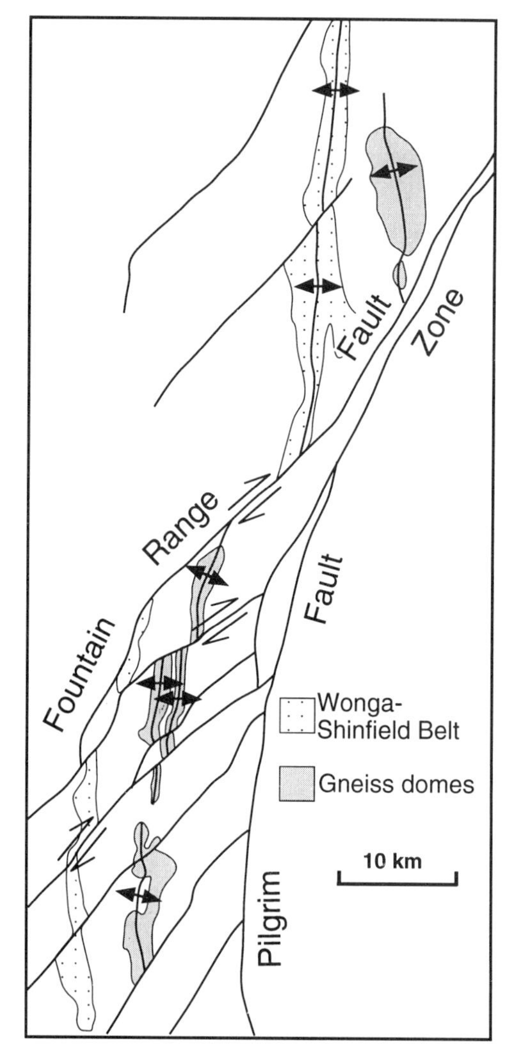

Fig. 13. Line drawing of the Fountain Range Fault and the Pilgrim–Corella Fault system in the Eastern Fold Belt. The Fountain Range Fault and related faults to the south display consistent right lateral strike-separation. These faults link into the Pilgrim–Corella Fault which in turn is inferred to have a synthetic sense of displacement of much larger magnitude.

Wrench faulting in the Mount Isa terrain

One of the most striking features in the map pattern of the Mount Isa terrain is the high density of faults, a large proportion of which show evidence for right-lateral strike-slip displacement (Fig. 12). The interval during which strike-slip faulting was active is loosely constrained, but has previously been considered to reflect the culmination of regional east–west shortening (Lister *et al.* 1986). Most strike-slip faults offset structures formed during regional shortening and peak metamorphism and predate the emplacement of the 1100 Ma Lakeview dolerite. We consider the possibility that the Mount Isa terrain evolved into a wrench system during the late stages of the Isan Orogeny.

A close examination of the fault pattern indicates that the strike-slip faults did not develop in a simple conjugate array. Rather, a characteristic feature is that most small faults feed into larger ones which in turn link into even larger ones, with intersection angles of less than 30° (Fig. 12). Where conjugate strike-slip faults developed, an inferred east-west oriented maximum principal stress axis bisects the obtuse angle between the conjugate faults, which is inconsistent with simple east–west shortening. This high conjugate angle may indicate that these conjugate faults rotated in response to kinematic constraints during continued east–west shortening. Alternatively, the high angle is consistent with the development of Riedel shears in a wrench setting.

In the Eastern Fold Belt, geometrical relations between strike-slip faults are best exemplified by the relation between the Fountain Range Fault and the much larger Pilgrim–Corella Fault Zone (Fig. 13). At its northeastern termination, the Fountain Range Fault merges with the Pilgrim–Corella Fault Zone making an intersection angle of about 30°. The Fountain Range Fault appears to have splayed from a prominent bend in the Pilgrim–Corella Fault Zone, suggesting that the two faults are kinematically related. In addition, detailed mapping shows that parts of this system are defined by numerous fault strands which coalesce, split, splay, and are marked by ribbon-like elongate fault blocks, often comprising quite different rock types.

Based on the strike separation of doubly plunging gneiss domes of the Wonga–Shinfield Belt, the Fountain Range Fault has accommodated approximately 30 km of right lateral displacement (Stewart 1987). The Fountain Range Fault splays from the Pilgrim Fault system (Fig. 13) and is in the correct orientation to be a synthetic fault to the latter. Given that the strike length of the Pilgrim–Corella Fault Zone is an order of magnitude greater than that of the Fountain Range Fault, then the Pilgrim–Corella Fault Zone may be a strike-slip fault across which there has been considerable displacement. Due to the absence of reliable piercing points, efforts to obtain estimates of displacement across this structure have been unsuccessful.

The significance of strike-slip faulting in the Mount Isa terrain is still unclear. It is important to note that the wrenching hypothesis, first proposed by Lister *et al.* (1986), is based primarily on a qualitative assessment of the fault pattern and its similarity with well studied wrench terrains. It is possible that a large number of strike-slip faults formed along reactivated basin-bounding structures across which large lateral displacements were not accommodated. Alternatively, strike-slip faults may very well have been governed by pre-existing structures, although large relative displacements may have occurred.

The Mount Isa terrain has been eroded to upper mid-crustal levels. If wrenching played an important role in the structural evolution of the region, then there are many lessons to be learned regarding the geometries and deformational processes at crustal levels at which both brittle and ductile structures interact. This terrain may provide valuable insight into the interaction between brittle seismogenic processes and associated ductile flow. It may also allow the determination of the factors that drive fluid systems, and provide insight into the links between magmatic processes and fault activity.

Metamorphism and metasomatism during the Isan Orogeny

Although much of the Mount Isa terrain comprises rocks of sub-greenschist to greenschist facies, most pressure-temperature and microstructural work has concentrated on narrow (10–15 km wide), northerly trending strips of upper greenschist to upper amphibolite facies rocks. Detailed studies have been conducted west of the Mount Isa Fault (Connors *et al.* 1992; Rubenach 1992), in the Mary Kathleen Fold Belt (Reinhardt & Rubenach 1988; Oliver *et al.* 1991; Reinhardt 1992*a, b*), and in the Cloncurry–Soldiers Cap Belt (Jaques *et al.* 1982; Loosveld 1988). These authors all noted a broad synchroneity between the metamorphic peak and the deformation peak, with maximum pressure–temperature conditions attained in the range 500–650°C at 3–4 kbar.

The geodynamic significance of widespread low-pressure high-temperature metamorphism is problematic due to the lack of an obvious heat source. Elevation of the geothermal gradient following rifting (e.g. as a result of the advection of magma) seems an unlikely source because of the 100 Ma interval between the last extension event and the metamorphism. Emplacement of mid-crustal melts has also been suggested (Rubenach 1992; Oliver *et al.* 1993); however no plutons of the appropriate age have been recognized despite the exhumation of up to 10 km of crustal section during later deformation, uplift and erosion. Mafic intrusions in the lower crust, inferred from broad gravity anomalies, may have been a heat source for the regional metamorphism.

There are some examples of small intrusions of the appropriate age. For instance, around the Sybella granite, there are abundant pegmatites within the sillimanite and sillimanite–K-feldspar metamorphic zones. These pegmatites have zircon U–Pb ages of 1565 ± 5 Ma and 1532 ± 7 Ma (Connors & Page 1995), overprinting relations indicate that the younger pegmatites intruded during the main phase of deformation. Although pegmatites alone could not have provided enough heat for the metamorphism, they may have been a partial heat source for this event. In addition, these tourmaline-bearing pegmatites may indicate the existence of a granite body at depth from which they were differentiated (Deer *et al.* 1966).

The role of fluids during metamorphism has been an important topic for the last two decades. One of the most notable features of the Mount Isa terrain, by world standards, is the intensity and extent of syn-metamorphic hydrothermal systems. These systems were responsible for the formation of giant ore deposits such as the Mount Isa copper ore body along with hundreds of smaller ore deposits, and a wide variety of areally extensive metasomatic alteration suites. The excellent exposure, coupled with the sheer number of operating and abandoned mines, makes the Mount Isa terrain one of the best places on Earth to study fluid flow processes and the products of fluid–rock interaction.

The most impressive zone of hydrothermally altered rocks is in the evaporitic metacarbonates of the Corella Formation of the Eastern Fold Belt (Oliver & Wall 1987; deJong & Williams 1995; Oliver 1995). High-temperature alteration by this syn-orogenic metasomatism is found in approximately 10% of all exposed Corella Formation and equivalent rocks east of the Kalkadoon–Leichhardt Block. Albitic alteration surrounds calcite veins and pods, with widespread but mainly sub-economic concentrations of chalcopyrite. This alteration is well displayed in hundreds of calcite quarries in the Mary Kathleen and Cloncurry regions.

Hydrothermal alteration is generally localized in shear zones and fracture arrays (Oliver *et al.* 1990). Stable isotopic results suggest the dominant component of the fluid was externally derived (Oliver *et al.* 1993), requiring large distances of fluid advection and an appropriate mechanism of dilatancy-driven fluid flow (Oliver

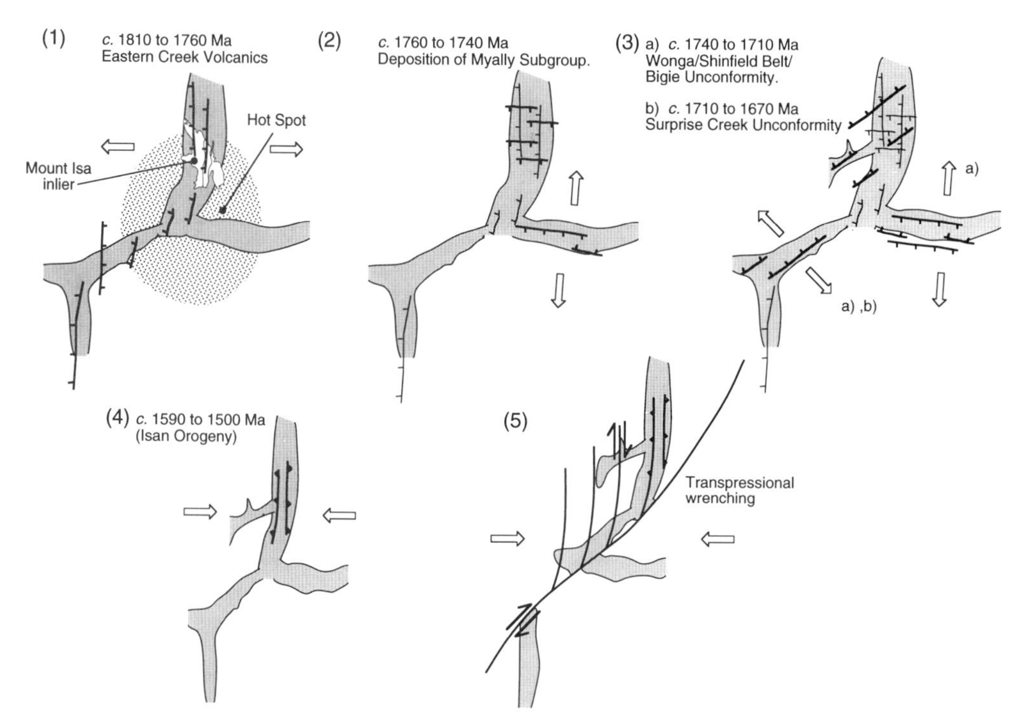

Fig. 14. The Mount Isa area may have originated close to the intersection of three rift systems, localized above a mantle plume. In such an environment each arm of the rift operated in turn, imposing different extensional geometries on the adjacent rift systems. This figure illustrates the sequential development (1 to 3) of rift-related structures within the Mount Isa terrain. During the Isan Orogeny (Stages 4 and 5) these rift arms controlled the orientation of thrusts, folds and finally wrench faults which dismembered the terrain during regional E–W shortening.

et al. 1990). Given the huge extent of the hydrothermal systems at this time, documented in the western, central and eastern parts of the terrain, the ultimate origin for this fluid is problematic. Stable isotope studies have not revealed a large meteoric component of the fluid, nor do they reveal a carbonate-dominated signature. Igneous signatures dominate, perhaps reinforcing the notion that voluminous granitoid plutons were crystallizing during the Isan Orogeny below the level of the exposed upper amphibolite-facies rocks.

At the end of the Isan Orogeny, the Eastern Fold Belt was intruded by voluminous granitic batholiths (e.g. the Williams Batholith *c.* 1500 Ma, Nisbet *et al.* 1983) with an estimated extent of 10 000 km^2. Wrenching, batholithic emplacement and the development of extensive fluid systems created numerous copper-gold deposits. Intrusion of these granites at the end of the Isan Orogeny produced a subsequent metamorphic event in the Eastern Fold Belt which marks the final thermal perturbations of the Isan Orogeny.

Tectonic synthesis

The geological evolution of the Mount Isa terrain, prior to the Isan Orogeny, has much in common with modern intraplate rift systems such as the Baikal rift system. The belts of the Mount Isa terrain formed as part of a large intracratonic extensional basin (Fig. 14), which, on the basis of simple reconstructions, may have been as much as 500 km in total breadth. This wide expanse of attenuated continental crust was probably extensively underplated as a result of mafic magmatism during the extensional process. Currently, the crust, which is in isostatic equilibrium, is approximately 55 km thick (Wellman 1976). Based on the assumption that 5–10 km of material were eroded in the late Proterozoic and Phanerozoic we conclude that the crust was thickened to 60–65 km during the Isan Orogeny.

The tectonics between approximately 1800 Ma and 1600 Ma were characterized by successive episodes of rifting. One explanation for these successive rifting episodes is that the

Mount Isa area was situated close to the intersection of three rift systems, localized above a mantle plume (Fig. 14). In such an environment it is possible to explain both the variation in stretching directions (because each arm of the rift operated in turn, imposing different extensional geometries on the adjacent rift systems), and the voluminous outpourings of igneous rocks (because lithospheric extension resulted in a high degree of partial melting). The voluminous out-pourings of continental tholeiites found throughout the Mount Isa terrain may have localised above an original 'hot spot'.

This geodynamic process helps explain the complex rifting history of the Mount Isa terrain and carries some predictive capabilities. The Western Fold Belt for example, is characterized by variably oriented arrays of normal faults, and rotational tilt blocks. While this belt is essentially a north–south-trending feature, its internal structure has been sequentially overprinted by east–west-trending, and northeast–southwest-trending normal faults. Geometries produced during the subsequent Isan Orogeny were governed by the compression direction relative to the orientation of original rift arms and their internal fault block geometries.

In the central part of the Western Fold Belt, the earliest rift border faults were oriented north–south and east–west (stages 1 & 2 in Fig. 14). Thereafter, an inferred northeast–southwest-trending rift arm may have developed to the south. This phase of rifting was of regional significance, creating the most pronounced extensional features in the Mount Isa terrane. The Leichhardt River Fault Trough was dissected during this phase of extension causing tilting of different fault blocks and the development of the Bigie Unconformity. The deeper parts of this extensional system are now exposed in the gneiss domes of the Wonga–Shinfield Belt of the Eastern Fold Belt. Renewed rifting in the northeast–southwest-oriented rift arm resulted in the development of northeast–southwest-striking normal faults, the development of the Surprise Creek Unconformity and the deposition of Cover Sequence 4.

During its rifting history, the Mount Isa terrain developed an extensional fault block mosaic of considerable complexity. This extensional architecture formed the underlying template upon which compressional structures were superimposed during the Isan Orogeny, resulting in geometrical variability and complexity. The extensional template had a pronounced influence on the orientations of thrusts, folds and wrench faults which dismembered the terrain during regional shortening. On a terrain-wide scale, compressional deformation is favoured by crust that has been thermally weakened by a preceding period of rifting (Ziegler 1982). Being thinned and weak, the crust underlying the Mount Isa terrain deformed preferentially over the more stable cratonic areas to the west. Once the crust was thickened, and the extension-related thermal anomaly waned, brittle processes dominated, and the Mount Isa terrain became a large-scale wrench system.

Conclusions

The Mount Isa terrain of northwest Queensland comprises lower to middle Proterozoic sediments, bimodal volcanic rocks and plutons that record two cycles of intracontinental rifting each followed by crustal shortening and metamorphism. It displays superb examples of rift basin development, basin inversion and strike-slip faulting, and offers an opportunity to examine the superposition of extensional and compressional geometries within a spectacularly exposed stratigraphic sequence.

At around 1900 Ma, a widespread extensional event in Australia led to the extrusion of voluminous bimodal volcanic rocks and deposition of rift-sag sequences (Etheridge *et al.* 1987) which were then deformed and metamorphosed during the Barramundi Orogeny (1870–1850 Ma). This interval of tectonism can been correlated globally with parts of Laurentia and Baltica, suggesting that these terrains may have occupied approximately contiguous positions throughout the Early Proterozoic. The Neoproterozoic supercontinent (Rodinia) comprised, in part, Australia, Antarctica, Laurentia and Baltica (Hoffman 1988, 1991; Moores 1991; Dalziel 1991). Similarities between the Barramundi Orogeny of Australia and the Trans-Hudson Orogeny of Laurentia suggest that parts of this supercontinent (i.e. northern Australia and Laurentia) were fused by as early as *c.* 2000 Ma. Although relatively few Barramundi age rocks are presently exposed, they form the crystalline basement of the Mount Isa terrain, and underlie much of the northern Australian craton (Plumb *et al.* 1980).

Following the Barramundi Orogeny the Mount Isa terrain, along with much of northern Australia, underwent a complex history of rifting and deposition between 1800 and 1600 Ma. Igneous and sedimentary rocks of this period were formed or deposited during at least four episodes of superimposed intracontinental rifting and associated thermal subsidence. Three major epochs of extensional tectonism are manifested by the development of the Leichhardt Rift, the Bigie Unconformity/ Wonga-Shinfield Belt, and deposition after the development of the Surprise Creek Unconformity. These rifting events may be explained in the context of a complex intracontinental rift system. Further extension and deposition were interrupted prior to the formation of oceanic crust by the compressional Isan Orogeny (1590–1500).

The main phase of the Isan Orogeny involved regional east–west shortening synchronous with low-pressure metamorphism reflecting a cycle of early heating and crustal thickening. Deformational features produced during the Isan Orogeny were superimposed on inherited extensional structures related to the earlier rifting history. Lithospheric thinning and asthenospheric rise at the onset of metamorphism may have caused an elevated geothermal gradient leading to the regional low-pressure high-temperature metamorphism. During the waning stages of the Isan Orogeny, an extensive network of strike-slip faults developed reflecting a change in deformational style related to crustal thickening and cooling.

Geometries produced during the Isan Orogeny cannot be directly linked to any master structure such as a plate suture or a through-going detachment surface. High-pressure metamorphic belts, ophiolites, magmatic arcs and intermediate volcanic rocks did not develop, suggesting that crustal shortening was not in response to an adjacent subduction zone. There is also little evidence of foreland basin development. The locus of structural development of the Mount Isa terrain seems to have been largely controlled by the reactivation of pre-existing structures during mid-Proterozoic orogenesis.

The Mount Isa region provides some of the world's best-exposed examples of structurally controlled fluid flow. Three temporally distinct episodes of metasomatism and associated mineralization are related to deformation as: pre-orogenic sediment-hosted Pb–Zn–Ag deposits; syn-orogenic alteration-breccia hosted Cu-mineralization; and late-post orogenic iron stone/black slate related Cu–Au mineralization.

This report is an outcome of collaborative research within the Australian Geodynamics Cooperative Research Centre. Critical reviews by P. Hoffman and C. Passchier, and comments by M. Ford and J.-P. Burg improved the manuscript. The research assistance of L. Richards is gratefully acknowledged. This publication is released with the permission of the director of the AGCRC.

References

Bell, R. T. & Jefferson, C. W. 1987. An hypothesis for an Australian-Canadian connection in the late Proterozoic and the birth of the Pacific Ocean. *Pacific Rim Congress '87*, 39–50.

Bell, T. H. 1983. Thrusting and duplex formation at Mount Isa, Queensland, Australia. *Nature,* **304**, 493–497.

Blake, D. H. 1987. Geology of the Mount Isa Inlier and environs, Queensland and Northern Territory. *Bureau of Mineral Resources Bulletin,* **225**, 83pp.

—— 1991*a*. *Geology of the Lake Mary Kathleen Zone* (1:50 000 scale map). Bureau of Mineral Resources, Canberra.

—— 1991*b*. *Geology of the Rifle Creek region* (1:50 000 scale map). Bureau of Mineral Resources, Canberra.

—— & Page, R. W. 1988. Early Proterozoic migmatic basement in the Kalkadoon-Leichhardt Belt of the Mount Isa Inlier, northwest Queensland. *Bureau of Mineral Resources Journal of Australian Geology and Geophysics,* **10**, 323–328.

—— & Stewart, A. J. 1992. Stratigraphic and tectonic framework, Mount Isa Inlier. *In*: Stewart, A. J. & Blake, D. H. (eds) *Detailed Studies of the Mount Isa Inlier*. Australian Geological Survey Organisation Bulletin, **243**, 1–11.

——, Etheridge, M. A., Page, R. W., Stewart, A. J., Williams, P. R. & Wyborn, L. A. I. 1990. Mount Isa inlier- Regional geology and mineralisation. *In*: Hughes, F. E. (ed.) *Geology of the mineral deposits of Australia and Papua New Guinea.* The Australasian Institute of Mining and Metallurgy Monographs, **14**, 915–925.

Borg, S. G. & DePaolo, D. J. 1994. Laurentia, Australia, and Antarctica as a Late Proterozoic supercontinent: Constraints from isotopic mapping. *Geology,* **22**, 307–310.

Burke, K. & Kidd, W. S. F. 1980. Volcanism on Earth through time. *In*: Strangeway, D. W. (ed.) *The Continental crust and its mineral deposits*. Geological Association of Canada, Special Papers, **20**, 503–522.

Butler, R. W. H. 1989. The influence of pre-existing basin structure on thrust system evolution in the Western Alps. *In*: Cooper, M. A. & Williams, G. D. (eds) *Inversion Tectonics*. Geological Society of London, Special Publications, **44**, 105–122.

Carter, E. K., Brooks, J. H. & Walker, K. R. 1961. *The Precambrian mineral belt of north-western Queensland*. Bureau of Mineral Resources Bulletin, **51**.

Coney, P. J., Edwards, A., Hine, R., Morrison, F. & Windrim, D. 1990. The regional tectonics of the Tasman orogenic system, eastern Australia. *Journal of Structural Geology,* **12**, 519–543.

Connors, K. A. & Page, R. W. 1995. Relationships between magmatism, metamorphism and deformation in the western Mount Isa Inlier, Australia. *Precambrian Research,* **71**, 131–153.

——, Proffett, J. M., Lister, G. S., Scott, R.,

OLIVER, N. H. S. & YOUNG, D. J. 1992. Geology of the Mount Novit Ranges, southwest of Mount Isa Mine. *In*: STEWART, A. J. & BLAKE, D. H. (eds) *Detailed Studies of the Mount Isa Inlier*. Australian Geological Survey Organisation Bulletin, **243**, 137–160.

DALZIEL, I. W. D. 1991. Pacific margins of Laurentia and East Antarctica-Australia as a conjecture rift pair: Evidence and implications for an Eocambrian supercontinent. *Geology*, **19**, 598–601.

—— 1992. Antarctica: A tale of two supercontinents? *Annual review of Earth Sciences*, **20**, 501–526.

DE GRACIANSKY, P. C., DARDEAU, G., LEMOINE, M. & TRICART, P. 1989. The inverted margin of the French Alps and foreland basin inversion. *In*: COOPER, M. A. & WILLIAMS, G. D. (eds) *Inversion Tectonics*. Geological Society, London, Special Publications, **44**, 7–104.

DE JONG, G. & WILLIAMS, P. J. 1995. A giant metasomatic system formed during exhumation of mid crustal Proterozoic rocks in the vicinity of the Cloncurry Fault, NW Queensland. *Australian Journal of Earth Sciences*, **42**, 281–290.

DEER, W. A., HOWIE, R. A. & ZUSSMAN, J. 1966. *An Introduction to the Rock Forming Minerals*. Longmans, London.

DERRICK, G. M. 1982. A Proterozoic rift zone at Mount Isa, Queensland, and implications for mineralisation. *Bureau of Mineral Resources Journal of Australian Geology and Geophysics*, **7**, 81–92.

——, WILSON, I. H. & SWEET, I. P. 1980. The Quilalar and Surprise Creek Formations- new Proterozoic units from the Mount Isa inlier: their regional sedimentology and application to regional correlation. *Bureau of Mineral Resources Journal of Australian Geology and Geophysics*, **5**, 215–223.

DUNNET, D. 1976. Mt Isa; reconstruction of a faulted ore body. *Philosophical Transactions of the Royal Society of London*, **A283**, 333–344.

ERIKSSON, K. A., SIMPSON, E. L. & JACKSON, M. J. 1993. Stratigraphical evolution of a Proterozoic syn-rift to post-rift basin: constraints on nature of lithospheric extension in the Mount Isa Inlier, Australia. *In*: FROSTICK, L. E. & STEEL, R. J. (eds) *Tectonic controls and signatures in sedimentary successions*. Special Publications of the International Association of Sedimentologists, **20**, 203–221.

ETHERIDGE, M. A. & WALL, V. 1994. Tectonic and structural evolution of the Australian Proterozoic. *Geological Society of Australia Abstracts*, **37**, 102.

——, RUTLAND, R. W. R. & WYBORN, L. A. I. 1987. Orogenesis and tectonic process in the Early to Middle Proterozoic of northern Australia. *In*: KRÖNER, A. (ed.) *Precambrian Lithospheric Evolution*. American Geophysical Union Geodynamic Series, **17**, 131–147.

FREEMAN, M. J., SHERGOLD, J. H., MORRIS, D. G. & WALTER, M. R. 1990. Late Proterozoic and Palaeozoic basins of Central and Northern Australia – regional geology and mineralisation. *In*: HUGHES, F. E. (ed.) *Geology of the Mineral Deposits of Australia and Papua New Guinea*. Australasian Institute of Mining and Metallurgy Monographs, **14**, 1125–1133.

GAAL, G. & GORBATSHEV, R. 1987. An outline of the Precambrian evolution of the Baltic Shield. *Precambrian Research*, **35**, 15–52.

GLIKSON, A. Y., DERRICK, G. M., WILSON, I. H. & HILL, R. M. 1976. Tectonic evolution and crustal setting of the middle Proterozoic Leichhardt River fault trough, Mount Isa region, northwestern Queensland. *Bureau of Mineral Resources Journal of Australian Geology and Geophysics*, **1**, 115–129.

GOWER, C. F. 1985. Correlations between the Grenville Province and the Sveconorwegian orogenic belt – implications for Proterozoic evolution of the southern margins of the Canadian and Baltic Shields. *In*: TOBI, A. C. & TOURET, J. L. R. (eds) *The Deep Proterozoic Crust in the North Atlantic Provinces*. NATO Advanced Study Institute Series C: Mathematical and Physical Sciences, **158**, 247–257.

GRATIER, J. & VIALON, P. 1980. Deformation pattern in a heterogeneous material: Folded and cleaved sedimentary cover immediately overlying a crystalline basement (Oisans, French Alps). *Tectonophysics*, **65**, 151–180.

HAMBLIN, W. K. 1965. Origin of 'reverse drag' on the downthrown side of normal faults. *Geological Society of America Bulletin*, **76**, 1145–1164.

HAYWARD, A. B. & GRAHAM, R. H. 1989. Some geometrical characteristics of inversion. *In*: COOPER, M. A. & WILLIAMS, G. D. (eds) *Inversion Tectonics*. Geological Society of London, Special Publications, **44**, 17–39.

HOFFMAN, P. F. 1988. United Plates of America, the birth of a craton: Early Proterozoic assembly and growth of Laurentia. *Annual review of Earth and Planetary Sciences*, **16**, 543–603.

—— 1989. Speculations on Laurentia's first gigayear (2.0 to 1.0 Ga). *Geology*, **17**, 135–138.

—— 1991. Did the breakout of Laurentia turn Gondwanaland inside-out? *Science*, **252**, 1409–1412.

—— & BOWRING, S. A. 1984. Short lived 1.9 Ga continental margin and its destruction, Wopmay Orogen, Northwest Canada. *Geology*, **12**, 68–72.

HOLCOMBE, R. J., PEARSON, P. J. & OLIVER, N. H. S. 1991*a*. Geometry of a middle Proterozoic extensional detachment surface. *Tectonophysics*, **191**, 255–274.

——, —— & SLIWA, R. 1991*b*. *Geology of the Mary Kathleen Area* (1:25 000 scale map). Bureau of Mineral Resources, Canberra.

HOUSEMAN, G. A. & HEGARTY, K. A. 1987. Did rifting on Australia's southern margin result from tectonic uplift? *Tectonics*, **6**, 515–527.

HUANG, W. 1994. *Structural and stratigraphic relations on the western flank of the Mitakoodi Culmination: a case study in the Roos Mine area south of the Corella Dam, Eastern Mount Isa Inlier, NW Queensland*. Australian Crustal Research Centre Technical Publications, **21**.

IDNURM, M. & GIDDINGS, J. W. 1995. Paleoproterozoic-Neoproterozoic North America-Australia link: new evidence from paleomagnetism. *Geology,* **23**, 149–152.

JACKSON, M. J., SIMPSON, E. L. & ERIKSSON, K. A. 1990. Facies and sequence stratigraphic analysis in an intracratonic, thermal-relaxation basin; the early Proterozoic, lower Quilalar Formation and Ballara Quartzite, Mount Isa Inlier, Australia. *Sedimentology,* **37**, 1053–1078.

JAQUES, A. L., BLAKE, D. H. & DONCHAK, P. J. T. 1982. Regional metamorphism in the Selwyn Range area, Northwest Queensland. *Bureau of Mineral Resources Journal of Australian Geology and Geophysics,* **7**, 181–196.

KEEN, C. E. 1988. Contrasting styles of lithospheric extension determined from crustal studies across rift basins, eastern Canada. *In*: PRICE, R. A. (ed.) *Origin and Evolution of Sedimentary Basins and Their Energy and Mineral Resources.* American Geophysical Union Geophysical Monographs, **48**, 37–42.

KLEIN, E. M., LANGMUIR, C. H., ZINDLER, A., STRANDIGEL, H. & HAMELIN, B. 1988. Isotope evidence of a mantle convection boundary at the Australian-Antarctic discordance. *Nature,* **333**, 623–629.

LE MESSURIER, P., WILLIAMS, B. T. & BLAKE, D. H. 1990. Tennant Creek Inlier- regional geology and mineralisation. *In*: HUGHES, F. E. (ed.) *Geology of the Mineral Deposits of Australia and Papua New Guinea.* The Australasian Institute of Mining and Metallurgy Monographs, **14**, 829–838.

LISTER, G. S. & ETHERIDGE, M. A. 1989. Towards a general model; detachment models for uplift and volcanism in the eastern highlands, their application to the origin of passive margin mountains. *In*: JOHNSON, R. W., KNUTSON, J. & TAYLOR, S. R. (eds) *Intraplate volcanism in Eastern Australia and New Zealand.* Australian Bureau of Mineral Resources, Division of Petrology and Geochemistry, 297–313.

——, THOMAS, A. & DUNN, J. 1986. Tectonic processes in the Mount Isa Inlier; the significance of transpressional strike-slip faulting. *Geological Society of Australia Abstracts,* **15**, 127–128.

LOOSVELD, R. J. H. 1988. *Structure and tectonothermal history of the eastern Mount Isa Inlier, Australia.* PhD Thesis, Australian National University, Canberra.

—— 1991. Structural Geology of the Central Soldiers Cap Group (1:100 000 scale map). Bureau of Mineral Resources, Canberra.

MARJORIBANKS, R. W, RUTLAND, R. W. R., GLEN, R. A. & LAING, W. P. 1980. The structure and tectonic evolution of the Broken Hill region, Australia. *Precambrian Research,* **13**, 209–240.

MCCLAY, K. R. & BUCHANAN, P. G. 1992. Thrust faults in inverted extensional basins. *In*: MCCLAY, K. R. (ed.) *Thrust Tectonics.* Chapman and Hall, London, 93–104.

——, INSLEY, M. W. & ANDERTON, R. 1989. Inversion of the Kechika Trough, Northeastern British Columbia, Canada. *In*: COOPER, M. A. & WILLIAMS, G. D. (eds) *Inversion Tectonics.* Geological Society of London, Special Publications, **44**, 235–257.

MOORES, E. M. 1991. Southwest U.S.-East Antarctic (SWEAT) connection: A hypothesis. *Geology,* **19**, 425–428.

MORSE, M. P., MILLIGAN, P. R. & RAJAGOPALAN, S. 1992. *Gravity Anomaly Map of Australia*, 1:5 000 000, AGSO.

MURRAY, C. G., SCHEIBNER, E. & WALKER, R. N. 1989. Regional geological interpretation of a digital coloured residual Bouguer gravity image of eastern Australia with a wavelength cut-off of 250 km. *Australian Journal of Earth Sciences,* **36**, 423–449.

MYERS, J. S., SHAW, R. D. & TYLER, I. M. 1994. Proterozoic Tectonic evolution of Australia. *Geological Society of Australia Abstracts,* **37**, 312.

NEEDHAM, R. S. & DE ROSS, G. J. 1990. Pine Creek Inlier – Regional geology and mineralisation. *In*: HUGHES, F. E. (ed.) *Geology of the Mineral Deposits of Australia and Papua New Guinea.* The Australasian Institute of Mining and Metallurgy Monographs, **14**, 727–737.

NEUDERT, M. K. & RUSSELL, R. E. 1981. Shallow water and hypersaline features from the Middle Proterozoic Mount Isa Sequence. *Nature,* **293**, 284–286.

NIJMAN, W., MIJNLIEFF, H. F. & SCHALWIJK, G. 1992*a*. The Hero Fan delta (Lower Mount Isa Group) and its structural control: deformation in the Hero/Western Fault Zone and Paroo Range compared, Proterozoic, Mount Isa Inlier, Queensland, Australia. *In*: STEWART, A. J. & BLAKE, D. H. (eds) *Detailed Studies of the Mount Isa Inlier.* Australian Geological Survey Organisation Bulletin, **243**, 75–110.

——, VAN LOCHEM, J. H., SPLIETHOFF, H. & FEIJTH, J. 1992*b*. Deformation model and sedimentation patterns of the Proterozoic of the Paroo Range, Mount Isa Inlier, Queensland, Australia. *In*: STEWART, A. J. & BLAKE, D. H. (eds) *Detailed Studies of the Mount Isa Inlier.* Australian Geological Survey Organisation Bulletin, **243**, 29–73.

NISBET, B. W., DEVLIN, S. P. & JOYCE, P. J. 1983. Geology and suggested genesis of cobalt-tungsten mineralization at Mount Cobalt, northwestern Queensland. *Proceedings of the Australasian Institute of Mining and Metallurgy,* **287**, 9–17.

O'DEA, M. G. & LISTER, G. S. 1995. The role of ductility contrast and basement architecture in the structural evolution of the Crystal Creek block, Mount Isa Inlier, Australia. *Journal of Structural Geology,* **17**, 949–960.

OLIVER, N. H. S. 1995. The hydrothermal history of the Mary Kathleen Fold Belt, Mount Isa Block, Queensland, Australia. *Australian Journal of Earth Sciences*, **42**, 267–279.

—— & WALL, V. J. 1987. Metamorphic plumbing system in Proterozoic calc-silicates, Queensland, Australia. *Geology,* **15**, 793–796.

——, VALENTA, R. K. & WALL, V. J. 1990. The effect of heterogeneous stress and strain on metamorphic fluid flow, Mary Kathleen, Australia,

and a model for large scale fluid circulation. *Journal of Metamorphic Geology*, **8**, 311–331.
——, HOLCOMBE, R. J., HILL, E. J. & PEARSON, P. J. 1991. Tectono-metamorphic evolution of the Mary Kathleen Fold Belt, northwest Queensland: a reflection of mantle plume processes? *Australian Journal of Earth Sciences*, **38**, 425–455.
——, CARTWRIGHT, I., WALL, V. J. & GOLDING, S. D. 1993. The stable isotope signature of kilometrescale fracture-dominated metamorphic fluid pathways, Mary Kathleen, Australia. *Journal of Metamorphic Geology*, **11**, 705–720.
PAGE, R. W. 1983*a*. Chronology of Magmatism, Skarn formation, and Uranium mineralisation, Mary Kathleen, Queensland, Australia. *Economic Geology*, **78**, 838–853.
—— 1983*b*. Timing of superposed volcanism in the Proterozoic Mount Isa Inlier, Australia. *Precambrian Research*, **21**, 223–245.
—— 1988. Geochronology of early to middle Proterozoic fold belts in northern Australia: a review. *Precambrian Research*, **40/41**, 1–19.
—— 1993. Geochronological results form the Eastern Fold Belt, Mount Isa Inlier. *Bureau of Mineral Resources Research Newsletter*, **19**, 4–5.
—— & BELL, T. H. 1986. Isotopic and structural responses of granite to successive deformation and metamorphism. *Journal of Geology*, **94**, 365–379.
—— & SWEET, I. P. Geochronology of basin phases in the Mount Isla Inlier, and correlation with McArthur Basin. *Australian Journal of Earth Sciences*, in press.
—— & WILLIAMS, I. S. 1988. Age of the Barramundi Orogeny in northern Australia by means of ion microprobe and conventional U-Pb zircon studies. *Precambrian Research*, **40/41**, 21–36.
——, SUN, SHEN-SU & CARR, G. 1994. Proterozoic sediment-hosted lead-zinc-silver deposits in northern Australia- U-Pb zircon and Pb isotopic studies. *Geological Society of Australia Abstracts*, **37**, 334–335.
PARKER, A. J. 1990*a*. Gawler Craton and Stewart shelf - regional geology and mineralisation. *In*: HUGHES, F. E. (ed.) *Geology of the Mineral Deposits of Australia and Papua New Guinea*. The Australasian Institute of Mining and Metallurgy Monographs, **14**, 999–1008.
—— 1990*b*. Precambrian provinces of South Australia - tectonic setting. *In*: HUGHES, F. E. (ed.) *Geology of the Mineral Deposits of Australia and Papua New Guinea*. The Australasian Institute of Mining and Metallurgy Monographs, **14**, 985–990.
PASSCHIER, C. W. 1986. Evidence for early extensional tectonics in the Proterozoic Mount Isa Inlier, Australia. *Geology*, **14**, 1008–1011.
—— & WILLIAMS, P. R. 1989. Proterozoic extensional deformation in the Mount Isa Inlier, Queensland, Australia. *Geological Magazine*, **126**, 43–53.
PEARSON, P. J., HOLCOMBE, R. J. & PAGE, R. W. 1992. Synkinematic emplacement of the Middle Proterozoic Wonga Batholith into a mid-crustal extensional shear zone, Mount Isa Inlier, Queensland. *In*: STEWART, A. J. & BLAKE, D. H. (eds) *Detailed studies of the Mount Isa Inlier*. Australian Geological Survey Organisation Bulletin, **243**, 289–328.
PLUMB, K. A. 1990. Subdivision and correlation of the Australian Precambrian. *In*: HUGHES, F. E. (ed.) *Geology of the Mineral Deposits of Australia and Papua New Guinea*. The Australasian Institute of Mining and Metallurgy Monographs, **14**, 27–32.
—— & WELLMAN, P. 1987. McArthur Basin, Northern Territory: mapping of deep troughs using gravity and magnetic anomalies. *Bureau of Mineral Resources Journal of Australian Geology and Geophysics*, **10**, 243–251.
——, DERRICK, G. M. & WILSON, I. H. 1980. Precambrian geology of the McArthur River-Mount Isa region, northern Australia. *In*: HENDERSON, R. A. & STEPHENSON, P. J. (eds) *The Geology and Geophysics of Northeastern Australia*. Geological Society of Australia, Queensland Division, 71–88.
——, DERRICK, G. M., NEEDHAM, R. S. & SHAW, R. D. 1981. The Proterozoic of Northern Australia. *In*: HUNTER, D. R. (ed.) *Developments in Precambrian Geology* **2**, Elsevier, Amsterdam, 205–307.
POWELL, C. M. & LI, Z. X. 1994. Neoproterozoic and Palaeozoic tectonic framework of Australia: from Rodinia to Pangea. *Geological Society of Australia Abstracts*, **37**, 355.
REINHARDT, J. 1991. *Geology of the Rosebud Syncline* (1:25 000 scale map). Bureau of Mineral Resources, Canberra.
—— 1992*a*. The Corella Formation of the Rosebud Syncline (central Mount Isa Inlier): deposition, deformation and metamorphism. *In*: STEWART, A. J. & BLAKE, D. H. (eds) *Detailed Studies of the Mount Isa Inlier*. Australian Geological Survey Organisation Bulletin, **243**, 229–255.
—— 1992*b*. Low-pressure, high-temperature metamorphism in a compressional tectonic setting; the Mary Kathleen fold belt, northeastern Australia. *Geological Magazine*, **129**, 41–57.
—— & RUBENACH, M. J. 1988. Temperature-time relationships across metamorphic zones: evidence from porphyroblast-matrix relationships in progressively deformed metapelites. *Tectonophysics*, **158**, 141–161.
RODGERS, J. 1990. Fold and thrust belts in sedimentary rocks. Part 1: typical examples. *American Journal of Science*, **290**, 321–359.
RUBENACH, M. J. 1992. Proterozoic low-pressure/high-temperature metamorphism and an anticlockwise P-T-t path for the Hazeldene area, Mount Isa Inlier, Queensland, Australia. *Journal of Metamorphic Geology*, **10**, 333–346.
SMITH, W. D. 1969. Penecontemporaneous faulting and its likely significance in relation to Mount Isa ore deposition. *Special Publications of the Geological Society of Australia*, **2**, 225–235.
STEWART, A. J. 1987. Evidence for early extensional tectonics in the Proterozoic Mount Isa inlier, Australia – Comment. *Geology*, **15**, 976–977.
—— 1991. *Geology of the Ballara-Mount Frosty*

Region (1:25 000 scale map). Bureau of Mineral Resources, Canberra.

—— 1992. Geology of the Horse's Head structure, north of Mount Isa, and its bearing on the 'roof thrust and imbricate stack' hypothesis, Mount Isa Inlier, Queensland. *In*: STEWART, A. J. & BLAKE, D. H. (eds) *Detailed Studies of the Mount Isa Inlier*. Australian Geological Survey Organisation Bulletin, **243**, 111–124.

—— & BLAKE, D. H. (eds) 1992. *Detailed Studies of the Mount Isa Inlier*. Australian Geological Survey Organisation Bulletin, **243**.

WELLMAN, P. 1976. Regional variation of gravity, and isostatic equilibrium of the Australian crust. *Bureau of Mineral Resources Journal of Australian Geology and Geophysics,* **1**, 297–302.

—— 1992*a*. Structure of the Mount Isa region inferred from gravity and magnetic anomalies. *Exploration Geophysics,* **23**, 417–422.

—— 1992*b*. Structure of the Mount Isa region inferred from gravity and magnetic anomalies. *In*: STEWART, A. J. & BLAKE, D. H. (eds) *Detailed Studies of the Mount Isa Inlier*. Australian Geological Survey Organisation Bulletin, **243**, 15–27.

WILLIAMS, G. D., POWELL, C. M. & COOPER, M. A. 1989. Geometry and kinematics of inversion tectonics. *In*: COOPER, M. A. & WILLIAMS, G. D. (eds) *Inversion Tectonics*. Geological Society of London, Special Publications, **44**, 3–15.

WYBORN, L. A. I. 1988. Petrology, geochemistry and origin of a major Australian 1880–1840 Ma felsic volcano-plutonic suite: A model for intracontinental felsic magma generation. *Precambrian Research,* **40/41**, 37–60.

——, PAGE, R. W. & MCCULLOCH, M. T. 1988. Petrology, geochronology and isotope geochemistry of the post-1820 Ma granites of the Mount Isa Inlier; mechanisms for the generation of Proterozoic anorogenic granites. *Precambrian Research,* **40/41**, 509–541.

ZIEGLER, P. A. 1982. Faulting and graben formation in western and central Europe. *Philosophical Transactions of the Royal Society of London,* **A305**, 113–143.

Contraction, extension and timing in the South Norwegian Caledonides: the Sognefjord transect

A. G. MILNES[1], O. P. WENNBERG[1], Ø. SKÅR[1] & A. G. KOESTLER[2]

[1]*Geological Institute, University of Bergen, Allégaten 41, N–5007 Bergen, Norway*

[2]*Geo-Recon AS., Munkedamsveien 59, N–0270 Oslo, Norway*

Abstract: The Sognefjord transect through the lower to middle Palaeozoic Caledonian mountain belt in southern Norway provides one of the best and most completely documented examples of late collisional tectonics in an Alpine-type orogen. It exposes a 250 km long cross-section from the cratonic foreland, in the east, through the heavily deformed continental margin (Baltica), to the remains of the Caledonian ocean (Iapetus) at the Norwegian west coast. Exceptionally detailed and complete structural data are available along the whole transect, together with good stratigraphic, radiometric, petrological, and geophysical control. In this synthesis, the structural data are analysed, in terms of the kinematics and relative age of the different deformation phases, and correlated along the whole transect. The analysis is then used, in conjunction with the other data, to carry out a retrodeformation, reconstructing the crustal geometry at different stages backward in time. The earliest of the present reconstructions (*c.* 410 Ma) marks the time of formation of the well-known West Norwegian eclogites, in an over-deepened root of Baltica which had developed in the ductile lower crust as a response to extreme crustal shortening. The brittle upper crust took up the shortening by the SE movement of a rigid sheet of Precambrian basement (Jotun complex) above the low-angle Jotunheimen contractional detachment, across a rigid wedge of the Baltic Shield. During the final stages of contraction (*c.* 410–395 Ma), the upper crust acted as an orogenic lid, against which the root 'collapsed' upwards by sub-vertical shortening and lateral E–W extension. During this process of inverted gravity spreading, the eclogites were carried upwards from 60–70 km to 40 km (exhumation phase 1, rate 2–3 mm a^{-1}) and retrograded within their deforming gneissic matrix. At the end of this phase, the strain field in the upper crust changed from contraction to extension, concomitant with a broad up-doming (base Devonian unconformity) which caused a further 10 km exhumation by 385 Ma (exhumation phase 2, 1 mm a^{-1}). This was followed by the main phase of crustal extension with the development of low-angle normal top-to-W or NW fault and shear zones, of which the Nordfjord–Sogn detachment was the most important (50 km of normal displacement). Exhumation in this phase took place by rapid uplift and erosion of the footwall of the detachment, causing the currently exposed eclogites in outer Sognefjord to rise the remaining 30 km (exhumation phase 3, 1.5 mm a^{-1}), to become juxtaposed against Devonian conglomerates on the hanging wall. The reconstructions confirm the general picture of eclogite exhumation in western Norway, and fill out some of the details. However, they do not support the idea that the process was due to extensional orogenic collapse caused by advective or convective lithospheric thinning. Although gravity played a significant role at various stages in the process, the main phase of crustal extension seems to have been mainly related to changes in Devonian plate motion.

The lower Paleozoic, Caledonian mountain belt in southern Norway is one of the best described collisional orogens. This is because of its good accessibility, excellent and deep exposure, long history of geological mapping and research, and, on a global basis, spectacular tectonic and petrologic features. One of the world's longest and deepest fjords, Sognefjord, together with its head valleys and mountains (Jotunheimen), provides a continuous, 250 km long transect through the whole belt, from the cratonic foreland (Baltica), in the east, to the remains of the Caledonian ocean (Iapetus), along the Norwegian west coast and beneath the rifted, mainly Mesozoic sediments of the northern North Sea. During the past 20 years, various research groups have studied different parts of this particular transect in great detail, particularly with respect to their structural amd metamorphic histories. Since coverage is now

From Burg, J.-P. & Ford, M. (eds), 1997, *Orogeny Through Time*, Geological Society Special Publication No. 121, pp. 123–148.

Table 1. *Terminology used in the text for designating different phases of orogenesis, with an overview of typical processes, together with Caledonide and Alpine nomenclature*

	Term	Processes	Scandinavian Caledonides	Central Alps
	Post-orogenic extension	Reorganization of plate motion, start of break-up of orogen due to external forces	Post-Caledonian extension	
Orogenic processes	Late orogenic extension	Gravitational instability, collapse of lithospheric and/or crustal roots and topographic highs	Late Caledonian extension	Late orogenic extension (partly due to arcuate shape)
Orogenic processes	Late collisional contraction	Suture zone inactive (deformed as passive marker), intracontinental deformation, crustal shortening/thickening	Late Scandian (Lower–Middle Allochthon)	Neo-Alpine (Helvetic nappes, Insubric line, etc.)
Orogenic processes	Early collisional contraction	Destruction of ocean, narrow zone of intercontinental deformation, formation of suture zone, major obduction	Early Scandian (suture = Upper–?Uppermost Allochthon)	Meso-Alpine (suture = Pennine zone)
Orogenic processes	Pre-collisional contraction	Closing of ocean, various subduction zones, arcs, local extension (back arc), local collisions/obductions	Early Caledonian including 'Finnmarkian'	Eo-Alpine
TIME ↑	Pre-orogenic extension	Rifting, break-up, spreading	Sparagmite basins, Iapetus ocean	Alpine Tethys, Piedmont ocean

practically complete, we have recently been concerned with the correlation and integration of the great diversity of maps, theses and papers along the whole cross-section. This has led to the compilation presented here on Fig. 1, and also to the realization that the resulting geometry and movement picture places tight constraints on the reconstruction and interpretation of the later stages of orogeny, i.e. the late collisional contractional phase and the late/post-orogenic phases of extension (for terminology, see Table 1).

Numerous models for the development of the South Norwegian Caledonides are to be found in recent publications (e.g. Séranne & Seguret 1987; Andersen & Jamtveit 1990; Andersen *et al.* 1991; Fossen 1992, 1993*b*; Chauvet *et al.* 1992; Andersen 1993; Dewey *et al.* 1993; Koenemann 1993; Chauvet & Séranne 1994; Wilks & Cuthbert 1994; see also Rey *et al.* this volume), but few take into account the depth of knowledge which is available across the whole belt, particularly along the Sognefjord transect. This paper is intended as a compilation and correlation of basic structural data which, it is hoped, will contribute to assessing the applicability of these models. They have often focused on the formation and exhumation of the well-known eclogites in western Norway. In this context, it is important to note that two different types of eclogite exist in the Caledonides, as in some other collision belts (Platt 1993). The eclogites occurring in the Caledonian allochthon are pre-collisional and give older formation ages (Ordovician, i.e. early Caledonian, see Table 1); they are related to subduction activity during the closing of the Iapetus ocean. This activity included local collisional events, but is essentially intra-oceanic (Mørk *et al.* 1988; Dallmeyer *et al.* 1991). The eclogites occurring in the parautochthonous Western Gneiss Complex (Griffin *et al.* 1985) formed during the collisional event and are younger (mid-Silurian to early Devonian, i.e. Scandian, see Table 1). They formed at a position which was at least 100 km continent-ward from the Baltica margin, where the Scandian collision started (see below). It is the formation and exhumation of the younger, Scandian eclogites which can be reconstructed in some detail on the basis of the present work.

In the following, we first give an overview of the tectono-stratigraphy of the South Norwegian Caledonides, before concentrating on the Sognefjord transect and the structural histories which have been documented for the different segments. Then, we attempt to correlate the different, structurally defined, deformation phases across the whole belt, taking into account geophysical, stratigraphic and radiometric data, and established upper crustal kinematics. Finally, we 'retrodeform' the whole transect, i.e. reconstruct the different stages in orogenic development by successively removing the deformation in a backward sequence (see Suppe 1980, 1985). This can be carried out successfully along the Sognefjord transect back to the time of Scandian eclogite formation. Finally, based on these results, we assess current models for eclogite exhumation and orogenic collapse in this area of spectacular late/post-orogenic extensional structures.

Tectono-stratigraphic units

The Sognefjord transect crosses a segment of the Caledonides with a relatively clear tectonostratigraphy (Fig. 2). It consists of a sequence of regionally flat-lying tectono-stratigraphic units which, from bottom to top, have been designated Autochthon/Parautochthon, Lower Allochthon, Middle Allochthon and Upper Allochthon (Roberts & Gee 1985). These were juxtaposed by overthrusting during the final contractional episode of the Caledonian mountain building, the Scandian orogeny (mid-Silurian to early Devonian, Table 1). The tectono-stratigraphic units are intersected and displaced by two major low-angle shear and fault zones, which represent a late to post-orogenic phase of crustal extension (early to late Devonian). The more easterly of these is known as the Lœrdal-Gjende fault where it crosses the Sognefjord transect (Battey 1965; Heim *et al.* 1977; Milnes & Koestler 1985; LGF on Fig. 2), and the Hardangerfjord shear zone further to the southwest (HFSZ on Fig. 2, see Fossen 1992). It marks the southeastern border of what has traditionally been referred to as the 'Faltungsgraben' (Goldschmidt 1912), a regional NE–SW-trending depression in southern Norway in which the Caledonian allochthon is preserved. Approximately in line with common usage, we will refer to the region to the southeast of the 'Faltungsgraben', where the allochthon overlaps onto the Baltic Shield, as the Caledonian foreland, and the region to the northwest of the 'Faltungsgraben' as the Caledonian hinterland. The latter is characterized in many areas by medium- to high-grade Caledonian metamorphism and polyphase ductile deformation. The western edge of the hinterland is intersected by the other late to post-orogenic extensional shear zone, with a prominent fault known as the Nordfjord–Sogn detachment (Norton 1987; NSD on Fig. 2). (The term detachment is used here to designate a low-angle fault with such a large displacement that it can only be estimated from regional considerations. A detachment may be extensional or contractional. Most detachments are accompanied by a zone of intense shearing or mylonitization which may be several kilometres thick, and when the whole zone of intense deformation related to movement on a detachment is addressed, we will use the term 'detachment zone'). This has recently found a prominent place in the literature as a clear example of extensional detachment tectonics (e.g. Hossack 1984; Norton 1987; Steel *et al.* 1985; Séranne & Seguret 1987; Andersen & Jamtveit 1990; Andersen *et al.* 1991; Dewey *et al.* 1993; Hartz *et al.* 1994; Chauvet & Séranne 1994; Wilks & Cuthbert 1994). The hanging wall of the Nordfjord–Sogn detachment, containing the erosional remnants of a large Devonian basin or group of basins, with coarse clastics lying unconformably on eroded stumps of Caledonian allochthon, is referred to in the literature as the 'upper plate'. The footwall (in the literature, the 'lower plate'), below a thick zone of mylonitized rocks, is built of gneisses showing Scandian high-grade metamorphism and poly-phase deformation, belonging to the Western Gneiss Complex (Fig. 2). These enclose the above-mentioned Scandian eclogites (Krogh 1980; Griffin & Mørk 1981; Griffin *et al.* 1985; Smith & Lappin 1989; Cuthbert & Carswell 1990).

The tectono-stratigraphic units of the Caledonian allochthon (Lower Allochthon, Middle Allochthon and Upper Allochthon) are groups of nappes with similar pre-orogenic origins. These units are characterized by rock associations which reflect different palaeogeographic locations within the pre-Caledonian plate tectonic configuration. For the Caledonides, two large continental masses are postulated (e.g. Ziegler 1985; Gower *et al.* 1990; Torsvik *et al.* 1992*b*; Rey *et al.* this volume), Baltica to the east and Laurentia to the west, separated by a largely oceanic area called Iapetus. In southern Norway, only remnants of the deformed Baltica margin and Iapetus are preserved, the former represented by the Autochthon/Parautochthon and the Lower and Middle Allochthon, the latter by the Upper Allochthon. The nearest exposed remnant of Laurentian basement is to be found on the Shetland Islands. The simplicity of this configuration on a continental scale, however, should not be translated into a simplistic view of the closing of Iapetus on a regional scale. This seems to have been a complex process, with shifting subduction zones, magmatic arcs and back-arc basins, local collisions involving continental fragments and arcs (leading to early Caledonian eclogite formation, see above), changing movement rates and directions, spread over a time period of 100 Ma (late Cambrian to mid-Silurian). These

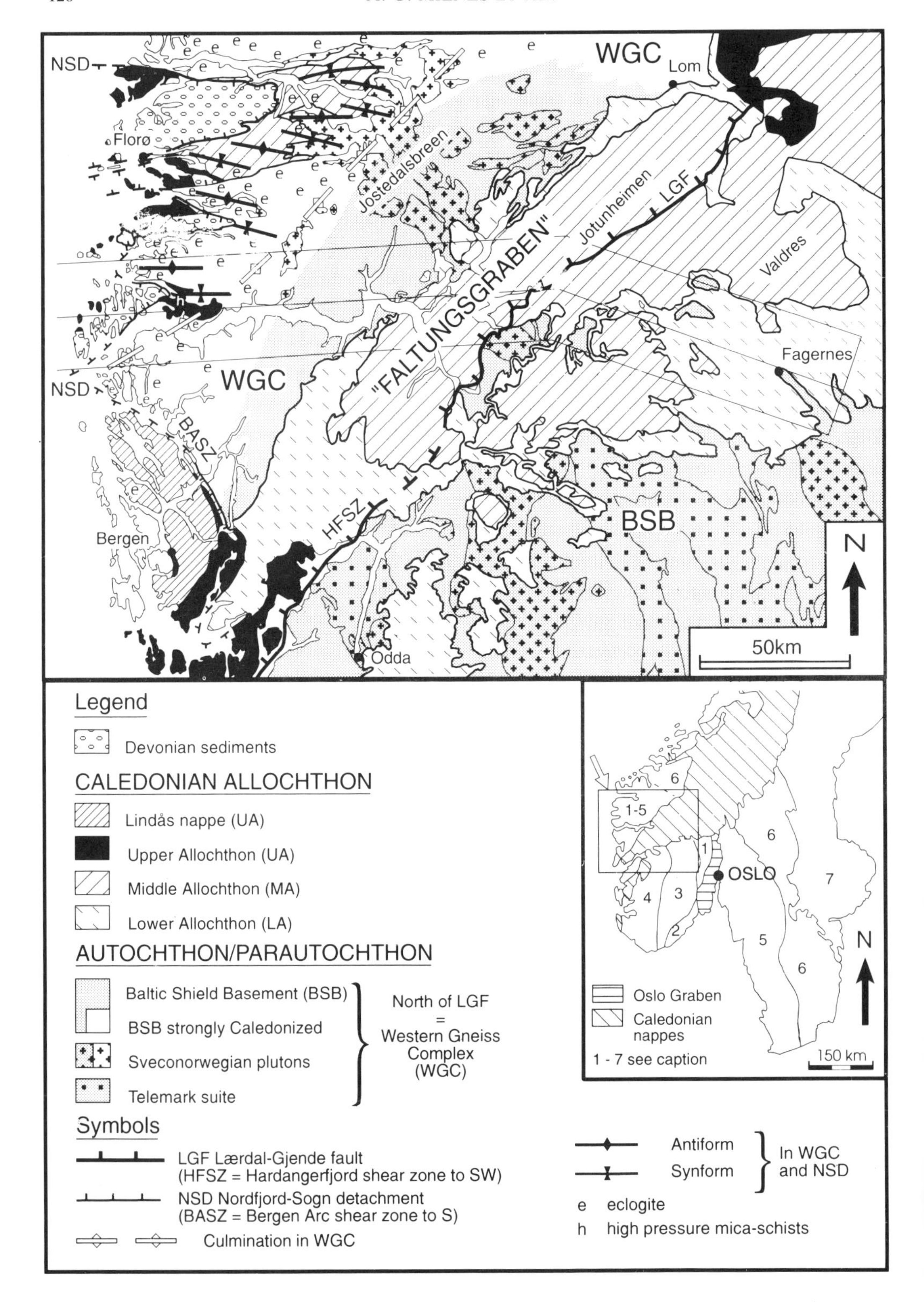

NSD
WGC
Lom
Florø
Jostedalsbreen
Jotunheimen
LGF
"FALTUNGSGRABEN"
Valdres
Fagernes
NSD
WGC
BASZ
HFSZ
BSB
Bergen
Odda
N
50km
Legend
Devonian sediments
CALEDONIAN ALLOCHTHON
Lindås nappe (UA)
Upper Allochthon (UA)
Middle Allochthon (MA)
Lower Allochthon (LA)
AUTOCHTHON/PARAUTOCHTHON
Baltic Shield Basement (BSB)
BSB strongly Caledonized
Sveconorwegian plutons
Telemark suite
North of LGF = Western Gneiss Complex (WGC)
Symbols
LGF Lærdal-Gjende fault (HFSZ = Hardangerfjord shear zone to SW)
NSD Nordfjord-Sogn detachment (BASZ = Bergen Arc shear zone to S)
Culmination in WGC
Antiform
Synform
In WGC and NSD
e eclogite
h high pressure mica-schists
1-5
OSLO
Oslo Graben
Caledonian nappes
1 - 7 see caption
150 km

early Caledonian complexities are particularly well-known from the northern parts of the chain (e.g. Bryhni 1988; Stephens 1988; Dallmeyer 1988; Stephens & Gee 1989; Dallmeyer *et al.* 1991), but are also described from the more fragmentary Upper Allochthon in the south (see below).

In the following, we briefly summarize the characteristics of the main tectono-stratigraphic units as they occur in southern Norway, in the region of the Sognefjord transect.

Autochthon/Parautochthon

The lowest unit in the tectono-stratigraphic pile consists mainly of Precambrian basement rocks which have undergone varying degrees of Caledonian overprinting. To the southeast of the 'Faltungsgraben', little Caledonization is discernible at the surface and the youngest pre-Caledonian events are the intrusion of granites in the age range 950–880 Ma (Gaal & Gorbatschev 1987). These were intruded post-tectonically with respect to the Precambrian 'Southwest Scandinavian Orogen' (Gaal & Gorbatschev 1987), which was formed during two phases, 1700–1500 Ma (Gothian) and 1250–900 Ma (Sveconorwegian). Sveconorwegian reworking of the Gothian orogenic phase is variable and concentrated in different sectors separated by major faults and mylonite zones (Fig. 2, inset, see Starmer 1993). One of these sectors contains the mono-metamorphic Telemark Suite of sediments and volcanics, deposited/extruded between the two orogenic phases and interfolded with their Gothian basement during the Sveconorwegian movements.

This Baltic Shield basement was deeply peneplained before the marine transgression in early Cambrian times, after which marine conditions reigned until mid-Silurian, when continental 'Old Red Sandstone' deposition commenced (Bassett 1985). Southeast of the 'Faltungsgraben', the present-day erosion level approximates to the pre-Cambrian peneplain, and occasional pockets of autochthonous cover sediments are preserved above a stratigraphic unconformity. Autochthonous cover remnants occur in places far to the northwest, under the nappe pile and even in the hinterland (Fig. 1, Kinnfokfjell formation). The basal decollement of the Lower Allochthon lies immediately above the basement–cover contact in most of the foreland region, and the Baltic Shield can be called truly autochthonous. Towards the northwest, where the shield becomes buried below an increasingly thick nappe pile, slicing and imbrication of the basement–cover contact justifies the term parautochthonous.

This term becomes increasingly appropriate as the Baltic Shield basement, still with pockets of *in situ* cover, emerges on the northwest side of the 'Faltungsgraben', where it is called the Western Gneiss Complex (Fig. 2). Not only is the slicing more prominent (Lutro 1989) but the whole basement becomes progressively more 'Caledonized' towards the west and north (i.e. progressively overprinted by increasingly intense shearing, folding and metamorphism, as described in Milnes *et al.* 1988). Nevertheless, a broad zone of only slightly Caledonized Precambrian basement lines the northwestern side of the 'Faltungsgraben' and forms the massif of Jostedalsbreen. Within this area, Precambrian geology very similar to that of the foreland can be observed. The Precambrian basement is composed of various migmatites, granitic gneisses, heterogeneous banded gneisses, amphibolites and paragneisses (Lutro 1986, 1987, 1988, 1989), with the main crust-forming events occuring between 1700 and 1450 Ma, i.e. Gothian (Gorbatschev 1985; Kullerud *et al.* 1986). Sveconorwegian reworking is slight, except for the intrusion of numerous post-tectonic granites, and associated aplites and pegmatites around 900–1000 Ma (Priem *et al.* 1973; Corfu 1980; Milnes *et al.* 1988). In the southern part of the Western Gneiss Complex, the ages of the felsic intrusives vary from 1300 to 870 Ma, with a possible orogenic event at 1050 Ma (Kullerud *et al.* 1986). This correlates well with the sectors to the south of the 'Faltungsgraben' (Gorbatschev 1985). In the northern part (north of Fig. 2), the Precambrian basement is correlated with the Trans-Scandinavian Granite-Porphyry Belt of southeast Sweden (Fig. 2, inset).

Fig. 2. Tectonic map of the Sognefjord region, showing the precise location of the Sognefjord transect (Fig. 1). The centre line of the strip was used to mark the main contacts on Fig. 1, and the true dips of the contacts were converted to apparent dips along this line. 'Off-section' in the text is used to designate information which is projected into the section from outside the marked strip. **Inset map**: Precambrian geology of the Baltic Shield. 1–5 Southwest Scandinavian Orogen (1700–900 Ma): 1, Kongsberg sector; 2, Bamble sector; 3, Telemark sector; 4, Rogaland-Vestagder sector; 5, Østfold sector and coeval sectors in Sweden. 6, Trans-Scandinavian Granite-Porphyry Belt (1810–1750 Ma); 8, Svecofennian Orogen (2000–1750 Ma), after Gower *et al.* (1990) and Tucker *et al.* (1990).

Because of the 'Faltungsgraben', the connection between the Baltic Shield basement and the Western Gneiss Complex is obscured over a distance of 20 km at its narrowest. However, the similarities between the Baltic Shield basement and the Precambrian geology of the Western Gneiss Complex, together with the geophysical character of the crust below the 'Faltungsgraben' (see below), make it improbable that there exists a hidden, fundamental separation between these two areas of Autochthon/Parautochthon, as has sometimes been suggested (e.g. Pedersen *et al.* 1988). In the following, we treat them together as a single tectonic unit.

Lower Allochthon

The Lower Allochthon in the area shown on Fig. 2 consists mainly of Vendian to lower Ordovician sediments with a stratigraphy correlatable with autochthonous sequences on the Baltic Shield to the southeast (Nickelsen *et al.* 1985). The sediments are involved in a typical thin-skinned fold-and-thrust belt, with the outer edge of deformation lying to the south of Oslo (Morley 1986, 1987). At the eastern edge of the Sognefjord transect, the unit consists of two duplex or imbricated nappe structures (Aurdal and Synnfjell duplexes, Hossack *et al.* 1985), which in their undeformed state correspond to an estimated original width of 190 km and are thought to have been transported southeastwards at least 50 km on the basal decollement. This implies that they represent the stripped-off cover of the Western Gneiss Complex. Towards the northwest, the coherent duplex structures grade into a narrow shear zone underlying the basal thrust of the Middle Allochthon (the Fortun–Vang nappe of Lutro & Tveten 1987 and Milnes & Koestler 1985). In this zone, all vestiges of stratigraphy are lost. The typical rock types of this zone (Vang and Fortun phyllites, Fig. 1) can be followed continuously across the 'Faltungsgraben' (Fig. 2). To the southwest and along the northwestern side of the 'Faltungsgraben', heavily deformed slices of basement (Baltic Shield type) and, occasionally, more exotic rocks (serpentinite, meta-gabbro, greenschists) occur. Because of the predominance of characteristic graphitic phyllites, dark fine-grained quartzites, light-coloured meta-arkoses and carbonate horizons, however, the Lower Allochthon as a unit can be mapped with a good deal of certainty (Fig. 2), in spite of its 'mélange-like' aspect. The Lower Allochthon corresponds to the 'late Caledonian detachment' of Milnes (1987) and the 'decollement zone' of Fossen (1992).

Middle Allochthon

The Middle Allochthon in southern Norway is represented by the erosional remnant of a large sheet of Precambrian basement and associated cover sequences which occupies the central part of the 'Faltungsgraben'. The basement forms the high mountain area at the head of Sognefjord (Jotunheimen) and has been called the Jotun Complex (Milnes & Koestler 1985). The main component in the complex is a pyroxene-granulite/anorthosite association which is intruded by, or was tectonically juxtaposed with, non-granulitic syenites and gabbros. These were together affected by an orogenic event at amphibolite facies, with granite intrusion, at about 900 Ma (Sveconorwegian; Schärer 1980; Koestler 1982; Corfu & Emmett 1992). In spite of overthrusting during the Scandian phase, the Jotun Complex is remarkably intact, with presumed Caledonian effects being confined to localized zones of low-grade mylonitization and cataclasis, and diffuse, weak, ductile-penetrative deformation of, for example, the 900 Ma granite, with indications of heating to lower amphibolite facies conditions. The bulk strain suffered by the Jotun Complex during transport is judged to be minimal (Milnes & Koestler 1985), except in a basal high-strain zone which varies in thickness from a few metres to 3 km.

In spite of intense mylonitization, this basal zone (the Turtagrø/Valdres zone, Milnes & Koestler 1985) contains recognizable remnants of stratigraphic basement–cover contacts, now inverted. The Jotun Complex cover rocks consist of a characteristic, basal quartz conglomerate (Bygdin conglomerate, Fig. 1) and characteristic current-bedded meta-arkoses with typical purplish clasts ('Valdres sparagmites', Fig. 1). These are better preserved towards the foreland, where they develop into a separate nappe structure (Valdres nappe). The inverted stratigraphy shows close similarities to that of the Lower Allochthon, including similar latest Precambrian, Cambrian and lower Ordovician formations (Mellsenn Group, Tya Series, Fig. 1: Heim *et al.* 1977; Hossack *et al.* 1981). One of the most southeasterly exposures of the Jotun Complex shows a steeply dipping but right-way-up basement/cover contact, with an undeformed basal quartz conglomerate (Hossack 1972).

All in all, the stratigraphic relationships indicate that the Jotun–Valdres nappe complex was derived from the Baltica continental margin immediately to the northwest of the one-time basement of the Lower Allochthon, i.e. the Western Gneiss Complex. However, on the northwest side of the 'Faltungsgraben', the strongly sheared basal zone (Turtagrø zone, Milnes & Koestler 1985; lower Jotun nappe, Lutro 1986) contains exotic elements which could represent mafic lavas, and mafic to ultramafic intrusives, in a 'melange-like' association with the units mentioned above. Because of the consistent indications of inversion, these are interpreted as remnants of tectonic units which lay on top of the Jotun Complex and its cover, before the overthrusting of the Middle Allochthon and the inversion in the basal zone took place (i.e. remnants of rocks similar to those now occurring in the Upper Allochthon). In general, the strongly sheared and mylonitized basal zone of the Middle Allochthon, together with the strongly sheared metasediments and basement slices in the underlying Lower Allochthon, can be grouped together as the Jotunheimen detachment zone. This is a major low-angle contractional shear zone separating two practically un-Caledonized Precambrian basement units, the Baltic Shield basement below and the Jotun complex above (Fig. 1). This is a complex zone in which both early and late contractional (top-to-SE) movements can be distinguished, together with a prominent later phase of reversed (top-to-NW) movement, which is interpreted to represent reactivation in an extensional tectonic regime (see later discussions). The total displacement on the Jotunheimen detachment is thought to be more than 300 km and its thickness is 1–5 km. The displacement estimate is based on the original width and position of the stratigraphy in the Lower Allochthon, as indicated above, plus the present width of the only slightly Caledonized part of the Middle Allochthon, as reconstructed below.

In addition to the main area of exposure of the Middle Allochthon, in the 'Faltungsgraben', two other types of occurrence have been identified in the hinterland. One occurs as strongly interfolded remnants within the Western Gneiss Complex in the footwall of the Nordfjord–Sogn detachment to the north of our transect (Fig. 2, see also Bryhni 1989; Andersen & Jamtveit 1990; Wilks & Cuthbert 1994). The other lies, relatively undeformed, in the hanging wall (Fig. 2). The hanging-wall occurrence (Dalsfjord nappe) is important because the equivalent of the Jotun complex basement (Dalsfjord complex, Fig. 1) is associated with a cover unit similar to the Valdres sparagmites (Høyvik group, Fig. 1). This in turn is unconformably overlain by a marine Lower Palaeozoic sequence containing, at the top, fossils of Wenlock age (Herland group, Fig. 1, Brekke & Solberg 1987; Andersen *et al.* 1990). This places a clear lower limit to the Scandian phase on the orogenic time-scale (see below).

Upper Allochthon

Units belonging to the Upper Allochthon are mainly exposed in the hanging wall of the Nordfjord–Sogn detachment in the present area (Fig. 2), tectonically overlying the Middle Allochthon. To the southwest (Bergen Arcs and Sunnhordland) and northeast (Trondheim Nappe Complex), however, large areas exist where Upper Allochthon lies variously on Autochthon/Parautochthon, Lower Allochthon and Middle Allochthon. In southern and western Norway, the Upper Allochthon is characterized by ophiolite complexes with island arc and marginal basin assemblages of early and late Ordovician age. These are unconformably overlain by clastic, late Ordovician–early Silurian sequences, intruded by plutons of various ages and affinities, with the youngest dated at 430 Ma (Andersen & Jansen 1987; Fossen & Austrheim 1988). The early Caledonian histories of these complexes have been studied in detail (e.g. Pedersen *et al.* 1988; Andersen & Andresen 1994). They record the complicated intra-oceanic evolution of Iapetus prior to the Scandian phase. During the latter phase, the whole complex was intensely redeformed, dismembered and obducted onto the margin of Baltica. The obduction melange (Sunnfjord melange, Fig. 1, see Andersen *et al.* 1990) is well exposed on the Sognefjord transect, where it overlies fossiliferous Wenlock sediments of the Baltica continental margin (Middle Allochthon, see above).

Sognefjord transect

Although the general outline of early Caledonian and Scandian orogenesis and late/post-orogenic extensional tectonics is well established, there have been few compilations of detailed structural data across the whole belt which could aid in advancing past the conceptual stage. Since the Valdres–Jotunheimen–Sognefjord–West Coast cross-section, here called the Sognefjord transect, has now been studied in detail, we felt the time was ripe for an attempt to compile and synthesize the structural data base and correlate it with the available stratigraphic, petrographic and radiometric data. The structural relations have been summarized as accurately as possible on the accompanying composite profile (Fig. 1), and the main sources of data are indicated as a series of panels (Fig. 1, panels 1–7), as detailed below. Within each panel, the geometry, kinematics and history of the Scandian deformation has been reported in detail, and is sufficiently well documented to be correlated with neighbouring panels. The results of structural correlation provide the main basis for the following retrodeformation. In addition to this, however, two other types of data provide important constraints on the reconstructions. Firstly, although there is as yet no deep reflection seismic profile along the transect, other types of geophysical data provide invaluable information on crustal structure. These will be discussed first. Secondly, the absolute dating of some of the events, by stratigraphic or radiometric means, provides an orogenic timetable which is sufficiently reliable to check the consistency and plausibility of the movement picture. This will be discussed after the relative sequence of deformational events has been established in the section on structural correlation.

Crustal structure

A recent compilation of geophysical data on Moho depth under Scandinavia (Kinck *et al.* 1993) shows a maximum Moho depth of about 40 km in the central part of the Sognefjord transect, decreasing westwards to less than 30 km just offshore. Further offshore, there is little published data and the indicated level (Fig. 1) is speculative. However, the combination of westward rising Moho and large Permo-Triassic normal faults with downthrow to the west (Færseth *et al.* 1995), together with the Devonian Nordfjord–Sogn detachment zone, also with a large downthrow to the west, means that the crust may be mainly composed of Upper Allochthon or higher units in the offshore part of the transect (Fig. 1).

Refraction seismic data were acquired along the transect during a recent experiment with ocean bottom seismometers (Iwasaki *et al.* 1994) and the results confirmed the general Moho level. The survey provided supplementary information which may have geological significance. Firstly, a mid-crustal velocity discontinuity was discovered under the eastern part of the profile. This sharp velocity increase at 20 km depth fades out in a westward direction (Fig. 1). In addition, the high velocity lower crust which underlies most of Scandinavia (P-wave velocities higher than 7 km s^{-1}, Korja *et al.* 1993) disappears westwards from about the same position (Fig. 1). Both these phenomena may be related to the 'Caledonization' of the outer edge of the Baltic Shield, a process which the retrodeformation is thought to illuminate (see below). In the upper parts of the Western Gneiss Complex, no lateral velocity variations were recognizable along the Sognefjord profile, but a marked increase in velocity was observed, where the Jotun Complex, with a high proportion of rocks with mafic composition, lies within the 'Faltungsgraben'. There, the high velocity body was detected down to a depth of 6–7 km, with no strong evidence for unusually high velocities in the deeper parts of the crust. This strengthens the geological interpretation that the Western Gneiss Complex is the direct westward extension of the Baltic Shield (Iwasaki *et al.* 1994).

Similar conclusions were reached during a reappraisal of the gravity and magnetic data across the 'Faltungsgraben' in the Jotunheimen–Valdres area by Skilbrei (1990). In spite of earlier interpretations which suggested a deep 'root' of high density material beneath the 'Faltungsgraben' (Smithson *et al.* 1974), the new modelling suggests that the Jotun Complex is less than 6 km thick at its deepest below the present surface. The relevant part of the present cross-section (Fig. 1) is thus constructed on the basis of these combined geophysical indications (refraction seismic, gravity, aeromagnetic).

An attempted reflection seismic profile using conventional marine instrumentation (Sellevoll & Mokhtari 1985) failed because of side reflections from the submerged fjord walls. However, analogous reflection seismic images for parts of the Sognefjord transect may be found on the 'Mobil Search' coast-parallel profiles ILP–10, ILP–11 and ILP–12 (Hurich & Kristoffersen 1988). The first segment of interest is the crossing of the Hardangerfjord shear zone, at the mouth of Hardangerfjord (Fig. 2: Hurich & Kristoffersen 1988; Deemer & Hurich 1991; Færseth *et al.* 1995). The Hardanger shear zone represents the southwestward continuation of the Lærdal-Gjende fault zone. A further image which may be similar to that which would appear on the Sognefjord transect is the prominent antiformal reflector array which is recorded on ILP–10 immediately south of the projected position of the Nordfjord–Sogn detachment (Færseth *et al.* 1995, NSD on their fig. 1). This lies on the offshore continuation of the culmination in the main Caledonian foliation in the Western Gneiss Complex along Sognefjord (Figs 1 & 2).

Structural correlation

On the scale of the whole transect (Figs 1 & 2), detailed structural information is available along a narrow strip, which in some parts must be supplemented with off-section data, along strike on both sides. For the purposes of correlation, we have subdivided this strip into three segments, within which a coherent deformational history can be established on the basis of available documentation. Correlation from one segment to the next is less secure and a working hypothesis has to be set up to make retrodeformation possible. To clarify the arguments, the structurally defined deformation phases in each segment will be prefixed with the segment name: Jotunheimen in the east (Fig. 1, panels 1–4), Sognefjord in the centre (Fig. 1, panels 5–6), and West Coast in the west (Fig. 1, panel 7).

Jotunheimen segment. A correlation and interpretation of Caledonian deformation phases in the Jotunheimen segment, across the 'Faltungsgraben', was carried out by Milnes & Koestler (1985). This work was completed in 1981 and confirmed and supplemented by later studies (e.g. Koestler 1983, 1988, 1989; Lutro 1986, 1987, 1988, 1989) and off-section data (e.g. Fossen 1993*a*; Fossen & Holst 1995). It showed that the structural history of the Jotunheimen detachment zone, the major zone of Caledonian deformation in this segment, could be described in terms of a sequence of phases (Milnes & Koestler 1985, Table 2), here called Jotunheimen-D1 to Jotunheimen-D6. The characteristics of these are summarized below and the general tectonic significance of each phase is added in parentheses (Milnes & Koestler 1985).

Jotunheimen-D1: imbrication and ductile shearing, incorporation of exotic fragments (tectonic burial of the Jotun Complex and its cover by the Upper Allochthon, including ophiolite obduction).

Jotunheimen-D2: main phase of ductile-penetrative deformation in the basal zone of the Middle Allochthon – ductile shearing in basement, isoclinal folding in cover, inversion of basement–cover contacts, main foliation–stretching lineation, pebble flattening–elongation (early stage of SE-directed thrusting in the Jotunheimen detachment zone).

Jotunheimen-D3: *en bloc* movement of Middle Allochthon on the basal thrust, main phase of ductile-penetrative deformation in the underlying meta-sediments of the Lower Allochthon, formation of the Lower Allochthon duplex structures ahead of the Middle Allochthon block (later stage of SE-directed thrusting in the Jotunheimen detachment zone, mainly accommodated by deformation in the Lower Allochthon).

Jotunheimen-D4: development of foreland-dipping cleavage and NW-vergent asymmetrical folds, superimposed on all earlier structures (reversed, NW-directed movement on the Jotunheimen detachment zone, 'mode I extension' of Fossen 1992).

Jotunheimen-D5: truncation and displacement of the Jotunheimen detachment zone by movement on the Lærdal–Gjende fault (low-angle, NW-dipping normal fault, with associated mylonites, cataclasites and gouges, 'mode II extension' of Fossen 1992).

Jotunheimen-D6: isolated major upright folds with NW–SE-trending axial planes, associated sporadic minor folding (regional significance unknown, may be several minor phases, age with respect to D5 not known – could be earlier).

The deformation phases Jotunheimen D1–D4 describe the structural history of the Jotunheimen detachment zone. Out on the foreland, the duplex structures described by Hossack *et al.* (1985) were formed during the Jotunheimen-D3 phase (main contractional deformation in the Lower Allochthon) and this places valuable controls on the movement picture (amount, direction and rate of displacement). The upper duplex structure has been traced northwestwards until it disappears under the Middle Allochthon, at which location the imbricates lose their coherence and are cut through by the 'reversed movement' cleavage of the Jotunheimen-D4 phase (see also Hossack 1976, fig. 6). This top-to-NW movement was relatively minor compared to the top-to-SE thrusting. Jotunheimen-D2 + D3 corresponds to more than 300 km of SE-directed shear displacement (see above), whereas Jotunheimen-D4 displacement must be only of the order of several kilometres (Milnes & Koestler 1985). Off-section to the southwest, Fossen & Holst (1995) estimate the amount of reversed movement (Jotunheimen-D4) in the same zone to be 20–36 km. The reversed movement in the detachment zone (Jotunheimen-D4) may mark the change to crustal extension at the end of Scandian contraction (Fossen 1992). However, it may partly correspond to the process suggested by Gee *et al.* (1994) for the formation of the Røragen Devonian basin to the north, i.e. extensional collapse at higher and inner levels in the nappe pile at the same time as thrusting was continuing at lower and outer levels.

With regard to depth of the Jotunheimen detachment zone at the time of movement, colour alteration of conodonts from the Lower Allochthon at the southeastern end of the

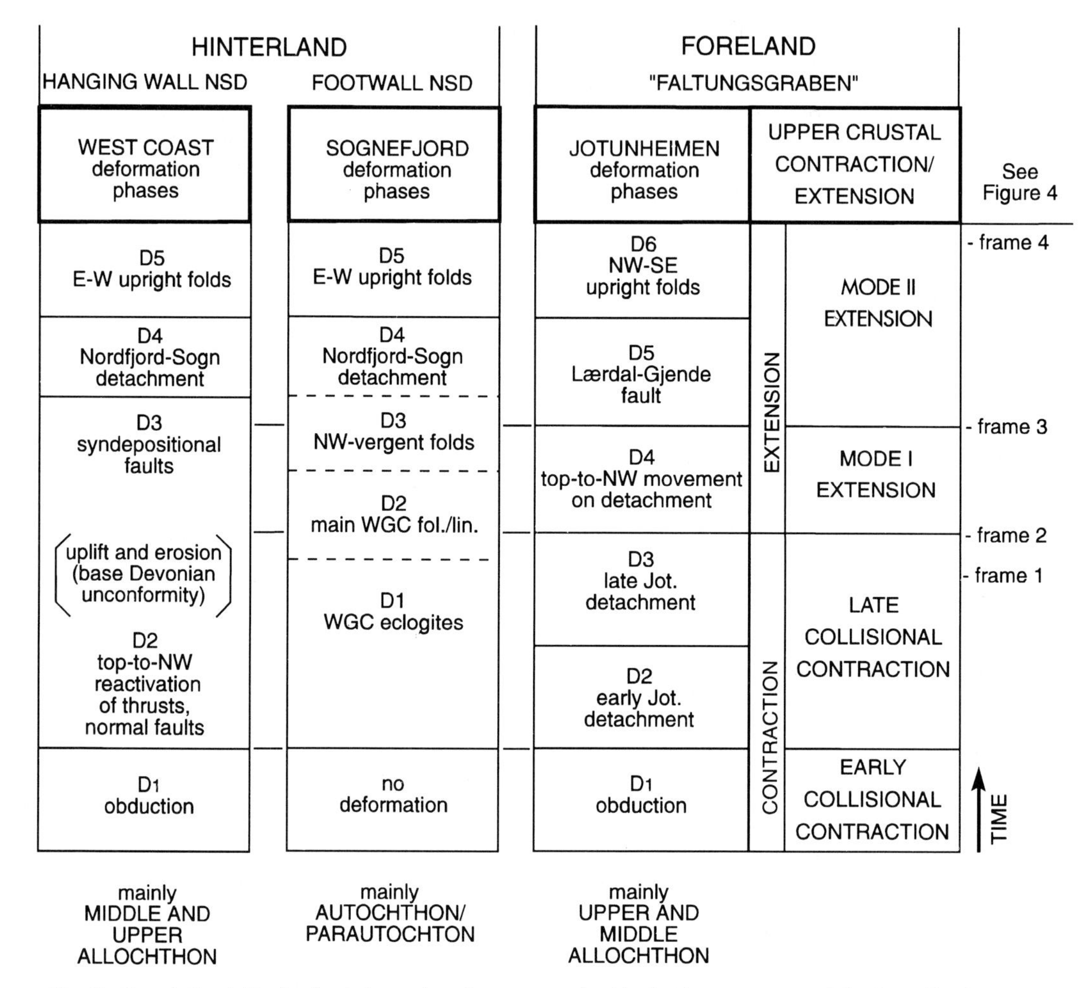

Fig. 3. Correlation table for the deformation phases recognized in the three segments of the Sognefjord transect, with a general subdivision based on orogenic processes in the rheological upper crust (see text, for discussion).

segment (Nickelsen *et al.* 1985) indicates that they lay at a maximum depth of 10–13 km, presumably at the end of overthrusting (Jotunheimen-D3). At the northwest end of the segment (Fig. 1, panel 4), metamorphic parageneses indicate maximum depths of 15–20 km for the same zone. This suggests a flat, gently NW dipping detachment zone at this time (end Jotunheimen-D3), with an average dip of about 5° over a distance of 100 km. There is nothing to suggest that any part of the Jotunheimen detachment zone reached depths deeper than those required for lower amphibolite facies metamorphism at any time during its history (see below, retrodeformation).

The sequence of structural phases established in the Jotunheimen segment (foreland and 'Faltungsgraben' along the Sognefjord transect) is summarized in Fig. 3. The change from crustal contraction to crustal extension is placed at the transition D3–D4, although, as noted above, there may be some overlap. Jotunheimen-D1 (obduction and overthrusting of Upper Allochthon on to the margin of Baltica, the future Middle Allochthon) is taken to represent the early collisional phase (early Scandian). During Jotunheimen-D2 and D3, the suture zone (Upper Allochthon) became inactive and was replaced by a deep crustal detachment surfacing some 100 km east of the Baltica margin (late collisional contraction, late Scandian). At the beginning of Jotunheimen-D4, contraction ceased and reverse (i.e. NW-directed) movement related to late orogenic or post-orogenic crustal extension started (mode I extension), to be followed by the development of a major, cross-cutting, low-angle normal fault during Jotunheimen-D5 (Lærdal–Gjende fault, mode II extension). This well defined sequence will be used in the following as a yardstick for the correlation of the deformation phases in the Sognefjord and West Coast segments, in the hinterland of the Caledonian orogen (Fig. 3).

Sognefjord segment. The structural history of the Sognefjord segment is well-known from the 'Sognefjord north shore log' (Dietler *et al.* 1985; Dietler 1987; Milnes *et al.* 1988; see Fig. 1, panel 5), and from detailed work in the footwall of the Nordfjord-Sogn detachment (Norton 1986, 1987; Bailey 1989; Andersen & Jamtveit 1990; Swensson & Andersen 1991; Andersen *et al.* 1994; Chauvet & Séranne 1994; Skår *et al.* 1994; see Fig. 1, panel 6). The 80 km long Sognefjord shore section exposes a continuous profile through the Western Gneiss Complex, ranging from un-Caledonized Precambrian granites and migmatites of the Baltic Shield basement, emerging from beneath the 'Faltungsgraben' in the east, to strongly Caledonized (originally Precambrian) gneisses, containing eclogite pods, forming the core of a regional culmination in the west (Fig. 2). This progressive Caledonization from east to west (in its original position, from top to bottom) of a 25 km thick slice of continental crust, was subdivided into three regimes for descriptive purposes (Milnes *et al.* 1988): (1) an easterly regime, mainly un-Caledonized, with isolated Caledonian shear zones; (2) a transitional regime, characterized by heterogeneous shears in an anastomosing network; and (3) a westerly regime, with more or less complete Caledonian overprinting, polyphase deformation and high-grade metamorphism. This has been simplified on Fig. 2 to an easterly, un-Caledonized or weakly overprinted zone (on Fig. 1, marked BSB, since it is interpreted to be the north-westward extension of the Baltic Shield basement, see earlier discussions), and a westerly, Caledonized zone (on Fig. 1, 'Caledonized' BSB), both containing part of the transitional regime. Within the easterly zone, the shear zones contain a mylonitic foliation and stretching lineation which were originally called S1/L1 (called now S2/L2 for the reasons given below). These can be followed through the transitional regime into the westerly zone, where they represent the main deformation phase. Remnants of earlier phases were originally grouped together as pre-D1, whether Precambrian or Caledonian (Milnes *et al.* 1988). In this paper, the Caledonian effects in this group are distinguished as D1 (see below).

The well defined culmination in the Western Gneiss Complex (Fig. 2), marking the change from E- to SE-dipping foliation and E-plunging lineations, to W–NW-dipping/plunging structures, lies in the footwall of the Nordfjord–Sogn detachment (Fig. 1, panel 6). Its western limb is characterized by progressively more intense shearing and mylonitization upwards towards the detachment, and progressively lower metamorphic grades (from amphibolite to greenschist facies). The western boundary of the Western Gneiss Complex is located in the lower part of this zone of mylonitization (the detachment zone). The formation of the culmination and the detachment-related shearing post-dates the D2 structures. The latter include major F2 folds with remnants of Upper Allochthon in their cores (Lavik mafics and Hyllestad schists, Fig. 1, panel 6) which show signs of pre-folding eclogite-facies metamorphism (Bailey 1989).

The mylonitic sequence in the upper part of the Nordfjord-Sogn detachment zone is known as the Askvoll Group (Fig. 1). It consists of meta-sediments, meta-volcanics and meta-gabbros which are thought to represent the Upper Allochthon (Solund–Stavfjord ophiolite complex, see Fig. 1), together with fragments of Precambrian basement (Swensson & Andersen 1991; Skår *et al.* 1994). The detachment zone contains abundant evidence of top-to-W or NW kinematics. The whole zone was later folded by upright E–W-trending open folds (e.g. Andersen & Jamtveit 1990; Swensson & Andersen 1991; Andersen *et al.* 1994; Chauvet & Séranne 1994) which die out eastwards (see Fig. 2).

The structural relations in the footwall of the Nordfjord–Sogn detachment can be combined with those of the Sognefjord shore section to give a well-defined deformational history for the Western Gneiss Complex, as follows.

Sognefjord-D1: in the west, foliations, lineations and folds within the eclogitic pods, some of which are certainly coeval with eclogite facies metamorphism (Andersen *et al.* 1994, stage 1); juxtaposition of Upper Allochthon against the Western Gneiss Complex and occasionally preserved remnants of penetrative deformation (Bailey 1989). In the east, the earliest Caledonian deformation is Sognefjord-D2.

Sognefjord-D2: towards the west, main ductile-penetrative deformation (foliation, E–W stretching lineation, major and minor isoclinal folding) at amphibolite facies (earlier eclogite-facies parageneses completely obliterated, except for the cores of some mafic bodies and some other remnants); towards the east, isolated shear zones in the Precambrian basement, increasing in importance and intensity westwards. This phase corresponds to D1 in Milnes *et al.* (1988), D2 in Bailey (1989), and stage 2 in Andersen *et al.* (1994). Sognefjord-D2 post-dates all pegmatites, migmatites and granites in the Sognefjord segment and is nowhere associated with signs of anatexis.

Sognefjord-D3: east of the culmination and across its hinge, tight to isoclinal, NW-vergent folding of S2/L2, still at amphibolite facies (Dietler 1987; Bailey 1989; see also Norton 1986, 1987), possibly coeval with the development of amphibolite facies mylonites in the lower part of the Nordfjord–Sogn detachment zone, where kinematic indicators show top-to-W or NW movement and the shear zones truncate D2 isoclines (Swensson & Andersen 1991; Andersen *et al.* 1994, early stage 3).

Sognefjord-D4: in the west, main movement on Nordfjord–Sogn detachment, intense mylonitization at increasingly lower metamorphic grade (Andersen *et al.* 1994, late stage 3), concentrated particularly in the upper part of the detachment zone, abundant kinematic indicators show top-to-W or NW; culmination probably developed during this period in response to movement in the detachment zone (see also Norton 1986, 1987).

Sognefjord-D5: open, upright E–W-trending major and minor folds to the north of Sognefjord, affecting the detachment zone mylonites and all earlier structures (Andersen & Jamtveit 1990; Swensson & Andersen 1991; Andersen *et al.* 1994; Chauvet & Séranne 1994; Fig. 2).

This sequence is now well established but its interpretation, particularly Sognefjord-D2/D3, has been problematic. Originally, both D2 and D3 were interpreted as related to a general top-to-NW shearing episode, at the time called 'back-thrusting' (Milnes *et al.* 1988, their phases D1/D2). For D2, this was based on the close association with D3, the latter showing obvious top-to-NW asymmetry, and not on the occurrence and consistence of D2 kinematic indicators, which were generally lacking. Since then, new data indicate that the bulk deformation field during D2 was probably irrotational (coaxial, or 'pure shear' deformation, Andersen *et al.* 1994, see also, Andersen & Jamtveit 1990; Dewey *et al.* 1993; Wilks & Cuthbert 1994). These authors show that this was related to subvertical crustal thinning and E–W extension after the extreme crustal thickening necessary for eclogite formation. The currently exposed eclogites along Sognefjord were formed at about 15 kbar and 600°C (Griffin *et al.* 1985; Bailey 1989). During Sognefjord-D2 they decompressed isothermally to about 10 kbar (Bailey 1989; Andersen & Jamtveit 1990; Dewey *et al.* 1993), corresponding to a shortening of the distance to the contemporary Earth's surface of about 20 km. This places important constraints on the retrodeformation carried out below; its relation to other crustal processes will be discussed further there. The earlier Caledonian phase, Sognefjord-D1, is related to the juxtaposition of the Upper Allochthon on the Western Gneiss Complex and its tectonic burial to the depths required for eclogite facies metamorphism. Phases post-dating Sognefjord-D2, particularly D3 and D4, which both show consistent top-to-W-NW shear sense, seem to be associated with crustal extension. This is particularly clear for Sognefjord-D4 (main shear episode in the Nordfjord–Sogn detachment zone). The folding during Sognefjord-D5 is taken to mark late to post-detachment N–S shortening (Chauvet & Séranne 1994).

Because of the large wedge of only slightly Caledonized Baltic Shield basement across the eastern part of the Sognefjord segment (Figs 1 & 2), correlation with the structural history of the Jotunheimen segment is indirect. The Jotun Complex (Middle Allochthon) is regarded as a part of Baltica, sub-adjacent to and immediately oceanward of, the Baltic Shield basement, and the Lower Allochthon is interpreted as the one-time cover of the outer part of the Baltic Shield basement (the Western Gneiss Complex), as indicated above. Hence, most of the overthrusting of the Middle and Lower Allochthon (i.e. the top-to-SE movement on the Jotunheimen detachment, Jotunheimen D2-D3) must have taken place before the juxtaposition of Upper Allochthon against the Western Gneiss Complex in outer Sognefjord (Sognefjord-D1). At the other end of the time span, the latest possible correlation for the NW-verging folds of Sognefjord-D3 is with movement on the Lærdal-Gjende fault (Jotunheimen-D5). This assumes that the folding represents the deep ductile equivalent of this major low-angle, normal, down-to-NW dislocation (see Fig. 1), but Sognefjord-D3 could also be related to the reversed movement on the Jotunheimen detachment, Jotunheimen-D4. However, the geometry and kinematics of the main movement on the Nordfjord-Sogn detachment (Sognfjord-D4) and the Lærdal-Gjende fault (Jotunheimen-D5) suggests coeval activity over much of these phases.

The conclusions drawn from this discussion are shown on the correlation chart (Fig. 3). For the retrodeformation, the main point is that the Sognefjord phases D1–D4 certainly occupy the same time bracket as Jotunheimen phases D3–D5, even though the detailed correlation within this time bracket remains unclear.

West Coast segment. The West Coast segment comprises the hanging wall of the Nordfjord-Sogn detachment (Fig. 1, panel 7). Its deformation history can be divided into phases which pre-date and post-date the base Devonian unconformity, which represents a period of uplift, erosion and then subsidence before the deposition of the coarse, middle Devonian conglomerates of the Solund basin (Nilsen 1968; Indrevær & Steel 1975; Séranne & Seguret 1987). The pre-unconformity history is best known off-section, particularly from the Askvoll area to the north, where the present day erosion level below the unconformity is much deeper. The geology of the Askvoll area has been studied in great detail, and the structural history is now very well documented (Brekke & Solberg 1987; Andersen *et al.* 1990; Skjerlie & Furnes 1990; Furnes *et al.* 1990; Osmundsen & Andersen 1994). The basement of the Devonian conglomerates in this area contains a well exposed obduction melange (Sunnfjord melange, Fig. 1), separating the Upper Allochthon (Solund–Stavfjord ophiolite complex, Fig. 1) from the presumed outermost fragment of Baltica (Dalsfjord complex, Middle Allochthon, Fig. 1). This fragment, composed of crystalline rocks similar to the Jotun complex, has a cover of presumed late Precambrian sediments (Høyvik group, Fig. 1) and unconformably overlying platform sediments of ?upper Ordovician to middle Silurian (Wenlock) age (Herland group, Fig. 1, see Brekke & Solberg 1987; Andersen *et al.* 1990). The shear zones associated with obduction and overthrusting show a complex history (Osmundsen & Andersen 1994), which is overprinted by, top-to-NW shearing, NW-vergent folding and down-to-NW normal faulting, all pre-dating the deposition of the Devonian clastics (Osmundsen & Andersen 1994). The unconformity was later affected by syn-depositional normal faulting related to the formation of the Devonian basins. The basin fills were then truncated by post-depositional movement on the Nordfjord–Sogn detachment, in places with strong deformation of the conglomerates adjacent to the fault (Séranne & Seguret 1987). The Devonian sequences were also affected by large, open, upright folds with E–W trends, some confined to the hanging wall (deformation during movement on the detachment), some affecting all units (post-detachment in age, see Fig. 2). For correlation purposes, we have subdivided this complex sequence of events into the following phases.

West Coast-D1: obduction and overthrusting with top-to-SE movement, in several phases (Osmundsen & Andersen 1994; Brekke & Solberg 1987); folding and imbrication of the overridden Herland Group sediments under greenschist facies conditions (Andersen *et al.* 1990, fig. 4).

West Coast-D2: reactivation of D1 shear zones with top-to-NW movement, development of NW-vergent asymmetrical folds (F3 folds of Brekke & Solberg 1987 and Osmundsen & Andersen 1994); development of normal faults which pre-date the base Devonian unconformity, in places reactivating D1 shear zones, still in greenschist facies (Osmundsen & Andersen 1994; Hartz *et al.* 1994).

(Uplift and erosion.)

West Coast-D3: syn-depositional normal faulting associated with basin formation (Osmundsen & Andersen 1994, possibly also Indrevær & Steel 1975), ?initial movement on Nordfjord–Sogn detachment.

West Coast-D4: main top-to-W or NW movement on Nordfjord–Sogn detachment, truncation of basin sequences, pebble deformation against fault, eastward tilting of bedding and shearing along fault zone (Séranne & Seguret 1987), open folds due to hanging-wall deformation.

West Coast-D5: folding about upright E–W-trending axial planes, related to N–S shortening (e.g. Torsvik *et al.* 1987; Chauvet & Séranne 1994).

The West Coast deformation phases can be correlated with those of the Sognefjord and Jotunheimen segments as shown in Fig. 3. The obduction phase D1 probably correlates with the signs of obduction-related deformation noted in the Jotunheimen detachment zone (Jotunheimen-D1). The deformation related to the main movement on the Nordfjord–Sogn detachment, West Coast-D4, and the subsequent and possibly related (Chauvet & Séranne 1994) E–W upright folding, West Coast-D5, are thought to correspond to the Sognefjord phases D4 and D5. In between, lies a sequence of extensional phases and an episode of uplift and erosion (down to Caledonian greenschist facies rocks) which are difficult to correlate in detail. The top-to-NW phase in the West Coast Upper Allochthon (West Coast-D2) which predates the base Devonian unconformity is clearly synchronous with the top-to-SE movement in the Jotunheimen detachment zone, if the above correlations are correct. This implies important syn-contractional extension in the Upper Allochthon, which could indicate that gravitational collapse (thinning and extension) coeval with nappe translation, as is known from the Caledonian allochthon in mid-Scandinavia (Gee 1978), was operative also along the Sognefjord transect. The top-to-NW phase which post-dates the base Devonian unconformity and pre-dates the main movement on the Nordfjord–Sogn detachment (West Coast D3) represents the rifting associated with the formation of the Devonian basins.

Contraction, extension and timing

The relative sequence of structural events, which in many cases represents an artificial subdivision of a continuous process, and the correlation of these events across the orogen (Fig. 3), provides the basis for an intepretation in terms of orogen-wide deformation and timing. The terms contraction and extension will be used in the first instance with reference to the horizontal strain field in the 'rheological upper crust'. This can be considered as that part of the crust which lies within the brittle regime (pressure-dependent deformation by fracture, frictional sliding and cataclastic flow, see Ranalli & Murphy 1987), with relatively high strength, or which can sustain stress, if ductile, over time periods of the order of 10 Ma (Rey *et al.* this volume). For the South Norwegian Caledonides, a low geothermal gradient can be assumed during collision and throughout subsequent extension, since there is an almost total absence of magmatic activity. For such conditions, the stiff upper crust is expected to be 15–20 km thick. This will overlie a weak 'rheological lower crust' (temperature and strain rate-dependent deformation by power-law creep, stress relaxation time less than 10 Ma, see Ranalli & Murphy 1987; Rey *et al.* this volume), which in turn overlies a stiff upper mantle, using the same criteria.

Within the rheological upper crust, horizontal contraction or extension relative to the Earth's surface at any particular time can usually be unambiguously deduced from the geometry and kinematics of the contemporaneous faults and shear zones (structural criteria in Wheeler & Butler 1994). The structural history of the Jotunheimen segment can be subdivided into a contractional phase and an extensional phase on this basis (Fig. 3). The Jotunheimen segment, together with the more fragmentarily exposed West Coast segment, spent all its orogenic life in the rheological upper crust. Contraction and extension, on this basis, describe the strain in the upper crust as a whole, i.e. the change in distance between two points on undeformed Baltica and undeformed Laurentia.

In the course of orogeny, rock masses may cross the boundary between the rheological upper crust and the rheological lower crust. For the periods of time when a rock mass lies in the rheological lower crust, pervasive strain in the weakly ductile matrix may not directly reflect upper crustal movement. In the Sognefjord transect, this applies particularly to the phase Sognefjord-D2 in the Western Gneiss Complex (Fig. 3). The significance of this phase only becomes clear after carrying out the retrodeformation (see below).

Finally, there is the question of absolute timing, i.e. the relation between the established sequence of structural events (relative ages), and stratigraphic and radiometric age data. This is summarized on Fig. 4. Because of the uncertainties inherent in the data, both in assigning radiometric ages to the stratigraphic boundaries and in interpreting radiometric ages from metamorphic rocks, we feel that a ± 10 Ma margin of error must be assumed, independent of the analytical accuracy achieved. Nevertheless, the age determinations give a consistent picture (Fig. 4). Ophiolite obduction on to the margin of Baltica started in mid-Wenlock times (425 Ma, Andersen *et al.* 1990), contemporaneous with the change from marine to continental sedimentation on the foreland (Bassett 1985). Most U/Pb and Nm/Sd ages on eclogites in the Western Gneiss Complex, interpreted as ages of formation, lie in the period 425–400 Ma, and Ar/Ar ages on muscovite and Rb/Sr ages on biotite, interpreted as cooling ages, in the range from 400 to 375 Ma (Fig. 4, e.g. Kullerud *et al.* 1986; Mørk & Mearns 1986; Jamtveit *et al.* 1991; Chauvet & Dallmeyer 1992). In addition, the U/Pb and Rb/Sr data from the Jotunheimen detachment zone (Schärer 1980) indicate that SE movement ceased at about 395 Ma, agreeing with the general experience that the achievement of 300 km or more of contraction by thrusting requires a time span of tens of millions of years. The start of sedimentation in the Devonian basins is not well constrained, but, on the basis of plant and fish fossils, is generally placed at the lower–middle Devonian boundary

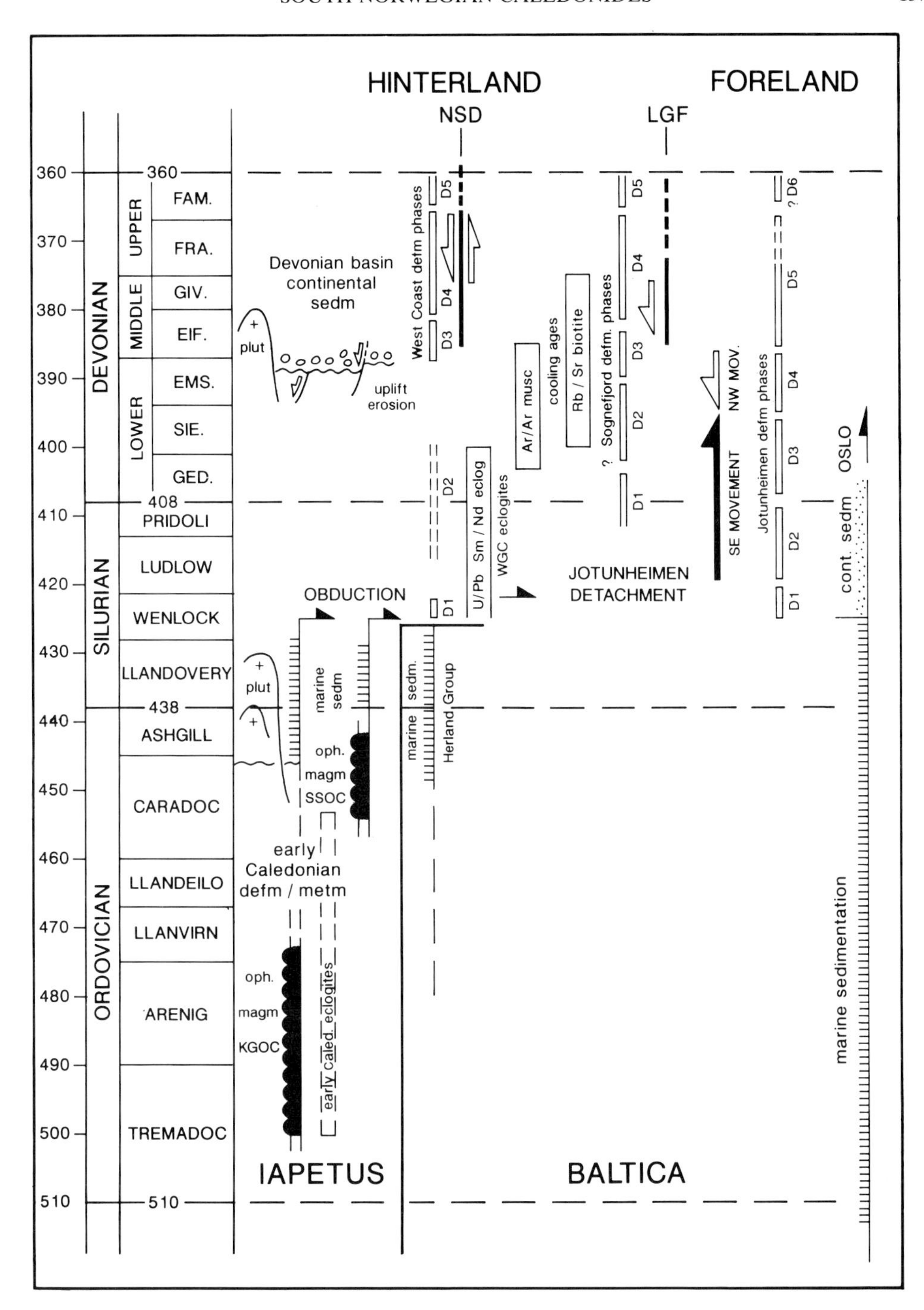

Fig. 4. Orogenic timetable for the South Norwegian Caledonides. The time-scale was constructed by taking average values from the following recent time-scales: DNAG (1983), IUGS (1989), Menning (1989) and Harland *et al.* (1990). All points are subject to an error of ±10 Ma. Main sources: Andersen *et al.* 1990; Andersen & Andresen 1994; Bassett 1985; Chauvet & Dallmeyer 1992; Furnes *et al.* 1989; Jamtveit *et al.* 1991; Kullerud *et al.* 1986; Mørk & Mearns 1986; Pedersen *et al.* 1988; Schärer 1980; Steel *et al.* 1985; Torsvik *et al.* 1986. KGOC, Karmøy–Gullfjellet ophiolite complex; SSOC, Solund–Stavfjord ophiolite complex. Deformation phases from Fig. 3 (for discussion, see text).

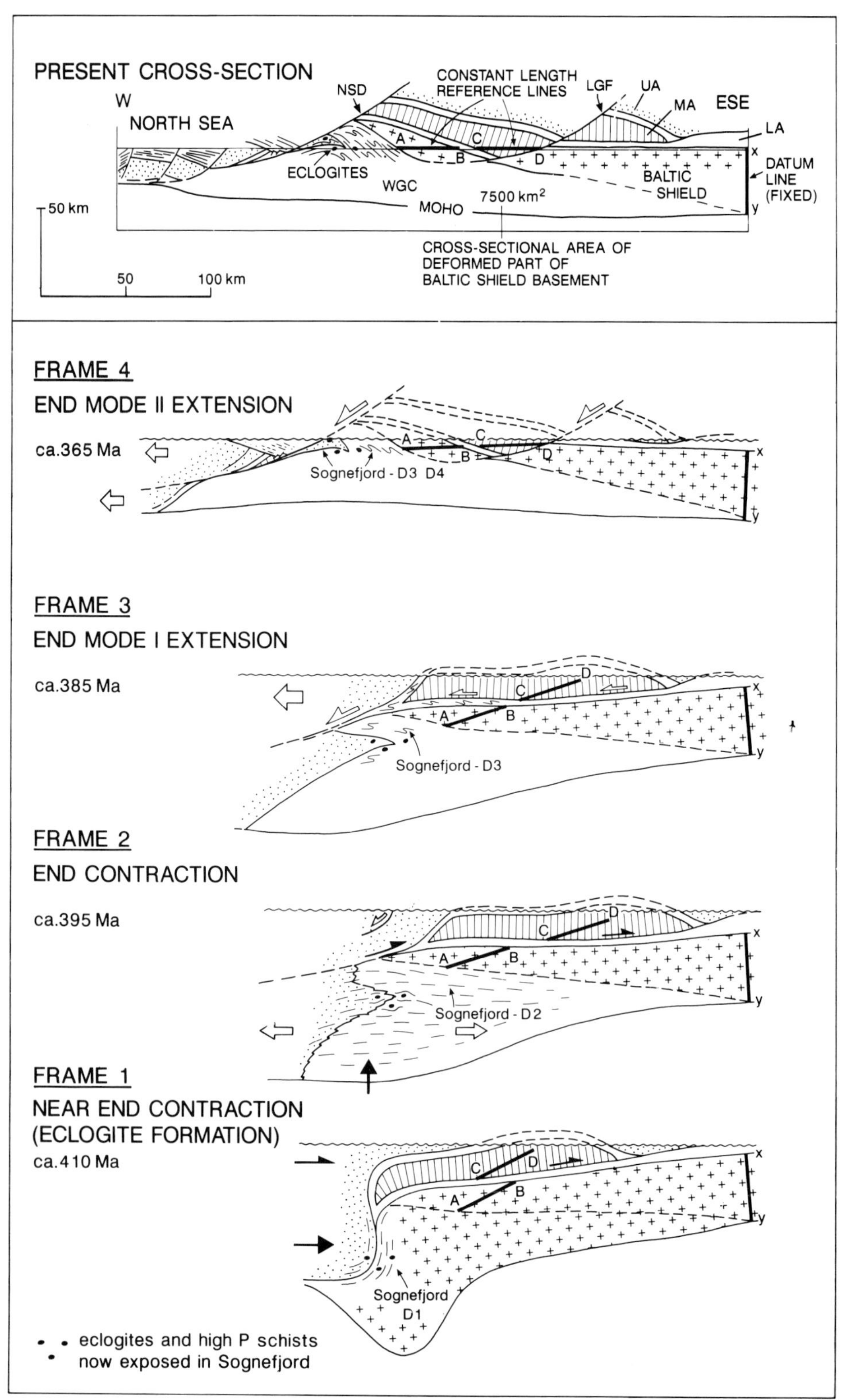

Fig. 5. Retrodeformation of the Sognefjord transect (see text, for explanation). For details of the simplified 'present cross-section', see Fig. 1.

(e.g. Kiær 1918; Jarvik 1949). A small pluton in the hanging wall of the Nordfjord–Sogn detachment has been reported to be of middle Devonian age (Furnes *et al.* 1989), with a wide margin of error, and is suggested to be contemporaneous with sporadic lavas interbedded with the Devonian conglomerates. These are the only signs of syn-extensional magmatism. Finally, paleomagnetic investigations suggest that the upright, E–W folding of the Devonian basins, the Nordfjord–Sogn detachment and the footwall mylonites are of late Devonian age (Torsvik *et al.* 1986, 1987). The thin zone of brittle deformation which caps the mylonites is a result of Permian and Upper Jurassic–Lower Cretaceous extensional reactivation (Torsvik *et al.* 1992*a*).

These data are plotted against the time-scale in Fig. 4, together with the approximate position of the deformation phases summarized in Fig. 3.

Retrodeformation

Reconstructing relationships backward in time by removing the effects of successively older events has been referred to as 'retrodeformation' (e.g. Suppe 1980, 1985), the tectonic equivalent of 'back-stripping' in basin analysis. It depends on a detailed knowledge of the deformational histories of the different exposed parts of the orogen, and the correlation of those histories across the whole transect, just as back-stripping relies on detailed knowledge of, and correlation between, well logs. Along the Sognefjord transect, the 'wells' are the columns in Fig. 3, and a correlation was established in the last section. The other basic elements of retrodeformation are balancing (e.g. Hossack 1979; Suppe 1980; Gibbs 1983) and depth control, i.e. the reconstruction of the Earth's surface at each stage in the process (see, for instance, Milnes & Pfiffner 1977; Milnes 1978; Pfiffner 1986). The basic assumptions used for balancing and depth control in the present case will be discussed first, before the results of the retrodeformation (Fig. 5) are outlined. The aim of retrodeformation is to reconstruct the orogenic structure as far back in time as possible, without recourse to conceptual models. In the case of the Sognefjord transect, this can be reasonably carried out to the time of formation of the Scandian eclogites in the Western Gneiss Complex.

Balancing

The following assumptions have been made in constructing the profiles on Fig. 5.

(1) The wedge of the Baltic Shield basement in the hinterland is assumed to be the northwestern edge of the un-Caledonized part of the crust, after its initial restoration by reversing the movement on the Lærdal–Gjende fault. At the eastern end of the transect, the crust is taken to be un-Caledonized throughout its thickness and a vertical line at that position is taken as a fixed datum line (Fig. 5, line XY). The shape of the crustal wedge is, of course, arbitrary: the lower surface is unlikely to be planar, as drawn, and there may have been Caledonian effects southeastwards from the datum line in the lower crust. The rest of the Baltic Shield basement is assumed to have undergone ductile-penetrative deformation of the type exposed in the Sognefjord segment. For balancing the crustal section, its area has been kept approximately constant throughout the process. Within the wedge, a reference line has been inscribed, along the present erosion surface, which remained practically undeformed throughout the orogeny (Fig. 5, line AB).

(2) The Jotun Complex has also remained only slightly Caledonized throughout orogenesis. However, it occurs today as erosional remnants of a much larger nappe whose thickness and horizontal extent are unknown. Here, the reconstruction of the original size and shape of the complex is based on the assumption that the fragment of Jotun-like basement in the hanging wall of the Nordfjord–Sogn detachment (Dalsfjord Complex) represents the westermost tip of the original nappe, disconnected from the main mass by movement on the detachment. Its original thickness is taken to be about 15 km from the present thickness preserved in the 'Faltungsgraben' (>10 km, Fig. 1) and from the estimate of the metamorphic temperatures reached in the Jotunheimen detachment zone giving maximum depths of about 20 km (see below). Another reference line (Fig. 5, line CD) is inscribed in the Jotun Complex at the present erosion level as a passive, practically undeformed marker.

Depth control

Part of our depth control is based on knowledge of metamorphic conditions during the different deformation phases. This knowledge is only rudimentary in many cases. However, the presence of phyllites in the Jotunheimen detachment zone (Fig. 1, Fortun and Vang phyllites) suggests that this zone should not be allowed to attain depths of more than 20 km at point B on line AB (Fig. 5), where the Baltic Shield basement is overlain by phyllites which reached the greenschist/amphibolite facies boundary. The depth of the Jotunheimen detachment is a major control on the whole retrodeformation, since the Lower Allochthon at the eastern end of the transect probably never even sank below 10 km (see section on Lower Allochthon, above). Since the Lower Allochthon contains the cover of the Western Gneiss Complex, it was obviously stripped off before the Western Gneiss Complex was eclogitized (present exposures representing depths of 60–70 km, see above). Further depth control on the Middle Allochthon is provided by the Ordovician–Silurian cover of the Dalsfjord Complex (Herland Group, Fig. 1), which has also experienced only low-grade metamorphic conditions.

The other type of depth control which has to be applied to rock masses which do not undergo internal deformation is that the erosion level at each stage in the retrodeformation must lie above the previous one, i.e. eroded parts cannot appear again in the cross-section at a later time. This is particularly important for the Jotun Complex, since parts must have been eroding away throughout much of these later stages of the orogeny (Fig. 5).

Results

The results of the exercise are shown in Fig. 5. At the top, the constructed present-day cross-section (Fig. 1) is presented in a simplified form. The retrodeformation starts with a late Devonian reconstruction (Fig. 5, frame 4, *c.* 365 Ma), after the end of movement on the Nordfjord–Sogn detachment and the Lærdal–Gjende fault. Since there is little information on the erosion level at this time, the erosion surface was drawn just slightly above the present level (Fig. 5, AB and CD).

The first step is the removal of the effects of the Mode II extension in the upper crust, including the removal of the coarse clastic sequences in the middle Devonian basins. This reconstructs conditions at the end of the Mode I extension phase (Fig. 5, frame 3, *c.* 385 Ma). Focusing on the Western Gneiss Complex, this removes the Sognefjord-D4 deformation and places the eclogites at a depth of 30 km by movement in the footwall of a crustal scale extensional detachment, with a displacement of about 50 km (Fig. 5). This represents strong footwall uplift which must have resulted in high mountain chains, since it was true uplift: the basal Devonian unconformity on the hanging wall, at the Earth's surface today, was near the Earth's surface during deposition and forms the Earth's surface at this stage of orogenic evolution. In contrast, the footwall loses 30 km of crust by erosion (or rather, gains it, during retrodeformation).

The next step is to remove the Mode I extension (Fig. 3) and to reconstruct conditions at the end of contraction (Fig. 5, frame 2, *c.* 395 Ma). This pushes the Middle Allochthon (Jotun Complex) to its farthest outreach over the foreland. For this step we have removed an estimated 20 km of top-to-NW movement on the Jotunheimen detachment (see Jotunheimen-D4). At deeper levels, the step removes the asymmetrical, top-to-W or NW folds (Sognefjord-D3) in the Western Gneiss Complex and considerable ductile thinning in the lower crust. This step in the retrodeformation has relatively little effect on eclogite depth. It shows that at the end of the contractional phase, the currently exposed eclogites lay at about 40 km and were almost completely retrograded to amphibolite facies, since the all-pervasive Sognefjord-D2 deformation was nearing completion.

The last step is the reconstruction of conditions at the time of eclogite formation (Fig. 5, frame 1, *c.* 410 Ma). This is more speculative, but still subject to important constraints. The meta-sediments in the Jotunheimen detachment, which is in its later phase of development (Jotunheimen-D3), must remain at shallow depths, as must the top of the rigid wedge of the Baltic Shield (Fig. 5, point B). But the eclogites must move downwards to 60–70 km at the same time as the main deformation in the surrounding gneisses (Sognefjord-D2) is removed. This leads to a deep root which can only be achieved by a process of downfolding of the thrust zone separating the Baltic Shield basement from the Upper Allochthon, after the Middle Allochthon has passed and after the Lower Allochthon has been stripped off. It is the destruction of this root which lifts the eclogites from 60–70 km to 40 km depth at this time and produces the Sognefjord-D2 deformation, with bulk subvertical shortening and bulk horizontal E–W extension. This process took place whilst the horizontal contraction of the rheological upper crust was continuing.

The retrodeformation (Fig. 5) is a kinematic reconstruction: it is not concerned with the forces involved, only with the probable geometry of the crustal segment at different points of time, based on the structural data and their interpretation in terms of strain and movement. Its interpretation in terms of driving forces and orogenic processes (dynamic models) is discussed in the following section.

Discussion of dynamic models

The Sognefjord transect through the South Norwegian Caledonides (Fig. 1) provides one of the best exposed and most completely documented examples of late collisional tectonics in a collisional orogen. Collision, in the sense of the final closing of the Iapetus ocean (suture formation) and the obduction of oceanic remnants onto the margin of Baltica, took place in mid-Silurian times, and was followed by intense deformation of the overridden continental crust, involving extreme crustal shortening (measured in hundreds of kilometres) and thickening. During collisional contraction, the obducted oceanic remnants were transported passively on the back of the Baltica-derived allochthon, uplifted and eroded over a broad area, and partly overridden and incorporated into the developing contractional detachment, the Jotunheimen detachment zone in the present transect. The early and late collisional phases, referred to together as the Scandian orogeny, are closely comparable to Meso- and Neo-Alpine orogenies which affected the European continental margin of the Central Alps (Table 1). The difference lies in the sizes of the structures and the distances of transport, which in the Scandinavian Caledonides are almost an order of magnitude larger. Related to this, the amount of crustal thickening was also exceptional, twice that in the present Himalaya. The overridden crust developed a deep root, probably extending down to 150 km in places, causing the formation of the well-known Scandian eclogites of western Norway, preserved within gneisses which belong to the outermost affected parts of the Baltica Shield basement (Fig. 2). The retrodeformation carried out above provides a framework for dynamic models of the later phase of the collisional process, which resulted in the formation and exhumation of the eclogites. The exhumation process, for the eclogites on the Sognefjord transect, from 60–70 km depth to the present-day surface, can be described in terms of three phases, as outlined below.

Eclogite formation took place after most of the Lower and Middle Allochthon had moved to a position east of the present eclogite localities (Fig. 5, frame 1) and the Upper Allochthon was juxtaposed directly against Baltic Shield basement. Subsequent shortening of the rheological lower crust was accomplished by the development of a deep root, mainly of Baltic Shield basement (becoming later the Western Gneiss Complex), but including parts of the Upper Allochthon, and, in other areas, remnants of the Lower/Middle Allochthon (Bryhni 1989; Andersen & Jamtveit 1990; Wilks & Cuthbert 1994). Deep root formation at this position is difficult to relate to the original (presumed W dipping) collision zone and may mark late collisional thickening at a new site, either due to a change in subduction polarity, or, possibly more likely, to the 'indentor'-type thickening seen in some finite-element models (e.g. Beaumont & Quinlan 1994, model 5) and in interpretations of the deep structure of the Alps (e.g. Argand 1924; Schmid *et al.* in press). Exhumation began at about 410 Ma (Fig. 5, frame 1), when the eclogites now exposed in outer Sognefjord, and their country rocks, lay at depths of 60–70 km (see *PT*t-paths in Bailey 1989 and Dewey *et al.* 1993).

Exhumation phase 1

During the first phase of eclogite exhumation (Fig. 5, from frame 1 to frame 2), intense deformation took place and all eclogitized rocks were retrograded to amphibolite facies with the exception of the cores of some mafic pods. The reconstruction suggests that this isothermal decompression event (Andersen & Jamtveit 1990; Dewey *et al.* 1993) procedes by 'inverted gravity spreading' of the root due to buoyancy (Ramberg 1980), resulting in subvertical shortening and subhorizontal E-W extension of a large part of the rheological lower crust (the Sognefjord-D2 phase). The top of this zone is marked today at the surface by the transition from strongly Caledonized to un-Caledonized Precambrian basement along the Sognefjord profile. At the same time, parts of the Upper Allochthon, and in other areas possibly parts of Lower and Middle Allochthon, become tightly folded into the Western Gneiss Complex. The rheological upper crust, however, is still undergoing contraction at this time, and acts as a rigid lid (Fig. 5, frame 2). The gravity-driven upward movement and lateral spreading of the lower crust is thought of as similar to the processes causing the overhangs around salt diapirs (Ramberg 1981). Similar effects have been observed in centrifuge models of orogens (Ramberg & Sjöström 1973) and the concept has been mentioned earlier in a Caledonide context (Ramberg 1980), but hitherto no specific examples have been given.

During this phase, the presently exposed eclogites, undergoing retrogression to amphibolite facies, rose from 60–70 km to about 40 km depth over a period of around 10 Ma (exhumation rate 2–3 mm a^{-1}). This process corresponds to that described by Platt (1993) under the heading 'exhumation by buoyancy forces' after crustal rocks are dragged down into the mantle. Platt remarks that 'buoyant return through the mantle may in fact be the only viable mechanism for the first stages in exhumation of ultra-high-pressure terrains . . .' (p. 125). Almost as interesting, however, is the corollary, that the 'Caledonization' of the Baltic Shield basement (Sognefjord-D2) is due to bulk coaxial deformation driven solely by gravity. Metamorphic conditions at the time were very high grade, but there are no signs of anatexis or magmatism (the widespread migmatites, granites and pegmatite veins in the Western Gneiss Complex all predate Sognefjord-D2, see above). The isotherms rose at the same rate as exhumation, with *P–T* conditions remaining below the eutectic, and the geothermal gradient increased from 10°C km^{-1} to 15°C km^{-1} during exhumation phase 1.

Exhumation phase 2

At the transition from upper crustal contraction to extension (Fig. 5, frame 2 to frame 3), bouyancy forces are envisaged as resulting in uplift and erosion of the stiff upper crust over a broad area due to horizontal stress relaxation, marked today by the base Devonian unconformity. During this phase, the all-pervasive, ductile–penetrative deformation in the lower crust (Sognefjord-D2) ended, cooling started, and the mineral parageneses (eclogite and amphibolite facies) became 'frozen in'. Exhumation slowed down to 1 mm a^{-1} (10 km in 10 Ma), and the gneisses enclosing the currently exposed eclogites slowly became part of the rheological upper crust, whilst thinning and spreading continued at depth. The first signs of upper crustal extension appear as top-to-NW movement on the Jotunheimen detachment and distributed asymmetrical folding in the Western Gneiss Complex (Fig. 5, frame 3). Since the crust is still overthickened, the plateau that developed at this time may have been very high and gravitational adjustments could have caused extension (mode I) in the upper crust concomitant with the thinning and spreading at depth.

Exhumation phase 3

In the final phase of exhumation (Fig. 5, frame 3 to frame 4), the presently exposed gneisses with their eclogite pods become an integral part of the rheological upper crust. The main phase of upper crustal extension begins, accommodated by movement on low-angle, cross-cutting (mode II) extensional detachments, of which the Nordfjord–Sogn detachment becomes the most long-lived and fundamental structure. At an early stage, detachment-related basins received large volumes of coarse clastic sediments from localized areas of high relief due to the uplift, tilting and erosion of the footwalls. However, sedimentation in these local basins ceased long before the main detachment became inactive and long before the eclogites reached the surface. Apart from the truncation of basin stratigraphy and deformation of basin fill by movement on the detachment, the best indication of this is the paucity of typical Western Gneiss Complex lithologies in the Devonian conglomerates (Cuthbert 1991). During the course of this phase, the footwall of the Nordfjord-Sogn detachment was strongly uplifted and eroded, and tilted *en bloc*, to accommodate the upper crustal extension. The eclogites were exhumed a further 30 km (at an exhumation rate of 1.5 mm a^{-1}), and the crust of the footwall assumed a normal thickness, which it has retained ever since. Conditions during this phase have been compared with those leading to the formation of metamorphic core complexes in the Basin and Range province, USA (e.g. Andersen & Jamtveit 1990; Dewey *et al.* 1993). Superficial resemblances do exist. However, there are no signs of the anomalously thin crust, the anomalously high heat flow and the widespread volcanic activity typical of the Basin and Range and similar regions (Platt & England 1994), so the comparison should not be taken too far.

The exhumation history outlined above agrees in general with earlier models i.e. an early phase of exhumation by vertical bulk coaxial thinning followed by a later phase of strongly localized footwall uplift along extensional detachments (Dewey *et al.* 1993). The present kinematic reconstruction has allowed some details to be filled out, and provides a dynamic model for the early phase: inverted gravity spreading of the crustal root due to buoyancy. Rutter & Brodie (1992) have shown *en passant* that such a process is mechanically plausible: they estimate that a Moho topography of only 10 km is necessary for buoyancy to overcome the creep strength of quartzo-feldspathic rocks at 500–600°C and a strain rate of *c.* $10^{-14} s^{-1}$.

The relatively good agreement at a crustal scale contrasts sharply with the divergent dynamic models which have been proposed for the southern Scandinavian Caledonides at a lithospheric scale (Rey *et al.* this volume). These fall into two groups.

(1) Lithospheric collapse models are based on the postulated occurrence of advective or convective lithospheric thinning by the detachment and sinking of part of the lithospheric root below the thickened crust (Andersen & Jamtveit 1990; Andersen *et al.* 1991; Fossen 1992; Andersen

1993; Dewey *et al.* 1993). The resultant upwelling of hot asthenosphere causes high heat flow, plateau uplift and volcanism at the surface, and the hot, weak, uplifted crustal welt collapses under gravity (Dewey 1988; Platt & England 1994). Thrusting (contraction) continues along the outer margins, towards the foreland, whilst extensional tectonics dominates in the region of high elevation (hinterland), as seen, for instance, in the western Mediterranean and surrounding mountain chains (Vissers *et al.* 1995). The processes are gravity-driven and not necessarily related to plate motion, which can continue to be convergent, as in present-day Tibet (England & Houseman 1989). The extension is an integral part of the orogenic process and is therefore referred to as late orogenic extension (Table 1).

Application to the South Norwegian Caledonides provides a conceivable framework for the eclogite exhumation and related features, but it meets several difficulties (Fossen 1992, 1993*b*; Wilks & Cuthbert 1994; Rey *et al.* this volume). One of the main ones is the lack of evidence for high heat flow, particularly the lack of evidence for appreciable Scandian magmatism (anatexis, intrusions, volcanism). Using crustal balancing arguments, Wilks & Cuthbert (1994) also show that the size of the postulated collapse-related fold-and-thrust belts and foreland basins would be out of all proportion to those actually observed. In addition, as illustrated in the present paper, the geometry and timing of the main contractional and extensional detachments in the Sognefjord transect are hard to relate to those expected from this mechanism. However, other transects through the Caledonides, such as the Central Scandinavian transect (e.g. Gee *et al.* 1985; Palm *et al.* 1991), show good evidence for syn-contractional extension, with simultaneous extensional thinning of the higher nappes in the hinterland and continued thrusting over the foreland (Gee 1978). The evidence indicates, therefore, that a high plateau did develop (exhumation phase 2, above) and that gravity spreading due to surface topography (Gee 1978), as well as due to Moho topography (exhumation phase 1, this paper), did take place during late collisional contraction in the South Norwegian Caledonides. However, it seems unlikely that the main extensional detachments (exhumation phase 3, above) were formed and driven solely by that process. It also seems unlikely, in agreement with the conclusions of Rey *et al.* (this volume), that the plateau uplift and extensional collapse was related to advective or convective lithospheric thinning, as envisaged in the papers cited above.

(2) Lithospheric extension models relate the extensional tectonics in southwestern Norway to a change in plate motion, caused by processes external to the orogen (the 'free space' model of Wilks & Cuthbert 1994; see also Fossen 1992; Séranne *et al.* 1991). Many authors interpret global plate motions during the Devonian as implying a regional sinistral wrench regime at the site of the Caledonian orogen (e.g. Ziegler 1985; Coward 1993; Rey *et al.* this volume). Séranne *et al.* (1991) relate the Devonian extensional tectonics in western Norway (our exhumation phase 3, Fig. 5, frames 3 and 4) to a large-scale releasing overstep within this regime, and Chauvet & Séranne (1994) suggest that the system of E–W upright folds along the west coast (West Coast-D5 and Sognefjord-D5, see Fig. 2) are syn-extensional in the same regime. Rey *et al.* (this volume) argue that the wrench regime was probably preceded by a period of pure extension (Early Devonian), giving an age, 395 Ma, which is identical to our estimate for the change from contraction to extension in the Sognefjord transect (Fig. 4). These models imply that the extensional tectonics is not directly related to the Scandian orogeny, i.e. that it is basically post-orogenic (Table 1). There seems to be no reason to negate that the upper crustal extension implied by the top-to-NW movement on the Caledonian thrusts (extension mode I) and the formation of large-scale cross-cutting extensional detachments (extension mode II) could be related to these changes in plate motion.

In conclusion, one can say that the southern Scandinavian Caledonides, as illustrated by the Sognefjord transect, exemplifies a collisional process which resulted in extreme crustal thickening. The orogenic welt so formed was strongly modified by gravitational forces caused by the buoyancy of the crustal root and possibly (by analogy with the central Caledonides) by collapse of the corresponding topographic high. This occurred particularly towards the end of plate convergence and coincided with the early part of the eclogite exhumation process. Crustal extension at a later stage was mainly a result of a change from convergent to divergent or transcurrent plate motion. The later part of the exhumation process was achieved by footwall uplift below a major extensional detachment which developed at this time. There is no evidence for advective or convective lithospheric thinning marking the change from contractional to extensional tectonics. In fact, perhaps the southern Scandinavian Caledonides exemplify what happens when detachment and sinking of part of the lithospheric root does not take place.

This paper was significantly improved by the thorough work of the reviewers, D. Gee, A. Pfiffner and C. Talbot. This was very much appreciated. We also wish to thank H. Fossen, N. Odling and R. Romer for useful comments on an early draft of the manuscript.

References

ANDERSEN, T. B. 1993. The role of extensional tectonics in the Caledonides of south Norway: discussion. *Journal of Structural Geology*, **15**, 1379–1380.

—— & ANDRESEN, A. 1994. Stratigraphy, tectonostratigraphy and the accretion of outboard terranes in the Caledonides of Sunnhordland, W. Norway. *Tectonophysics*, **231**, 71–84.

—— & JAMTVEIT, B. 1990. Uplift of deep crust during orogenic extensional collapse: a model based on field studies in the Sogn-Sunnfjord region of West Norway. *Tectonics*, **9**, 1097–1112.

—— & JANSEN, Ø. J. 1987. The Sunnhordland Batholith, W. Norway: regional setting and internal structure, with emphasis on the granitoid plutons. *Norsk Geologisk Tidsskrift*, **67**, 159–183.

——, SKJERLIE, K. P. & FURNES, H. 1990. The Sunnfjord Melange, evidence of Silurian ophiolite accretion in the West Norwegian Caledonides. *Journal of the Geolological Society, London*, **147**, 59–68.

——, JAMTVEIT, B., DEWEY, J. F. & SWENSSON, E. 1991. Subduction and eduction of continental crust: major mechanisms during continent-continent collision and orogenic extensional collapse, a model based on the south Norwegian Caledonides. *Terra Nova*, **3**, 303–310.

——, OSMUNDSEN, P. T. & JOLIVET, L. 1994. Deep crustal fabrics and a model for the extensional collapse of the southwest Norwegian Caledonides. *Journal of Structural Geology*, **16**, 1191–1203.

ARGAND, E. 1924. La tectonique de l'Asie. *Proceediings International Geological Congress*, 13rt Session **1**, 171–372.

BAILEY, D. E. 1989. *Metamorphic evolution of the crust of southwestern Norway: an example from Sognefjord*. PhD thesis, Oxford University.

BASSETT, M. G. 1985. Silurian stratigraphy and facies development in Scandinavia. *In*: GEE, D. G. & STURT, B. A. (eds) *The Caledonide Orogen – Scandinavia and Related Areas*. Wiley, Chichester, 283–292.

BATTEY, M. H. 1965. Layered structure in rocks of the Jotunheimen complex, Norway. *Mineralogical Magazine*, **34**, 35–51.

BEAUMONT, B. & QUINLAN, G. 1994. A geodynamic framework for interpreting crustal-scale seismic-reflectivity patterns in compressional orogens. *Geophysical Journal International*, **116**, 754–783.

BREKKE, H. & SOLBERG, P. O. 1987. The geology of Atløy, Sunnfjord, western Norway. *Norwegian Geological Survey Bulletin*, **410**, 73–94.

BRYHNI, I. 1988. Early Palaeozoic metamorphism in the Scandinavian Caledonides. *In*: HARRIS, A. L. & FETTES, D. J. (eds) *The Caledonian–Appalachian Orogen*, Geological Society, London, Special Publications, **38**, 135–140.

—— 1989. Status of the supracrustal rocks in the Western Gneiss Region, S. Norway. *In*: GAYER, R. A. (ed.) *The Caledonide Geology of Scandinavia*. Graham and Trotman, London, 221–228.

CHAUVET, A. & DALLMEYER, R. D. 1992. Ar/Ar mineral dates related to Devonian extension in the southwestern Scandinavian Caledonides. *Tectonophysics*, **210**, 155–177.

—— & SÉRANNE, M. 1994. Extension-parallel folding in the Scandinavian Caledonides: implications for late-orogenic processes. *Tectonophysics*, **238**, 31–54.

——, KIENAST, J. R., PINARDON, J. L. & BRUNEL, M. 1992. Petrological constraints and *PT* path of Devonian collapse tectonics within the Scandian Mountain Belt. *Journal of Geological Society, London*, **149**, 383–400.

CORFU, F. 1980. U–Pb and Rb–Sr systematics in a poly-orogenic segment of the Precambrian shield, central southern Norway. *Lithos*, **13**, 305–323.

—— & EMMETT, T. 1992. U–Pb age of the Leirungsmyran gabbroic complex, Jotun Nappe, southern Norway. *Norsk Geologisk Tidsskrift*, **72**, 369–374.

COWARD, M. P. 1993. The effect of Late Caledonian and Variscan continental escape tectonics on basement structure, Paleozoic basin kinematics and subsequent Mesozoic basin development in NW Europe. *In*: PARKER, J. R. (ed.) *Petroleum Geology of Northwest Europe: Proceedings of the 4th Conference*. Geological Society, London, 1095–1108.

CUTHBERT, S. J. 1991. Evolution of the Devonian Hornelen Basin, west Norway: new constraints from petrological studies of metamorphic clasts. *In*: MORTON. A. C., TODD, S. P. & HAUGHTON, P. D. W. (eds) *Developments in Sedimentary Provenance Studies*. Geological Society, London, Special Publications, **57**, 343–360.

—— & CARSWELL, D. A. 1990. Formation and exhumation of medium-temperature eclogites in the Scandinavian Caledonides. *In*: CARSWELL, D. A. (ed.) *Eclogite Facies Rocks*. Blackie (Glasgow), 180–204.

——, HARVEY, M. A. & CARSWELL, D. A. 1983. A tectonic model for the metamorphic evolution of the Basal Gneiss Complex, Western South Norway. *Journal of Metamorphic Geology*, **1**, 63–90.

DNAG 1983. *Decade of North American Geology time scale*. *Geology*, **11**, 503–504.

DALLMEYER, R. D. 1988. Polyphase tectonothermal evolution of the Scandinavian Caledonides. *In*: HARRIS, A. L. & FETTES, D. J. (eds) *The Caledonian–Appalachian Orogen*. Geological Society, London, Special Publications, **38**, 365–379.

——, ANDREASSON, P. G. & SVENNINGSEN, O. 1991. Initial tectonthermal evolution within the Scandinavian Caledonide acrretionary prism: constraints from 40Ar/38Ar mineral ages within the Seve Nappe Complex, Sarek Mountains, Sweden. *Journal of Metamorphic Geology*, **9**, 203–218.

DEEMER, S. J. & HURICH, C. A. 1991. Comparison of coincident high-resolution wide-aperture and CDP profiling along the southwest coast of Norway. *In*: *Continental Lithosphere: Deep Seismic Reflections*. American Geophysical Union (Washington DC), Geodynamics Series, **22**, 435–442.

DEWEY, J. F. 1988. Extensional collapse of orogens. *Tectonics*, **7**, 1123–1139.

——, RYAN, P. D. & ANDERSEN, T. B. 1993. Orogenic uplift and collapse, crustal thickness, fabrics and metamorphic phase changes: the role of eclogites. *In*: PRITCHARD, H. M. *et al.* (eds) *Magmatic Processes and Plate Tectonics*. Geological Society, London, Special Publications, **76**, 325–343.

DIETLER, T. N. 1987. *Strukturgeologische und tektonische Entwicklung der Western Gneiss Complexes im Sognefjord-Querschnitt, westliches Norwegen*. Doctoral thesis, ETH Zürich.

——, KOESTLER, A. G. & MILNES, A. G. 1985. A preliminary structural profile through the Western Gneiss Complex, Sognefjord, southwestern Norway. *Norsk Geologisk Tidsskrift*, **65**, 233–235.

ENGLAND, P. & HOUSEMAN, G. A. 1989. Extension during continental convergence, with application to the Tibetan plateau. *Journal of Geophysical Research*, **94**, 17561–17579.

FOSSEN, H. 1992. The role of extensional tectonics in the Caledonides of south Norway. *Journal of Structural Geology*, **14**, 1003–1046.

—— 1993*a*. Structural evolution of the Bergsdalen Nappes, Southwest Norway. *Norwegian Geological Survey Bulletin*, **424**, 23–50.

—— 1993*b*. The role of extensional tectonics in the Caledonides of south Norway: reply. *Journal of Structural Geology*, **15**, 1381–1383.

—— & AUSTRHEIM, H. 1988. Age of the Krossnes Granite. *Norwegian Geological Survey Bulletin*, **413**, 61–65.

—— & HOLST, T. B. 1995. Northwest-verging folds and the northwestward movement of the Caledonian Jotun Nappe, Norway. *Journal of Structural Geology*, **17**, 3–15.

FURNES, H., PEDERSEN, R. B., SUNDVOLL, B., TYSSELAND, M. & TUMYR, O. 1989. The age, petrography, geochemistry and tectonic setting of the late Caledonian Gåsøy Intrusion, west Norway. *Norsk Geologisk Tidsskrift*, **69**, 273–289.

——, SKJERLIE, K. P., PEDERSEN, R. B., ANDERSEN, T. B., STILLMAN, C. J., SUTHRENS, R. J., TYSSELAND, M. & GARMAN, L. B. 1990. The Solund-Stafjord Ophiolite Complex and associated rocks, west Norwegian Caledonides: geology, geochemistry and tectonic environment. *Geological Magazine*, **127**, 209–224.

FÆRSETH, R. B., GABRIELSEN, R. H. & HURICH, C. A. 1995. Influence of basement in structuring of the North Sea basin, offshore Southwest Norway. *Norsk Geologisk Tidsskrift*, **75**, 105–119.

GAAL, G. & GORBATSCHEV, R. 1987. An outline of Precambrian evolution of the Baltic Shield. *Precambrian Research*, **35**, 15–52.

GEE, D. G. 1978. Nappe displacement in the Scandinavian Caledonides. *Tectonophysics*, **47**, 393–419.

——, GUÉZOU, J.-C., ROBERTS, D. & WOLFF, F. C. 1985. The central-southern part of the Scandinavian Caledonides. *In*: GEE, D. G., STURT & B. A. (eds) *The Caledonide Orogen – Scandinavia and Related Areas*. Wiley, Chichester, 109–133.

——, LOBKOWICZ, M. & SINGH, S. 1994: Late Caledonian extension in the Scandinavian Caledonides – the Røragen Detachment revisited. *Tectonophysics*, **231**, 139–155.

GIBBS, A. D. 1983. Balanced cross-section construction from seismic sections in areas of extensional tectonics. *Journal of Structural Geology*, **5**, 153–160.

—— 1987. Deep seismic profiles in the northern North Sea. *In*: BROOKS, J. & GLENNIE, K. W. (eds) *Petroleum Geology of North-West Europe*. Graham & Trotman, London, **2**, 1025–1028.

GOLDSCHMIDT, V. M. 1912. Die kaledonische Deformation der sfdnorwegischen Urgebirgstafel. *Skrifler Norsk Videnskaps-Akademi. Oslo, Mat. naturvitensk. Kl.*, **19**, 1–11.

GORBATSCHEV, R. 1985. Precambrian basement of the Scandinavian Caledonides. *In*: GEE, D. G. & STURT, B. A. (eds) *The Caledonide Orogen-Scandinavia and Related Areas*. Wiley, Chichester, 197–212.

GOWER, C. F., RYAN, A. B. & RIVERS, T. 1990. Mid-Proterozoic Laurentia-Baltica: overview of its geological evolution and a summary of the contributions made by this Volume. *In*: GOWER, C. F., RIVERS, T. & RYAN, B. (eds) *Mid-Proterozoic Geology of the Southern Margin of Proto-Laurentia-Baltica*. Geological Association of Canada, Special Papers, **38**, 1–20.

GRIFFIN, W. L. & MØRK, M. B. E. 1981. *Eclogites and basal gneisses in Western Norway*. Uppsala Caledonide Symposium, Excursion Guide B1.

——, AUSTRHEIM, H., BRASTAD, K., BRYHNI, I., KRILL, A. G., KROGH, E. J., MØRK, M. B. E., QVALE, H. & TØRUDBAKKEN, B. 1985. High-pressure metamorphism in the Scandinavian Caledonides. *In*: GEE, D. G. & STURT, B. A. (eds) *The Caledonide Orogen–Scandinavia and Related Area*. Wiley, Chichester, 783–801.

HARLAND, W. B., ARMSTRONG, R. L., COX, A. V., CRAIG, L. E., SMITH, A. G. & SMITH, D. G. 1990. *A Geologic Time Scale 1989*. Cambridge University Press, Cambridge, UK.

HARTZ, E., ANDRESEN, A. & ANDERSEN, T. B. 1994. Structural observations adjacent to a large-scale extensional detachment zone in the hinterland of the Norwegian Caledonides. *Tectonophysics*, **231**, 123–137.

HEIM, M. 1979. *Struktur und Petrographie des Jotun-Valdres-Deckenkomplexes und der ihn unterlagenden kaledonischen Deformationszone im Gebiet des östlichen Vangsmjøsi (zentrales Süd-norwegen*. Doctoral thesis, ETH Zürich.

——, SCHÄRER, U. & MILNES, A. G. 1977. The nappe complex in the Tyin-Bygdin-Vang region, central southern Norway. *Norsk Geologisk Tidsskrift*, **57**, 171–178.

HOSSACK, J. R. 1972. The geological history of the Grønsennknipa nappe, Valdres. *Norwegian Geological Survey Bulletin*, **281**, 1–26.

—— 1976. Geology and structure of the Beito window, Valdres. *Norwegian Geological Survey Bulletin*, **327**, 1–33.

—— 1979. The use of balanced cross-sections in the calculation of orogenic contraction: a review. *Journal of the Geological Society, London*, **136**, 705–711.

—— 1984. The geometry of the listric faults in the Devonian basins of Sunnfjord, Western Norway. *Journal of the Geological Society, London*, **141**, 629–637.

——, KOESTLER, A. G., LUTRO, O., MILNES, A. G. & NICKELSEN, R. P. 1981. *A traverse from the foreland through the Jotun-Valdres thrust sheets, Jotunheimen*. Uppsala Caledonide Symposium, Excursion Guide B3.

——, GARTON, M. R. & NICKELSEN, R. P. 1985. The geological section from the foreland up to the Jotun thrust sheet in the Valdres area, south Norway. *In*: GEE, D. G. & STURT, B. A. (eds) *The Caledonide Orogen–Scandinavia and Related Areas*. Wiley, Chichester, 443–456.

HURICH, C. A. & KRISTOFFERSEN, Y. 1988. Deep structure of the Caledonide orogene in southern Norway: new evidence from marine seismic reflection profiling. *In*: KRISTOFFERSON, Y. (ed.) *Progress in Studies of the Lithosphere in Norway*. Norwegian Geological Survey Special Publications, **3**, 96–101.

IUGS 1989. Global Stratigraphic Chart. *Episodes*, **12**.

INDREVÆR, G. & STEEL, R. J. 1975. Some aspects of the sedimentary and structural history of the Ordovician and Devonian rocks of the westernmost Solund islands, West Norway. *Norwegian Geological Survey Bulletin*, **317**, 23–32.

IWASAKI, T., SELLEVOLL. M. A., T., KANAZAWA, T., VEGGELAND, T. & SHIMAMURA, H. 1994. Seismic refraction crustal study along the Sognefjord, south-west Norway, employing ocean-bottom seismometers. *Geophysical Journal International*, **119**, 791–808.

JAMTVEIT, B., CARSWELL, D. A. & MEARNS, E. W. 1991. Chronology of the high-pressure metamorphism of Norwegian garnet peridotites/pyroxenites. *Journal of Metamorphic Geology*, **9**, 125–139.

JARVIK, E. 1949. On the Middle Devonian Crossopterygians from the Hornelen field in western Norway. *Universitetet i Bergen Aarbok 1948, Naturvidenskaplige Række*, **8**, 1–48.

KINCK, J. J., HUSEBYE, E. S. & LARSSON, F. R. 1993. The Moho depth distribution in Fennoscandia and the regional tectonic evolution from Archean to Permian times. *Precambrian Research*, **64**, 23–51.

KIÆR, J. 1918. Fiskerester fra den devonske sandsten paa Norges vestkyst. *Bergens Museums Aarbok, Naturvidenskaplige Række*, **7**, 1–17 [in Norwegian with English summary].

KLEMPERER, S. L. 1988. Crustal thinning and nature of extension in the northern North Sea from deep seismic profiling. *Tectonics*, **7**, 803–821.

KOENEMANN, F. H. 1993. Tectonics of the Scandian Orogeny and the Western Gneiss Region in southern Norway. *Geologische Rundschau*, **82**, 696–717.

KOESTLER, A. G. 1982. A Precambrian age for the Ofredal granodiorite intrusion, Central Jotun Nappe, Sogn, Norway. *Norsk Geologisk Tidsskrift*, **62**, 225–228.

—— 1983. *Zentralkomplex und NW-Randzone der Jotundecke, West-Jotunheimen*. Doctoral thesis, ETH Zürich.

—— 1988. Heterogeneous deformation and mylonitization of a granulite complex, Jotun-Valdres Nappe Complex, central South Norway. *Geological Journal*, **23**, 1–13.

—— 1989. *Hurrungane, berggrunnskart 1517 IV*, 1:50 000. Norwegian Geological Survey, bedrock maps.

KORJA, A., KORJA, T., LUOSTO, U. & HEIKKINEN, P. 1993. Seismic and geoelectric evidence for collisional and extensional events in the Fennoscandian Shield – implications for Precambrian crustal evolution. *Tectonophysics*, **219**, 129–152.

KROGH, E. J. 1980. Geochemistry and petrology of glaucophane-bearing eclogites and associated rocks from Sunnfjord, western Norway. *Lithos*, **13**, 355–380.

KULLERUD, L., TØRUDBAKKEN, B. O. & ILEBEKK, S. 1986. A compilation of radiometric age determinations from the Western Gneiss Region, South Norway. *Norwegian Geological Survey Bulletin*, **406**, 17–42.

LUTRO, O. 1986. *Lustrafjord, berggrunnskart 1417 I*, 1 : 50 000. Norwegian Geological Survey bedrock maps.

—— 1987. *Møkrisdalen, berggrunnskart 1418 II*, 1 : 50 000. Norwegian Geological Survey, bedrock maps.

—— 1988. *Beskrivelse. Bergrunnskart 1417 I, Lustrafjorden*, 1 : 50 000. Norwegian Geological Survey, Skrifter, **83**, 1–38.

—— 1989. Bergartsbeskrivelse: berggrunnskart 1418 II, Møkrisdalen, 1 : 50 000. *Norwegian Geological Survey, Skrifter*, **94**, 2–24.

—— & TVETEN, E. 1987. *Geologisk kart over Norge, berggrunnskart Årdal 1 : 250 000, foreløpig utgave*. Norwegian Geological Survey, bedrock maps, provisional edition.

MENNING, M. 1989. A synopsis of numerical time scales 1917–1986 and a new time scale for Central Europe. *Episodes*, **12**, 3–5.

MILNES, A. G. 1978. Structural zones and continental collision, Central Alps. *Tectonophysics*, **47**, 369–392.

—— 1987. The Lower Allochthon in southern Norway: an exhumed analog of the southern Appalachian deep detachment? *In*: SCHAER, J-P. & RODGERS, J. (eds) *The Anatomy of Mountain Ranges*. Princeton University Press, 59–64.

—— & KOESTLER, A. G. 1985. Geological structure of Jotunheimen, southern Norway (Sognefjell-Valdres cross-section. *In*: GEE, D. G. & STURT, B. A. (eds) *The Caledonide Orogen – Scandinavia and Related Areas*. Wiley, Chichester, 457–474.

—— & PFIFFNER, O. A. 1977. Structural development

of the Infrahelvetic complex, eastern Switzerland. *Eclogae Geologicae Helvetiae*, **70**, 83–95.
——, DIETLER, T. N. & KOESTLER, A. G. 1988. The Sognefjord north shore log – a 25 km depth section through Caledonized basement in Western Norway. *In*: KRISTOFFERSON, Y. (ed.) *Progress in Studies of the Lithosphere in Norway*. Norwegian Geological Survey Special Publication, **3**, 114–121.
MORLEY, C. K. 1986. The Caledonian thrust front and palinspastic reconstructions in the southern Norwegian Caledonides. *Journal of Structural Geology*, **8**, 753–765.
—— 1987. Lateral and vertical changes of deformation style in the Osen-Røa thrust sheet, Oslo region. *Journal of Structural Geology*, **9**, 331–343.
MØRK, M. B. E. & MEARNS, E. W. 1986. Sm-Nd isotopic systematics of a gabbro-eclogite transition. *Lithos*, **19**, 255–267.
——, KULLERUD, K.& STABEL, A. 1988. Sm-Nd dating of Seve eclogites, Norrboten, Sweden – evidence for early Caledonian (505 Ma) subduction. *Contributions to Mineralogy and Petrology*, **99**, 344–351.
NICKELSEN, R. P., HOSSACK, J. R., GARTON, M. & REPETSKY, J. 1985. Late Precambrian to Ordovician stratigraphy and correlation in the Valdres and Synnfjell thrust sheets of the Valdres area, southern Norwegian Caledonides; with some comments on sedimentation. *In*: GEE, D. G. & STURT, B. A. (eds) *The Caledonide Orogen – Scandinavia and Related Areas*. Wiley, Chichester, 369–378.
NILSEN, T. H. 1968. The relationship between sedimentation and tectonics in the Solund Devonian District of southwestern Norway. *Norwegian Geological Survey Bulletin*, **259**, 1–108.
NORTON, M. G. 1986. Late Caledonian extension in western Norway: a response to extreme crustal thickening. *Tectonics*, **5**, 195–204.
—— 1987. The Nordfjord-Sogn detachment, W. Norway. *Norsk Geologisk Tidsskrift*, **67**, 93–106.
OSMUNDSEN, P. T. & ANDERSEN, T. B. 1994. Caledonian compressional and late-orogenic extensional deformation in the Staveneset area, Sunnfjord, Western Norway. *Journal of Structural Geology*, **16**, 1385–1401.
PALM, H., GEE, D. G., DYRELIUS, D. & BJÖRKLUND, L. 1991. *A reflection seismic image of Caledonian structure in Central Sweden*. Swedish Geological Survey, Series Ca **75**.
PEDERSEN, R. B., FURNES, H. & DUNNING, G. 1988. Some Norwegian ophiolite complexes reconsidered. *In*: KRISTOFFERSON, Y. (ed.) *Progress in Studies of the Lithosphere in Norway*. Norwegian Geological Survey Special Publication, **3**, 80–85.
PFIFFNER, O. A. 1986. Evolution of the north Alpine foreland basin in the Central Alps. *In*: ALLEN, P. A. & HOMEWOOD, P. (eds) *Foreland Basins*. International Association of Sedimentologists, Special Publications, **8**, 219–228.
PLATT, J. P. 1993. Exhumation of high-pressure rocks: a review of concepts and processes. *Terra Nova*, **5**, 119–133.
—— & ENGLAND, P. C. 1994. Convective removal of lithosphere beneath mountain belts: thermal and mechanical consequences. *American Journal of Science*, **294**, 307–336.
PRIEM, H. N. A., BOELRIJK, N. A. I. M., HEBEDA, E. H., VERDUMEN, E. A. T. & VERSCHURE, R. H. 1973. A note on the geochronology of the Hestbrepiggan granite in West Jotunheimen. *Norwegian Geological Survey Bulletin*, **289**, 31–35.
QUALE, H. 1982. *Jotundekkets anorthositter: Geologi, mineralogi og geokjemi*. Norwegian Geological Survey Report, **1560/32**.
RAMBERG, H. 1980. Diapirism and gravity collapse in the Scandinavian Caledonides. *Journal of the Geological Society, London*, **137**, 261–270.
—— 1981. *Gravity, Deformation and the Earth's Crust*. Academic Press, London.
—— & SJÖSTRÖM, H. 1973. Experimental geodynamical models relating to continental drift and orogenesis. *Tectonophysics*, **19**, 105–132.
RANALLI, G. & MURPHY, D. C. 1987. Rheological stratification of the lithosphere. *Tectonophysics*, **132**, 281–295.
REY, P., BURG, J.-P., & CASEY, M. 1997. The Scandinavian Caledonides and their relationship to the Variscan belt. *This volume*.
ROBERTS, D. & GEE, D. G. 1985. An introduction to the structure of the Scandinavian Caledonides. *In*: GEE, D. G. & STURT, B.A. (eds) *The Caledonide Orogen – Scandinavia and Related Areas*. Wiley, Chichester, 55–68.
RUTTER, E. H. & BRODIE, K. H. 1992. Rheology of the lower crust. *In*: FOUNTAIN, D. M., ARCULUS, R. & KAY, R. (eds) *The Geology of the Lower Continental Crust*. Elsenvier, Amsterdam, 201–267.
SCHÄRER, U. 1980. U-Pb and Rb-Sr dating of a polymetamorphic nappe terrain: the Caledonian Jotun nappe, southern Norway. *Earth and Planetary Science Letters*, **49**, 205–218.
SCHMID, S. M., FROITZHEIM, N., PFIFFNER, O. A., SCHÖNBORN, G. & KISSLING, E. Geophysical–geological transect and tectonic evolution of the Swiss-Italian Alps. *Tectonics*, in press.
SELLEVOLL, M. A. & MOKHTARI, M. 1985. *The Coast-Fjord test project*. Seismological Observatory, University of Bergen, Seismology Series, **6**.
SÉRANNE, M. & SEGURET, M. 1987. The Devonian Basins of western Norway: tectonics and kinematics of an extending crust. *In*: COWARD, M. P. *et al.* (eds) *Continental Extensional Tectonics*. Geological Society, London, Special Publications, **28**, 537–548.
——, CHAUVET, A. & FAURE, J.-L. 1991. Cinématique de l'estension tardi-orogénique (Dévonien) dans les Calédonides Scandinaves et Britanniques. *Comptes Rendus de l'Academie de Sciences, Paris, Serie II*, **313**, 1305–1312.
SKILBREI, J. R. 1990. *Structure of the Jotun Nappe Complex, southern Norwegian Caledonides, ambiguity of gravity modelling, and reinterpretation*. Norwegian Geological Survey, Rapport, **89.169**.
SKJERLIE, K. P. & FURNES, H. 1990. Evidence for a

fossil transform fault in the Solund-Stavfjord Ophiolite Complex, West Norwegian Caledonides. *Tectonics*, **9**, 1631–1648.

SKÅR, Ø., FURNES, H. & CLAESSON, S. 1994. Middle Proterozoic magmatism within the Western Gneiss Region, Sunnfjord, Norway. *Norsk Geologisk Tidsskrift*, **74**, 114–126.

SMITH, D. C. & LAPPIN, M. A. 1989. Coesite in the Straumen kyanite-eclogite pod, Norway. *Terra Nova*, **1**, 47–56.

SMITHSON, S. B., RAMBERG, I. B. & GRØNLIE, G. 1974. Gravity interpretation of the Jotun nappe of the Norwegian Caledonides. *Tectonophysics*, **22**, 205–222.

STARMER, I. C. 1993. The Sveconorwegian Orogeny in southern Norway, relative to deep structures and events in the North Atlantic Proterozoic Supercontinent. *Norsk Geologisk Tidsskrift*, **73**, 109–132.

STEEL, R., SIEDLECKA, A. & ROBERTS, D. 1985. The Old Red Sandstone basins of Norway and their deformation: a review. *In*: GEE, D. G. & STURT, B. A. (eds) *The Caledonide Orogen – Scandinavia and related Areas*. Wiley, Chichester, 293–315.

STEPHENS, M. B. 1988. The Scandinavian Caledonides: a complexity of collisions. *Geology Today*, **4**, 1, 20–26.

STEPHENS, B. S. & GEE, D. G. 1989. Terranes and polyphase accretionary history in the Scandinavian Caledonides. *In*: DALLMEYER, R. D. (ed.) *Terranes in the Circum-Atlantic Paleozoic Orogens*. Geological Society of America Special Papers, **230**, 17–30.

SUPPE, J. 1980. A retrodeformable cross-section of northern Taiwan. *Geological Society of China, Proceedings*, **23**, 46–55.

—— 1985. *Principles of Structural Geology*. Prentice-Hall, Engelwood Cliffs, NJ.

SWENSSON, E. & ANDERSEN, T. B. 1991. Contact relationships between the Askvoll group and the basement gneisses of the Western Gneiss Region (WGR), Sunnfjord, Western Norway. *Norsk Geologisk Tidsskrift*, **71**, 15–27.

TORSVIK, T. H., STURT, B. A., RAMSAY, D. M., KISCH, H. J. & BERING, D. 1986. The tectonic implications of Solundian (Upper Devonian) magnetization of the Devonian rocks of Kvamshesten, western Norway. *Earth and Planetary Science Letters*, **80**, 337–347.

——, ——, —— & VETTI, V. 1987. The tectonomagnetic signature of the Old Red Sandstone and pre-Devonian strata in the Håsteinen area, Western Norway, and implications for the later stages of the Caledonian Orogeny. *Tectonics*, **6**, 305–322.

——, ——, SWENSSON, E., ANDERSEN, T. B. & DEWEY, J. 1992*a*. Paleomagnetic dating of fault rocks: evidence for Permian and Mesozoic movements and brittle deformation along the extensional Dalsfjord Fault, western Norway. *Geophysical Journal International*, **109**, 565–580.

——, SMETHURST, M. A., VAN DER LOO, M., TRENCH, A., ABRAHAMSEN, N. & HALVORSEN, E. 1992*b*. Baltica: a synopsis of Vendian–Permian paleomagnetic data and their paleotectonic implications. *Earth Science Reviews*, **33**, 133–152.

TUCKER, R. D., KROGH, T. E. & RÅHEIM, A. 1990. Proterozoic evolution and age-province boundaries in the central part of the Western Gneiss Region, Norway: results of U-Pb dating of accessory minerals from Trondheimsfjord to Geiranger. *In*: GOWER, C. F., RIVERS, T. & RYAN, B. (eds) *Mid-Proterozoic Geology of the Southern Margin of Proto-Laurentia-Baltica*. Geological Association of Canada, Special Papers, **38**, 149–173.

VISSERS, R. L. M., PLATT, J. P. & VAN DER WAAL, D. 1995. Late orogenic extension of the Betic Cordillera and the Alboran Domain: a lithospheric view. *Tectonics*, **14**, 786–803.

WENNBERG, O. P. & MILNES, A. G. 1994. Interpretation of kinematic indicators along the northeastern margin of the Bergen Arc System: a preliminary field study. *Norsk Geologisk Tidsskrift*, **74**, 166–173.

WHEELER, J. & BUTLER, R. W. H. 1994. Criteria for identifying structures related to true crustal extension in orogens. *Journal of Structural Geology*, **16**, 1023–1027.

WILKS, W. J. & CUTHBERT, S. J. 1994. The evolution of the Hornelen Basin detachment system, western Norway: implications for the style of late orogenic extension in the southern Scandinavian Caledonides. *Tectonophysics*, **238**, 1–30.

ZIEGLER, P. A. 1985. Late Caledonian framework of western and central Europe. *In*: GEE, D. G. & STURT, B. A. (eds) *The Caledonide Orogen – Scandinavia and Related Areas*. Wiley, Chichester, 3–18.

Tectonics of the southeastern Australian Lachlan Fold Belt: structural and thermal aspects

DAVID R. GRAY

Australian Geodynamics Cooperative Research Centre, Department of Earth Sciences, Monash University, Melbourne, Vic. 3168, Australia

Abstract: The Lachlan Fold Belt of southeastern Australia is a 700 km wide belt of deformed, Palaeozoic deep and shallow marine sedimentary rocks, cherts, and mafic volcanic rocks. Characterized by large areas of chevron-folded turbidite sequences, linked contractional and strike-slip faults, and superposed thrust-belts of different age and vergence, it has large volumes of granite and low pressure/high temperature metamorphic rocks. Surface structural elements suggest that it formed by massive telescoping and strike-slip translation within a continental margin sediment prism along the former margin of Gondwanaland during the mid-Palaeozoic. Migrating, sporadic compressional and extensional events over the 100 Ma deformational history of the belt produced a crustally thickened, high geothermal gradient orogen, involving non-craton directed thrusting in an inferred convergent margin setting. Compressional periods were marked by eastward-migrating deformation zones involving detachment-related folding and reverse faulting in the largely turbidite dominated sequences. High-grade metamorphic rocks are not part of a metamorphic hinterland, but are largely confined to a shear zone-bounded crustal wedge in the central part of the fold-belt. Extension was marked by localized development of rift basins and half grabens accompanied by extensive granitic magmatism and silicic volcanism. Subsequent basin collapse was by reactivation/inversion of the extensional faults and folding of the basinal sequence. Magmatic activity and high temperature/low pressure metamorphism are linked to a convergent margin setting over much of the Middle Palaeozoic, but with major convergence in the Silurian.

The Lachlan Fold Belt is an enigmatic part of the north–south-trending composite Palaeozoic Tasman Orogenic Belt along the eastern margin of Australia (Fig. 1). Part of a Phanerozoic orogenic system that formed along the eastern margin of Gondwanaland in the Palaeozoic, the Lachlan Fold Belt contains large tracts of deformed deep-marine sedimentary rocks (quartz-rich turbidites), cherts, and mafic volcanic rocks of Cambrian to Devonian age. It has a complex Early Cambrian to Early Carboniferous (*c*. 200 Ma) amalgamational and deformational history with inferred periods of compression (Late Ordovician–Early Silurian, Late Silurian, Early Devonian, Mid-Devonian, and mid- Early Carboniferous) and extension (Late Silurian, Early Devonian, Late Devonian) tectonics (e.g. Cas 1983; Coney *et al.* 1990). Showing little evidence of continent–continent collision, the Lachlan Fold Belt has been described as an 'accretionary continental margin orogen' that provides an unmodified example of Palaeozoic Circum-Pacific tectonics (Coney 1992). It has an upper crustal framework of linked thrust belts (Figs 2 & 3). Recent deep crustal seismic studies suggest that the steep faults at the present erosion level flatten with depth and link into a common detachment (see Gray *et al.* 1991*a*; Leven *et al.* 1992).

The character of the Lachlan Fold Belt is quite distinct and arguably somewhat different to other orogenic belts, such as the Alps, Appalachians and North American Cordillera (Coney *et al.* 1990; Coney 1992). It shows:

(1) the same lithotectonic assemblages, general structural style and level of exposure along the entire *c*. 1000 km length and across the *c*. 700 km breadth of the belt;
(2) dimensions comparable to the North American Cordillera (see Fig. 1);
(3) a dominant lithotectonic assemblage, mostly part of an Ordovician quartz-rich turbidite fan with palinspastic dimensions comparable to the Bengal Fan (e.g. Fergusson & Coney 1992*a*);

From Burg, J.-P. & Ford, M. (eds), 1997, *Orogeny Through Time*, Geological Society Special Publication No. 121, pp. 149–177.

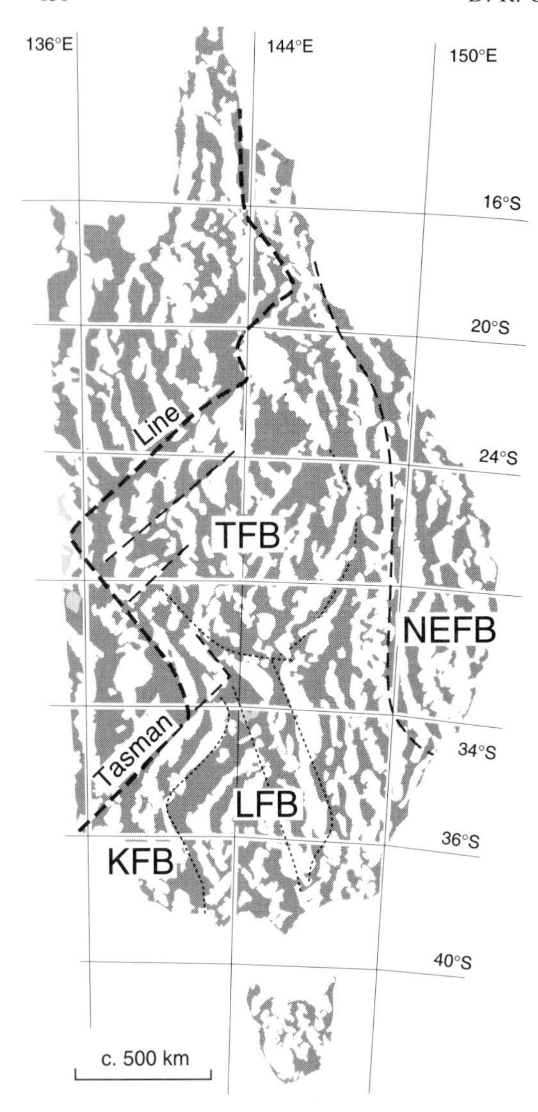

Fig. 1. Residual Bouger gravity map of eastern Australia in greyscale (from Murray *et al.* 1989, fig.5) with major subdivisions of the Tasman Orogenic Belt shown. Structural trends within, and relationships between, the major structural zones and fold belts are shown by shaded regions which are the relatively positive domains. The Tasman Line approximately delineates the Early Palaeozoic cratonic margin. More than two thirds of the Tasman Orogenic Belt lies beneath younger sedimentary cover basins. KFB, Kanmantoo Fold Belt; LFB, Lachlan Fold Belt: TFB, Thomson Fold Belt; NEFB, New England Fold Belt.

(4) no preservation of a shelfal-platform carbonate sequence;
(5) extraordinarily complex patterns of the timing of deformation (see Glen 1992; Fergusson & Coney 1992*b*);
(6) widespread granite plutons which account for almost one-third of the present outcrop (see Chapple *et al.* 1989);
(7) a minimum of 50% shortening in a belt at least 700 km wide (Gray 1988; Gray & Willman 1991*a,b; Fergusson & Coney 1992b*);
(8) no deep-seated crustal rocks brought-up on large-scale ramping thrusts.

There are many difficulties interpreting the geology of the Lachlan Fold Belt. For example, it is unclear how to reconcile the upper crustal structure, the extensive middle to lower crustal melting to produce the large volumes of granite, the high *T*–low *P* metamorphism, and the complex deformational pattern within the context of plate tectonics. Most aspects of the geology require a major convergent plate boundary somewhere to the east of the present Lachlan exposures. However, the effects of tectonic processes such as aborted 'A'-type intraplate subduction leading to orogen obduction (e.g. Woodward 1995), extensional roll-back due to sinking thickened lithosphere, leading to lithospheric delamination and crustal melting, are still being debated. The thin-skinned nature and marked upper crustal shortening have been explained by massive intraplate failure of a vast, quasi-oceanic–continental lithosphere (Fergusson & Coney 1992*b*). The extent and importance of granites during development of the Lachlan Fold Belt is also contentious. Certainly, the generation of large volumes of granite (up to 30% areal extent) have been explained previously through anomalous thermal behaviour in the mantle unrelated to plate subduction (Chappell *et al.* 1988), multiple subducting slabs (magmatic underplating) (Collins & Vernon 1992), and lithospheric delamination (Looseveld & Etheridge 1990; Cox *et al.* 1991; Collins 1994; Collins & Vernon 1994).

This paper attempts to summarize the important geological, structural and geochronological aspects of the Lachlan Fold Belt that help to explain the thermo-mechanical evolution of this orogenic belt. The role of thick- versus thin-skinned deformation is assessed and linked to the likelihood of a regionally extensive mid-crustal decollement underlying the presently exposed orogenic belt. The thermal evolution of the belt is related to granitic magmatism and crustal metamorphism, and the crustal structure linked to processes of lithospheric thickening and interpreted in terms of large scale crustal rheology. The timing and kinematics of deformation provide the basis for development of tectonic evolutionary models.

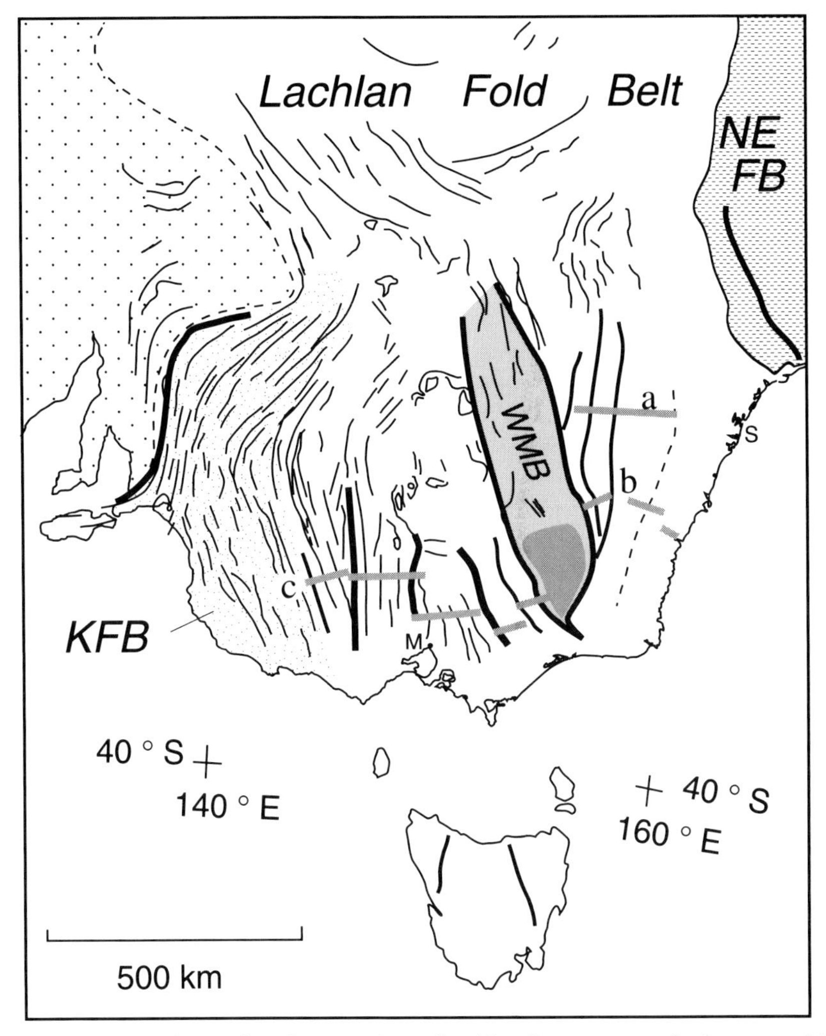

Fig. 2. Simplified geological map of southeastern Australia with sedimentary cover basins removed. The major geological elements outboard of the Precambrian craton (coarse stipple) are the Kanmantoo Fold Belt (KFB, fine stipple), the much larger Lachlan Fold Belt (blank with structural form lines) including the Wagga Metamorphic Belt (WMB, grey shading; heavier shading, higher grade Omeo belt), and the New England Fold Belt (NEFB, fine horizontal dashes). The superimposed fine lines are structural and aeromagnetic trend lines and the heavy lines are major fault traces. a, b and c indicate the positions of the tectonic profiles in Fig. 3. M, Melbourne; S, Sydney.

Geological framework

The Lachlan Fold Belt consists of three separate and distinct subprovinces (Figs 3 & 4), each with differences in rock type, metamorphic grade, structural history and geological evolution (Table 1). The western and central subprovinces are dominated by a turbidite succession consisting of quartz-rich sandstones and black shales. The eastern subprovince consists of mafic volcanics, volcaniclastic rocks and limestone, as well as quartz-rich turbidites and extensive black shale in the easternmost part (VandenBerg & Stewart 1992). The nature of the basement to the turbidite succession is controversial. In the western subprovince Cambrian mafic volcanic rocks of oceanic affinities underlie the quartz-rich turbidite succession, whereas in the eastern subprovince the oldest rocks observed are Ordovician shoshonitic volcanics and a Late Cambrian–Early Ordovician chert–turbidite–mafic volcanic sequence.

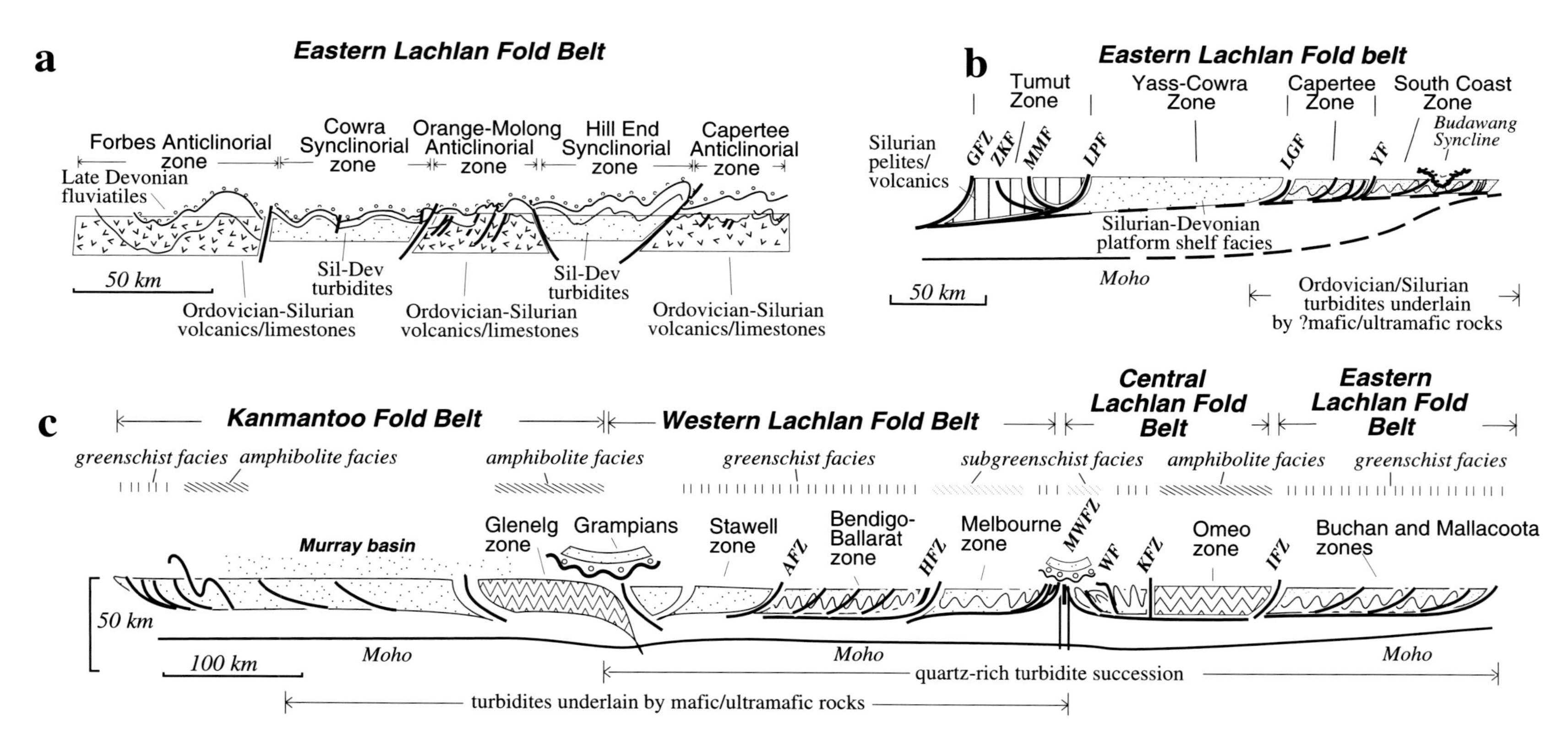

Fig. 3. Schematic west–east structural profiles showing the main geological, structural and metamorphic aspects of the Lachlan Fold Belt (vertical approximately equals horizontal scale). Locations of the sections are shown on Fig. 2. (**a**) Eastern Lachlan Fold Belt (33°S latitude). (**b**) Eastern Lachlan Fold Belt at 36°S latitude. GFZ, Gilmore Fault Zone; KFZ, Killimicat Fault Zone; MMF, Mooney Mooney Fault; LPF, Long Plain Fault; LGF, Lake George Fault; YF, Yarralaw Fault. (**c**) Lachlan Fold Belt at 37°S latitude. AFZ, Avoca Fault Zone; HFZ, Heathcote Fault Zone; MWFZ, Mt Wellington Fault Zone; WF, Wonnangatta Fault; KFZ, Kiewa Fault Zone; IFZ, Indi Fault Zone.

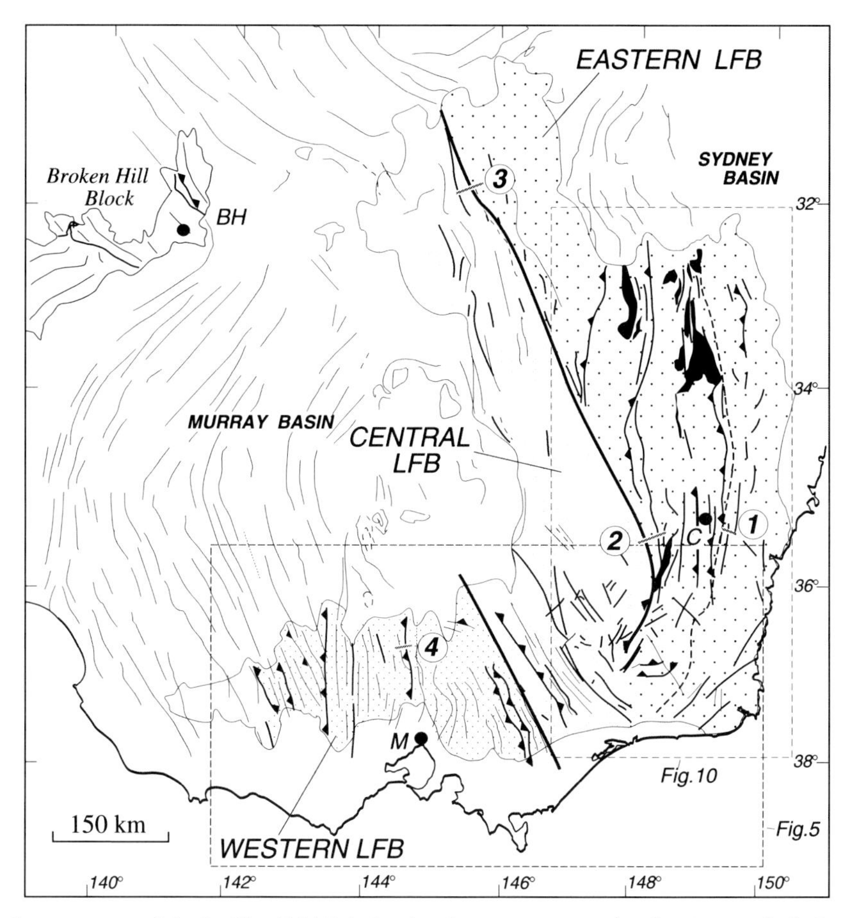

Fig. 4. Structure map of the Lachlan Fold Belt showing the western, central and eastern geological subprovinces, the cover sequences of the Murray and Sydney basins (grey shaded regions), structural and aeromagnetic trend lines (medium lines), major faults (heavy lines with barbs: contraction faults; heavy lines without barbs: strike-slip faults), and the distribution of the Ordovician volcanic rocks (black areas). Detailed map regions of Figs. 5 and 10 are shown. Circled numbers indicate positions of seismic traverses: 1, Tibbinbilla–Braidwood; 2, Tumut; 3, Cobar; 4, Heathcote. M, Melbourne; C, Canberra; BH, Broken Hill.

Structurally the Lachlan Fold Belt consists of a simple sequence of upright chevron folds and steep faults. However, within this sequence there are a number of fault-bounded structural zones which show differences in structural trends, the timing and nature of deformation and tectonic vergence (Fig. 3). These zones show no simple accretionary trends and the fold belt is not dominated by thrust-belts verging towards the craton (i.e. westwards), like other orogenic systems (e.g. Appalachians). The western subprovince consists of an east-vergent thrust system with alternating zones of north-west- and north-trending structures (Fig. 3c). The central subprovince is dominated by north-west-trending structures and consists of a south-west vergent thrust-belt linked to a fault-bounded metamorphic complex (Morand & Gray 1991). The eastern subprovince is dominated by a north-south structural grain and east-directed thrusting associated with inverted extensional basins in the west and an east-vergent thrust system in the most eastern part which links into an accretionary complex (Fig. 3a, b).

The central subprovince clearly separates regions of different lithostratigraphy which may have been different parts of a complex plate margin. These regions were subsequently juxtaposed along the regional fault systems both within and bounding the central belt (Fergusson *et al.* 1986; Morand & Gray 1990; VandenBerg & Stewart 1992; Glen 1992). Facies analysis of the Ordovician sequences (VandenBerg &

Table 1 *Subprovinces of the Lachlan Fold Belt (modified from Rutland 1976)*

	Western	Central	Eastern
Main facies	Quartz-rich turbidite continental margin sequence	Quartz-rich turbidite continental margin sequence	Platform carbonates and clastics with rhyolites and dacitic tuffs Chert, melange, basalt (South Coast, NSW)
Age of Sequence	Ordovician to Early Carboniferous	Ordovician to Late Silurian	Ordovician to Early Carboniferous
Initial Record	Cambrian basic volcanics	Tremadoc chert	Ordovician andesitic volcanics
Main Deformation	Early Silurian to Late Silurian	Late Ordovician to Early Silurian	Mild Early Devonian with more intense Early Carboniferous phase
Terminal folding	Late Early Devonian	Mid-Silurian	Early Carboniferous
Tectonic vergence	East-directed thrusting	Overall strike-slip with southeast directed thrusting	East-directed thrusting
Metamorphic Grade	Greenschist facies or lower	Up to amphibolite facies	Greenschist facies or lower
Main plutonism	Late Devonian	Late Silurian	Late Carboniferous

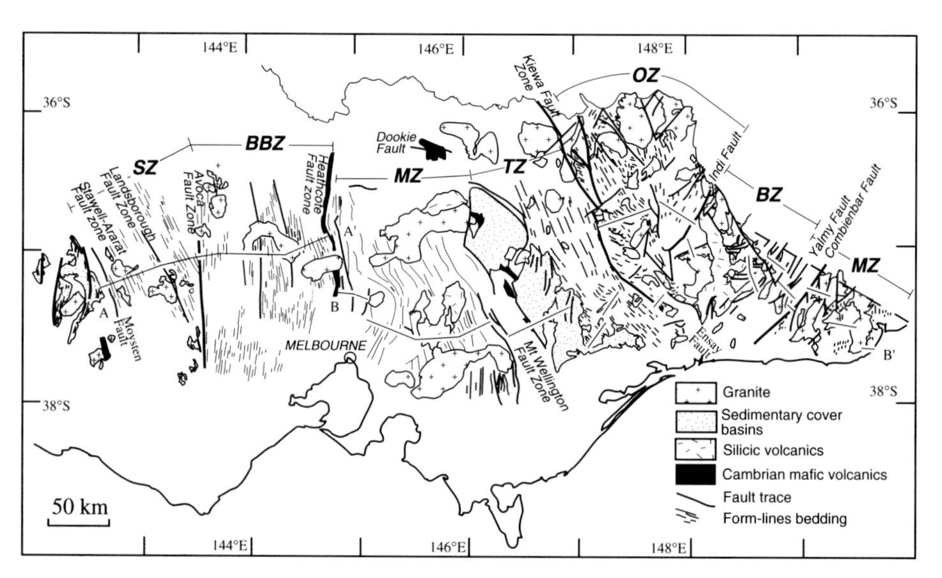

Fig. 5. Structure map of Victoria with trend lines (thin lines), major fault traces (heavy lines), outcrop patterns of the Cambrian 'greenstones' (black regions) and granites (regions designated by crosses) (Based on Gray 1988). Structural zones are designated by SZ, Stawell Zone; BBZ, Bendigo-Ballarat Zone; MZ, Melbourne Zone; TZ, Tabberabbera Zone; OZ, Omeo Zone; BZ, Buchan Zone; MZ, Mallacoota Zone.

Stewart 1992) suggests that approximately 400 km dextral displacement may have taken place on the leading fault of the Mt Wellington Fault Zone and/or Wonnangatta Fault (MWFZ & WF: Fig. 2c).

The Western Subprovince

The Western subprovince (Figs 5, 6, 7 & 8) is made up of a deformed turbidite sequence cut by a series of major strike-parallel, west-dipping,

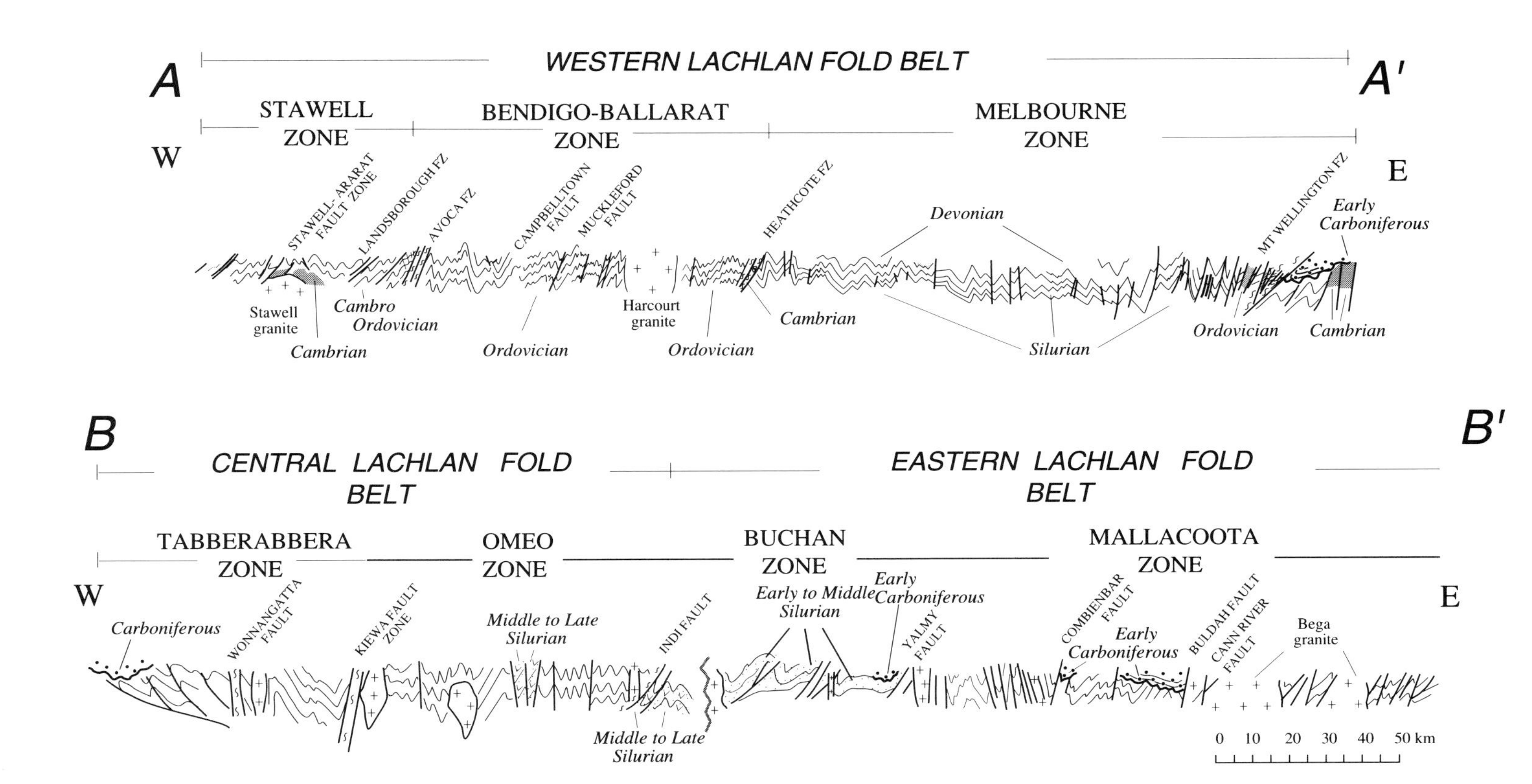

Fig. 6. Structural profile across the southern part of the Lachlan Fold Belt (vertical equals horizontal scale). Geological zones, major faults, bedding form lines, and granite bodies are shown. Chevron folding and steep faults define the structural style at the present level of exposure.

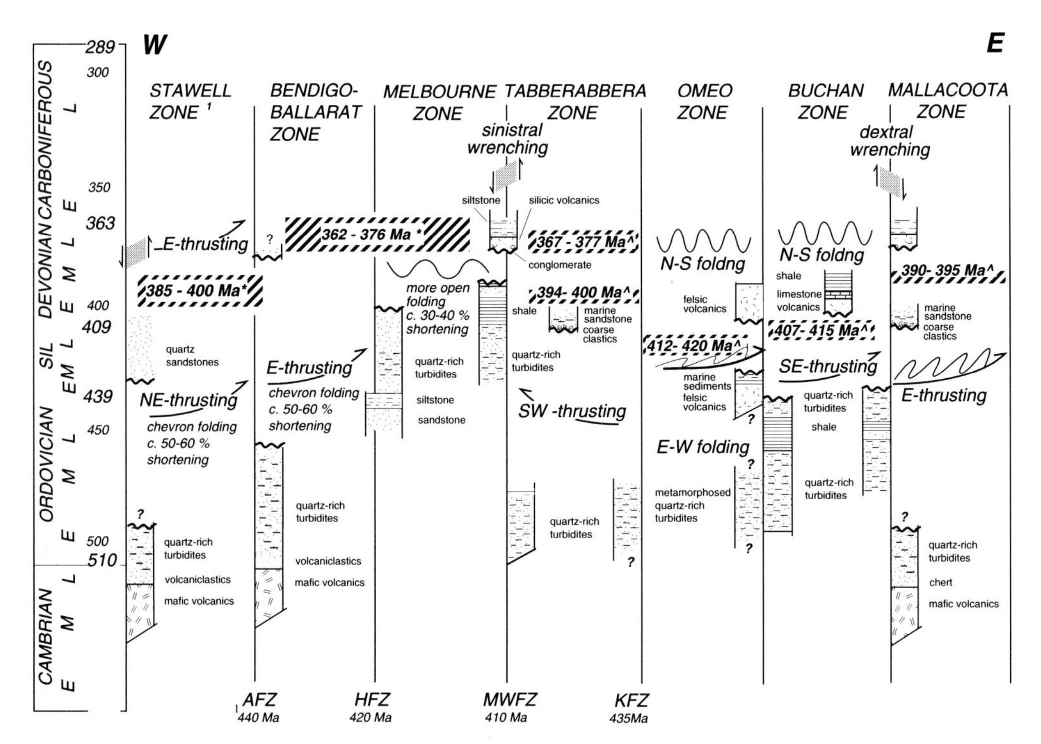

Fig. 7. Time–space plot showing the relationships between sedimentation, deformation and plutonism for the southern Lachlan Fold Belt (based on time scale of Harland *et al.* 1990). Subdivisions are the major geological zones of Victoria arranged from west to east (see Fig. 3). Ages and the nature of sediments, directions of thrusting, and age ranges of granitic intrusives (oblique slashes) are shown. Note the general eastward younging of the sedimentation and the deformation in the the western-most zones (i.e. the Stawell, Bendigo–Ballarat, and Melbourne Zones). Post-tectonic granitoids and clastic and volcanic overlap sequences constrain the deformation in this western part of the fold belt to be pre-Late Devonian in age. A major change occurs across the Mt Wellington Fault Zone between the Melbourne and Tabberabbera Zones. AFZ: Avoca Fault zone; HFZ, Heathcote Fault zone; MWFZ, Mt Wellington fault zone; KFZ, Kiewa Fault zone.
[1]Ages are crystallization ages of white mica in phyllites from major fault zones (see Bucher *et al.* 1996*a*)
*Granite age data from Richards and Singleton (1981) and Gray (1990).

reverse faults which link into an inferred major mid-crustal detachment fault to form an east-vergent thrust system (e.g. Cox *et al.* 1991; Gray & Willman 1991*a, b*; Gray *et al.* 1991*b*; Glen 1992; Fergusson & Coney 1992*a*). These major faults are steeply dipping, up to 2 km wide zones of strongly deformed rocks with intense development of obliquely trending crenulation cleavages associated with generally, but variably steeply plunging meso- and micro-folds. The faults expose Cambrian metavolcanics (tholeiitic pillow basalts, boninites, andesites and rare ultramafics), cherts and volcaniclastics in their immediate hanging walls. Although the Cambrian rocks are all fault-bounded, they form basement to the younger clastic succession, have MORB affinities and probably represent oceanic crust (Crawford & Keays 1978; Crawford *et al.* 1984) (Fig. 6).

The turbidites consist of chevron-folded sandstone and mudstone layers which reflect up to 65% shortening above this mid-crustal detachment fault (Gray & Willman 1991*b*). Regional scale anticlinoria and synclinoria with a fanning to axial surface slaty cleavage, are cut by quartz veins and steeply east- and west-dipping reverse faults (Gray *et al.* 1991*a, b*). The main fabric development, related to mica growth in both slate and psammite, is late in chevron fold development after fold lock-up and attainment of maximum tectonic thickening (Gray & Willman 1991*b*; Yang & Gray 1994). Overall tectonic shortening ranges from 30 to 50% (due largely to folding) in the eastern segment, and from 60 to 70 % in the other zones due to folding, cleavage development and faulting (Fig. 5).

Hanging-wall stratigraphy indicates that de-

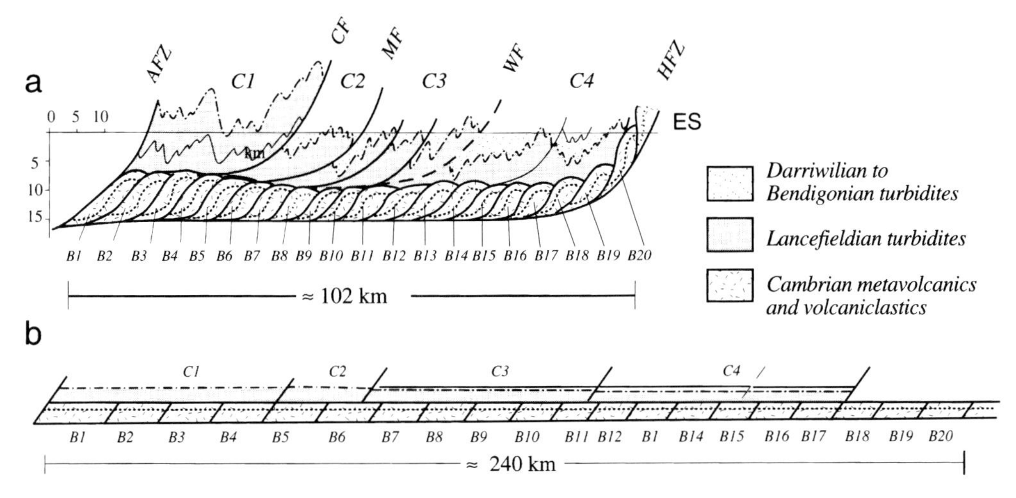

Fig. 8. Typical geometry of the western Lachlan Fold Belt shown by schematic and restored cross sections across the Bendigo–Ballarat zone. (**a**) Interpretative cross section showing leading imbricate-fan geometry with cover thrust-sheets (C1 to C4) in the quartz-rich turbidite succession overlying stacked, basement horses (B1 to B20) within Cambrian mafic volcanics. Deformed length between selected reference points is 102 km. (**b**) Restored section showing length of the sedimentary cover sequence relative to the Cambrian mafic volcanics. Undeformed length is 240 km implying 57% shortening due to contraction faulting and chevron-folding within the turbidite package. Retro-deformation was by both area and line length balancing of fault blocks C1-C4 (Woodward *et al.* 1986).

tachments occur at the base of the Ordovician turbidite sequence (Lancefieldian biostratigraphic stage) and within the Cambrian mafic volcanics of the western Lachlan Fold Belt (Gray & Willman 1991*a, b*) (Fig. 8). Deep crustal seismic profiling (Gray *et al.* 1991*a*) (Fig.9) and microseismicity studies (Gibson *et al.* 1981) indicate the detachment to be at approximately 15–17 km depth shallowing to the east where it is exposed as the Mt Wellington Fault Zone. This sole or detachment fault beneath the Melbourne Zone is a major crustal discontinuity which separates fold-thrust zones of different age and vergence (Fergusson *et al.* 1986; Murphy & Gray 1992). In the Melbourne Zone there is an eastward transition from upright to inclined folds to overturned as the Mt Wellington Fault Zone is approached. There is also a marked eastwards increase in the intensity of deformation, as reflected by widespread development of cleavage and faults, and a zone of moderately to shallowly dipping transposition layering, containing strongly deformed slices of Cambrian mafic volcanic rocks and highly cleaved slices of Ordovician black shale (Gray 1995).

The Central Subprovince

The Central subprovince (Figs 5, 6 & 7) is dominated by the fault-bounded, NNW-trending Wagga/Omeo Metamorphic Belt (Fig. 2) of greenschist to upper amphibolite facies metamorphic rocks (Morand 1990). The subprovince consists of complexly deformed Ordovician metasedimentary rocks dominated by Early to Mid-Silurian deformation, and is intruded by granites of Silurian and Devonian age (see Morand & Gray 1990; Glen 1992; Fergusson & Coney 1992*b*, fig. 5) (Fig. 7, Tabberabbera and Omeo Zones). Much of the complex, which represents the main locus of Palaeozoic low-pressure metamorphism in the Lachlan Fold Belt, is bounded by major mylonitic strike-slip faults, especially the western and eastern margins (Kancoona, Kiewa, Ensay Faults on the west and the Gilmore Fault in the east). Part of the southeastern margin is bounded by a major reverse fault (Indi Fault) (Morand & Gray 1990) (Fig. 5). For example, the Kiewa fault zone along the western margin is a 1.5 km wide, steeply west-dipping mylonite zone, with S–C fabrics indicating an early dextral strike-slip movement, overprinted by post-Mid-Devonian faults with sinistral strike-slip movement (Morand & Gray 1991).

External to the subprovince on the southeast side is a low metamorphic grade, northnorth-west-trending belt of chevron-folded and faulted Ordovician turbidites (Tabberabbera Zone; Fig. 5) (Fergusson 1987). A 2 km wide melange zone

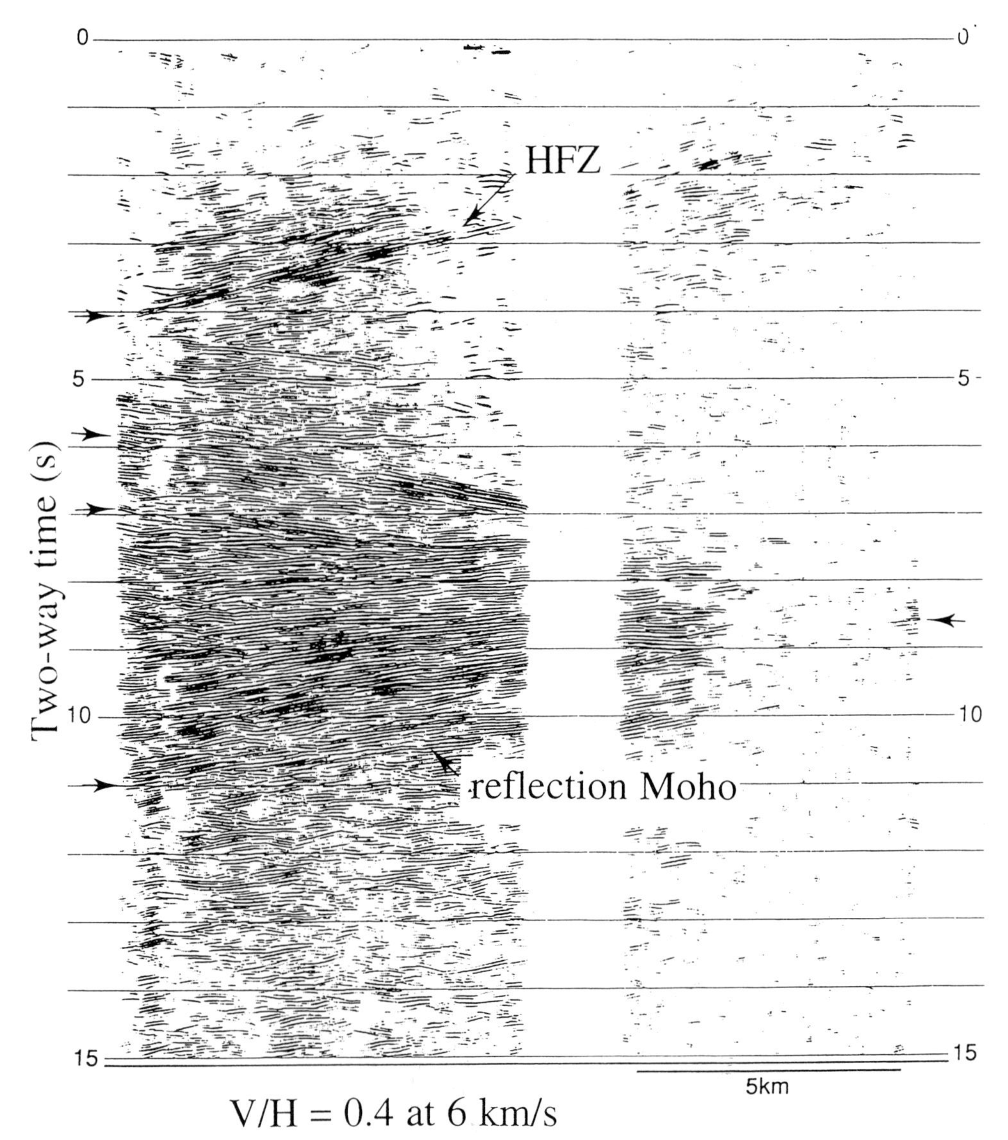

Fig. 9. Unmigrated 15 s TWTT stacked profile north of Heathcote, western Lachlan Fold Belt (see Gray *et al.*, 1991*b*) assumes 6 km s^{-1} as an average crustal velocity (Gibson *et al.* 1981). HFZ, Heathcote Fault Zone. Arrows indicate positions of inferred detachments; detachment faults have been defined where truncations occur between packets of reflections. Line located on Fig. 4 (line 4).

(Wonnangatta Fault) separates thick-skinned structures, dominated by folds with tighter wavelengths (*c.* 10 km) and larger amplitudes (*c.* 10 km) and polydeformation, from a thin-skinned zone with only one set of fold structures cut by north-dipping reverse faults (Fergusson 1987; Fergusson & Coney 1992*b*, fig. 5).

The Eastern Subprovince

The Eastern subprovince (Figs 4, 10, 11, 12, 13) is characterised by a series of anticlinorial and synclinorial zones bounded by both east- and west-dipping reverse faults. Folding is more open (interlimb angles of 70–120° with shorten-

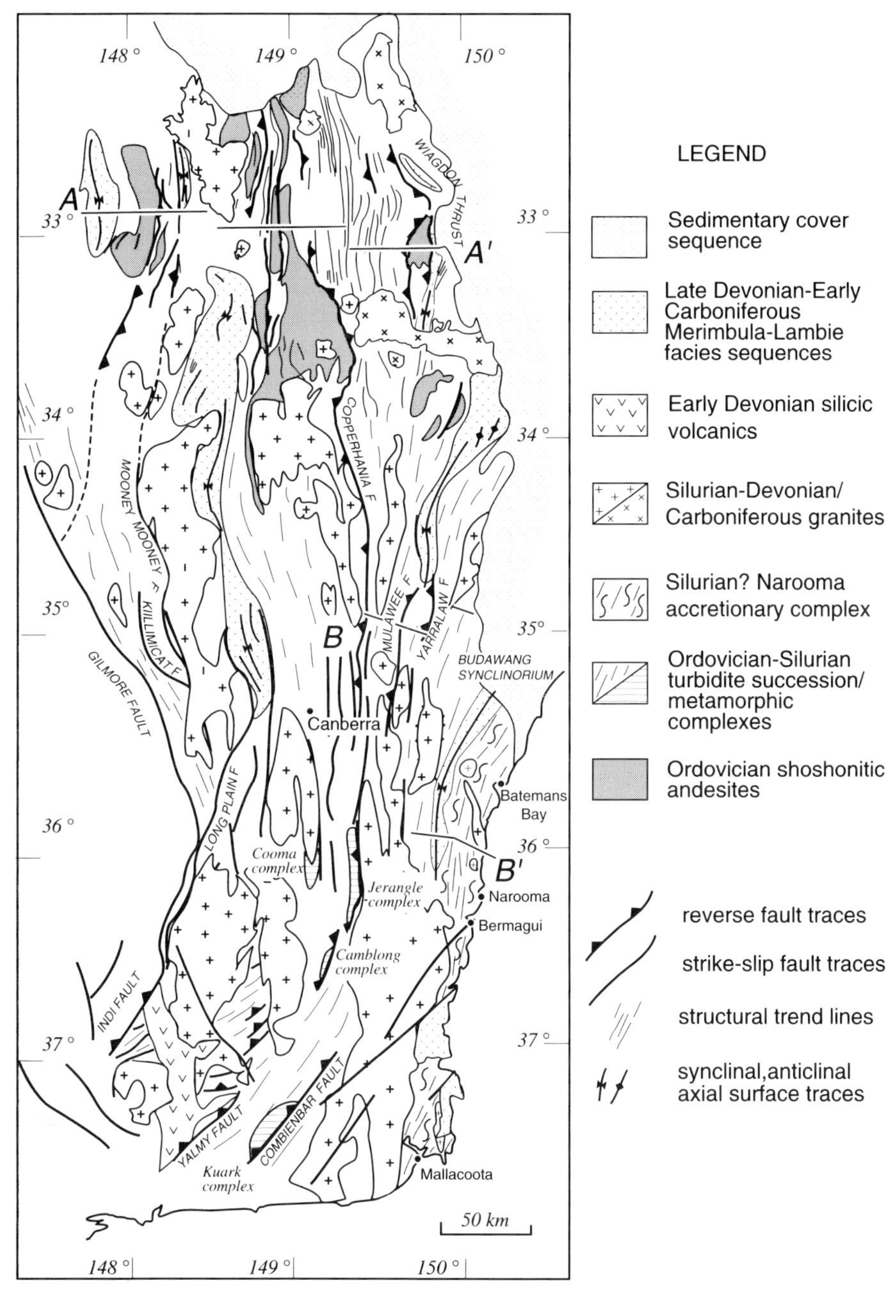

Fig. 10. Structural map of the eastern Lachlan Fold Belt showing structure trend lines (thin lines), major fault traces (heavy lines), outcrop patterns of the Ordovician shoshonitic andesitic volcanics (dark grey regions), granites (regions designated by crosses), and the eastern belt of metamorphic complexes (horizontal slashes). The Narooma accretionary complex (see Miller & Gray in press *a, b*) is shown. A–A′ and B–B′ are structural profiles shown in Fig. 11.

ings of 24–33%; Fergusson & Coney 1992*a*), with an overall east vergence and an eastwards increase in degree of cleavage development (Fig. 11a). The regional folds become tight and inclined towards the Capertee Anticlinorial Zone (Fig. 11a). Inversion of a series of former Mid-Silurian to Late Devonian extensional basins represented by alternating troughs (turbi-

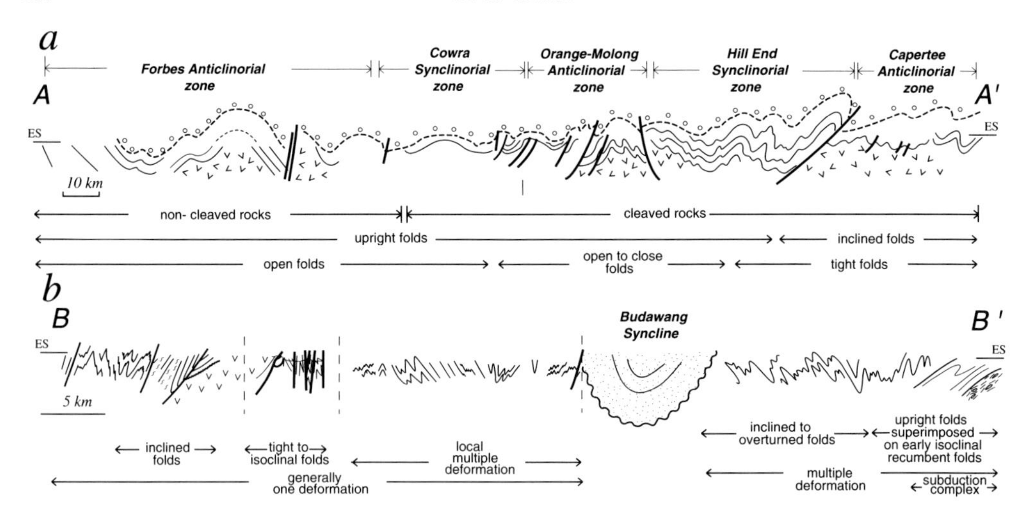

Fig. 11. Structural profiles across the eastern Lachlan Fold Belt located in Fig. 10 (vertical equals horizontal scale). Geological zones, major faults, bedding form lines, and granite bodies are shown. (**a**) A–A′: northern profile showing anticlinoria (dominated by Ordovician andesites and shallow water deposits) and synclinoria (dominated by turbidites). (based on Powell 1984, fig. 204) (**b**) B–B′: southern profile in the turbidite dominated eastern part of the eastern belt. Composite profile based on Fergusson & VandenBerg (1990) and Powell (unpublished). Stipple pattern (**b**) /open circles (**a**): Late Devonian-Early Carboniferous cover sequence ('Lambie' facies). V pattern, Ordovician shoshonitic volcanics. Form lines (no pattern), Ordovician–Silurian turbidites.

dites) and highs (carbonates/shoshonitic volcanic rocks) has controlled the present distribution of synclinorial and anticlinorial zones respectively (Fig. 11a; Powell 1984, p. 311).

The eastern part of the subprovince is characterized by the chevron-folded turbidites of the Bungonia–Delegate Thrust belt (Fergusson & VandenBerg 1990) which grades into the coastal facies of Powell (1984), including a broken formation (melange) of an Early Silurian? accretionary complex within a Late Cambrian–Early Ordovician chert–basalt–turbidite sequence (Fig. 11b; Miller & Gray in press *a*, *b*). Several isolated, north–south-elongated low pressure/high temperature metamorphic complexes (e.g. Cooma, Jerangle, Cambalong and Kuark Complexes) occur within this thrust-belt (Fig. 10). They are associated with foliated granitoids and have very narrow, north–south-trending metamorphic zones defined by biotite, knotted schist (cordierite) and sillimanite/migmatite (K feldspar) (see Glen 1992, fig. 9).

Basin inversion in the north is largely due to east-directed thrusting along earlier formed extensional faults and their associated detachment (Glen 1992; Glen *et al.* 1992). This deformation, as recorded by K–Ar ages and $^{40}Ar/^{39}Ar$ spectra of metamorphic micas in cleaved slates, was earliest in the west; 395–400 Ma late Early Devonian in the Cobar Basin (Glen *et al.* 1992), and 344–355 Ma Late Devonian–Early Carboniferous in the Hill End 'Trough' (Cas *et al.* 1976), although this was preceded by uplift and broad folding in the Middle Devonian (Powell 1984). The age of thrusting within the eastern part (Bungonia–Delegate Thrust belt) is considered to range from Early Silurian to Mid-Devonian based on age constraints of granites (cf. pre-Early Devonian: Fergusson & VandenBerg 1990). The deformed Ordovician through Early Devonian sequence of the eastern Lachlan Fold Belt is overlain by Late Devonian molasse (Lambie facies) which has been folded into a series of open, meridional folds in the Early Carboniferous (Fig. 10). The intensity of this Early Carboniferous deformation, as recorded by fold tightness in Lambie facies rocks, increases dramatically eastwards towards the developing New England Fold Belt (see Powell 1984, fig. 222).

Crustal structure

Crustal thicknesses in the Lachlan Fold Belt range from 40 to 52 km under the eastern highlands of Australia, from 35–36 km in the western subprovince (central and northeast Victoria) to 26 km in the eastern subprovince (central NSW) (Finlayson *et al.* 1980; Gibson *et

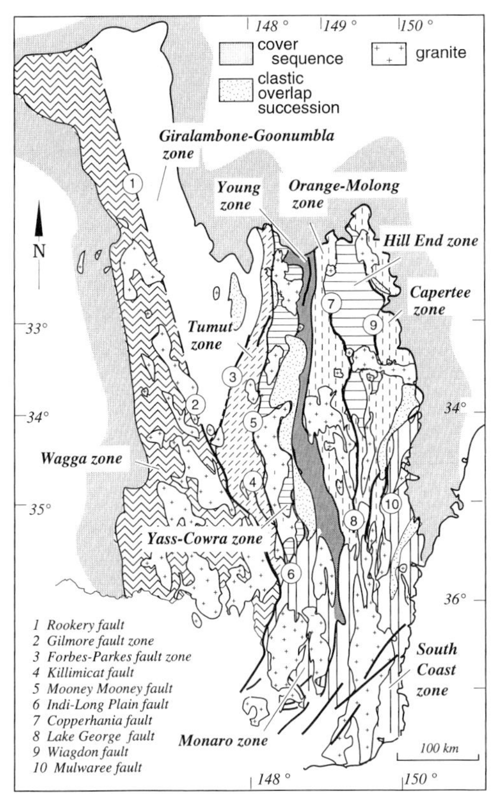

Fig. 12. Geological zone map of the eastern Lachlan Fold Belt. Cover sequences, clastic overlap successions and granites are shown. Major faults are numbered. The South Coast Zone correlates with the Bungonia–Delegate thrust-belt of Fergusson & VandenBerg (1990). Compare with geological and structural history of zones (see Fig. 13).

al. 1981). The southeastern part of the eastern subprovince is characterized by an upper crust with P-wave velocities between 5.6 and 6.3 km s^{-1}, a lower crust with velocities between 6.7 and 7.4 km s^{-1} and a low velocity zone at a depth of about 15–20 km (Finlayson *et al.* 1979, 1980). In the western subprovince the upper crustal velocities range between 5.3 and 5.9 km s^{-1} with a change to 6.3 km s^{-1} at a depth of 17 km (Gibson *et al.* 1981).

Limited deep crustal seismic reflection profiling in parts of the Lachlan Fold Belt Australia (Pinchin 1979; Leven *et al.* 1992; Gray *et al.* 1991*a*) has provided some 'snapshots' of the nature and geometry of the lower crust. Petrologic interpretations (Finlayson 1982, fig. 2; Griffin & O'Reilly 1987, fig. 2b; Clemens 1988, fig. 1) have inferred a lower crust dominated by subhorizontal alternating mafic and felsic layers, but structural interpretations of seismic profiles suggest the existence of a structurally complex lower crust cut by shear zones. The Tibbinbilla–Braidwood seismic traverses (Fig. 4, #1) have extremely short traverse lengths (<10 km) and were shot above granites, but show intracrustal reflections (particularly at 10–12 km and a pronounced low-velocity zone at 22–28 km) with a depth to the reflection Moho of 41 km (Pinchin, 1979). The Tumut seismic traverse (Fig. 4, #2) shows that steep faults exposed at the surface (e.g. Killimicat and Mooney Mooney faults, Fig. 3b) correspond to east-dipping, listric shaped, non-reflective zones (Leven *et al.* 1992). These are truncated by a series of strong, continuous, gently west-dipping reflectors interpreted to be a mid-crustal detachment at approximately 20 km depth (Leven *et al.* 1992). The Cobar seismic traverses (Fig. 4, #3) show a Moho depth of 30–33 km and a mid-crustal detachment fault which varies in depth from 25 to 12–16 km (Glen *et al.* 1994).

The Heathcote seismic traverse (Gray *et al.* 1991*a*) in the western subprovince of the Lachlan Fold Belt (Fig. 4, #4) shows a depth to the reflection Moho of approximately 36 km (Fig. 8). The steeply dipping Heathcote fault-zone appears as a moderately dipping reflection which flattens with depth to approximately 15 km. The lower crust appears structurally complex and has been interpreted as a westward-dipping duplex or crustal stacking-wedge developed on a ramp which originates at the crust-mantle boundary (Gray *et al.* 1991*a*). The lower crustal region (*c.* 18–33 km depth, Fig. 9), has low angle (<20° on non-migrated sections) packets of largely east-dipping discontinuities which were interpreted as deep crustal shear zones.

Timing of upper crustal deformation

Timing of deformation has been generally constrained by the age of the youngest sedimentary rocks and the ages of granitic intrusions within the respective structural zones (see Figs 7 & 13). Age of deformation maps (Fig. 14) reflect the broad patterns of deformation but do not show the problems of overprinting deformations due to later 'orogenic' pulses (Glen 1992, figs 5, 6 & 7). Recent synopses of these deformations and their effects are given in Fergusson & Coney (1992*b*) and are only briefly treated here. Four major periods of deformation (Fig. 14) can be delineated based on the timing of the major deformational 'event' which has affected the regions indicated.

(1) *Early Silurian*: earliest deformation co-

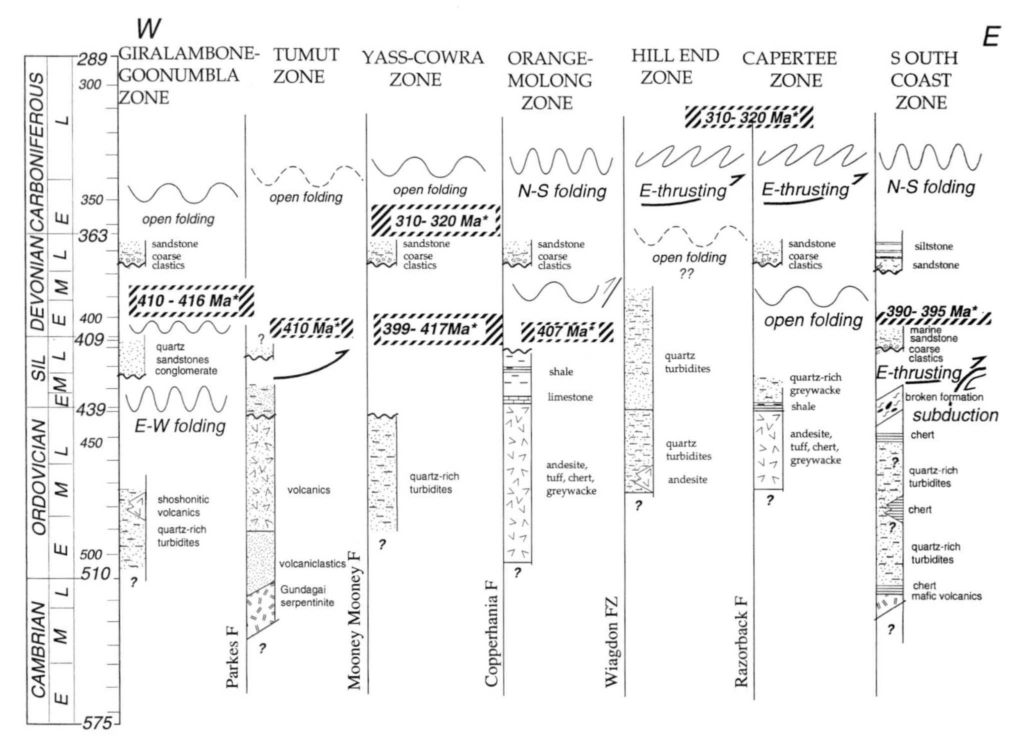

Fig. 13. Time–space plot showing the relationships between sedimentation, deformation and plutonism for the eastern Lachlan Fold Belt (based on time scale of Harland *et al.* 1989). Subdivisions are the major geological zones of New South Wales arranged from west to east (see Fig. 3). Ages and the nature of sediments, directions of thrusting, and age ranges of granitic intrusives (oblique slashes) are shown.
[1]Data from; *Data from Evernden & Richards (1962), Richards & Singleton (1981) and Gray (1990).

inciding with the central Lachlan Fold Belt (see Fergusson & Coney 1992*a*, fig. 4), but recently recognized to be the most widespread deformation within the Lachlan Fold Belt now incorporating the western part (see Bucher *et al.* 1996). In the central subprovince it is characterized by east–west-trending folds and related thrusts, and high temperature/low pressure metamorphism. The east–west-trending folds have been related to transpression resulting from oblique convergence (Powell 1983). The western Lachlan Fold Belt is dominated by an eastwards migrating east vergent fold-thrust belt.

(2) *Mid-Silurian to Early Devonian*: main effects preserved as an east vergent fold-thrust belt in the eastern (see Fergusson & Coney 1992*a*, fig. 6) and western subprovinces of the Lachlan Fold belt (Gray *et al.* 1996).

(3) *Early Devonian*: main effect in the Melbourne Zone of the western Lachlan Fold Belt (see Fergusson & Coney 1992*a*, fig. 7). Widespread effects occur in the eastern and central subprovinces with conjugate strike-slip faulting and reactivation of older strike-slip faults (e.g. Morand & Gray 1991).

(4) *Early Carboniferous*: the final regionally extensive deformation, which folded the Late Devonian-Early Carboniferous (Lambie facies) cover sequences throughout the Lachlan Fold Belt. The main effects were in the northeasternmost part of the eastern Lachlan Fold Belt (see Powell 1984, fig. 222; Fergusson & Coney 1992*a*, fig. 8; Glen 1992, fig. 5).

The traditional approach has been to consider Lachlan Fold Belt evolution as the result of a series of orogenies, originally recognized by regional and localized unconformities (Powell 1983; Coney *et al.* 1990; Glen 1992; Fergusson & Coney 1992*a*). These are:

(1) Benambran Event (Early Silurian: *Llandovery c.* 439–435 Ma)
(2) Quidongan Event (late Early Silurian: *Wenlock c.* 430–424 Ma)
(3) Bowning Event (Early Devonian: *Lochkovian c.* 409–396 Ma)
(4) Bindian Event (late Early Devonian: *Emsian c.* 390–386 Ma)

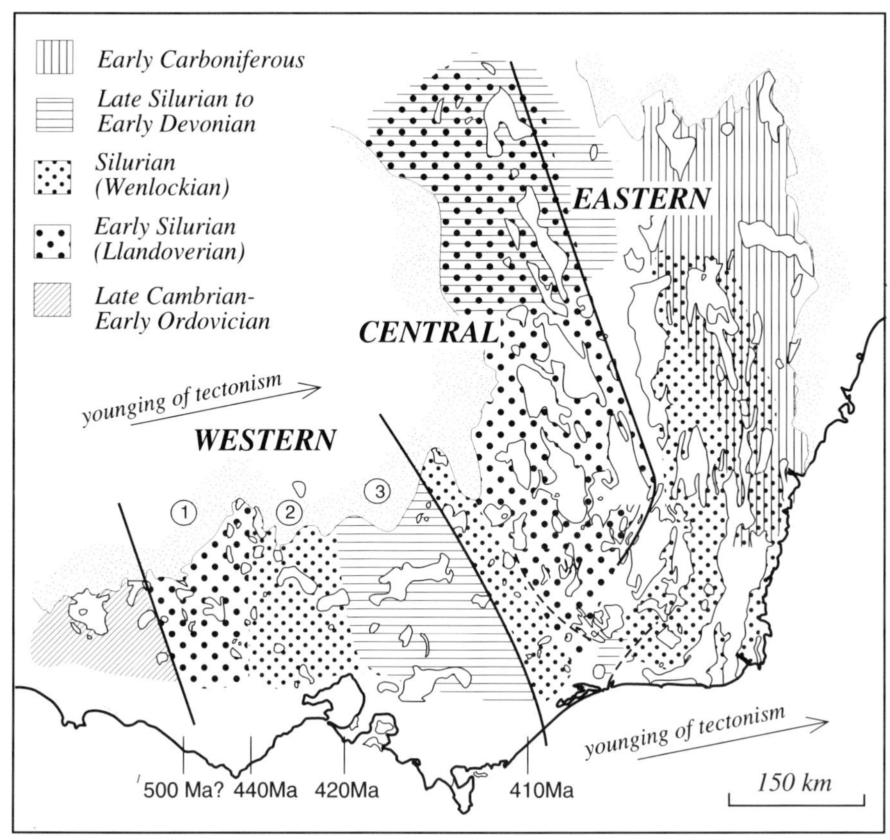

Fig. 14. Age of deformation map for the Lachlan Fold belt showing the temporal and spatial relationships of the major deformational events. Deformation ages are poorly constrained to time periods of 10–30 Ma (see Fig. 6) largely by ages of the post-tectonic granites, ages of the youngest sedimentary rocks affected by the deformation, and by local to regional unconformities; based on data from Richards and Singleton (1981), Gray (1990) and Cas (1983). [1]Ages shown for the western subprovince are $^{40}Ar/^{39}Ar$ metamorphic ages of white micas in slates and phyllites from fault zones (see Bucher *et al.* 1996); circled numbers 1–3 reflect the eastwards younging of the deformation across structural zones within the western subprovince.

(4) Tabberabberan Event (Middle Devonian: *Givetian c.* 381–377 Ma)
(5) Kanimblan Event (Early Carboniferous: *early Visean c.* 360–340 Ma)

(after Crook & Powell 1976, fig.2). This approach assumes each orogeny represents a fold-belt wide event and does not necessarily allow for either localised or wide ranging, diachronous deformations. As in other orogenic belts, this concept of orogeny is perhaps outmoded, in that localised unconformities do not necessarily reflect fold-belt wide major deformational events. Fold-belt evolution is better linked to a series of tectonothermal events which can be correlated over large parts of the fold-belt (Glover *et al.* 1983). At this point only limited geochronological and thermochronological data are available for the Lachlan Fold Belt (Cas *et al.* 1976; Glen *et al.* 1992; Bucher *et al.* 1996).

Regional deformation kinematics

Variations in the age, style and intensity of regional deformation, as well as the kinematics shown by movement indicators in fault zones (Gray & Willman 1991*a*, fig. 18) and changes in structural vergence across the fold-belt, reflect the structural history of the fold belt and must relate to the overall tectonic development. How this relates to the existing framework of orogenic episodes constitutes a major problem. It necessitates a revised approach to orogenic belt evolution, particularly in the light of thin-skinned deformation where the deformation is

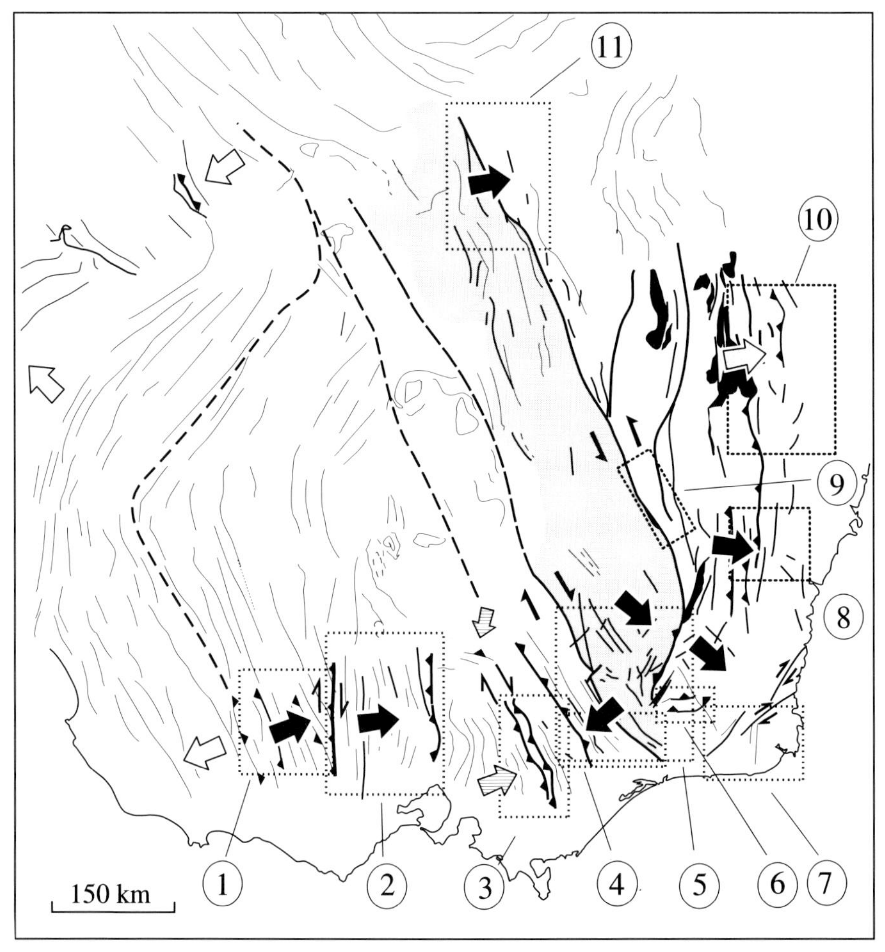

Fig. 15. Vergence map for the Lachlan Fold Belt showing structural and aeromagnetic trends (thin lines), fault traces (heavy lines), the Wagga Metamorphic Belt (shaded region), and the tectonic vergence (arrows) for individual structural blocks. Vergence here is the tectonic transport direction as determined from regional fold asymmetry and overall fault dip based on structural data for the regions outlined from the following publications (1) Wilson *et al.* (1992); (2) Gray & Willman (1991*a*, *b*); (3) Murphy & Gray (1992), Gray (1995); (4) Fergusson (1987); (5) Morand (1990), Morand & Gray (1991); (6) Glen & VandenBerg (1987); (7) Wilson *et al.* (1992); (8) Fergusson & VandenBerg (1990); (9) Stuart-Smith (1990); (10) Powell (1983); (11) Glen (1988). Arrows with dense stipple fill, Cambrian timing; black arrows, Silurian deformation; horizontal slash arrow, Late Silurian to Mid-Devonian deformation; arrow with coarse stipple fill, Carboniferous deformation. Heavy dashed lines are inferred major tectonic boundaries.

clearly diachronous at the scale of a thrust-belt (see above).

Regional vergence is shown by overall fault dip and fold asymmetry into major fault zones and defines a complex pattern centred about the Wagga Metamorphic belt (shaded area, Fig. 15). Vergence vectors and bulk regional shortening orientations for individual structural zones across the fold-belt, are largely east-directed away from the craton. In the western subprovince northwest-trending aeromagnetic /structural trends are associated with an eastnortheast-directed transport with a progression in deformation from Early Silurian (area 1, Fig. 14), Late Silurian to Early Devonian (area 2, Fig. 14) to late Early Devonian (area 3, Fig. 14). This suggests diachronous development of detachment-related chevron-folding (Geiser 1988) accompanied by a bulk eastwards propagation of a mid-crustal detachment fault. Detachment related folding also occurs in the eastern subpro-

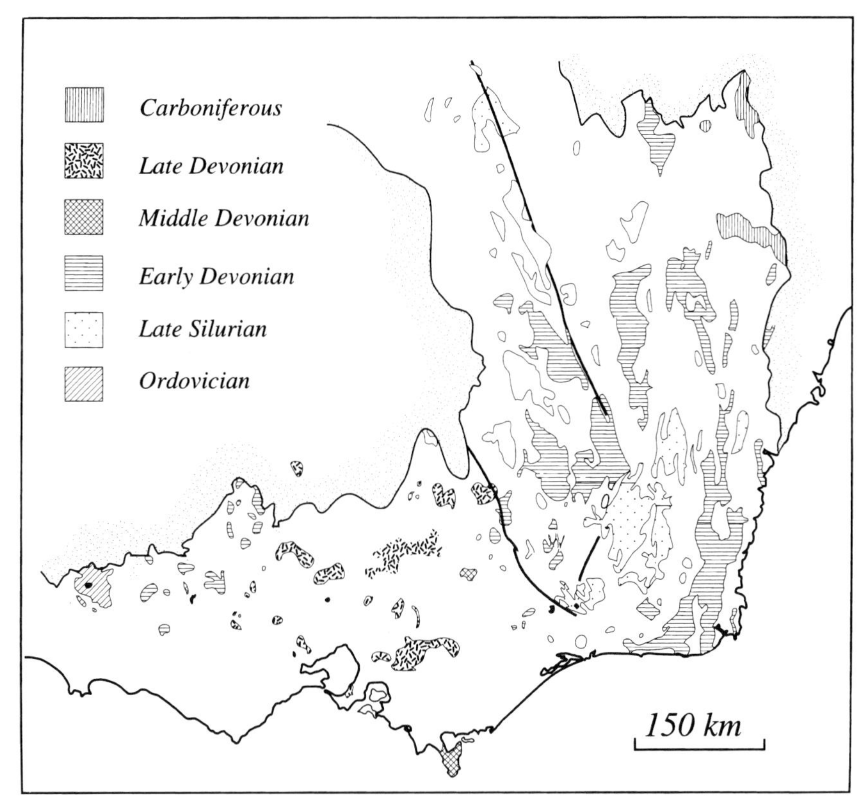

Fig. 16. Granite age map showing the distribution and ages of granites for the Lachlan Fold Belt; based on data from Everden & Richards (1962), Richards & Singleton (1981), Shaw *et al.* (1982) and Gray (1990). Granites that have not been radiometrically dated have no fill.

vince of the fold-belt which has a pronounced north-south structural grain, shows east-directed thrusting and a generalised eastwards younging of tectonism (Fergusson & Vanden-Berg 1990).

In summary, two separate eastward-migrating zones of Silurian deformation converted the Ordovician continental margin into the present fold-belt (Fergusson & Coney 1992*a*; Collins 1994). (1) Central/Eastern subprovinces: deformation began in the Wagga-Omeo Metamorphic Belt in the Early Silurian and progressed eastwards through the Devonian (Fergusson & Coney 1992*a*). (2) Western subprovince: deformation began at the western margin of the belt in the Early Silurian and also moved eastward, eventually terminating against the 'cold' central Lachlan Fold Belt in the Early Devonian (Fergusson *et al.* 1986; Gray & Willman 1991; Gray & Mortimer 1996, fig. 13).

The central subprovince of the Lachlan Fold Belt shows the earliest deformation within the fold belt and has strong control on vergence zones and the overall kinematic evolution of the fold-belt (see Fig. 15). Vergence zones are bounded by the faults separating the central subprovince from the western and eastern parts respectively. Structures across the western subprovince show vergence to the east (Figs 3c & 14) and are related to a Silurian to Early Devonian event (Bucher *et al.* 1996), whereas those in the central subprovince (Tabberabbera zone) to the east of the Mt Wellington Fault Zone show vergence to the southwest (Figs 3c & 14) and are related to an Early Silurian deformational event (Fergusson 1987). There should be another major crustal discontinuity separating zones of different age and vergence between the Kanmantoo and the western Lachlan Fold Belt, but this boundary is somewhat contentious (Wilson *et al.* 1992; Glen 1992) and may underlie some of the younger sedimentary cover sequences (Figs 2 & 3c).

The central and western subprovinces were not joined till the late Early Devonian with 'docking' movement along the Mt Wellington

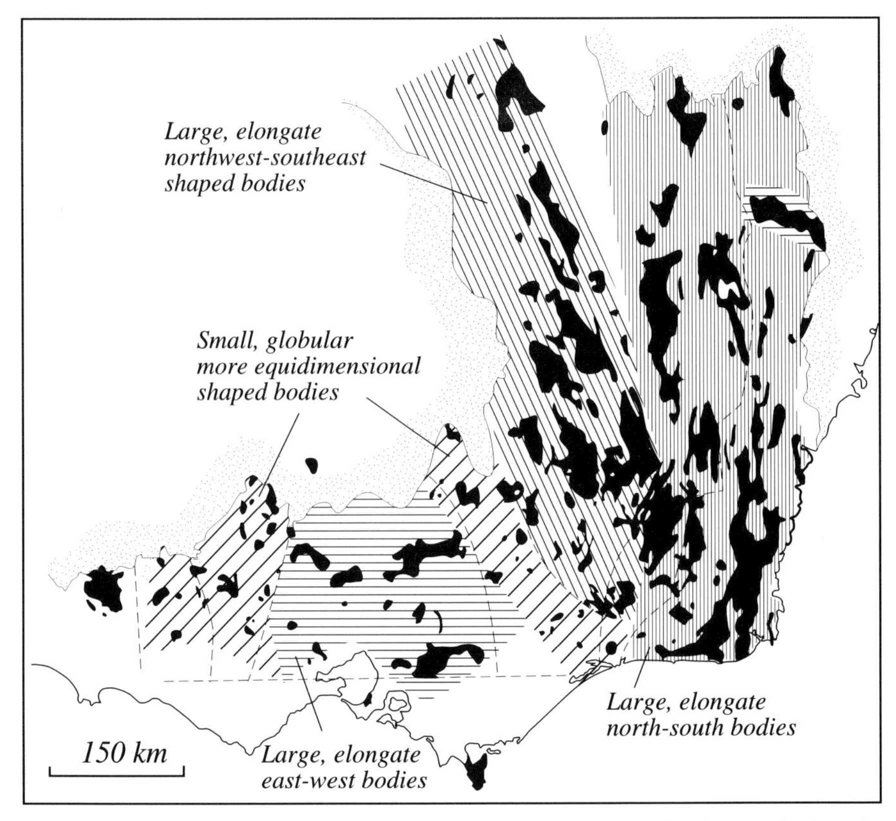

Fig. 17. Granite outcrop pattern map showing five distinct provinces based on the shape and orientation of granitic bodies in black; dashed lines indicate boundaries of granite geochemistry domains and are equivalent to the 'basement terranes' of Chappell *et al.* (1988). These crustal provinces potentially reflect the states of stress in the mid- to lower crust at the time of granite emplacement.

Fault Zone. The easternmost part of the western subprovince was an eastwards-prograding continental margin sediment prism until the late Early Devonian (VandenBerg & Stewart 1992; Gray & Mortimer 1996, fig. 13) when there was a change in provenance and sediment input from the east (Powell pers. comm., 1994). Docking effects are shown by localized development of east–west-trending folds and north-over-south thrusting (Gray & Mortimer 1996).

Thermal evolution

The thermal evolution of the Lachlan Fold Belt is documented by the age, spatial and volume distribution of granites and related volcanism, and by the nature and extent of the metamorphism and regions of crustal extension (Morand 1990; Gray & Cull 1992).

Granites

Granites cover up to 36% of the exposed Lachlan Fold Belt with maximum development in the central and eastern subprovinces (Figs 16 & 17). 'Regional aureole', 'contact aureole' and 'subvolcanic field' associations, as well as S and I type granites based on geochemistry/mineralogy have been recognised (White *et al.* 1974; White & Chappell 1983; Chappell *et al.* 1988). In the western Lachlan Fold Belt granites are either 'contact aureole' types with narrow (1–2 km wide) contact aureoles, or 'subvolcanic' types associated with cauldron-collapse rhyolites and ash flows of similar composition. The 'regional aureole' types are less common and are associated with low *P* metamorphism, migmatites and K–feldspar–cordierite–andalusite–sillimanite gneisses of the Wagga–Omeo Metamorphic Belt

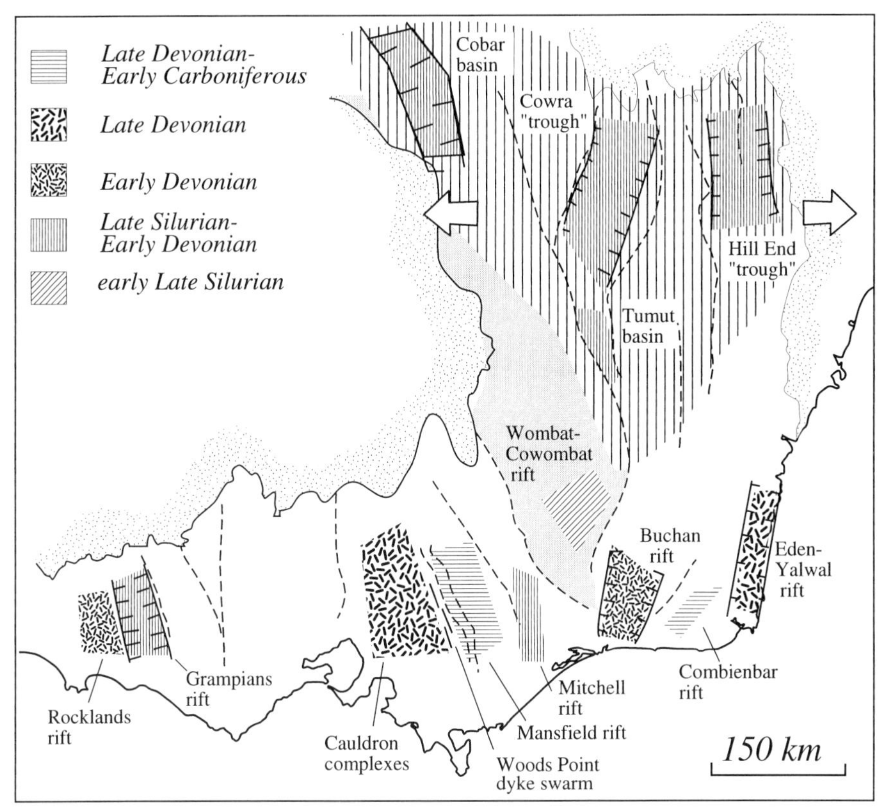

Fig. 18. Map showing the timing and geographic distribution of extensional elements in the Lachlan Fold Belt, including sedimentary accumulations in rift basins, regions of silicic volcanism, and dyke swarms. Arrows indicate major E–W crustal extension in the northeast portion of the fold belt (highlighted by the wider-spaced vertical slash pattern) in the Late Silurian to Early Devonian. Dashed lines are major fault traces in the exposeed parts of the fold belt.

in the central Lachlan Fold Belt and the Cooma, Cambalong and Kuark belts of the eastern Lachlan Fold Belt (Fig. 10).

Granite cooling ages clearly vary across the Lachlan Fold Belt (Fig. 16). In the eastern subprovince granitic magmatism is predominantly Late Silurian to Early Devonian, where granites are syntectonic, with emplacement facilitated by thrust-faulting (Burg & Wilson 1988; Paterson *et al.* 1990). Within the eastern subprovince there is an apparent eastwards younging in the age of plutonism which correlates with a slow (6 mm a^{-1}) eastwards migrating ?mantle heat source (Powell 1983, fig. 9). In the central subprovince granitic magmatism is dominantly Early Devonian, with some pluton emplacement in strike-slip fault zones in the Late Silurian (Morand 1992). The western subprovince has different character to the rest of the fold-belt and shows two main periods of granitic magmatism (Early Devonian and Late Devonian). The Late Devonian granites are post-tectonic and associated with volcanic cauldron complexes and ignimbrite flows.

The shape distribution of the granites (Fig. 17) suggests five major granite provinces, which presumably reflect the mode and timing of emplacement relative to tectonism, and the state of stress at the particular level of emplacement. Elongate NNW-trending granites (Early Devonian ages) define the Wagga–Omeo Metamorphic Belt of the central subprovince. North–south-trending granites (Late Silurian to Early Devonian ages) of the eastern Lachlan Fold Belt are elongated parallel to the regional structural grain. Many of these granites are syn-tectonic with major mylonites developed along their eastern margins (White *et al.* 1974; Paterson *et al.* 1990). In the western Lachlan Fold Belt post-tectonic granites of the central Victorian Magmatic province (Late Devonian age) are east–west-trending, have elongated form and were probably emplaced by stoping or piston uplift (Fig. 17). The remaining granites (Early Devonian ages) are smaller and more equant in shape. Granite spacing fits a diapiric emplacement model requiring a source depth of 12–24 km from a 35 km thick crust (Rickard & Ward 1981).

Metamorphism

The Lachlan Fold Belt is dominated by greenschist to sub-greenschist facies metamorphism, with high grade rocks largely confined to fault-bounded metamorphic complexes (Vallance 1967; Smith 1969; Morand 1990). The Wagga–Omeo Metamorphic Belt of the central Lachlan Fold Belt (Fig. 2) and a more easterly grouping of the discontinuous, north–south-trending Cooma, Cambalong, Jerangle and Kuark metamorphic complexes of the eastern Lachlan Fold Belt (Fig. 10), display high temperature–low pressure metamorphism characterized by andalusite–sillimanite assemblages (Morand 1990; Glen 1992, table 2). Metamorphic zones include chlorite, biotite, cordierite, andalusite–K-feldspar and sillimanite–K-feldspar, with local development of migmatites within the sillimanite–K-feldspar zone (Morand 1990). Such assemblages are typical of contact metamorphism but occur on a regional scale in the high-grade part of the Wagga–Omeo Metamorphic Belt (Fig. 2).

The largest belt of high-grade metamorphism occurs as the NNW-trending zone of greenschist to upper amphibolite facies metamorphism of the Wagga-Omeo Metamorphic Belt in the central Lachlan Fold Belt. Associated with large volumes of Silurian and Devonian granite, the highest grade metamorphism is in the southern Omeo part of the complex (Fig. 2). Peak metamorphic conditions were $T \approx 700°C$ and $P \approx 3.5$ kbar (Morand 1990). Erosional unroofing of the metamorphic complex in the Mid–Late Silurian, documented in the Wombat and Cowombat rifts (Fig. 18) constrains the metamorphism to be Early Silurian, and necessitates shallow overburden with geothermal gradients on the order of 65°C km^{-1} (Morand 1990). Wrenching on the NW-trending marginal shear zones combined with thrusting along the leading edge, led to emplacement of the Wagga–Omeo Metamorphic Complex as a SE-moving crustal wedge in the Late Silurian to Early Devonian (see Morand & Gray 1991, fig. 13). This exposed the deepest levels of the complex in the south.

The much smaller Cooma, Jerangle, Cambalong and Kuark Metamorphic complexes within the eastern Lachlan Fold Belt have very narrow, north–south-trending metamorphic zones and are associated with foliated granitoids (see Glen 1992, fig. 9). These complexes have been considered to represent 'hot-spots' in the regional metamorphism (Flood & Vernon 1978), or have been recently interpreted to be a series of fault-emplaced metamorphic complexes representative of a 10–11 km deep crustal layer below the eastern part of the fold-belt (Glen 1992). Granodiorite in the Cooma Complex is considered to be derived from *in situ* melting of the sediments now preserved within the metamorphic envelope (Pidgeon & Compston 1965). Both the granodiorite and the metamorphic envelope are considered to have been both diapirically emplaced into the upper crust (Flood & Vernon 1978).

The dominant metamorphic assemblages are white mica – chlorite in pelites for both the eastern (Smith 1969; Offler & Prendergast 1985; Farrell & Offler 1989) and western subprovinces of the Lachlan Fold Belt (Wilson *et al.* 1992; Gray 1995). Metamorphic conditions are characteristic of the greenschist facies, with *P–T* conditions of 330–450°C and 1–7 kbar (Farrell & Offler 1989) and $450 \pm 10°C$ and 1.7 ± 0.7 kbar implying burial depths <7 km (Wilson *et al.* 1992). It would appear therefore that the rocks in both the western and eastern subprovinces of the Lachlan Fold Belt have been deformed without significant cover. Burial prior to crustal thickening must range between 5 and 10 km, with uplift (< 10 km) in the Mid–Late Devonian causing a change from deep marine to continental fluvial facies sedimentation (Figs 7 & 13).

Crustal extension: a thermal response

The Lachlan Fold Belt has undergone extension at different times throughout its 250 Ma developmental history (e.g. Cas 1983; Coney *et al.* 1990). Extensional periods were marked by localized development of rift basins and half grabens accompanied by extensive granitic magmatism and silicic volcanism (Fig. 18). Regions of former crustal extension within the Lachlan Fold Belt are now preserved as elongate zones of sedimentary rocks forming cover 'basins', and silicic volcanic rocks (e.g. Early Devonian Rocklands Rhyolite of the Rocklands 'Rift' and Snowy River Volcanics of the Buchan 'Rift'). They represent structural/topographic remnants of inverted extensional basins which were generally bounded by north-south trending extensional faults.

Half-graben sequences are typified by the Cobar Basin, an overlap–cover basin to the central Lachlan Fold Belt (Glen *et al.* 1994) and the succession of 'troughs' and 'highs' in the northern part of the eastern Lachlan Fold Belt in New South Wales (Glen 1992). Other regions have been interpreted as strike-slip basins, such as the Tumut Basin (Stuart-Smith, 1991), the Mansfield and Combienbar Basins (Powell 1984). Ages of sedimentary rocks within the rifts provide an approximate timing for extension or pull-apart (Fig. 18):

Wombat and Cowombat rifts: late Early to early Late Silurian (*c.* 30 Ma)
Tumut Basin: Early to Late Silurian (*c.* 30 Ma) (Stuart-Smith *et al.* 1992)
Cobar Basin: Late Silurian–Early Devonian (*c.* 30 Ma)
Cowra Trough: Late Silurian–Early Devonian (*c.* 30 Ma)
Hill End Trough: Late Silurian–Early Devonian (*c.* 30 Ma)
Grampians Basin: Late Silurian–Early Devonian (*c.* 30 Ma) (Wilson *et al.* 1992)
Mitchell Basin: Late Silurian–Early Devonian (*c.* 30 Ma)

The northwest part of the exposed eastern Lachlan Fold Belt underwent regional extension after the Early Silurian deformation in the Wagga–Omeo Metamorphic Belt (central Lachlan Fold Belt). Short lived (<30 Ma) rift basins (e.g. Cowombat and Wombat 'rifts': Fig. 18) developed within the uplifted metamorphic complex and were subsequently deformed in the Late Silurian. Several meridional rift basins developed east of the metamorphic belt in the Late Silurian to Early Devonian (e.g. Cobar Basin, Cowra and Hill End 'Troughs': Fig. 18).

Other regions show Early to Late Devonian bimodal volcanism, including the Woods Point mafic dyke swarm (387 ± 14 Ma, K/Ar – Richards & Singleton 1981) and Early to Late Devonian silicic volcanism (Rocklands Rhyolite, western Victoria, 397 Ma, Rb/Sr – Gray 1990; Lake Mountain Rhyodacite, 367 Ma, Rb/Sr – Gray 1990; Snowy River volcanics, eastern Victoria, Early Devonian, Talent 1965; Eden Rhyolite, Late Givetian–Early Frasnian *c.* 377–370 Ma, Fergusson *et al.* 1979) (see Fig. 18).

Thermo-mechanical evolution

The timing relationships between deformation, metamorphism and plutonism vary across the Lachlan Fold Belt. The eastward-migrating deformation fronts, the mixed character of the granites including the post-tectonic granites of the western belt, and the largely syn-tectonic, and eastwards younging granites in the eastern belt are the key features. In a fold-belt dominated by such large volumes of granite, deformation may have been localised along axes of magmatic activity. Thermal softening was probably important in the development of the Wagga–Omeo Metamorphic Belt and parts of the eastern belt, where granites are syn-thrusting (e.g. Paterson *et al.* 1992) and the timing of plutonism and deformation is closely linked. In the western Lachlan Fold Belt, however, thermally enhanced deformation must be less important because granites are post-tectonic (Gray & Cull 1992).

The cause of metamorphism is uncertain. The large volume of granite, the high *T*–low *P* nature of the metamorphism and high heat flow have been linked to magmatic processes involving underplating of basaltic magma (e.g. Chappell *et al.* 1988; Clemens 1988; Collins & Vernon 1992). The cause of underplating remains controversial with explanations due to (1) a mantle pertubation (Collins 1994, p. 144), (2) a moving mantle plume (Wyborn 1992), (3) flip and change in the position of subduction zones from some distant trench (Collins & Vernon 1994; Collins 1994), (4) change in the dip of a subduction zone to the east (Powell 1983; Fergusson 1987), and (5) crustal thickening accompanied by convective thinning or delamination of mantle lithosphere (Looseveld & Etheridge 1990; Cox *et al.* 1991; Gray & Cull 1992, p. 458; Collins 1994). The high geothermal gradient, particularly that required for the Wagga–Omeo Metamorphic complex, has been previously related to a back-arc environment (Powell 1983; Morand 1990) and to arc magmatism (Collins 1994; Collins & Vernon 1994).

The role of plate convergence in the evolution of the Lachlan Fold Belt has been questioned on geochemical grounds (e.g. Chappell *et al.* 1992; Wyborn 1992), yet regionally extensive high temperature–low pressure metamorphism is typical of magmatic arcs (e.g. Barton & Hansen 1989). Granite production can be linked to subduction and roll-back models (e.g. the Hercynian event in the Appalachians: Sinha & Zietz 1982), but in the Lachlan Fold Belt two subduction zones are necessary to explain the two zones of eastward migrating deformation and granite plutonism. There does not appear to be a simple outwards thermal migration from a core of high grade metamorphic rocks (magmatic arc) (Fig. 15). None of the previously proposed models however, really explain the upper crustal vergence zones and the kinematics of Lachlan Fold Belt crustal evolution. Structural requirements necessitate some form of underthrusting in a convergent plate setting. That the granites young to the east, as does the deformation which is linked to E-vergence, suggests roll-back associated with W-dipping subduction. The presence of the two eastwards-migrating belts either necessitates subsequent strike-slip duplication within the fold belt or two W-dipping subduction zones.

Tectonothermal models must also explain the extensional phases, an integral part of the fold-belt evolution (e.g. Cas 1983). Extensional zones have been linked to periods of wrenching with development of strike-slip basins (e.g. Powell 1983; Stuart-Smith *et al.* 1992), a post-thickening response involving 'extensional collapse' of the crust up to 30 Ma after compressional orogenesis (Collins 1994), and roll-back of a W-dipping subduction zone with extensional pull-apart in the overriding plate (this paper). There is no single 'regional' phase of extension as required by crustal thinning models (e.g. Sandiford & Powell 1986). The observed geological complexity is also difficult to relate solely to a thermal pertubation arising from extensional thinning of the lithosphere. A 'Basin and Range' analogue has been applied by Cas (1983, fig. 23) to explain the Late Devonian–Early Carboniferous bimodal volcanism of the Lachlan Fold Belt. This phase of volcanism is related to crustal extension (<10%; Cas 1983), and it immediately post-dates the late Early to Mid-Devonian folding event which was a response to docking and amalgamation of the western and central subprovinces of the Lachlan Fold Belt (Fergusson *et al.* 1986).

Another aspect which must be explained is the preservation, firstly of the Late Devonian volcanic cauldron complexes and exposure of the Late Devonian granites which were intruded at pressures of about 1.7 kbar (Clemens & Wall 1982) in the western Lachlan Fold Belt, and secondly of the high level Ordovician Cu–Au porphyry systems associated with high K, shoshonitic mafic lavas in the eastern Lachlan Fold Belt. Their preservation indicates that there has only been a small amount of uplift and erosion (<3 km), particularly since the Early Carboniferous. This therefore (1) implies isobaric cooling after uplift, and (2) may suggest the presence of a higher density, structurally thickened, mafic lower crust (Coney *et al.* 1990). The nature of this lower crust however, still remains controversial based on geochemical and isotopic grounds (e.g. McCulloch & Chappell 1982; Gray 1984, 1990; Chappell *et al.* 1988)

Development of upper crustal structure

The positions, transport direction, and magnitude of thin-skinned fold–thrust belts relative to 'stable' craton, in this case Gondwanaland, reflect some form of collision in a convergent plate margin setting. Thrust-belts develop on the leading edges of actively driven plates where they verge toward, and against the motion of these plates (Coney 1973). Vergence away from the craton is a major distinguishing feature of the Lachlan Fold Belt, in contrast to the Kanmantoo and New England Fold Belts which together with the Lachlan Fold Belt make up the Tasman Orogenic Belt along the former margin of Gondwana (Fig. 1). The non-craton directed thrusts can be related to (1) thrusting over and against a small plate (former oceanic plateau) perhaps comparable in size to the southern portion of the present fold-belt (Scheibner 1985), (2) underthrusting of a 'continental slab' beneath a continental margin sequence (Crawford *et al.* 1984; Fergusson *et al.* 1986; Gray *et al.* 1991*b*; Fergusson & Coney 1992*a*), and (3) in this paper to imbrication above a W-dipping subduction zone for much of the Silurian and Early Devonian.

The interaction of hot and cold crust in the collisional process may also be important. The presence of cold crust may act as a 'backstop' causing transmission of stress across blocks (microplates) and the development of synthetic and antithetic (back) thrusts (Hamilton 1988). Mixed vergence around the Wagga–Omeo Metamorphic Belt reflects its role, and that of its substrate, as a buttress during the later (Devonian) compression (Scheibner 1992; Collins & Vernon 1994). During the buttressing event major faults bounding the metamorphic belt were reactivated, accompanied by superimposed conjugate brittle-faulting and localized development of north-trending crenulation cleavages (Morand & Gray 1991). In the Early Carboniferous intraplate compressional effects occurred within the Lachlan Fold Belt due to collision outboard in the developing New England orogen. Stress transfer resulted in the development of localized zones of en echelon folds within the Late Devonian cover sequence (Mansfield and Combienbar rifts: Fig. 18) (see Powell 1984, fig. 222A).

Strike-slip faults

The Lachlan Fold Belt has certainly been affected by strike-slip faulting (e.g. Stuart-Smith 1990; Gray & Willman 1991*a*; Morand & Gray 1991; Glen 1992). Arguments for such fault movements are based on shear sense criteria within shear zones (Stuart-Smith 1990; Morand & Gray 1991; Willman & Gray 1991*a*), the extreme undeformed width of the original turbidite fan (Fergusson & Coney 1992*b*), facies changes in the Ordovician sequence across major faults (VandenBerg & Stewart 1992), and the juxtaposition of different metamorphic and plutonic belts (Figs 15 & 16).

Within the Lachlan Fold Belt regional strike-slip faults are northwest-trending with major displacements on faults bounding the central subprovince (e.g. Morand & Gray 1991). The main locus of movement is inferred to be along the leading edge of the Mt Wellington Fault Zone which separates the western and central subprovinces of the Lachlan Fold Belt (Fergusson *et al.* 1986; VandenBerg & Stewart 1992). This fault is probably responsible for duplication of the strike-width of the fold belt in the late Early Devonian, resulting in the extreme width for the palinspastically restored Lachlan turbidite blanket in the Ordovician (see Fergusson & Coney 1992, fig. 11c). At this time major dislocations probably formed between the now hidden northern segment of the Lachlan Fold Belt and the exposed southern portion (Fig. 1). Other faults show evidence for considerably less displacement (<100 km) (Gray & Willman 1991*a*).

The significance of strike-slip faults in the tectonic evolution of the Lachlan Fold Belt is problematical. Strike-slip faults are associated with oblique subduction in modern tectonic settings (Karig *et al.* 1980; Beck 1983) where arc-parallel strike-slip faults develop along zones of weakness in the overriding plate. These faults occur within the magmatic arc, but can also form in the forearc at the ocean-continent boundary (e.g. Mentawai Fault: Diament *et al.* 1992). Strike-slip faults are however, also important in intracontinental deformation, particularly in areas of active collision (e.g. Southern Tibet and Central China: Tapponnier & Molnar 1977; Tapponnier *et al.* 1982; Armijo *et al.* 1989).

Tectonic setting

The progressive west to east younging of upper crustal segments in the Tasman Orogenic Belt (see Table 1) has led to concepts of large-scale continental accretion for eastern Australia. Most tectonic evolutionary models require arc-continent collisional interaction along the Gondwana continental margin in the Palaeozoic (Oversby 1971; Solomon & Griffiths 1972; Scheibner 1974; Crook 1974, 1980; Cas *et al.* 1980; Powell 1983), although there is some dispute about the existence of an 'arc' in the Ordovician based on the shoshonitic nature and geochemistry of the Ordovician volcanic rocks (Wyborn 1992). Cratonisation was either by back-arc accretion in a marginal sea setting (Cas *et al.* 1980; Powell 1983, 1984, p. 290) or in a convergent continental margin setting, involving some form of plate interaction to the east of its present geographic coordinates. It has been proposed that the Lachlan Fold Belt formed as a collage of amalgamated terranes prior to their accretion to the continental edge of Australia (Scheibner 1985; Fergusson *et al.* 1986; Leitch & Scheibner 1987; Coney *et al.* 1990). The lack of exotic terranes and the overall facies similarity however, suggests limited movement on faults within this continental margin sediment prism. At present crustal levels there is little evidence for juxtaposition of continental and oceanic fragments (oceanic plateaux, submarine arc and ocean floor) with the disrupted continental margin sequences (greywackes/chert–melange belts), particularly in the style of west-Pacific type belts.

As in most orogenic belts problems have related to the recognition of palaeogeographic elements in the deformed rock record and to their assembly into a coherent framework defining a plate tectonic setting. In the Lachlan Fold Belt major disputes relate to the tectonic setting in the Ordovician and Cambrian periods (continental rift or oceanic setting?) and the nature of the basement to the voluminous quartz-rich turbidite succession (continental, oceanic or mixed?). Observations from modern fan systems indicate that large parts of such fans accumulate on oceanic rather than continental crust (e.g. Curray *et al.* 1982). The presence of Cambrian MORB tholeites and boninites suggest an oceanic setting associated with a possible island arc(s) for the Cambrian (Crawford & Keays 1978; Crawford *et al.* 1984). The recognition of an accretionary complex (Powell 1984; Bischoff & Prendergast 1987; Miller & Gray in press *a, b*) within Late Cambrian and Early to Late Ordovician chert–turbidite sequences supports a subduction setting.

For the Ordovican most workers have argued for an island arc setting; the western subprovince representing a marginal sea back-arc basin, and the eastern subprovince part of a disrupted, segmented, and imbricated island arc sequence represented by the Ordovician volcanics. Varying effects inboard of the arc have been related to changes in the rate and dip of subduction, and to plate margin setting (subduction to a intracontinental transform setting); and to movement of a triple junction, as proposed for western North America (e.g. Atwater 1970). For the Lachlan Fold Belt most models argue for ocean–continent collisional setting with local obduction associated with deeper continental

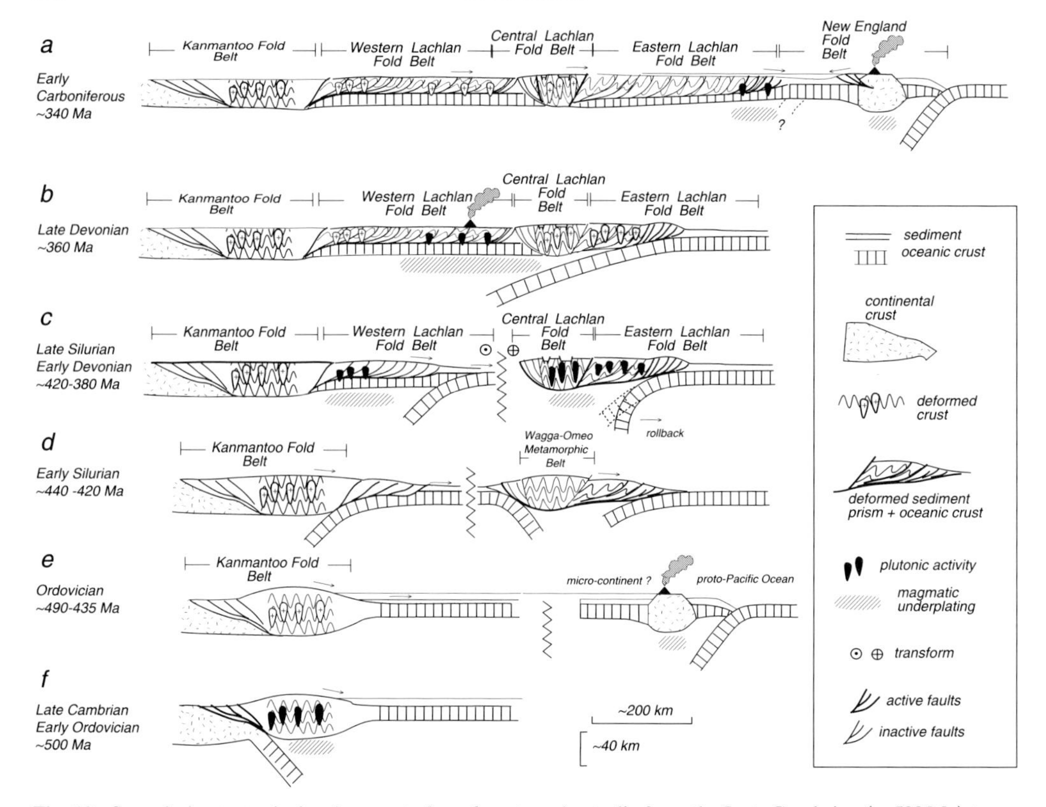

Fig. 19. Speculative tectonic development of southeastern Australia from the Late Cambrian (*c.* 500 Ma) to the early Carboniferous (*c.* 340 Ma) based on a geological reconstruction along 37°S latitude. The cartoons utilise upper crustal vergence to indicate directions of underthrusting and positions of subduction zones. The 'zigzag' line indicates a break in section (see c, d, e) which becomes a dextral transform fault in the Early Devonian to juxtapose the western and central/eastern Lachlan Fold Belt segments.

underthrusting (e.g. Fergusson *et al.* 1986; Fergusson & VandenBerg 1990; Fergusson & Coney 1992*a*, fig. 10).

The shoshonitic rather than andesitic character of the Ordovician volcanics, the lack of associated calc-akaline volcanics, and the lack of recognition of a former curvilinear arc system has led to questioning of a subduction-related origin (Wyborn 1992). Shoshonites occur on both continental and oceanic substrates, and are found in subduction settings when subduction has ceased either during rifting of older arcs or arc-polarity reversals (Morrison 1980). In the eastern Lachlan Fold belt these volcanics have now been linked to localized volcanic centres within an abyssal plain covered by Ordovician quartz-rich turbidites (Fergusson & Coney 1992*b*; VandenBerg & Stewart 1992). Inclusion of Late Ordovician strata within the chert–mélange sequence of the Narooma accretionary complex (Bischoff & Prendergast 1987; Stewart & Glen 1991) suggests that subduction in the eastern belt may not have started until the Early Silurian (Miller & Gray in press *a*, *b*).

The complicated time and space variations in the deformation and magmatic pattern (Figs 14 & 16) require a complex oceanic plate setting where the western and eastern subprovinces must represent different parts of such a setting. The reasons are as follows:

(1) The easternmost part of the western subprovince (Melbourne 'trough') was in an open east-facing ocean setting until the late Early Devonian (VandenBerg & Stewart 1992; Powell pers. comm. 1994).

(2) Silurian–Devonian: an extensional region existed in part of the eastern subprovince (Cowra and Hill End 'troughs'), whereas the western subprovince was still a continental margin sediment-prism dominated by quartz-rich turbidites. The central subprovince underwent high temperature/low pressure metamor-

phism and accompanying deformation in the Early Silurian, and erosional unroofing and extensional collapse by the mid-Silurian.

(3) Cambrian–Ordovician: an oceanic setting existed in the western subprovince with mostly turbidites deposited on oceanic crust in a convergent margin setting. There is some evidence of subduction with the presence of Cambrian high Ti-andesites (boninites: Crawford & Keays 1978; Crawford *et al.* 1984). The eastern subprovince at this time was dominated by a shallow marine shoshonitic basalt–limestone–volcaniclastic facies (Molong Province: VandenBerg & Stewart 1992).

(4) Deformation pulses and the tectonic vergences for individual structural belts are centred about the central subprovince.

Lachlan Fold Belt history

The tectonic models presented in this paper (Fig. 19) require an ocean–continent collisional setting, with local geological complexities due to: (1) changes from a convergent to a transform margin (normal to oblique convergence) in the Early Devonian, (2) changes in subduction zone dip in the Late Devonian and (3) variations in the distribution of continental and oceanic crust.

The Cambrian was dominated by an oceanic setting with a number of submarine arcs, perhaps modified by intra-arc rifting (Crawford & Keays 1978; Crawford *et al.* 1984). In the Late Cambrian–Early Ordovician subduction-related convergence produced an east-vergent thrust-belt and associated metamorphism and plutonism (500 Ma) inboard at the craton margin (Kanmantoo Fold Belt) (Fig. 19f). Outboard there was little change in setting into the Ordovician. This period was dominated by a massive influx of turbidites from the west and southwest (present coordinates) (Fig. 19e). Initiation of W-dipping subduction in the Early Silurian (both eastern and western subprovinces) (Fig. 19d) led to a continent–ocean collisional interaction (Chilean setting) of a micro-plate with development of the Wagga–Omeo Metamorphic Belt in the central/eastern subprovince, and east-directed thrusting in the western Lachlan Fold Belt. Most interpretations have the western subprovince in a back-arc position at this time (e.g. Powell 1983). The Tabberabbera Zone reflects either rear-arc thrusting or alternatively limited E-dipping subduction during development of the central Lachlan Fold Belt (Fig. 19d) (Gray & Miller 1995). The southeastern part of the Tabberabbera Zone could therefore represent part of an accretionary complex developed above an Early Silurian E-dipping subduction zone along the eastern margin of the Wagga–Omeo Metamorphic Belt.

In the Late Silurian to Early Devonian east-directed thrusting was taking place in the western and eastern subprovinces respectively (Fig. 19c). An inferred change in plate motion led to oblique convergence with eventual docking and amalgamation of the western and central/eastern subprovinces along the Mt Wellington Fault Zone in the late Early Devonian (Fig. 19c). This collisional interaction led to a change in the angle of subduction and magmatic underplating inboard of the continental margin leading to a period of silicic volcanism and plutonism in the western subprovince and broad continental style sedimentation (Lambie facies) across the eastern subprovince (Fig. 19b).

In the Early Carboniferous (Fig. 19a) collisional interaction of the developing New England Fold Belt with the northern part of the eastern Lachlan Fold Belt, deformed this part of the eastern subprovince and may have led to cessation of subduction along an inner subduction zone, or to a stepping of subduction to the east in present geographic coordinates. By this time the Lachlan Fold Belt had been accreted to the Gondwana margin.

Conclusions

In any orogenic belt, models of crustal evolution must explain the diachronous nature of the deformation, plutonism and volcanism, metamorphism and changes in sedimentation. The observed surface geology of the Lachlan Fold Belt and the complex magmatic history recorded by granite plutonism and silicic volcanism is linked with long lived subduction in the Silurian to Early Devonian along the former Gondwana margin. Despite the lack of a classic 'suture' in the Lachlan Fold Belt, surface structures have been used to infer collisional processes. Previously, the Cambrian and Ordovician volcanics in the western and eastern subprovinces have been used to define the plate margin setting.

The Lachlan Fold Belt is unusual in that there is no craton-verging thrust-belt and metamorphic hinterland typical of thrust-belts such as the Appalachians. The wedge-shaped Wagga–Omeo Metamorphic Belt which is located in the central subprovince dominates the gross form of the fold-belt, controlling fold-belt symmetry shown by faults within the various linked thrust-systems (Fig. 3) and the overall kinematic evolution of the fold-belt (Fig. 15). Thrust slices or windows exposing Proterozoic basement within the Lachlan Fold Belt have not been recognized indicating that deeper levels of the crust have not been exhumed. Crustal-scale decoupling is probably associated with a fold-belt-wide detachment which links different thrust and strike-slip fault systems. Different crustal thickening processes have operated at different crustal levels. At upper to mid- crustal levels deformation is dominated by chevron-folding within turbidites and duplexing within the mafic volcanics which underlie the turbidites of the western subprovince. Extensive magmatic underplating at the base of the crust with intrusions into the mid- and lower crust have generated the high T/low P metamorphic rocks and the large volumes of granitoids exposed at the present level of erosion.

Thermal and structural aspects of the Lachlan Fold Belt appear to be associated with W-dipping subduction zones initiated in the Early Silurian. The timing and location of both compressional tectonic events and magmatic activity have been linked to changes in the convergent margin geometry, changes in subduction zone dip, and changing positions of ancient magmatic 'arc' systems. The juxtaposition of the eastern and central subprovinces with the western subprovince in the late Early Devonian led to an extremely wide fold-belt and produced an apparent complexity in the timing of deformation, plutonism and deformation kinematics.

The research was supported by Australian Research Council Grants E8315666 (awarded to DRG), A38315675 (awarded to R. Cas and D.R.G.), and A38930542 (awarded to C.J. Wilson and D.R.G.), and Monash University Special Research Funds. This paper is the culmination of 10 years research where many of the ideas have evolved during interaction and discussions with a number of geologists in particular C. Fergusson, V. Morand, D. Forster, C. Wilson, R. Gregory, D. Durney, R. Cas, C. Willman, F. VandenBerg, N. Woodward, P. Coney, C. Elliott, B. Hobbs, I. Nicholls, I. Stewart, J. Miller and various Honours students. Completion of the text for the manuscript was undertaken as part of my time commitment to the Australian Geodynamics Co-operative Research Centre. The paper was published with permission of the Director, Australian Geodynamics Cooperative Research Centre (AGCRC). I thank M. O'Dea, S. Zakowski, G. Price (AGCRC Director), J.-P. Burg and M. Ford for helpful comments on the manuscript, and C. Fergusson and L. Glover III for constructive reviews.

References

Armijo, R., Tapponnier, P. & Tonglin, H. 1989. Late Cenozoic right-lateral strike-slip faulting in southern Tibet. *Journal of Geophysical Research,* **94**, 2787–2838.

Atwater, T. 1970. Implications of plate tectonics for the Cenozoic tectonic evolution of western North America. *Geological Society of America Bulletin,* **81**, 3513–3536.

Barton, M. D. & Hansen, R. B. 1989. Magmatism and the development of low-pressure metamorphic belts: implications from the western U.S. and thermal modelling. *Geological Society of America Bulletin,* **101**, 1051–1065.

Beck, M. E. 1983. On the mechanism of tectonic transport in zones of oblique subduction. *Tectonophysics,* **93**, 1–11.

Bischoff, G. C. O. & Prendergast, E. I. 1987. Newly discovered Middle and Late Cambrian fossils from the Wagonga Beds of New South Wales, Australia. *Neues Jahrbuch fur Geologie und Palaontologie,* **175**, 39–64.

Bucher, M. B., Foster, D. A. & Gray, D. R. 1996. Timing of cleavage development in the western Lachlan Fold Belt: new constraints from ^{40}Ar/^{39}Ar geochronology. *Geological Society of Australia Abstracts* **41**, 66.

Burg, J. P. & Wilson, C. J. L. 1988 A kinematic analysis of the southernmost part of the Bega Batholith. *Australian Journal of Earth Sciences Bulletin,* **35**, 1–13.

Cas, R. A. F. 1983. *Palaeogeographic and tectonic development of the Lachlan Fold Belt of southeastern Australia.* Geological Society of Australia Special Publication, **10**.

——, Flood, R. H. & Shaw, S. E. 1976. Hill End Trough: new radiometric ages. *Search,* **7**, 205–207.

——, Powell, C. MC. A. & Crook, K. A. W. 1980. Ordovician palaegeography of the Lachlan Fold Belt: A modern analogue and tectonic constraints. *Journal of the Geological Society of Australia,* **27**, 19–31.

Chappell, B. W., White, A. J. R. & Hine, R. 1988. Granite provinces and basement terranes in the Lachlan Fold Belt, southeastern Australia. *Australian Journal of Earth Sciences Bulletin,* **35**, 505–524.

Clemens, J. D. 1988. Volume and composition relationships between granites and their lower crustal sources: An example from central Victoria, Australia. *Australian Journal of Earth Sciences Bulletin,* **35**, 445–449.

—— & Wall, V. J. 1981. Crystallisation and origin of some peraluminous (S-type) granitic magmas. *Canadian Mineralogist,* **19**, 111–132.

Collins, W. J. 1994. Upper- and middle-crustal response to delamination: an example from the Lachlan Fold Belt, eastern Australia. *Geology,* **22**, 143–146.

—— & Vernon, R. H. 1992. Palaeozoic arc growth, deformation and migration across the Lachlan Fold Belt, southeastern Australia. *Tectonophysics,* **214**, 381–400.

—— & —— 1994. Rift-drift-delamination model of continental evolution: Palaeozoic tectonic development of eastern Australia. *Tectonophysics,* **235**, 249–275.

Coney, P. J. 1973. Plate tectonics of marginal foreland

thrust-fold belts. *Geology,* **1**, 131–134.
—— 1992. The Lachlan Fold Belt of eastern Australia and Circum-Pacific tectonic evolution. *Tectonophysics,* **214**, 1–25.
——, EDWARDS, A., HINE, R. MORRISON, F. & WINDRUM, D. 1990, The regional tectonics of the Tasman orogenic system, eastern Australia: *Journal of Structural Geology,* **12**, 519–543.
COX, S. F., ETHERIDGE, M. A., CAS, R. A. F. & CLIFFORD, B. A. 1991. Deformational style of the Castlemaine area, Bendigo-Ballarat zone: Implications for evolution of crustal structure in central Victoria. *Australian Journal of Earth Sciences,* **38**, 151–170.
CRAWFORD, A. J. & KEAYS, R. R. 1978. Cambrian greenstone belts in Victoria: marginal sea-crust slices in the Lachlan Fold belt of southeastern Australia. *Earth Planet. Science Letters,* **41**, 197–208.
——, CAMERON, W. E. & KEAYS, R. R. 1984. The association bonninite low Ti-andesite-tholeiite in the Heathcote greenstone belt, Victoria: Ensimatic setting for the early Lachlan Fold belt. *Australian Journal of Earth Sciences,* **31**, 161–175.
CROOK, K. A. W. 1974. Kratonisation of west Pacific-type geosynclines. *Journal of Geology,* **87**, 24–36.
—— 1980. Fore-arc evolution in the Tasman Geosyncline. *Journal of the Geological Society of Australia,* **27**, 215–232.
—— & POWELL, C. MC. A. 1976. The evoultion of the southeastern part of the Tasman geosyncline. *Excursion Guide 17A*, 25th International Geological Congress, Sydney, 122p.
CURRAY, J. R., EMMEL, F. J., MOORE, D. G. & RAITT, R. W. 1982. Structure, tectonics and geological history of the northeastern Indian Ocean. *In*: NAIRN, A. E. M. & STEHLI, F. G. (eds) *The Indian Ocean, The ocean basin and margins,* **6**, Plenum Press, New York, 399–450.
DIAMENT, M. & 9 others 1992. Mentawai fault zone off Sumatra: a new key to the geodynamics of western Indonesia. *Geology,* **20**, 259–262.
EVERNDEN, J. F. & RICHARDS, J. R. 1962. Potassium-argon ages in eastern Australia. *Journal of the Geological Society of Australia,* **9**, 1–50.
FARRELL, T. R. & OFFLER, R. 1989. Greenschist facies metamorphism of a rift basin sequence, Breadalbane, N.S.W. *Australian Journal of Earth Sciences,* **36**, 337–349
FERGUSSON, C. L. 1987. Early Paleozoic back-arc deformation in the Lachlan Fold Belt, southeastern Australia: implications for terrane translations in eastern Gondwanaland. *In*: LEITCH, E. C. & SCHEIBNER, E. (eds) *Terrane accretion and orogenic belts.* American Geophysical Union Geodynamics Series **19**, 39–56.
—— & CONEY, P. J. 1992*a*. Convergence and intraplate deformation in the Lachlan Fold Belt of southeastern Australia. *Tectonophysics,* **214**, 417–439.
—— & —— 1992*b*. Bengal Fan. *Geology,* **20**, 1047–1049.
—— & VANDENBERG, A. H. M. 1990. Middle Palaeozoic thrusting in the eastern Lachlan Fold Belt, southeastern Australia. *Journal of Structural Geology,* **12**, 577–589.
——, GRAY, D. R. & CAS, R. A. F. 1986. Overthrust terranes in the Lachlan Fold Belt, southeastern Australia. *Geology,* **14**, 519–522.
FINLAYSON, D. M. 1982. Geophysical differences in the lithosphere between Phanerozoic and Precambrian Australia. *Tectonophysics,* **84**, 287–312.
——, PRODEHL, C. & COLLINS, C. D. N. 1979. Explosion seismic profiles and implications for crustal evolution in southeastern Australia: *Bureau of Mineral Resources Journal of Australian Geology and Geophysics,* **4**, 243–252.
——, COLLINS, C. D. N. & DENHAM, D. 1980. Crustal structure under the Lachlan Fold Belt, southeastern Australia: *Physics of the Earth and Planetary Interiors,* **21**, p. 321–342.
FLOOD, R. H. & VERNON, R. H. 1979. The Cooma Granodiorite, Australia: an example of *in situ* crustal anatexis? *Geology,* **6**, 81–84.
GEISER, P. A. 1988. Mechanisms of thrust propagation: some examples and implications for the analysis of overthrust terranes. *Journal of Structural Geology,* **10**, 829–845.
GIBSON, G., WESSON, V. & CUTHBERTSON, R. 1981. Seismicity of Victoria to 1980. *Journal of the Geological Society of Australia,* **28**, 341–356.
GLEN, R. A. 1992. Thrust, extensional and strike-slip tectonics in an evolving Palaeozoic orogen- a structural synthesis of the Lachlan Orogen of southeastern Australia. *Tectonophysics,* **214**, 341–380.
—— & VANDENBERG, A. H. M. 1987. Thin-skinned tectonics in part of the Lachlan Fold Belt near Delegate, southeastern Australia. *Geology,* **15**, 1070–1073.
——, DALLMEYER, R. D. & BLACK, L. P. 1992. Isotopic dating of basin inversion- the Palaeozoic Cobar Basin, Lachlan Orogen, Australia. *Tectonophysics,* **214**, 249–268.
——, DRUMMOND, B. J., GOLEBY, B. R., PALMER, D. & WAKE-DYSTER, K. D. 1994. Structure of the Cobar basin, New South Wales, based on seismic reflection profiling. *Australian Journal of Earth Sciences Bulletin,* **41**, 341–352.
GLOVER, L., SPEER, J. A., RUSSELL, G. S. & FARRAR, S. S. 1983. Ages of regional metamorphism and ductile deformation in the central and southern Appalachians. *Lithos,* **16**, 223–245.
GRAY, C. M. 1984. An isotopic mixing model for the origin of granitic rocks in southeastern Australia. *Earth and Planetary Science Letters*, **70**, 47–60.
—— 1990. A strontium isotopic traverse across the granitic rocks of southeastern Australia: Petrogenetic and tectonic implications. *Australian Journal of Earth Sciences,* **37**, 331–349.
GRAY, D. R. 1988. Structure and Tectonics. *In*: DOUGLAS, J. G. & FERGUSON, J. A. (eds) *Geology of Victoria* (Second Edition). Victorian Division Geological Society of Australia Publication, 1–36.
—— 1995. Thrust kinematics and transposition fabrics from a basal detachment zone, eastern Australia. *Journal of Structural Geology,* **17**, 1637–1654.

—— & CULL, J. P. 1992. Thermal regimes, anatexis, and orogenesis: relations in the western Lachlan Fold Belt, southeastern Australia. *Tectonophysics,* **214**, 441–461.

—— & MILLER, J. M. 1995. Implications of a subduction complex for the tectonic evolution of the Lachlan Fold Belt. *Geological Society of Australia Abstracts,* **40**, 55–56.

—— & MORTIMER, L. 1996. Implications of overprinting deformations and fold interference patterns in the Melbourne Zone, Lachlan Fold Belt. *Australian Journal of Earth Sciences,* **43**, 103–114.

—— & WILLMAN, C. E. 1991*a*. Deformation in the Ballarat Slate belt, central Victoria and implications for the crustal structure across SE Australia. *Australian Journal of Earth Sciences,* **38**, 171–201.

—— & —— 1991*b*. Thrust-related strain gradients and thrusting mechanisms in a chevron-folded sequence, southeastern Australia. *Journal of Structural Geology,* **13**, 691–710.

——, WILSON, C. J. L. & BARTON, T. J. 1991*a*. Intracrustal detachments and implications for crustal evolution within the Lachlan Fold belt, southeastern Australia. *Geology,* **19**, 574–577.

——, GREGORY, R. T. & DURNEY, D. W. 1991*b*. Rock-buffered fluid-rock interaction in deformed quartz-rich turbidite sequences, eastern Australia. *Journal of Geophysical Research,* **96**, 19681–19704.

——, FOSTER, D. A. & BUCHER, M. 1996. Revised Lachlan Fold belt orogenic patterns based on a new $^{40}Ar/^{39}Ar$ dataset. *Geological Society of Australia Abstracts,* **41**, p.164.

GRIFFIN, W. L. & O'REILLEY, S. Y. 1987. Is the continental Moho the crust-mantle boundary? *Geology,* **15**, 241–244.

HAMILTON, W. B. 1988. Plate tectonics and island arcs. *Geological Society of America Bulletin,* **100**, 1503–1527.

HARLAND, W. B., ARMSTRONG, R. L., COX, A. V., CRAIG, L. E., SMITH, A. G. & SMITH, D. G. 1990. *A Geologic Time Scale* 1989. Cambridge University Press, Cambridge.

KARIG, D. E., MOORE, G. F., CURRAY, J. R. & LAWRENCE, M. B. 1980. Morphology and shallow structure of the lower trench slope of Nias Island, Sunda Arc. *In*: HAYES, D. E. (ed.) *The Tectonic and Geologic Evolution of southeast Asian seas and Islands*. American Geophysical Union Monograph, **23**, 179–208.

LEITCH, E. C. & SCHEIBNER, E. 1987. Stratotectonic terranes of the Eastern Australian Tasmanides. *In*: LEITCH, E. C. & SCHIEBNER, E. (eds) *Terrane Accretion and Orogenic Belts*. American Geophysical Union Geodynamics Series, **19**, 1–19.

LEVEN, J. H., STUART-SMITH, P. G., MUSGRAVE, R. J., RICKARD, M. J. & CROOK, K. A. W. 1992. A geophysical transect across the Tumut Synclinorial Zone, N.S.W. *Tectonophysics,* **214**, 239–248.

LOOSEVELD, R. J. H. & ETHERIDGE, M. A. 1990. A model for low pressure facies metamorphism during crustal thickening. *Journal of Metamorphic Geology,* **8**, 257–267.

MCCULLOCH, M. T. & CHAPPELL, B. W. 1982. Nd isotopic characteristics of S- and I- type granites. *Earth & Planetary Science Letters,* **58**, 51–64.

MILLER, J. & GRAY, D. R. in press *a*. Sediment accretion in a Palaeozoic accretionary complex, eastern Australia. *Journal of Structural Geology.*

—— & —— in press *b*. Subduction-related deformation and the Narooma anticlinorium, eastern Lachlan Fold Belt. *Australian Journal of Earth Sciences.*

MORAND, V. J. 1990. Low- pressure regional metamorphism in the Omeo Metamorphic Complex, Victoria, Australia. *Journal of Metamorphic Geology,* **8**, 1–12.

—— 1992. Pluton emplacement in a strike-slip fault zone: the Doctors Flat Pluton, Victoria, Australia. *Journal of Structural Geology,* **14**, 205–213.

—— & GRAY, D. R. 1991. Major fault zones related to the Omeo Metamorphic Complex, northeastern Victoria. *Australian Journal of Earth Sciences Bulletin,* **38**, 203–221.

MORRISON, G. W. 1980. Characteristics and tectonic setting of the shoshonite rock association. *Lithos,* **13**, 97–108.

MURPHY, N. C. & GRAY, D. R. 1992. East-directed overthrusting in the Melbourne zone, Lachlan Fold Belt. *Australian Journal of Earth Sciences Bulletin,* **39**, 37–53.

MURRAY, C. G., SCHEIBNER, E. & WALKER, R. N. 1989. Regional geological interpretation of a digital coloured residual Bouger gravity image of eastern Australia with a wavelength cut-off of 250 km. *Australian Journal of Earth Sciences,* **37**, 493–497.

OFFLER, R. & PRENDERGAST, E. 1985. Significance of illite crystallinity and b_0 values of K-white micas in low-grade metamorphic rocks, North Hill End Synclinorium, New South Wales, Australia. *Mineralogical Magazine,* **49**, 357–364.

OVERSBY, B. 1971, Palaeozoic plate tectonics in the southern Tasman geosyncline. *Nature Physical Science,* **234**, 45–47.

PATERSON, S. R., TOBISCH, O. T. & MORAND, V. J. 1990. The influence of large ductile shear zones on the emplacement and deformation of the Wyangala Batholith, SE Australia. *Journal of Structural Geology,* **12**, 639–650.

PIDGEON, R. T. & COMPSTON, W. 1965. The age and origin of the Cooma granite and its associated metamorphic zones, N.S.W. *Journal of Petrology,* **6**, 193–222.

PINCHIN, J. 1979. Intracrustal seismic reflections from the Lachlan Fold Belt near Canberra: *Bureau of Mineral Resources Journal of Australian Geology and Geophysics,* **5**, 305–309.

POWELL, C.MC.A. 1983. Tectonic relationship between the Late Ordovician and Late Silurian palaeogeographies of southeastern Australia. *Journal of the Geological Society of Australia,* **30**, 353–373.

—— 1984. Uluru regime. *In*: VEEVERS, J. J. (ed.) *Phanerozoic Earth History of Australia*. Oxford Geological Sciences Series **2**, 290–337.

——, EDGECOMBE, D. R., HENRY, N. M. & JONES, J.

G. 1977. Timing of regional deformation in the Hill Trough: a reassessment. *Journal of the Geological Society of Australia,* **23**, 407–422.

RICHARDS, J. R. & SINGLETON, O. P. 1981. Palaeozoic Victoria, Australia : igneous rocks, ages and their interpretation. *Journal of the Geological Society of Australia,* **28**, 395–421.

RICKARD, M. J. & WARD, P. 1981. Palaeozoic crustal thickness in the southern part of the Lachlan Orogen deduced from volcano and pluton-spacing geometry. *Journal of the Geological Society of Australia,* **28**, 395–421.

RUTLAND, R. W. R. 1976. Orogenic evolution of Australia. *Earth Science Reviews,* **12**, 161–196.

SANDIFORD, M. & POWELL, R. 1986. Deep crustal metamorphism during continental extension: modern and ancient examples. *Earth & Planetary Science Letters,* **19**, 151–158.

SCHEIBNER, E. 1974. A plate tectonic model of the Palaeozoic history of New South Wales. *Journal of the Geological Society of Australia,* **20**, 405–426.

—— 1985. Suspect terranes in the Tasman Fold Belt System, eastern Australia. *In*: HOWELL, D. G. (ed.) *Tectonostratigraphic Terranes of the Circum-Pacific Region*. Circum-Pacific Council for Energy and Mineral Resources, Earth Science Series, **1**, 493–514.

—— 1992. Influence of detachment-related passive margin geometry on subsequent active margin dynamics: applied to the Tasman Fold Belt system. *Tectonophysics,* **214**, 211–237.

SHAW, S., FLOOD, R. H. & RILEY, G. H. 1982. The Wologorong batholith, New South Wales, and the extension of the I-S line of the Siluro-Devonian granitoids. *Journal of the Geological Society of Australia,* **29**, 41–48.

SINHA, A. K. & ZIETZ, I. 1982. Geophysical and geochemical evidence for a Hercynian magmatic arc, Maryland to Georgia. *Geology,* **10**, 593–596.

SMITH, R. E. 1969. Zones of progressive regional burial metamorphism in part of the Tasman Geosyncline, eastern Australia. *Journal of Pertrology,* **10**, 144–163.

SOLOMON, M. & GRIFFITHS, J. R. 1972. Tectonic evolution of the Tasman Orogenic zone, eastern Australia. *Nature Physical Science,* **237**, 3–6.

STEWART, I. R. & GLEN, R. A. 1991. New Cambrian and Early Ordovician ages from the New South Wales Souh Coast. *Quarterly Notes N.S.W. Geological Survey,* **85**, 1–8.

STUART-SMITH, P. G. 1990. The emplacement and fault history of the Coolac Serpentinite, Lachlan Fold Belt, southeastern Australia. *Journal of Structural Geology,* **12**, 621–638.

——, HILL, R. I., RICKARD, M. J. & ETHERIDGE, M. A. 1992. The stratigraphy and deformation history of the Tumut region: implications for the development of the Lachlan Fold Belt. *Tectonophysics,* **214**, 211–237.

TAPPONNIER, P. & MOLNAR, P. 1977. Active faulting and tectonics in China. *Journal of Geophysical Research,* **82**, 2905–2930.

——, PELTZER, G., LE DAIN, A. Y., ARMIJO, R. & COBBOLD, P. 1982. Propagating extrusion tectonics in Asia: new insights from simple experiments with plasticene. *Geology,* **10**, 611–616.

TALENT, J. A. 1965. The stratigraphic and diastrophic evolution of the central and eastern Victoria in Middle Palaeozoic times. *Proceedings of the Royal Society of Victoria,* **79**, 179–195.

VALLANCE, T. G. 1967. Palaeozoic low-pressure regional metamorphism in southeastern Australia. *Meddelelser fra Dansk Geologisk Forening,* **17**, 494–503.

—— 1969. Plutonic and metamorphic rocks. *In*: *Geology of New South Wales.*, PACKHAM, G. H. (ed.), *Journal of the Geological Society of Australia,* **16**, 180–200.

VANDENBERG, A. H. M. & STEWART, I. R. 1992. Ordovician terranes of the southeastern Lachlan Fold Belt: stratigraphy, structure and palaeogeographic reconstruction. *Tectonophysics,* **214**, 159–176.

WHITE, A. J. R. & CHAPPELL, B. W. 1983. Granitoid types and their distribution in the Lachlan Fold Belt, southeastern Australia. *Geological Society of America Memoir,* **159**, 21–34.

——, —— & CLEARY, J. R. 1974. Geologic setting and emplacement of some Australian Palaeozoic batholiths and implications for emplacement mechanisms. *Pacific Geology,* **8**, 159–171.

WILSON, C. J. L., WILL, T. M., CAYLEY, R. A. & CHEN, S. 1992. Geologic framework, tectonic evolution and displacement history in the Stawell Zone, Lachlan Fold Belt, Australia. *Tectonophysics,* **214**, 93–127.

WOODWARD, N. B. 1995. Thrust systems in the Tamworth Zone, southern New England Orogen, New South Wales. *Australian Journal of Earth Sciences,* **42**, 407–422.

——, GRAY, D. R. & SPEARS, D. B. 1986. Including strain data in balanced sections. *Journal of Structural Geology,* **8**, 313–324.

WYBORN, D. 1992. The tectonic significance of Ordovician magmatism in the eastern Lachlan Fold Belt. *Tectonophysics,* **214**, 177–192.

YANG, X. & GRAY, D. R. 1994. Strain, cleavage and microstructure variations in sandstone: implications for stiff layer behaviour in chevron folding. *Journal of Structural Geology,* **16**, 1353–1365.

The Scandinavian Caledonides and their relationship to the Variscan belt

P. REY[1], J.-P. BURG[2] & M. CASEY[2]

[1]*Department of Geology, Monash University, Clayton, VICT 3168, Australia*

[2]*Geologisches Institut, ETH-Zentrum, Sonneggstrasse 5, CH-8092 Zürich, Switzerland*

Abstract: The main events that mark the contraction and extension histories of the Scandinavian Caledonides and the European Variscides are summarized. It is shown that continental subduction may have developed similarly large and asymmetric thrust systems in both orogens. However, while continent–continent collision developed in the Variscides, extension began in the Scandinavian Caledonides marking the end of continental subduction. This led extensional tectonics to affect two continental crusts with contrasting rheology and therefore led to contrasting extensional modes. We argue that plate divergence, responsible for extension in the Scandinavian Caledonides, was triggered by the Variscan collision between Laurasia and Gondwana. In contrast, horizontal buoyancy forces acting on a thermally softened thickened crust are more likely to have been responsible for extension in the Variscan belt.

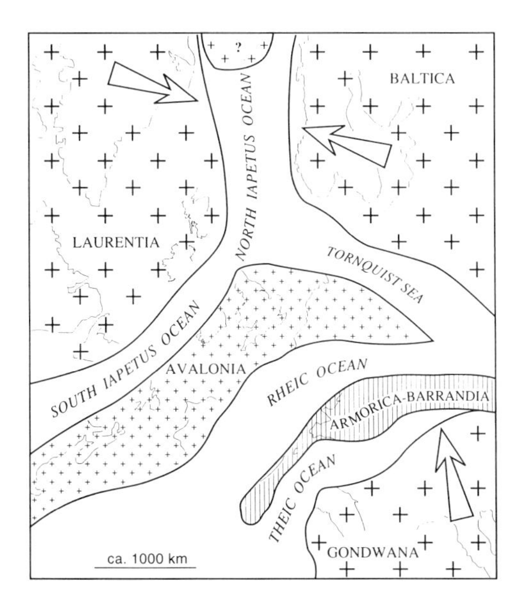

Fig. 1. Bulk distribution of continental masses in Silurian times (*c.* 430 Ma) and approximate directions of relative closure vectors after Ziegler (1986) and Scotese & McKerrow (1990). The width of the oceanic domains and the amount of intracontinental shortening (e.g. in Baltica, see Soper *et al.* 1992) are not established. The eastern extension of Armorica–Barrandia includes the Léon region according to Balé & Brun (1986) and the Central Iberian Zone according to Ballèvre *et al.* (1992).

From a tectonic point of view, Paleozoic history is essentially that of the assembly of Pangaea, the unique super continent that began to break up in the Mesozoic (Smith *et al.* 1973; Bambach *et al.* 1980). There is faunal (e.g. Cocks & Fortey 1982), sedimentological (e.g. Noblet & Lefort 1990) and paleomagnetic (e.g. Van der Voo 1982) evidence that in Europe an older (Caledonian) orogenic belt resulted from the frontal collision of Baltica and Laurentia (Fig. 1), followed in time by a younger (Variscan) orogeny when Gondwana and Laurussia (Laurentia together with Baltica, Fig. 1) joined. The Uralides formed later when Siberia collided with the earlier-formed Laurussia–Gondwana continental mass in the Latest Palaeozoic (Scotese & McKerrow 1990; Puchkov this volume). During this assemblage of Pangaea in Palaeozoic times, the Scandinavian Caledonides and the Variscides came about by similar but independent plate tectonic histories. Convergence between Laurentia and Baltica during the Ordovician and early Silurian (500–430 Ma) closed the latest Precambrian–Cambrian North Iapetus Ocean. The last deep-water marine sedimentation and the peak of the Caledonian orogeny indicate that collision occurred in the Mid-Silurian to Early Devonian (425–400 Ma, e.g. Harland & Gayer 1972; Robert & Sturt 1980; Gee 1982; Milnes *et al.* this volume). During the same period, the northward motion

From Burg, J.-P. & Ford, M. (eds), 1997, *Orogeny Through Time*, Geological Society Special Publication No. 121, pp. 179–200.

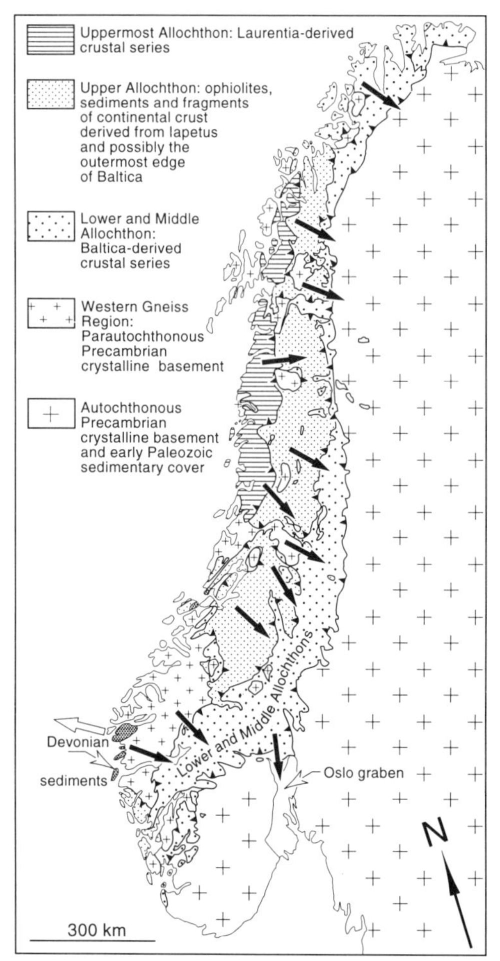

Fig. 2. Main structural elements of the Scandinavian Caledonides, simplified after Stephens & Gee (1989). Black arrows are average thrust transport directions (e.g. Soper *et al.* 1992) and empty arrow is the Devonian extension direction (e.g. Chauvet & Séranne 1989; Fossen 1992).

of Gondwana closed the Rheic and Theic oceanic domains. In the Early Devonian (*c.* 395 Ma) Gondwana collided with the recently assembled Laurussia (Bard *et al.* 1980; Matte 1986; Franke 1989*b*). Caledonian east–west contraction stopped when north–south Early Variscan contraction began (directions are given in a present-day reference frame). From this time onward, the two orogens evolved differently. The Scandinavian Caledonides underwent extension (Hossack 1984; McClay *et al.* 1986; Norton 1986; Séguret *et al.* 1989) while the Variscides were recording early collision. Synchronism between extension in the Caledonides and early contraction in the Variscides suggests that the Early Devonian evolution of the two belts may not be totally disconnected. Later on, from Early Devonian to Early–Mid-Carboniferous (390–330 Ma), the Variscides recorded active continent–continent collision (e.g. Matte 1986; Pin & Peucat 1986; Franke 1989*b*). This was followed by post-thickening extension from Mid-Carboniferous to Permian (330–300 Ma, Costa 1990; Malavieille 1993; Burg *et al.* 1994).

In this paper, we compare contraction and subsequent extension in the Scandinavian Caledonides with those of the Variscides. We argue that both their contractional and extensional structures differ, although they are reported to have followed a similar geodynamic evolution. We follow some authors (e.g. Séranne *et al.* 1991; Fossen 1992) in arguing that the origin of extension in the Scandinavian Caledonides should be found in changing boundary conditions, namely that the nappe pile slid back during the Devonian into a space created by divergence between Laurentia and Baltica (the free space model of Wilks & Cuthbert 1994). In conclusion, we propose a plate tectonic model in which extension in the Scandinavian Caledonides is related to the Early Variscan contraction.

Contractional tectonics

The Scandinavian Caledonides

The Scandinavian Caledonides (Fig. 2) are a sub-linear, 1800 km long orogenic belt with a rather simple structure. Five tectonic units are identified (e.g. Kulling 1972; Bryhni & Sturt 1985). Their relationships show that the Caledonides are essentially a wedge of nappes which become successively more far-travelled upward in the pile resulting from the southeastward imbrication of the outer margin of Baltica during Ordovician–Silurian times. The wedge is tectonically overridden by elements of Laurentian affinities (e.g. Dallmeyer & Gee 1986; Stephens & Gee 1989).

(1) The Autochthon–Parautochthon consists of the Precambrian Baltic Shield covered by Late Precambrian to Ordovician sediments. To the east, i.e. the foreland, the Caledonian overprint is restricted to thin-skinned fold and thrust tectonics immediately below the contact zone with the overlying Lower Allochthon. To the west, i.e. the hinterland, the Precambrian Baltic Shield re-appears as a window in the so-called Western Gneiss Region. In southern Norway, Caledonian parageneses record a

northwestward increase in pressure ranging from 15 to 30 kbar (Griffin *et al.* 1985; Smith & Lappin 1989; Andersen & Jamtveit 1990), which implies burial of the Baltic basement to depths of 70–100 km. Strain was concentrated in mylonitic zones that anastomose around large blocks in which Precambrian structures are preserved (Austrheim 1987; Milnes *et al.* 1988, this volume). These shear zones also include exotic fragments of garnet peridotite.

(2) The Lower Allochthon, (the Décollement Zone of Fossen 1992), developed mostly within the Upper Precambrian – Upper Silurian sediments. It is a high strain shear zone of thin-skinned thrusts in which basement slivers, derived from the western margin of the Baltic Shield, have been locally imbricated.

(3) The Middle Allochthon comprises flat-lying nappes of Precambrian basement and its psammitic cover. It has been translated over 300 km southeastward above the Precambrian shield (cf. Milnes *et al.* this volume). These nappes, derived from Baltica, were affected by generally low-grade metamorphism. They are remarkably large in areal extent (several thousands of square kilometres) in comparison to their thickness (4–6 km, Bryhni & Sturt 1985).

(4) The Upper Allochthon consists of variably metamorphosed and deformed packages, including ophiolitic fragments overlain by low-grade sediments and island-arc-related volcanic and plutonic rocks of mostly Ordovician age (Dunning & Pedersen 1988; Stephens & Gee 1989). They build a dominantly sedimentary wedge of far-travelled nappes in which the Caledonian strain intensity and metamorphic grade increase upward and westward. Granulite-facies rocks, probably of Precambrian age, are reported in the Upper Allochthon which also contains retrogressed Caledonian eclogites and garnet peridotites.

(5) The Uppermost Allochthon is composed of schists, marbles, gneisses and granites that may represent the eastern edge of Laurentia, including a basement and cover complex (Stephens & Gee 1985). They have been intruded by several synorogenic gabbro and granitoid intrusions.

In the tectonic models proposed (e.g. Williams 1984; Dallmeyer & Gee 1986; Andersen *et al.* 1991*a*), Baltica's margin (the Autochthon and Parautochthon together with the Lower and Middle Allochthons) began to subduct westward beneath a volcanic arc (now dismembered in the Upper Allochthon) in Late Cambrian times. During the Early Ordovician, an accretionary wedge developed associated with subduction and imbrication of the outer margin of Baltica, and concomitant high-grade metamorphism. Until the earliest Silurian, however, ophiolites formed in back-arc basins in other parts of Iapetus and marine deposition continued on parts of the outer edge of Baltica until the Wenlock (Milnes *et al.* this volume, and reference therein). The complete closure of Iapetus was associated with emplacement of an imbricated nappe complex onto the Baltic Shield. Caledonian contraction culminated and migrated towards the foreland during the Silurian (Bockelie & Nystuen 1985). Except for the Upper Silurian to Lower Devonian molasse of the Ringerike Group preserved in the Oslo Graben, there are only very limited clastic sedimentary records of this event in the present foreland of the Scandinavian Caledonides. No significant flexural basin developed parallel to the Caledonian belt as could be expected in a collision orogen. Instead, the nappe complex overlaps the autochthonous domain consisting of the thin succession of Early Palaeozoic sediments deposited on the Precambrian basement (e.g. Roberts & Gee 1985) all along the Scandinavian Caledonian front.

The discovery of eclogites in the Caledonides of NE Greenland (Gilotti 1993), could modify this view of the Scandinavian Caledonides if these high-pressure rocks are revealed to be Caledonian. In this case, a microplate lying in between Baltica and Laurentia (see Fig. 9) might explain the occurrence of eclogites on both sides of the North Iapetus ocean.

The Variscan Belt

The Variscan Belt of Europe (Fig. 3) resulted from the Early Devonian to Mid-Carboniferous collision of Laurussia and Gondwana, between which smaller Precambrian continental blocks (Avalonia, Armorica and its eastern equivalent, Barrandia, in Bohemia) were squeezed (e.g. Bard *et al.* 1980; Behr *et al.* 1984; Matte 1986, 1991; Franke 1989 *a, b*). The Gondwana–Laurussia convergence closed two oceanic domains of Cambrian to Ordovician age (Pin 1990): the Rheic ocean, between Avalonia and Armorica–Barrandia, and the Theic (McKerrow & Ziegler 1972) ocean (or Prototethys, including the Galicia–Massif Central ocean of Matte 1991) between Armorica–Barrandia and Gondwana (Fig. 1). Their respective width is a point of contention (Neugebauer 1989). The collision of a Gondwana promontory (Iberia) with Laurussia s.l. (Baltica, Laurentia and Avalonia) formed the Ibero-Armorican arc, a salient orocline in the western part of the belt (Fig. 3, Matte 1986; Burg *et al.* 1987). By the end of

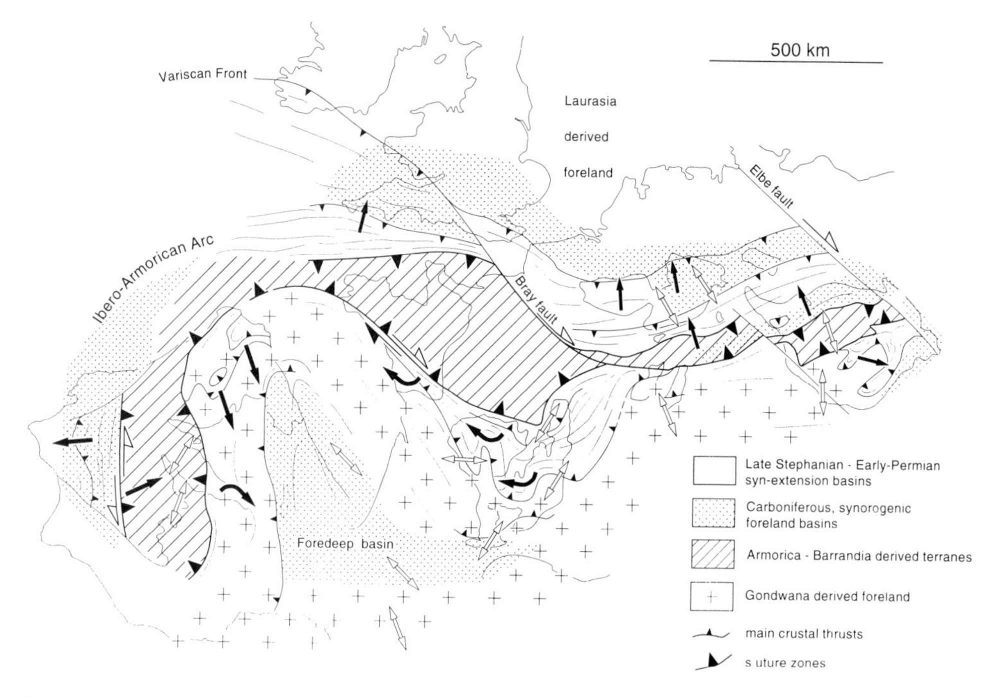

Fig. 3. Main structural elements of the Variscides (sutures as from Bard *et al.* 1980; Martínez-Catalán 1990; Ballèvre *et al.* 1992; Edel & Weber 1995) Stipple, Carboniferous, synorogenic clastic sedimentation in foreland basins after Franke (1989*a*). Black arrows, transport directions as from Matte (1986) and Burg *et al.* (1987). White arrows, extension directions as from Burg *et al.* (1994).

convergence, in Carboniferous times, the Variscan belt was characterized by the accretion of three main terranes (Avalonia, Armorica–Barrandia and Gondwana) with a bilateral symmetry centred on Armorica–Barrandia. Based on the direction of large scale thrust tectonics and the migration of the deformation and metamorphic events toward the forelands, it is inferred that the Rheic and Theic oceanic lithospheres were essentially consumed by subduction zones of opposite dip beneath the dorsal Armorica–Barrandia blocks (e.g. Matte 1986; Franke 1989*a, b*).

Subduction of oceanic and continental lithosphere is marked by high-pressure metamorphism (eclogites, blue schists, white schists and granulites) between 440 and 390 Ma (mostly zircon U-Pb data, e.g. Peucat & Cogné 1977; Gebauer & Grünenfelder 1979; Duthou *et al.* 1981; Pin & Lancelot 1982; Ducrot *et al.* 1983; Pin & Vielzeuf 1983; Paquette *et al.* 1987). In the Early Devonian (*c.* 400 Ma) most of the oceanic lithospheres were subducted (Franke & Engel 1986; Matte 1986). We consider that the Variscan high-pressure metamorphism is probably older than 380 Ma. However, it has been argued that many eclogites yield Sm–Nd whole rock–garnet ages that spread in the range of 433–290 Ma. Because garnet has a high Sm/Nd ratio compared to whole-rocks and other minerals usually used for computing whole rock-mineral Sm–Nd isochrons, the slope of the isochron, and thus the resulting age, are mainly controlled by the isotopic ratios of garnet. Sm–Nd whole rock–garnet ages may either represent the time when the garnet crystallized in a magmatic or metamorphic rock, or the time when the garnet closed to Sm–Nd diffusion. The closure temperature for Sm–Nd diffusion in garnet is estimated to be 600 ± 30°C (Mezger *et al.* 1992). If the maximum temperatures attained during the eclogite metamorphic event responsible for the crystallization of the investigated garnets exceeded this closure temperature, then Sm–Nd whole rock–garnet ages are cooling ages.

Several Sm–Nd ages are consistent with the U–Pb and Ar–Ar data where available. For example, the eclogite facies metamorphism in the Münchberg Massif is dated at 395–380 Ma, whichever isotopic system is used (Gebauer & Grünenfelder 1979; Kreuzer *et al.* 1989; Stosch & Lugmair 1990). This consistency in the geochronological dates supports the pre-Early to Mid-Devonian age of high-pressure meta-

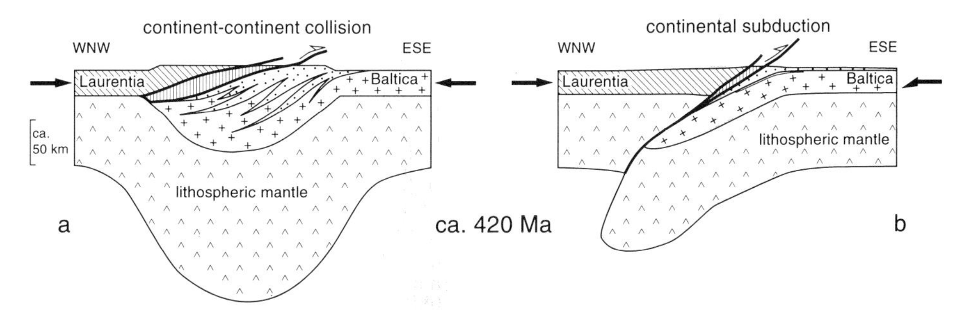

Fig. 4. Thickening models of the continental lithosphere in the Scandinavian Caledonides (**a**) adapted from Andersen & Jamtveit (1990) and (**b**) modified after Andersen *et al.* (1991*a*).

morphism. In other (numerous) cases, Sm–Nd whole rock–garnet isochrons obtained on high-pressure rocks contradict existing geological observation and other geochronological data. For example, in the Orlica–Snienik dome (eastern Bohemian Massif), Sm–Nd ages of high-pressure assemblages (352–327 Ma; Brueckner *et al.* 1989) are younger than K–Ar ages (380 Ma) obtained on the same rocks (Bakun-Czubarow 1968). In the Black Forest, Sm–Nd ages on eclogite (337–332 Ma; Kalt *et al.* 1994) are younger than Ar–Ar ages (357 Ma; Boutin 1992) obtained on eclogites found in a similar setting in the nearby Vosges Massif (see also Schmädicke *et al.* 1995). The Sm–Nd data obtained on the eclogites from the Mariankse Lazne Complex (Northwest Bohemian Massif) are even more confusing, since they show an age gradient from the garnet core (433 ± 12 Ma) to the garnet rims (375 ± 5 Ma; Beard *et al.* 1991). The young eclogite ages as age of metamorphism are locally ruled out by field observations, such as granulites, presumably dated at 308 Ma by the Sm–Nd method and yet occurring as pebbles in the Westphalian (310–300 Ma) molasse (Franke 1993; Schmädicke *et al.* 1995) or Sm–Nd age of 322 Ma on eclogite boudins embedded in low-pressure anatectic gneisses dated at 336–329 Ma (Kalt *et al.* 1994). Sm–Nd ages ranging between 370 Ma and 340 Ma may represent post-eclogitic cooling ages rather than the high-pressure crystallization.

Deformation of the colliding continental crusts is recorded by large, forelandward thrust imbrication (Burg & Matte 1978; Tollmann 1982; Burg *et al.* 1987; Ledru *et al.* 1989). Large eclogitic and granulitic crustal nappes were syn-kinematically retrogressed during thrusting, while intermediate pressure metamorphism developed in their relative autochthonous and parautochthonous domains (Burg *et al.* 1984). Related ages span from Early Devonian to Early–Mid-Carboniferous (390–330 Ma; Bernard-Griffiths *et al.* 1977; Gebauer & Grünenfelder 1979; Pin 1979; van Breemen *et al.* 1982; Lafon 1986; Pin & Peucat 1986; Kröner *et al.* 1988; Costa 1991–1992). Deformation, metamorphism, magmatism and synorogenic flysch sedimentation migrated from the suture zones toward both forelands, on the Gondwana margin to the south and on the southern Avalonia margin to the north. Migration is related to continuing collision that involved progressively more external domains as the belt widened (Matte *et al.* 1990). Intra-continental deformation due to long-lasting collision has been partly accommodated by widespread strike-slip faulting (Arthaud & Matte 1977), eventually producing lateral escape of crustal blocks (Matte 1986) and tightening of the Ibero–Armorican Arc (Burg *et al.* 1987). A set of dextral shear zones, delineated by strong gravimetric and magnetic gradients (Edel & Weber 1995) was responsible for the segmentation and eastward shift of the eastern parts of Avalonia and Armorica–Barrandia (Fig. 3).

Continental convergence in the Scandinavian Caledonides and the Variscides: a comparison

Convergence of continental lithospheres can be accommodated by continental subduction or by continent–continent collision (Fig. 4). Commonly, continent–continent collision follows continental subduction. During continental subduction, as during oceanic subduction, contractional structures are localized along the Wadatti–Benioff zone. Both the subducted and overlying crusts suffer limited deformation, which depends on the ratio between subduction and convergence rates. In contrast, during continent–continent collision, deformation migrates away from the plate boundaries so that both continental crusts are strongly affected by intra-crustal deformation. In both the Scandinavian Caledonides and the Variscides, continent–continent collision is inferred to have followed continental subduction. We emphasize, however, that continental subduction with only minor intracontinental deformation better explains the particular geometry of the Caledonian contractional structures.

Crustal scale cross-sectional models of the Scandinavian Caledonides vary according to the relative proportion of two lithospheric thickening models used. (i) Horizontal pure shear (Andersen & Jamtveit 1990; Andersen *et al.* 1994) and thick-skinned tectonic models (Norton 1986) assume that pressures of 16–30 kbar recorded in the eclogites witness at least 100 km crustal thickening caused by whole crustal imbrication during continent–continent collision (Fig. 4a). (ii) Asymmetrical thickening models (Fig. 4b) assume that the eclogitic pressures were reached during subduction of the Baltica continental lithosphere below Laurentia (Séranne *et al.* 1989; Andersen *et al.* 1991*a*; Fossen 1992; Wilks & Cuthbert 1994). The horizontal pure shear model predicts the building of a crustal wedge several tens of kilometres thick, and the relative pervasive distribution of deformation within the collided continental margins. In contrast, continental subduction predicts that deformation is mainly restricted to the Wadatti–Benioff zone and that the internal deformation of the subducted continental crust is less important than that involved in pure shear models.

Both models may explain the westward increase of both Caledonian deformation and metamorphic grade in the Scandinavian Caledonides. However, because elevation and crustal thickness are controlled by the ratio of the force needed to drive convergence and gravitational forces (England & McKenzie 1983), the crust may hardly thicken homogeneously more than 75 km. As pointed out by Andersen *et al.* (1991*a*), pressures equivalent to 100 km are likely to be related to subduction of a continental slab below Moho levels rather than to a homogeneous crustal thickening. Therefore, there is no need to invoke continental collision to explain the Silurian high-pressure metamorphism. During continental subduction the fluid-rich sedimentary layers may act as a décollement zone. The subduction-related thrust complexes give rise to large thin nappes, involving sedimentary cover and crustal slivers detached from the subducted basement. Because of the lid effect of the overriding upper plate, thin nappes may be translated without excessive folding over hundreds of kilometres above the subducting crust. This is reproduced by thermal and mechanical modelling showing that forces (buoyancy of crustal material and friction at the plate contact) acting upon a subducting continental lithosphere lead to the development of décollement zones parallel to the rheological layering of the subducted crust (Van den Beukel 1992). In the Scandinavian Caledonides the main rheological discontinuity of the subducted plate is defined by the contact between the crystalline basement of Baltica and its Lower Palaeozoic sedimentary cover. Therefore it is expected, and actually observed, that thrusts develop parallel to the basement mainly (but not only) within the sedimentary cover. A continental slab reaching a depth of 100 km with a dipping angle of 20° is consistent with the *c.* 300 km relative translation of the sedimentary cover on the subducted basement. Such large-scale, shallow dipping décollements parallel to the subducting crust may also explain some of the relatively flat seismic reflectors of the Scandinavian upper crust (Hurich *et al.* 1989), and the apparent decrease of Caledonian overprint toward the Moho since the lower crust seems less reflective than the upper and middle crust. Therefore, we believe that continental subduction better explains the geometry of the Caledonian nappes, and the localisation and distribution of contractional décollement zones.

The near absence of molassic basins parallel to the Scandinavian Caledonides is somehow intriguing because they are a common feature of continent–continent collision. Some would argue that these basins have been eroded. However, this is difficult to accept because foreland basins develop on low-elevation areas of the downward flexure of the crust near the mountain range. Foredeep basins have a low topography and therefore survive the erosion of the mountain belt. We contend that continental subduction has taken place without involving the development of molassic basins.

This model suggests that shear was the governing strain regime during contraction. This contrasts with the coaxial constrictional fabrics of the eclogites explained by a near vertical non-rotational and constrictional horizontal shortening during continental subduction (Andersen *et al.* 1991*a*) or by non-rotational vertical shortening during extensional collapse (Andersen & Jamtveit 1990; Andersen *et al.* 1994). Recent investigations in the Bergen Arc (Boundy *et al.* 1992; Rey *et al.* 1996) conflict with this view. In Holsnøy, strongly deformed eclogites occur in anastomosing, 30–150 m thick shear zones that are laterally continuous over several kilometres. Away from the shear zones,

Table 1 *Comparison of extension characteristics between the European Variscan Belt and the Scandinavian Caledonides*

Extension features and related events	European Variscan belt	Scandinavian Caledonides
End of the high-*P* event	420–410 Ma	420–410 Ma
Beginning of extension	*c*. 330 Ma	*c*. 395 Ma
Direction of extension	Variable in space and time	Same as contraction
Low-angle ductile normal faults	Do not necessarily overprint thrusts	Often overprint contractional structures
Sedimentary basins	Uniformly distributed	Localized above one unique detachment
Associated thermal regime	High-geothermal gradient	Medium to low geothermal gradient
Exhumed terranes	Anatectic gneisses, migmatites, granites	Eclogitized Precambrian formations
Associated metamorphic event	Low-*P*–high-*T* metamorphism	amphibolite facies or lower
Plutonism	Voluminous granitic plutons	Rare plutons
In the foreland	Contractional structures	No proven post–410 Ma sediments
Thermal subsidence	Yes	No

Precambrian structures and metamorphic assemblages are remarkably well-preserved (Austrheim 1987). The eclogite-facies shear zones exhibit a pronounced millimetre-scale layering defined by alternating omphacite–garnet- and kyanite–zoisite-rich layers and a strong shape-fabric foliation defined by aligned omphacite, kyanite, zoisite and phengite. The mineral lineation on this foliation is defined by the preferred orientation of zoisite, omphacite, kyanite and phengite, parallel to the stretching lineation defined by rod-shaped mineral aggregates (mainly garnet and omphacite) and elongated relict corona structures. The shear zones generally have a north to northeast dip of about 10° to 30°. In thin section, the geometrical relationships between S and C planes (both bearing eclogitic assemblages), the asymmetry of the omphacite-bearing crystallization tails around garnet, and the right-stepping overlaps of garnet and zoisite grains indicate a top to the ENE sense of shear consistent with the kinematics inferred from macroscopic features, such as sigmoid-type mineral clusters. Although these features indicate that the shear zones have a normal component in their present orientation, the shear zone orientation during metamorphism and deformation was likely to have been different. If these shear zones were dipping to the west before being rotated into their present attitude, the top to the ENE sense of shear is consistent with the view that eclogites recrystallized during the westward subduction of Baltica below Laurentia.

In the Variscides, structural models emphasize the large-scale thrusts associated with continental subduction (pressures as high as 20 kbar are recorded, e.g. Ballèvre *et al.* 1989), that succeeded synthetic oceanic subduction (Matte *et al.* 1986; Franke 1989*b*; Matte 1991). As a point of consensus in their convergence histories, we note that oceanic and subsequent continental subduction is marked by high-pressure metamorphism of oceanic and continental rocks in both the Scandinavian Caledonides and the Variscides. However, contractional structures differ significantly between the two belts. A major difference is that, in the Variscan belt, continent–continent collision continued for several tens of millions of years after the high-pressure metamorphism. This intracontinental deformation is inferred from (1) the development of molassic foreland basins, (2) the migration through time of deformation from the hinterland to the foreland, (3) the homogeneous distribution of deformation and metamorphism through the nappe piles (no pre-Variscan feature seems preserved in the hinterland), (4) the development of a syn-convergence arcuate structure (the Ibero-Armorican arc), (5) the development of strike-slip faults parallel to the strike of the belt, and (6) the variability through time and space of the direction of displacement. All these features may be explained by the continent–continent collision model and are, at best, poorly represented in the Scandinavian Caledonides.

Post-thickening extension: Scandinavian Caledonides versus Variscides

Contrasting features (Table 1) suggest that post-thickening extension in the Scandinavian Caledonides and in the Variscides do not result

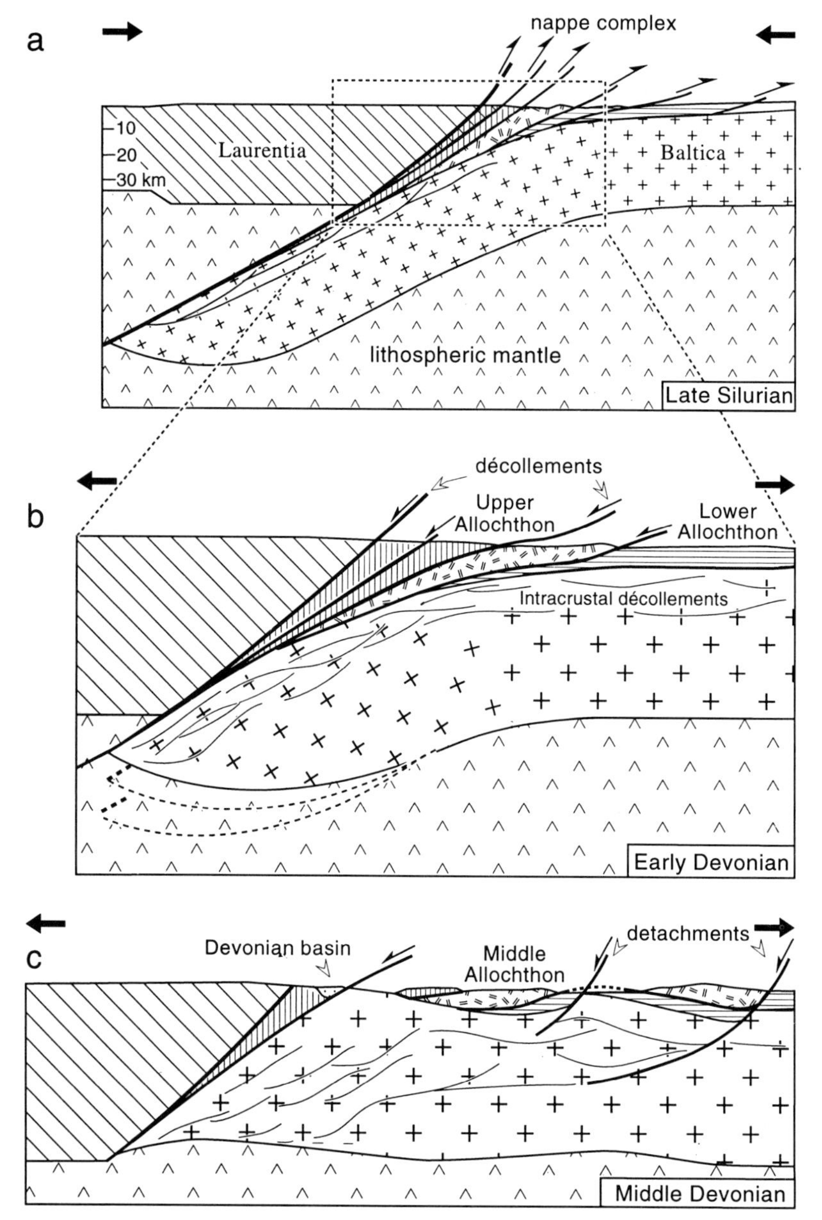

Fig. 5. Sketch of the structural evolution of the Scandinavian Caledonides discussed in this article. (**a**) The cross section illustrates the relationships between Laurentia, Baltica and the nappe complex at the end of the convergence. Following Fossen (1992), extension developed both extensional décollements (Mode I, section b) and detachments (Mode II, section c). (**b**) At first extension exhumed progressively the continental slab whereas former thrusts, reactivated in extensional décollements, rotate to horizontal. The reversed subduction (eduction) is consistent with reversal of relative plate motion, while the subducted lithosphere remains too strong to fail (see analogue experiment 7 of Shemenda 1993). (**c**) New detachments developed when sliding along décollements was no longer possible because they became too close to the horizontal.

from similar thermo-mechanical regimes. In the following section we argue that eduction of the subducted continental crust (i.e. back movement of the continental slab) may explain the extensional features of the Scandinavian Caledonides. In contrast, extensional features of the Variscan belt indicate that the lithosphere was thick and weak when extension occurred.

Extension could have been triggered by either a reduction of the compressional driving forces or the sudden increase of the horizontal buoyancy forces possibly related to the thinning of the lithospheric mantle. These contrasting models are based on the following significant differences.

Initiation and duration of extension

West-verging normal ductile shear began in the Early Devonian (*c.* 395 Ma, Hossack 1984; Lux 1985; McClay *et al.* 1986; Chauvet & Dallmeyer 1992) or even during the Late Silurian as suggested by Wilks & Cuthbert (1994) for the Scandinavian Caledonides, and by Hartz & Andersen (1995) for the east-verging extensional detachment of the East Greenland Caledonides. Therefore, extension followed closely after the high-pressure metamorphic event dated between 450 and 400 Ma (Krogh *et al.* 1974; Griffin & Brueckner 1980; Gebauer *et al.* 1985; Kullerud *et al.* 1986; Mørk & Mearns 1986), and probably started whilst the Baltica continental slab was still involved in high-pressure metamorphic conditions. The close temporal relationship (perhaps even overlap) between extension and the high-pressure event strongly supports a rapid switch from contraction to extension. This is also supported by the fact that upper crustal contraction ceased as soon as extension began (Fossen 1992; Milnes *et al.* 1996), and extension continued into Late Devonian times, probably dying out by the end of the Devonian (Milnes *et al.* this volume). The time span between high pressure metamorphism and extension may be less than 10 to 20 Ma.

In contrast, extension of the Variscides took place from Mid-Carboniferous to Permian times (330–280 Ma, Ménard & Molnar 1988; Burg *et al.* 1994), more than 70 Ma after high pressure metamorphism.

Controlling effect of pre-extensional structures

In the Scandinavian Caledonides, former thrust zones controlled the location of extensional décollement faults (Milnes & Koestler 1985; Chauvet & Séranne 1989; Fossen 1992; Milnes *et al.* this volume). These décollements were localized on the west margin of Baltica, close to the boundary with the remains of Iapetus and probably at the boundary of Laurentia (Fig. 5b). The extension direction was approximately parallel to, but with an opposite sense to that of the earlier contraction (Norton 1986; Séguret *et al.* 1989; Fossen 1992), and remained consistent in time and space. A second mode of extension (Mode II of Fossen 1992; see also Milnes *et al.* this volume) occurred partly during and partly after the décollement mode. It involved stretching of Baltica and the development of new detachment faults that cut through the nappe pile. According to our view, these new detachments developed in response to the rotation of the décollement zones into a horizontal attitude. This rotation, related to the eduction of the subducted continental crust (Fig. 5c), led to decreased shear stresses acting on the décollement. Slip along the horizontal décollements thus became more difficult than on higher-angle detachment faults.

In contrast, normal faults are periodically distributed throughout the internal zones of the Variscides where they do not necessarily overprint former thrust zones. The relationship between extension and contraction directions is complex, in particular because the direction of post-thickening extension varies in space and time (Fig. 3; Burg *et al.* 1994).

Extension-related structures

One of the best expressions of post-thickening collapse due to thermal softening of the lithosphere is the development of low-angle detachment zones that control (1) the exhumation of anatectic domes in their footwall and (2) the deposition of continental basins in the hanging wall. This coeval association of detrital basin, detachment and anatectic dome is known as a Metamorphic Core Complex (Coney & Harms 1984). There is no syn-extension anatectic dome in the Scandinavian Caledonides. Instead, exhumation of the Precambrian basement was mainly controlled by the reactivation of the former thrusts and the development of detachment faults. As a result, post-thickening extension did not penetratively affect the contractional nappe stack: earlier Caledonian contractional structures, and even Precambrian structures, are relatively well-preserved throughout the belt in comparison to the Variscan belt.

Conversely, post-thickening extension of the Variscides produced coeval migmatitic domes and intracontinental, widely distributed coal-bearing basins in a Metamorphic Core Complex-type geometry (Ménard & Molnar 1988; Malavieille *et al.* 1990; Van Den Driessche & Brun 1991–1992). Contractional structures are generally obliterated in the migmatitic and granitic hinterland, which indicates that the crust was partially molten and weakened when extension occurred. In the internal part of the belt, the

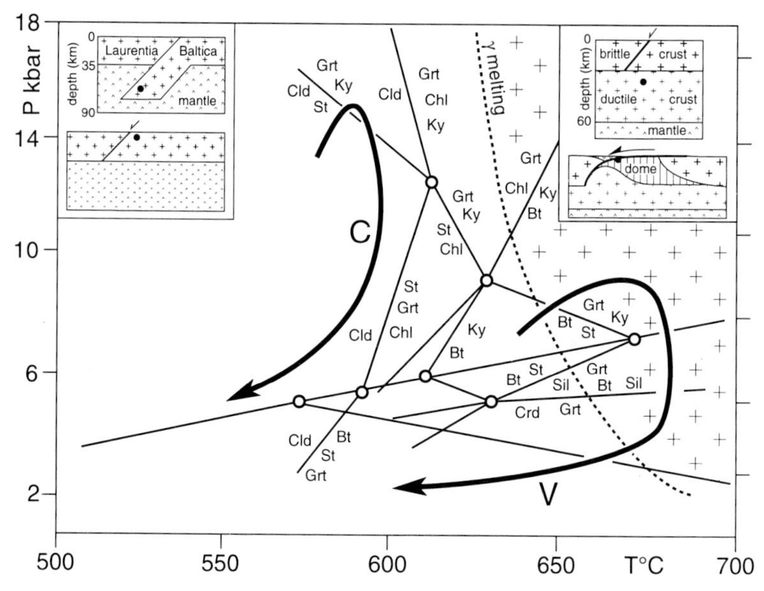

Fig. 6. Representative *P–T–t* trajectories recorded by metapelites in the Scandinavian Caledonides (C, after Chauvet *et al.* 1992) and the Variscides (V, after Rey *et al.* 1991–1992). Note the *c.* 100°C difference between the two retrograde paths. In the Variscides, exhumation of high-pressure metapelites occurred above their solidus. During late-orogenic extension a large volume of anatectic melts was produced that reduced the strength of the crust. Conversely, no extensive partial melting occurred in the Scandinavian Caledonides. During late-orogenic extension, exhumation of high-pressure rocks developed under a relatively cold geotherm. Inserts show related extension systems in the Scandinavian Caledonides (top left) and in the southern Variscides (top right).

upper crustal layer is deeply eroded. As a consequence, the association of coeval basins and Metamorphic Core Complexes is principally found near the external zone of the belt.

Syn-extensional metamorphism and plutonism

Exhumed Caledonian and Variscan rock units present distinct thermal histories.

In the Scandinavian Caledonides, eclogitic rocks sampled within ductile normal shear zones below the Devonian basins experienced nearly 10 to 20 kbar isothermal decompression (Fig. 6) within a period of *c.* 20 Ma (Andersen & Jamtveit 1990; Chauvet *et al.* 1992; Wilks & Cuthbert 1994). Extension-related ductile structures developed under lower amphibolite to greenschist-facies metamorphism (Norton 1987; Chauvet *et al.* 1992), with a complete absence of migmatites and associated granites (Stephens 1988; Milnes *et al.* this volume). This provides further evidence that extension exhumed rocks that were under high-pressure facies conditions, and strongly supports a rapid switch from contraction to extension.

In contrast, the Variscan extension exhumed rocks that were involved in medium-pressure amphibolite- to granulite-facies conditions. Indeed, 80% of the internal zones in the Variscides consists of migmatites and granites coeval with widespread high-*T*/low–*P* metamorphic conditions. *P–T–t* paths of the deepest rocks are characterized by high-temperature near-isothermal, 9–4 kbar decompression before cooling (Fig. 6). Temperatures as high as 700–800°C at depths equivalent to 3–6 kbar were common (e.g. Gardien 1990; Rey *et al.* 1991–1992). They imply a thermal gradient (50–90°C km^{-1}) that would produce unrealistic Moho temperatures (>2000°C) with a conductive geotherm. Accordingly, advective heat transport by pervasive flow in the deeper parts of the continental crust is inferred to maintain a temperature around 800–1000°C. Large volumes of granitic plutons, dated between 340 and 290 Ma, as well as low-*P* – high-*T* metamorphic conditions dated at about 330–310 Ma, support the hypothesis that both extensive melting and pervasive flow of the middle and lower crust were spatially and temporally associated with post-thickening extension (Rey 1993; Costa & Rey 1995). This high-*T*–low-*P* environment during Variscan extension is certainly a major difference with the extensional metamorphic conditions associated with the Scandinavian Caledonides.

Basins and foreland deformation

In the Scandinavian Caledonides, Devonian basins developed mainly in limited areas above a single main normal fault (the Norfjord–Sogn Detachment, Norton 1987; Chauvet & Séranne 1989). Away from this fault Devonian sediments are rare.

In the Variscides, Carboniferous basins are distributed throughout the internal zones of the orogenic belt (Ménard & Molnar 1988; Burg *et al.* 1994). During extension, mainly gravity-driven contractional structures developed in external domains of the belt (Ahrendt *et al.* 1983; Weber & Behr 1983; Echtler 1990; Meilliez & Mansy 1990). This contrasts with the Caledonian belt where evidence for contemporaneous contraction in the foreland and extension in the hinterland are rare and questionable (Fossen 1992, 1993; Andersen 1993).

Deep crust

A low-*P*–high-*T* granulite-facies metamorphism took place in the lower crust of the Variscan belt in Late Carboniferous times (Pin & Vielzeuf 1983), while the orogen was undergoing post-thickening extension. The high *T–P* gradient is associated with intrusion of mantle-derived mafic intrusions crystallized as layered mafic–ultramafic complexes, that may have played a key role in the origin of the high seismic reflectivity of the Variscan lower crust (Rey 1993). Indeed, geometric relationships between seismic reflectors and regional geological structures indicate that the 'layering' of the lower crust was acquired during post-thickening extension (Rey 1993; Costa & Rey 1995). There is no direct information on the Scandinavian lower crust. However, deep seismic profiles indicate that the lower crust is poorly reflective compared to the reflectivity of the lower Variscan crust (Rey 1993 and references therein).

Post-extensional thermal history

Variscan extensional tectonics ended with deposition of Permian sediments and was followed during the Early and Mid-Mesozoic by the formation of intra-continental basins such as the Aquitaine, Paris and London Basins. The development of these basins was controlled by thermal subsidence during which the lithospheric mantle recovered a normal thickness. Extension in the Scandinavian Caledonides ended, at the latest, in the Late Devonian (Milnes *et al.* this volume). As pointed out by Séguret *et al.* (1989), the absence of sedimentation between the Devonian and the Permian indicates that no thermal subsidence followed extension in the Scandinavian Caledonides, and as a consequence, that the lithospheric mantle may have preserved a constant thickness throughout Caledonian times. The absence of plutonic activity, often related to mantle delamination, supports this hypothesis.

Origin of post-thickening extensional tectonics

We conclude that post-thickening extensional tectonics in the Scandinavian Caledonides did not occur under a thermal and mechanical regime comparable to that of the Variscides. We will now argue that post-thickening extension in the latter was a mechanical consequence of the thermal evolution of the lithosphere during convergence. In contrast, Devonian extension in the Scandinavian Caledonides was not a direct consequence of the thermal and mechanical evolution expected in a lithosphere thickened by continental collision.

Rheological discussion

From the lack of intrusions in the Caledonides and their abundance in the Variscides, it is probable that the lithosphere in the Caledonides was cooler than that of the Variscides. As shown by Kusznir & Park (1987) the extensional strength of the lithosphere is very strongly dependent on the temperature gradient. These authors show that an initially distributed stress will be relaxed by creep in those parts of the crust that deform in a ductile manner (e.g. lower crust) and that the load is then transferred to the non-deforming parts (e.g. upper crust). The rheology of rocks is a function of temperature (Kirby 1985), which is expressed by:

$$\mu_{eff}=\frac{1}{2C_1\sigma_0^{n-1}}\exp(E_a/RT) \quad (1)$$

(Turcotte & Schubert 1982, p. 339, eqn 7–250) where μ_{eff} is the effective viscosity, C_1 is the pre-exponential term of power-law creep, σ_0 is the stress, n is the exponent of power-law creep, E_a is the activation energy, R the gas constant and T the absolute temperature.

The temperature dependence of rock rheology in turn gives a temperature dependence to the stress relaxation time τ_r

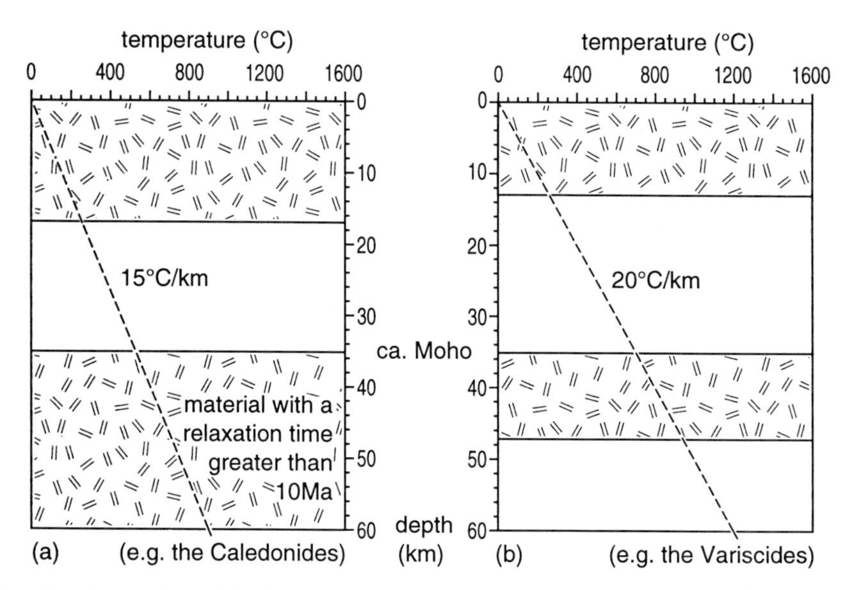

Fig. 7. Load bearing portions of the lithosphere for a temperature gradient of (a) 15°C km^{-1} and (b) 20°C km^{-1}. The patterned areas represent material that can sustain stress over a period of 10 Ma.

$$\tau_r = \frac{3\mu_{\text{eff}}}{E} \quad (2)$$

(Turcotte & Schubert 1982, p. 339, eqn 7–251), where E is the elasticity modulus. The relaxation time is a measure of the time over which a rock can sustain a stress and therefore, a useful concept in considering the redistribution of tectonic load in the lithosphere. In the Caledonides and in the Variscides, the lithospheric failure takes a period of time in the order of 10 Ma, therefore the strength of the lithosphere can be determined as the proportion of it that can sustain a load over this period of time. For this rheological discussion, we take a crust made up of 35 km of wet quartz and an upper mantle of dry olivine, and use data from Kirby (1985). Figure 7 illustrates the proportions of the crust that can sustain a load for linear temperature gradients of 15 and 20°C km^{-1}, the latter being chosen to reach the minimum melting temperature of crustal rocks at a Moho depth.

The precise strength of the lithosphere depends on the precise mechanism of failure in the non-ductile parts, but it can be seen from the figure that the lower temperature gradient gives a load bearing lithosphere twice that of the higher gradient. Thus it is a tenable hypothesis that the Variscides with a high-temperature gradient suffered extension caused by buoyancy forces, whilst the lithosphere of the Scandinavian Caledonides with a low geothermal gradient was strong enough to resist an equivalent load.

Models of post-convergence extension

Variscan belt. The bulk history that we envision for the Variscides (Fig. 8) can be summarized as follows. After a long-lasting and probably slower continental collision (as it is seen between India and Asia, Molnar & Tapponnier 1975), the crustal root of the Variscides was partially melted and the lithospheric mantle was more or less transformed into asthenosphere by thermal relaxation (Gaudemer *et al.* 1988) or mantle delamination (Bird 1979; Houseman *et al.* 1981). This transformation resulted in uplift and partial melting of fertile asthenosphere and subsequent intrusion of mafic magmas into the lower crust (McKenzie & Bickle 1988), which also contributed in the heating of the crust (Fig. 8c). Heating and production of large volumes of granitic melt were responsible for a dramatic drop of crustal strength. Thermal thinning of the lithospheric mantle may also have triggered a higher topographic relief of the belt. The decrease in crustal strength, and the rise in elevation induced lateral pressure gradients in the lower crust by variations in crustal thickness (Bird 1991). Subsequent pervasive flow which, in turn, induced formation of extensional de-

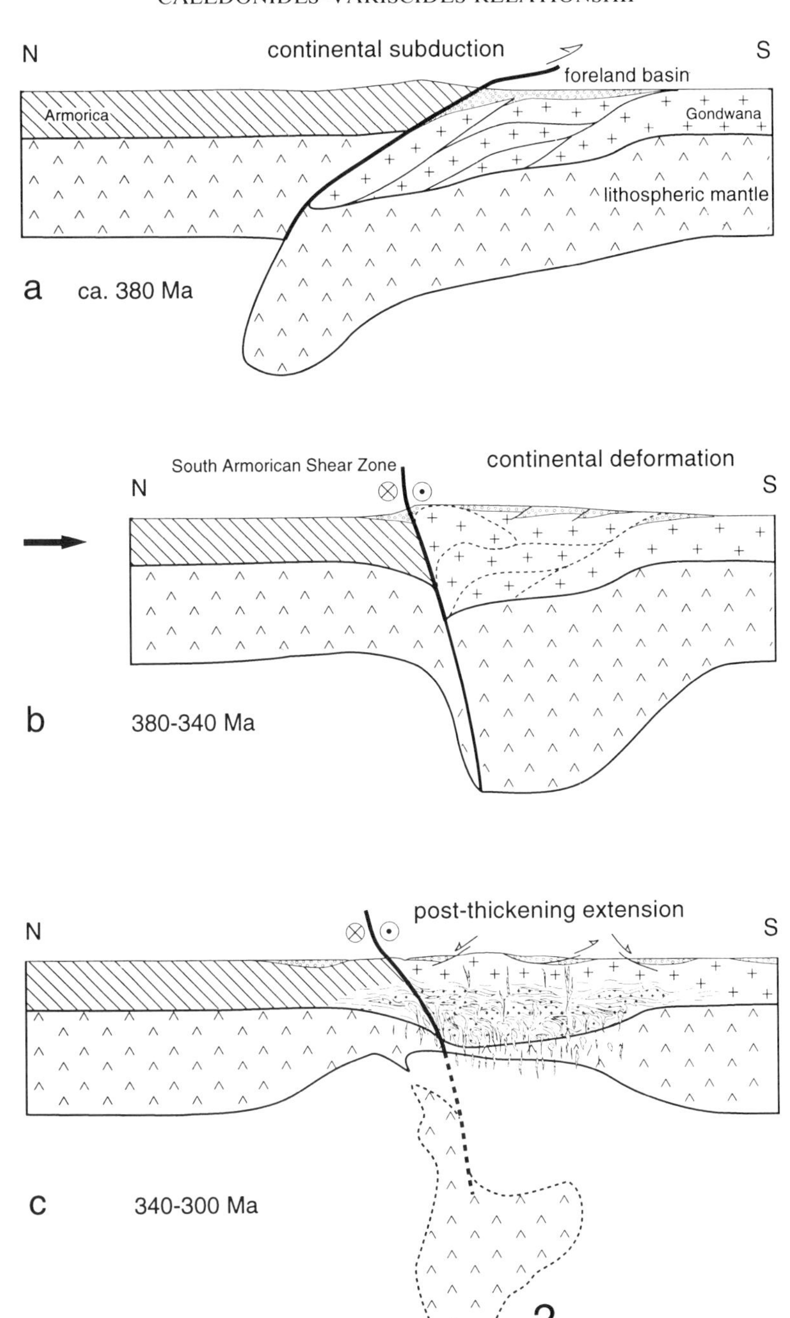

Fig. 8. Sketch of the structural evolution and post-thickening extension across the southern branch of the Variscides. (**a**) Cross section illustrating the relationships between Gondwana, Armorica and the nappe complex at the collision stage. (**b**) Further convergence maintained a thick crust while plate boundaries and steepened major thrust systems evolved into complex transpressive systems. (**c**) After long-lasting continental convergence (hyper-collision), the crustal root was partially melted and the lithospheric mantle was more or less transformed into asthenosphere by thermal relaxation or mantle delamination (?). This resulted in partial melting of fertile asthenosphere and intrusion of mafic magmas into the lower crust. The subsequent decrease in crustal strength allowed the thickened crust to recover a normal thickness within a few tens of million years.

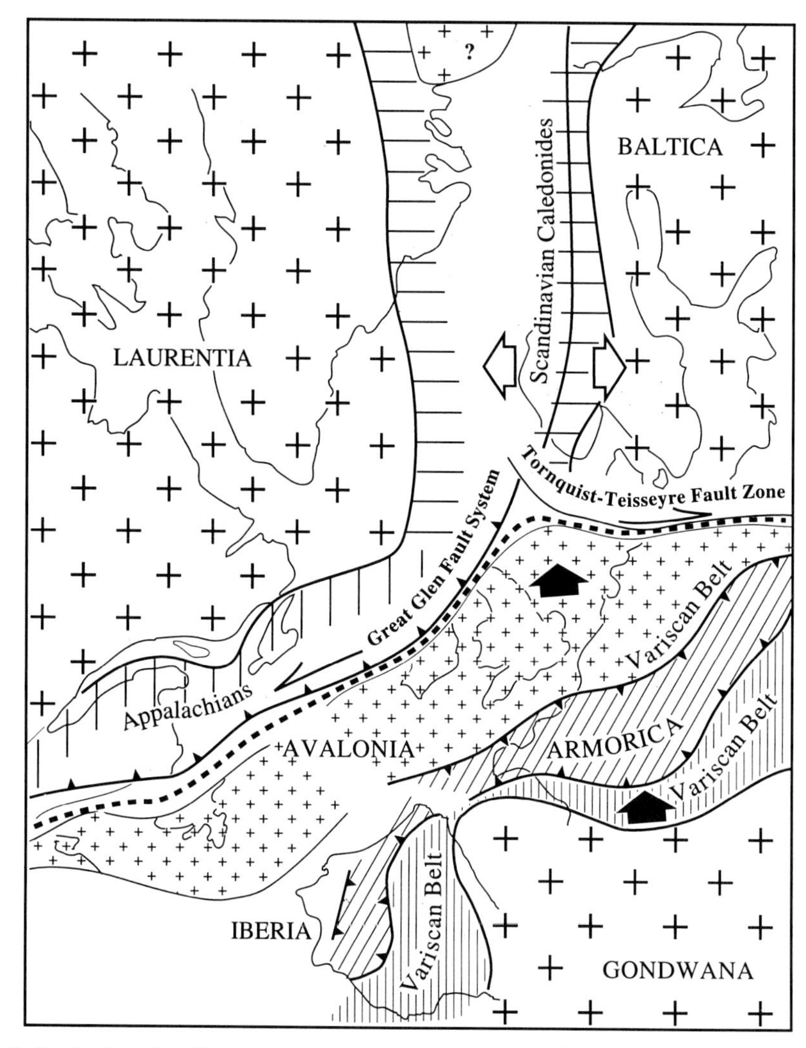

Fig. 9. Bulk distribution of continental masses in Devonian times (*c.* 390 Ma) and approximate directions of relative movement after Ziegler (1986) and Scotese & McKerrow (1990). Ibero-Armorican Arc opening angle after Perroud (1982). Continental boundaries as in Fig. 1.

tachment faults in the upper crust, accounts for a redistribution of crustal material from the thickened internal zone toward the external part of the belt. Synchronously, the rise in elevation induced a topographic gradient responsible for gravitational spreading and marginal thrusting toward the foreland (Dewey 1988). These mechanical processes permitted the thickened crust to recover a normal (*c.* 30 km) thickness in a few tens of million years. But the recovery of the lithospheric mantle thickness was controlled by slower thermal processes that involved thermal subsidence of most of Europe during the Mesozoic.

Scandinavian Caledonides. It is usually inferred that in this orogen the lithosphere doubled in thickness during early collision by continental subduction of Baltica under Laurentia. Many models then assume that gravitational collapse of the orogen was initiated in the Devonian by the catastrophic detachment and subsidence of the heavy and cold lower mantle of the thickened lithosphere (Norton 1986; Séranne & Séguret 1987; Séranne *et al.* 1989; Andersen & Jamtveit 1990; Klaper 1991; Chauvet *et al.* 1992). This delamination resulted, for buoyancy reasons, in the rapid exhumation (about 20 Ma) of the eclogitic crust (Fig. 5, Andersen &

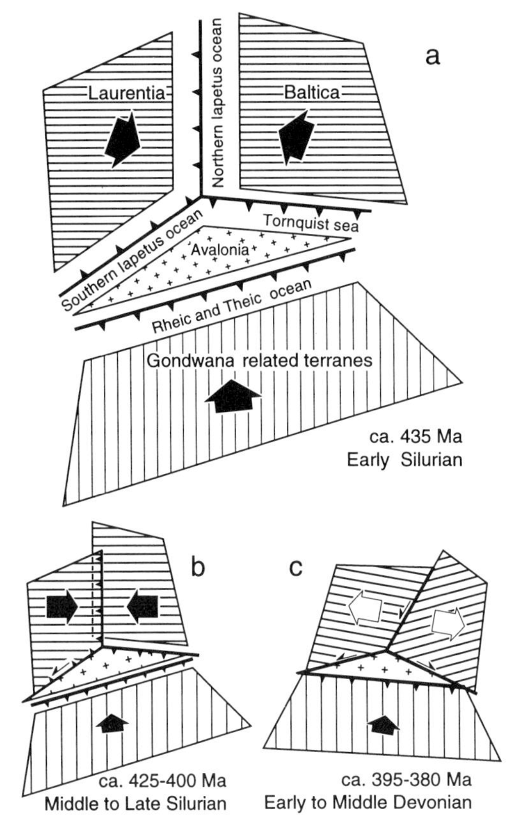

Fig. 10. Sketch to represent the spatial relationships between the Variscides and the Scandinavian Caledonides, and the interference involved to create extension in the Caledonides during early collision in the Variscides, followed by wrench-dominated tectonics suggesting clockwise rotation of Gondwana during the Silurian–Early Devonian times. This model is consistent with the Silurian closure of Iapetus described by Soper *et al.* (1992).

Jamtveit 1990; Andersen *et al.* 1991*a*; Fossen 1992). Extension of the orogenic belt is necessary to exhume the deepest rocks by a footwall rolling hinge mechanism below the backsliding nappe pile as suggested by the present day disposition of isobars (Norton 1986).

The internally driven decoupling model, where horizontal buoyancy forces play a governing role, fails to explain (i) why continental convergence ceased when the thermal boundary layer was removed and (ii) the absence of spreading-related contraction in the foreland during uplift and exhumation of the hinterland (Fossen 1992; Wilks & Cuthbert 1994). Furthermore, the lack of both significant crustal partial melting and thermal subsidence suggest that the driving forces of extensional tectonics were not related to either horizontal buoyancy forces combined with thermal softening of the crust or to isostatic rebound associated with a mantle delamination process. As an alternative model, plate divergence has been proposed to explain the origin of extension (Séranne *et al.* 1991; Fossen 1992; Rey 1993; Chauvet & Séranne 1994; Wilks & Cuthbert 1994; Milnes *et al.* this volume). We follow these authors in arguing that the origin of much of the extension can be found in changing boundary conditions, namely that the nappe pile slid back during the Devonian into a space created by divergence between Laurentia and Baltica (the free space model of Wilks & Cuthbert 1994). A consensus is not reached about the origin of extensional tectonics (Andersen 1993; Fossen 1993), and the origin of the plate divergence is poorly documented. How and why was extension externally imposed?

The collision of Avalonia with Laurasia has been proposed to explain the north–south compression coeval with east–west extension, which is deduced from folds contemporaneous with, and parallel to the late-orogenic extension (Séranne *et al.* 1991; Chauvet & Séranne 1994). However, Avalonia did not collide northwards with the recently assembled Laurasia in the Early Devonian. It underthrust Laurentia transpressively in the Silurian, rotating anticlockwise (Soper & Woodcock 1990; Soper *et al.* 1992), while turbidite sequences were synchronously deposited on both sides of the suture (Kneller 1991). We contend that Gondwana, was more likely to have been the plate that caused divergent plate motion between Baltica and Laurentia, inducing extension. In the following we propose that a relationship existed between the Early Variscan collision and Late Caledonian extension.

A Lower Palaeozoic palinspastic reconstruction shows a triple junction centred on the Avalonian block situated between Laurentia, Baltica and Gondwana (Figs 9 and 10a; Soper & Hutton 1984; Soper & Woodcock 1990). East–west closure of the North Iapetus Ocean was taking place during Cambrian–Silurian times, while northward translation of Gondwana was closing the Rheic and Theic oceanic domains to the south. Silurian ages of high-pressure metamorphism in both the Caledonian and Variscan belts, as well as the Late Ordovician age of calc-alkaline arc-magmatism in eastern Britain (Noble *et al.* 1993) and the Silurian age of unmetamorphosed ophiolites in southwest Poland (Oliver *et al.* 1993), suggest that the triple junction was stable as long as convergence was accommodated by subduction of the oceanic lithospheres. At the end of the Silurian, whilst the Scandinavian Caledonides reached continent–continent collision after continental subduction, the boundaries between the Laurentia and Baltica plates and between Laurasia and Gondwana were nearly perpendicular (Fig. 10b). In the Early Devonian, the Rheic and Theic oceans were closed and Variscan continental collision began. The pulling apart of Laurentia and Baltica in a tension gash manner at a lithospheric scale may then have occurred in the Early Devonian (Fig. 10c). In the meantime, Silurian basins were inverted by N–S compression in the British Isles (Soper & Woodcock 1990). This would suggest that north–south Variscan collision was responsible for east–west Caledonian divergence in Devonian times. Two conjugate fault zones accommodated the northward impingement of Gondwana, the sinistral Great Glen Fault System to the west (Fig. 9; e.g. Soper & Hutton 1984) and the dextral Tornquist–Teisseyre Fault Zone to the east (Figs 9 and 10). Furthermore, the northward impingement of Gondwana may also explain the Acadian Orogeny (Devonian deformation in Britain and in the Appalachians) as proposed by Soper *et al.* (1992). This transitory extension ceased, to give way to the predominantly wrenching regime recognized in the Caledonides through the Mid- and Late Devonian (e.g. Soper & Hutton 1984; Hutton 1987; Chauvet & Séranne 1994) and possibly as late as the earliest Carboniferous (Harland & Gayer 1972). This relative movement suggests that the northward impingement against Laurasia changed in direction, possibly because of a clockwise rotation of Gondwana (Fig. 10c see also the model developed by Hutton 1987). This rotation may have lasted until Namurian times, a period for which a rotation of Europe is seen by palaeomagnetic investigations (Edel 1987). It would be consistent with the ductile kinematic pattern described in the nappe pile of the arcuate Variscides in Western Europe (Burg *et al.* 1987) and would provide an explanation for a slower convergence between Gondwana and Laurasia before the establishment of Variscan orogenic collapse.

Conclusions

The Scandinavian Caledonides and the Variscides apparently document two different modes of convergence as well as two different modes of post-orogenic extension. First order differences such as the time delay between contraction and extension, the gross *P–T–t* paths of deeply subducted rocks, the establishment of high thermal gradients, the amount of granites and the distribution of extensional basins suggest that post-thickening extension in the Variscides was a consequence of gravitational collapse due to the thermal softening of the thickened crust. Conversely, extension in the Scandinavian Caledonides was probably controlled by a plate divergence motion which took place while Baltica was still being dragged below Laurasia.

In the Variscides, continent–continent collision is inferred from (1) the development of molassic foreland basins, (2) the migration through time of deformation from the hinterland to the foreland, (3) the homogeneous distribution of deformation and metamorphism through the nappe piles (no pre-Variscan features are preserved in the hinterland), (4) the development of syn-convergence arcuate structures (Ibero-Armorican arc) (5) the development of strike-slip faults parallel to the strike of the belt and (6) the variability through time and space of the direction of displacement. Gravitational instabilities related to the viscous flow of the lower crust and lithospheric mantle delamination may explain a number of Carboniferous events in the waning stages of the Variscan history. These are: (1) extraction of granitic magmas from the melted lower crust, (2) generation of widespread low-*P*–high-*T* metamorphic conditions, (3) mantle-derived mafic intrusion in the lower crust, (4) the seismic fabric of the lower crust, (5) deposition of Upper Carboniferous basins above low-angle ductile normal faults, (6) development of contractional structures in the foreland, and (7) a waning thermal subsidence.

The Early Paleozoic subduction of the Baltic margin beneath Laurentia explains: (1) the uniform pressure increase toward the northwestern margin of the belt, (2) the thinness of the Caledonian nappes, (3) the low geothermal gradient, (4) the absence of a foreland basin and (5) the absence of penetrative deformation within the subducted Baltica basement. The post-thickening east–west extension was responsible for the exhumation of the deeply subducted crust. This was triggered by external forces at the plate boundaries, which may explain: (1) the short delay between extension and high-*P* metamorphism, (2) the distribution of the Devonian basins above a single detachment, (3) the absence of thermal relaxation, and (4) the absence of long-lasting thermal subsidence. Based on palinspastic reconstruction and timing of tectonometamorphic events in both the Scandinavian Caledonides and the Variscan belt, it is proposed that east–west Caledonian extension was initiated during the north–south early-Variscan continental collision, rather than by gravitational collapse.

This work has been supported by the INSU-CNRS (ATP-ECORS 891705) for the field study in the Variscan belt and by the NSF grant EAR–9003956 for the field study in the Caledonian belt. An early version of this work profited from helpful comments provided by H. Austrheim, A. Chauvet, S. Costa, B. John, D. Fountain and M. Séranne. The final version has been improved by the reviews of A. G. Milnes, N. J. Soper, and an anonymous reviewer, all of them are gratefully thanked. We thank M. Ford for acting as editor for this article.

References

AHRENDT, H., CLAUER, N., HUNZIKER, J. C. & WEBER, K. 1983. Migration of folding and metamorphism in the Rheinisches Schiefergebirge deduced from K-Ar and Rb-Sr age determinations. *In*: MARTIN, H. & EDER, W. (eds) *Intracontinental Fold Belts*. Springer Verlag, 323–338.

ANDERSEN, T. B. 1993. The role of extensional tectonics in the Caledonides of south Norway: Discussion. *Journal of Structural Geology*, **15**, 1379–1380.

—— & JAMTVEIT, B. 1990. Uplift of deep crust during orogenic extensional collapse: A model based on field studies in the Sogn-Sunnfjord region of Western Norway. *Tectonics*, **9**, 1097–1111.

——, ——, DEWEY, J. F. & SWENSSON, E. 1991*a*. Subduction and eduction of continental crust: major mechanisms during continent-continent collision and orogenic extensional collapse, a model based on the south Norwegian Caledonides. *Terra Nova*, **3**, 303–310.

——, NIELSEN, P., RYKKELID, E. & SØLNA, H. 1991*b*. Melt-enhanced deformation during emplacement of gabbro and granodiorite in the Sunnhordland batholith, west Norway. *Geological Magazine*, **128**, 207–226.

——, OSMUNDSEN, P. T. & JOLIVET, L. 1994. Deep crustal fabrics and a model for the extensional collapse of the southwest Norwegian Caledonides. *Journal of Structural Geology*, **16**, 1191–1203.

ARTHAUD, F. & MATTE, P. 1977. Late Paleozoic strike-slip faulting in southern Europe and northern Africa: Result of a right-lateral shear zone between the Appalachians and the Urals. *Geological Society of America Bulletin*, **88**, 1305–1320.

AUSTRHEIM, H. 1987. Eclogitization of the lower crustal granulites by fluid migration through shear zones. *Earth and Planetary Science Letters*, **81**, 221–232.

BAKUN-CZUBAROW, N. 1968. Geochemical characteristic of eclogites from environs of Nowa Wies in the region of Śnieznik Klodzki. *Archiv für Mineralogie*, **28**, 248–382.

BALÉ, P. & BRUN, J.-P. 1986. Les complexes métamorphiques du Léon (NW Bretagne) : un segment du domaine éo-hercynien sud armoricain translaté au Dévonien. *Bulletin de la Société géologique de France*, **8 II**, 471–477.

BALLÈVRE, M., PARIS, F. & ROBARDET, M. 1992. Corrélations ibéro-armoricaines au Paléozoïque: une confrontation des données paléobiogéographiques et tectonométamorphiques. *Comptes Rendus de l'Académie des Sciences de Paris*, **315 II**, 1783–1789.

——, PINARDON, J.-L., KIÉNAST, J.-R. & VUICHARD, J.-P. 1989. Reversal of Fe-Mg partitioning between garnet and staurolite in eclogite-facies metapelites from the Champtoceaux Nappe (Brittany, France). *Journal of Petrology*, **30**, 1321–1349.

BAMBACH, R. K., SCOTESE, C. R. & ZIEGLER, A. M. 1980. Before Pangea: the geographies of the Paleozoic World. *American Scientist*, **68**, 26–38.

BARD, J. P., BURG, J.-P., MATTE, P. & RIBEIRO, A. 1980. La chaîne hercynienne d'Europe occidentale en termes de tectonique des plaques. *Mémoire du B.R.G.M.*, **108**, 233–246.

BEARD, B. L., MEDARIS, L. G., JOHNSON, C. M., MISAR, Z. & JELINEK, E. 1991. Nd and Sr isotope geochemistry of Moldanubian eclogites and garnet peridotites, Bohemian Massif, Czechoslovakia. Second Eclogites Field Symposium, Spain. *Terra Nova*, **3 (suppl. 6)**, 4.

BEHR, H.-J., ENGEL, W., FRANKE, W., GIESE, P. & WEBER, K. 1984. The Variscan Belt in Central Europe: Main structures, geodynamic implications, open questions. *Tectonophysics*, **109**, 15–40.

BERNARD-GRIFFITHS, J., CANTAGREL, J.-M. & DUTHOU, J.-L. 1977. Radiometric evidence for an Acadian tectonometamorphic event in Western Massif Central Français. *Contribution to Mineralogy and Petrology*, **61**, 199–212.

BIRD, P. 1979. Continental delamination and the Colorado Plateau. *Journal of Geophysical Research*, **84**, 7561–7571.

—— 1991. Lateral extrusion of lower crust from under high topography, in the isostatic limit. *Journal of Geophysical Research*, **96**, 10,275–10,286.

BOCKELIE, J. F. & NYSTUEN, J. P. 1985. The southeastern part of the Scandinavian Caledonides. *In*: GEE, D. G. & STURT, B. A. (eds) *The Caledonide Orogen – Scandinavia and Related Areas*. J. Wiley & Sons Ltd, New York, 69–88.

BOUNDY, T. M., FOUNTAIN, D. M. & AUSTRHEIM, H. 1992. Structural development and petrofabrics of eclogite facies shear zones, Bergen Arcs, Western Norway : implications for deep crustal deformational processes. *Journal of Metamorphic Geology*, **10**, 127–146.

BOUTIN, R. 1992. *Histoire de deux segments de la chaîne Varisque (le Plateau d'Aigurande, Massif Central Français, et les Vosges) à travers une étude* ^{40}Ar–^{39}Ar. Thèse de Doctorat, University of Strasbourg.

BRUECKNER, H. K., MEDARIS, L. G. & BAKUN-CZUBAROW, N. 1989. Nd and Sr age and isotope patterns from Hercynian eclogites and garnet pyroxenites of the Bohemian Massif and the East Sudetes. *Third International Eclogite Conference, Würzburg*, 4.

Bryhni, I. & Sturt, B. A. 1985. Caledonides of southwestern Norway. *In*: Gee, D. E. & Sturt, B. A. (eds) *The Caledonide Orogen – Scandinavia and Related Areas*. John Wiley & Sons Ltd, New York, 89–107.

Burg, J.-P. & Matte, P. 1978. A cross section through the french Massif central and the scope of its Variscan geodynamic evolution. *Zeitschrift der Deutschen Geologischen Gessellschaft,* **129**, 429–460.

——, Balé, P., Brun, J.-P. & Girardeau, J. 1987. Stretching lineations and transport direction in the Ibero-Armorican arc during the Siluro-Devonian collision. *Geodinamica Acta,* **1**, 71–87.

——, Leyreloup, A., Marchand, J. & Matte, P. 1984. Inverted metamorphic zonation and large-scale thrusting in the Variscan Belt: an example in the French Massif Central. *In*: Hutton, D. H. W. & Sanderson, D. J. (eds) *Variscan tectonics of the North Atlantic Region*. Geological Society, London, Special Publications, **14**, 47–61.

——, Van Den Driessche, J. & Brun, J.-P. 1994. Syn- to post-thickening extension in the Variscan Belt of Western Europe: Mode and structural consequences. *Géologie de la France,* **3**, 33–51.

Chauvet, A. & Dallmeyer, R. D. 1992. $^{40}Ar/^{39}Ar$ dates related to Devonian extension in the southwestern Scandinavian Caledonides. *Tectonophysics,* **210**, 155–177.

—— & Séranne, M. 1989. Microtectonic evidence of Devonian extensional westward shearing in Southwest Norway. *In*: Gayer, R. (ed.) *The Caledonide geology of Scandinavia*. Graham & Trotman, London, 245–254.

—— & —— 1994. Extension-parallel folding in the Scandinavian Caledonides: implications for late-orogenic processes. *Tectonophysics,* **238**, 31–54.

——, Kiénast, J.-R., Pinardon, J.-L. & Brunel, M. 1992. Petrological constraints and PT path of Devonian collapse tectonics within the Scandian mountain belt (Western Gneiss Region, Norway). *Journal of the Geological Society of London,* **149**, 383–400.

Cocks, L. R. M. & Fortey, R. A. 1982. Faunal evidence for oceanic separations in the Palaeozoic of Britain. *Journal of the Geological Society of London,* **139**, 465–478.

Coney, P. J. & Harms, T. A. 1984. Cordilleran metamorphic core complexes: Cenozoic extensional relics of Mesozoic compression. *Geology,* **12**, 550–554.

Costa, S. 1990. *De la collision continentale à l'extension tardi-orogénique: 100 millions d'années d'histoire varisque dans le Massif Central Français. Une étude chronologique par la méthode 40Ar–39Ar*. Thèse de Doctorat, University of Montpellier II.

—— 1991–1992. East–west diachronism of the collisional stage in the french Massif Central: Implications for the European Variscan Orogen. *Geodinamica Acta,* **5**, 51–68.

—— & Rey, P. 1995. Lower crustal rejuvenation and crustal growth during post-thickening collapse: Insights from a crustal cross section through a metamorphic core complex. *Geology,* **23**, 905–908.

Dallmeyer, R. D. & Gee, D. G. 1986. $^{40}Ar/^{39}Ar$ mineral dates from retrogressed eclogites within the Baltoscandian miogeosyncline: implications for a polyphase Caledonian orogenic evolution. *Geological Society of America Bulletin,* **97**, 26–34.

Dewey, J. F. 1988. Extensional collapse of orogens. *Tectonics,* **7**, 1123–1139.

Ducrot, J., Lancelot, J. R. & Marchand, J. 1983. Datation U-Pb sur zircons de l'éclogite de la Borie (Haut Allier, France) et conséquences sur l'évolution anté-hercynienne de l'Europe occidentale. *Earth and Planetary Science Letters,* **62**, 385–394.

Dunning, G. R. & Pedersen, R. B. 1988. U/Pb ages of ophiolites and arc-related plutons of the Norwegian Caledonides: implications for the development of Iapetus. *Contributions to Mineralogy and Petrology,* **98**, 13–23.

Duthou, J.-L., Piboule, M., Gay, M. & Dufour, E. 1981. Datations radiométriques Rb-Sr sur les ortho-granulites des Monts du Lyonnais (Massif Central français). *Comptes Rendus de l'Académie des Sciences de Paris,* **292**, 749–7526.

Echtler, H. 1990. Geometry and kinematics of recumbent folding and low-angle detachment in the Pardailhan nappe (Montagne Noire, Southern French Massif Central). *Tectonophysics,* **177**, 109–123.

Edel, J.-B. 1987. Paleopositions of the western Europe Hercynides during the Late Carboniferous deduced from paleomagnetic data: consequences for 'stable Europe'. *Tectonophysics,* **139**, 31–41.

—— & Weber, K. 1995. Cadomian terranes, wrench faulting and thrusting in the central Europe Variscides: geophysical and geological evidence. *Geologische Rundschau,* **84**, 412–432.

England, P. & McKenzie, D. P. 1983. Correction to: A thin viscous sheet model for continental deformation. *Geophysical Journal of the Royal Astronomical Society,* **73**, 523–532.

Fossen, H. 1992. The role of extensional tectonics in the Caledonides of south Norway. *Journal of Structural Geology,* **14**, 1033–1046.

—— 1993. The role of extensional tectonics in the caledonides of south Norway: Reply. *Journal of Structural Geology,* **15**, 1381–1383.

Franke, W. 1989*a*. Tectonostratigraphic units in the Variscan belt of central Europe. *In*: Dallmeyer, R. D. (ed.) *Terranes in the Circum-Atlantic Palaeozoic Orogens*. Geological Society of America, Special Papers, **230**, 67–90.

—— 1989*b*. Variscan plate tectonics in Central Europe – current ideas and open questions. *Tectonophysics,* **169**, 221–228.

—— 1993. The Saxonian Granulites: a metamorphic core complex? *Geologische Rundschau,* **82**, 505–515.

—— & Engel, W. 1986. Synorogenic sedimentation in the Variscan Belt of Europe. *Bulletin de la Société Géologique de France,* **8/2**, 25–33.

GARDIEN, V. 1990. *Evolutions P–T et structures associées dans l'est du Massif Central Français : Un exemple de l'évolution thermomécanique de la chaîne paléozoique*. Thèse de Doctorat, University of Grenoble I.

GAUDEMER, Y., JAUPART, C. & TAPPONNIER, P. 1988. Thermal control on post-orogenic extension in collision belts. *Earth and Planetary Science Letters*, **89**, 48–62.

GEBAUER, D. & GRÜNENFELDER, M. 1979. U/Pb Zircon and Rb/Sr mineral dating of eclogites and their country rocks, example : Münchberg gneiss Massif Northeast Bavaria. *Earth and Planetary Science Letters*, **42**, 35–44.

——, LAPPIN, M. A., GRUNENFELDER, M. & WYTTENBACH, A. 1985. The age and origin of some Norwegian eclogites: a U-Pb zircon and R.E.E. study. *Chemical Geology*, **52**, 227–248.

GEE, D. G. 1982. The Scandinavian Caledonides. *Terra Cognita*, **2**, 89–96.

GILOTTI, J. A. 1993. Discovery of a medium-temperature eclogite province in the Caledonides of North-East Greenland. *Geology*, **21**, 523–526.

GRIFFIN, W. L. & BRUECKNER, H. K. 1980. Caledonian Sm/Nd ages and a crustal origin for Norwegian eclogites. *Nature*, **285**, 319–321.

——, AUSTRHEIM, H., BRASTAD, K., BRYHNI, I., KRILL, A. G., KROGH, E. J., MØRK, M. B. E., QVALE, H. & TØRUDBAKKEN, B. 1985. High-pressure metamorphism in the Scandinavian Caledonides. *In*: GEE, D. E. & STURT, B. A. (eds) *The Caledonide Orogen – Scandinavia and Related Areas*. John Wiley & Sons Ltd, New York, 783–801.

HARLAND, W. B. & GAYER, R. A. 1972. The Arctic Caledonides and earlier oceans. *Geological Magazine*, **109**, 289–314.

HARTZ, E. & ANDERSEN, A. 1995. Caledonian sole thrust of central East Greenland: A crustal scale Devonian extensional detachment. *Geology*, **23**, 637–640.

HOSSACK, J. R. 1984. The geometry of listric growths faults in the Devonian basins of Sunnfjord, W Norway. *Journal of the Geological Society, London*, **141**, 629–637.

HOUSEMAN, G. A., MCKENZIE, D. P. & MOLNAR, P. 1981. Convective instability of a thickened boundary layer and its relevance for the thermal evolution of continental convergent belts. *Journal of Geophysical Research*, **86**, 6115–6132.

HURICH, C. A., PALM, H., DYRELIUS, D. & KRISTOFFERSEN, Y. 1989. Deformation of the Baltic continental crust during Caledonides intracontinental subduction: Views from seismic reflection data. *Geology*, **17**, 423–425.

HUTTON, D. H. W. 1987. Strike-slip terranes and a model for the evolution of the British and Irish Caledonides. *Geological Magazine*, **124**, 405–425.

KALT, A., HANEL, M., SCHLEIDER, H. & KRAMM, U. 1994. Petrology and geochronology of eclogites from the Variscan Schwarzwald (F.R.G.). *Contributions to Mineralogy and Petrology*, **115**, 287–302.

KIRBY, S. H. 1985. Rock mechanics observations pertinent to the rheology of the continental lithosphere and the localization of strain along shear zones. *Tectonophysics*, **119**, 1–27.

KLAPER, E. M. 1991. Eclogitic shear zones in a granulite-facies anorthosite complex: field relationships and an emplacement scenario – an example from the Bergen Arcs, western Norway. *Schweizerische Mineralogische und Petrographische Mitteilungen*, **71**, 231–241.

KNELLER, B. C. 1991. A foreland basin on the southern margin of Iapetus. *Journal of the Geological Society, London*, **148**, 207–210.

KREUZER, H., SEIDEL, E., SCHÜSSLER, U., OKRUSCH, M., LENZ, K. L. & RASCHKA, H. 1989. K-Ar geochronology of different tectonic units at the northwestern margin of the Bohemian Massif. *Tectonophysics*, **157**, 149–178.

KROGH, T. E., MYSEN, B. O. & DAVIS, G. L. 1974. A Palaeozoic age for the primary minerals of a Norwegian eclogite. *Annual Report of the Geophysics Laboratory, Carnegie Institute, Washington*, **73**, 575–576.

KRÖNER, A., WENDT, I., LIEW, T. C., COMPSTON, W., TODT, W., FIALA, J., VANKOVA, V. & VANEK, J. 1988. U-Pb zircon and Sm-Nd model ages of high-grade Moldanubian metasediments (Bohemian Massif, Czechoslovakia). *Contributions to Mineralogy and Petrology*, **99**, 257–266.

KULLERUD, L., TØRUDBAKKEN, B. O. & ILEBEKK, S. 1986. A compilation of radiometric age determinations from the Western Gneiss Region, South Norway. *Norges geologiske Underssøkelse Bulletin*, **406**, 17–42.

KULLING, O. 1972. The Swedish Caledonides. *In*: STRAND, T. & KULLING, O. (eds) *The Scandinavian Caledonides*. Wiley-Interscience, London, 147–285.

KUSZNIR, N. J. & PARK, R. G. 1987. The extensional strength of the continental lithosphere: its dependence on geothermal gradient, crustal composition and thiskness. *In*: COWARD, M. P., DEWEY, J. F. & HANCOCKS, P. L. (eds) *Continental Extensional Tectonics*, Geological Society, London, Special Publications, **28**, 35–52.

LAFON, J. M. 1986. *Géochronologie U-Pb appliquée à deux segments du Massif Central français : le Rouergue oriental et le Limousin central*. Thèse de Doctorat, University of Montpellier.

LEDRU, P., LARDEAUX, J. M., SANTALLIER, D., AUTRAN, A., QUENARDEL, J. M., FLOC'H, J. P., LEROUGE, G., MAILLET, N., MARCHAND, J. & PLOQUIN, A. 1989. Où sont les nappes dans le Massif central français? *Bulletin de la Société géologique de France*, **8**, 605–618.

LUX, D. R. 1985. K/Ar ages from the Basal Gneiss Region, Stadtlandet area, western Norway. *Norsk Geologisk Tidsskrift*, **65**, 277–286.

MALAVIEILLE, J. 1993. Late orogenic extension in mountain belts: insights from the Basin and Range and the Late Paleozoic Variscan Belt. *Tectonics*, **12**, 1115–1130.

——, GUIHOT, P., COSTA, S., LARDEAUX, J.-M. & GARDIEN, V. 1990. Collapse of a thickened

Variscan crust in the French Massif Central : Mont-Pilat extensional shear zone and Saint-Etienne Upper Carboniferous basin. *Tectonophysics,* **177**, 139–149.
MARTÍNEZ-CATALÁN, J. R. 1990. A non-cylindrical model for the northwest Iberian allochthonous terranes and their equivalents in the Hercynian belt of Western Europe. *Tectonophysics,* **179**, 253–272.
MATTE, P. 1986. Tectonics and plate tectonics model for the Variscan belt of Europe. *Tectonophysics,* **126**, 329–374.
—— 1991. Accretionary history and crustal evolution of the Variscan belt in Western Europe. *Tectonophysics,* **196**, 309–337.
——, MALUSKI, H., RAJLICH, P. & FRANKE, W. 1990. Terrane boundaries in the Bohemian Massif: Result of large-scale Variscan shearing. *Tectonophysics,* **177**, 151–170.
——, RESPAUT, J.-P., MALUSKI, H., LANCELOT, J. R. & BRUNEL, M. 1986. La faille NW-SE du Pays de Bray, un décrochement ductile dextre hercynien: déformation à 330 Ma d'un granite à 570 Ma dans le sondage Pays de Bray 201. *Bulletin de la Société géologique de France,* **8**, 69–77.
MCCLAY, M. R., NORTON, M. G., CONEY, P. & DAVIS, G. H. 1986. Collapse of the Caledonian orogen and the Old Red Sandstone. *Nature,* **323**, 147–149.
MCKENZIE, D. P. & BICKLE, M. 1988. The volume and composition of melt generated by extension of the lithosphere. *Journal of Petrology,* **29**, 625–680.
MCKERROW, W. S. H. & ZIEGLER, A. M. 1972. Paleozoic oceans. *Nature (Physical Sciences),* **240**, 92–94.
MEILLIEZ, F. & MANSY, J.-L. 1990. Déformation pelliculaire différenciée dans une série lithologique hétérogène: le Dévono-Carbonifère de l'Ardenne. *Bulletin de la Société géologique de France,* **6**, 177–188.
MÉNARD, G. & MOLNAR, P. 1988. Collapse of a Hercynian Tibetan Plateau into a Late Palaeozoic European Basin and Range Province. *Nature,* **334**, 235–237.
MEZGER, K., ESSENE, E. J. & HALLIDAY, A. N. 1992. Closure temperature of the Sm-Nd system in metamorphic garnets. *Earth and Planetary Science Letters,* **113**, 397–409.
MILNES, A. G. & KOESTLER, A. G. 1985. Geological structure of Jotunheim, southern Norway (Sognefjell-Valdres cross-section. *In*: GEE, D. G. & STURT, B. A. (eds) *The Caledonide Orogen – Scandinavia and Related Areas.* John Wiley & Sons Ltd, New York, 457–474.
——, DIETLER, T. N. & KOESTLER, A. G. 1988. The Sognefjord north shore log.- A 25 km depth section through Caledonized basement in Western Norway. *Norwegian Geological Survey Special Publication,* **3**, 114–121.
——, WENNBERG, O. P., SKÅR, Ø. & KOESTLER, A. G. 1997. Contraction, extension and timing in the South Norwegian caledonides – The Sognefjord transect. *This volume.*
MOLNAR, P. & TAPPONNIER, P. 1975. Cenozoic tectonics of Asia: Effects of a continental collision. *Science,* **189**, 419–426.
MØRK, M. B. E. & MEARNS, E. W. 1986. Sm-Nd isotope systematics of a gabbro-eclogite transition. *Lithos,* **19**, 255–267.
NEUGEBAUER, J. 1989. The Iapetus model: a plate tectonic concept for the Variscan belt of Europe. *Tectonophysics,* **169**, 229–256.
NOBLE, S. R., TUCKER, R. D. & PHARAOH, T. C. 1993. Lower Paleozoic and Precambrian igneous rocks from eastern England, and their bearing on Late Ordovician closure of the Tornquist Sea: Constraints from U-Pb and Nd isotopes. *Geological Magazine,* **130**, 835–846.
NOBLET, C. & LEFORT, J.-P. 1990. Sedimentological evidence for a limited separation between Armorica and Gondwana during the Early Ordovician. *Geology,* **18**, 303–306.
NORTON, M. G. 1986. Late Caledonide extension in Western Norway: A response to extreme crustal thickening. *Tectonics,* **5**, 195–204.
—— 1987. The Nordfjord-Sogn detachment, W. Norway. *Norsk Geologisk Tidsskrift,* **67**, 93–106.
OLIVER, G. J. H., CORFU, F. & KROGH, T. E. 1993. U-Pb ages from SW Poland: Evidence for a caledonian suture zone between Baltica and Gondwana. *Journal of the Geological Society of London,* **150**, 355–369.
PAQUETTE, J.-L., BALÉ, P., BALLÈVRE, M. & GEORGET, Y. 1987. Géochronologie et géochimie des éclogites du Léon: nouvelles contraintes sur l'évolution géodynamique du Nord-ouest du Massif Armoricain. *Bulletin de Minéralogie,* **110**, 683–696.
PERROUD, H. 1982. Contribution à l'étude paléomagnétique de l'Arc Ibéro-Armoricain. *Bulletin de la Société géologique et minéralogique de Bretagne,* **C14**, 1–114.
PEUCAT, J.-J. & COGNÉ, J. 1977. Geochronology of some blueschists from Ile de Groix, France. *Nature,* **268**, 131–132.
PIN, C. 1979. *Géochronologie U-Pb et microtectonique des séries métamorphiques anté-stéphaniennes de l'Aubrac et de la région de Marvejols (Massif Central).* Thèse de 3ième cycle, University of Montpellier.
—— 1990. Variscan oceans: Ages, origins and geodynamic implications inferred from geochemical and radiometric data. *Tectonophysics,* **177**, 215–227.
—— & LANCELOT, J. 1982. U/Pb dating of an early Paleozoic bimodal magmatism in the French Massif Central and of its further metamorphic evolution. *Contributions to Mineralogy and Petrology,* **79**, 1–12.
—— & PEUCAT, J.-J. 1986. Ages des épisodes de métamorphismes paléozoïques dans le Massif Central et le Massif Armoricain. *Bulletin de la Société géologique de France,* **8**, 461–469.
—— & VIELZEUF, D. 1983. Granulites and related rocks in Variscan Median Europe : a dualistic interpretation. *Tectonophysics,* **93**, 47–74.
PUCHKOV, V. N. 1997. Structure and geodynamics of the Uralian orogen. *This volume.*

REY, P. 1993. Seismic and tectonometamorphic characters of the lower continental crust in Phanerozoic areas: A consequence of post-thickening extension. *Tectonics,* **12**, 580–590.

——, BURG, J.-P. & CARON, J.-M. 1991–1992. Middle and Late Carboniferous extension in the Variscan Belt: structural and petrological evidences from the Vosges massif (Eastern France). *Geodinamica Acta,* **5**, 17–36.

——, FOUNTAIN, D. M. & BOUNDY, T. 1996. Kinematic indicators in eclogite-facies shear zones. *In*: SNOKE, A. W., TULLIS, J. A. & TODD, V. R. (eds) *Atlas of Mylonitic and Fault-Related Rocks*. Princeton University Press, Princeton, in press.

ROBERT, D. & STURT, B. A. 1980. Caledonian deformation in Norway. *Journal of the Geological Society of London,* **137**, 241–250.

ROBERTS, D. & GEE, D. G. 1985. An introduction to the structure of the Scandinavian Caledonides. *In*: GEE, D. G. & STURT, B. A. (eds) *The Caledonide Orogen – Scandinavia and Related Areas*. John Wiley & Sons Ltd, New York, 55–68.

SCHMÄDICKE, E., MEZGER, K., COSCA, M. A. & OKRUSCH, M. 1995. Variscan Sm-Nd and Ar-Ar ages of eclogite facies rocks from the Erzgebirge, Bohemian Massif. *Journal of Metamorphic Geology,* **13**, 537–552.

SCOTESE, C. R. & MCKERROW, W. S. 1990. Revised World maps and introduction. *In*: MCKERROW, W. S. & SCOTESE, C. R. (eds) *Palaeozoic Palaeogeography and Biogeography*. Geological Society, London, Memoirs, **12**, 1–21.

SÉGURET, M., SÉRANNE, M., CHAUVET, A. & BRUNEL, M. 1989. Collapse basin: A new type of extensional sedimentary basin from the Devonian of Norway. *Geology,* **17**, 127–130.

SÉRANNE, M. & SÉGURET, M. 1987. The Devonian basin of western Norway: tectonics and kinematics of an extending crust. *In*: COWARD, M. P., DEWEY, J. F. & HANCOCKS, P. L. (eds) *Continental Extensional Tectonics*. Geological Society, London, Special Publications, **28**, 537–548.

——, CHAUVET, A. & FAURE, J.-L. 1991. Cinématique de l'extension tardi-orogénique (Dévonien) dans les Calédonides Scandinaves et Britanniques. *Comptes Rendus de l'Académie des Sciences de Paris,* **313 II**, 1305–1312.

——, ——, SÉGURET, M. & BRUNEL, M. 1989. Tectonics of the Devonian collapse-basins of western Norway. *Bulletin de la Société géologique de France,* **8**, 489–499.

SHEMENDA, A. I. 1993. Subduction of the lithosphere and back arc dynamics: insights from physical modeling. *Journal of Geophysical Research,* **98**, 16,167–16,185.

SMITH, A. G., BRIDEN, J. C. & DREWRY, G. E. 1973. Phanerozoic world maps. *Special Papers in Palaeontology,* **12**, 1–42.

SMITH, D. C. & LAPPIN, M. A. 1989. Coesite in the Straumen kyanite-eclogite pod, Norway. *Terra Nova,* **1**, 47–56.

SOPER, N. J. & HUTTON, D. H. W. 1984. Late Caledonian sinistral displacements in Britain: implications for a three-plate collision model. *Tectonics,* **3**, 781–794.

—— & WOODCOCK, N. H. 1990. Silurian collision and sediment dispersal patterns in southern Britain. *Geological Magazine,* **127**, 527–542.

——, STRACHAN, R. A., HOLDSWORTH, R. E., GAYER, R. A. & GREILING, R. O. 1992. Sinistral transpression and the Silurian closure of Iapetus. *Journal of the Geological Society of London,* **149**, 871–880.

STEPHENS, M. B. & GEE, D. B. 1985. A tectonic model for the evolution of the eugeosynclinal terranes in the central Scandinavian Caledonides. *In*: GEE, D. G. & STURT, B. A. (eds) *The Caledonide Orogen – Scandinavia and related areas*. John Wiley & Sons Ltd, New York, 953–978.

—— & —— 1989. Terranes and polyphase accretionary history in the Scandinavian Caledonides. *Geological Society of America, Special Paper,* **230**, 17–30.

STEPHENS, W. E. 1988. Granitoid plutonism in the Caledonian orogen of Europe. *In*: HARRIS, A. L. & FETTES, D. J. (eds) *The Caledonian–Appalachian Orogen*. Geological Society, London, Special Publications, **38**, 389–403.

STOSCH, H. G. & LUGMAIR, G. W. 1990. Geochemistry and evolution of MORB type eclogites from the Münchberg Massif, southern germany. *Earth and Planetary Science Letters,* **99**, 230–249.

TOLLMANN, A. 1982. Großräumiger variszischer Deckenbau im Moldanubikum und neue Gedanken zum Variszikum Europas. *Geotektonische Forschungen,* **64**, 1–91.

TURCOTTE, D. L. & SCHUBERT, G. 1982. *Geodynamics: Applications of Continuum Physics to Geological Problems*. John Wiley & Sons, New York.

VAN BREEMEN, O., AFTALION, M., BOWES, D. R., DUDEK, A., MISAR, Z., POVONDRA, P. & VRANA, S. 1982. Geochronological studies of the Bohemian Massif, Czechoslovakia, and their significance in the evolution of central Europe. *Earth Sciences,* **73**, 89–108.

VAN DEN BEUKEL, J. 1992. Some thermomechanical aspects of the subduction of continental lithosphere. *Tectonics,* **11**, 316–329.

VAN DEN DRIESSCHE, J. & BRUN, J.-P. 1991–1992. Tectonic evolution of the Montagne Noire (french Massif Central) : a model of extensional gneiss dome. *Geodinamica Acta,* **5**, 85–99.

VAN DER VOO, R. 1982. Pre-Mesozoic paleomagnetism and plate tectonics. *Annual Review of Earth and Planetary Science,* **10**, 191–220.

WEBER, K. & BEHR, H. J. 1983. Geodynamic interpretation of the Mid-European Variscides. *In*: MARTIN, H. & EDER, W. (eds) *Intracontinental Fold Belts*. Springer Verlag, 427–469.

WILKS, W. J. & CUTHBERT, S. J. 1994. The evolution of the Hornelen Basin detachment system, western Norway: implications for the style of late orogenic extension in the southern Scandinavian Caledonides. *Tectonophysics,* **238**, 1–30.

WILLIAMS, H. 1984. Miogeoclines and suspect terranes of the Caledonian – Appalachian orogen: tectonic patterns in the North Atlantic region. *Canadian Journal of Earth Sciences,* **21**, 887–901.

ZIEGLER, P. A. 1986. Geodynamic model for the Palaeozoic crustal consolidation of western and central Europe. *Tectonophysics,* **126**, 303–328.

Structure and geodynamics of the Uralian orogen

V. N.PUCHKOV

Institute of Geology, ul. K. Marx 16/2, Ufa 450 000, Bashkiria, Russia

Abstract: The Urals are a Late Palaeozoic orogenic belt. The relicts of earlier orogens are traced in its basement. In particular, the Late Vendian pre-Uralian orogen is reconstructed and identified as a part of the Late Precambrian Cadomian orogen. The Uralian orogeny was preceded by Late Cambrian–Early Ordovician epicontinental rifting and formation of the Paleo-Uralian ocean whose remnants are Palaeozoic ophiolites. Calc-alkaline volcanites and plutons, typical of active margins, are widely developed in the eastern Urals. The Uralian foldbelt results from oblique collision between the East European (Laurussia) passive margin and the active margin on the Kazakhstanian continent. Collision began in the south of the Urals and moved, wave-like, to the north. The eastern and northern parts of the Urals have been affected by the Middle Jurassic Cimmerian intracontinental (intra-Pangaea) shortening. The Uralian–Cimmerian mountain belt was eroded and partially inundated by seas in the Late Jurassic–Early Cretaceous times and has been reactivated since the Oligocene in response to a recent intracontinental shortening.

The Urals are a Late Palaeozoic foldbelt that also experienced Mid-Jurassic Cimmerian deformation in its eastern and northern parts. The north–south-trending mountain range, approximately 2000 km long, is the geographic Europe–Asia boundary and is commonly divided into the Polar, Cis-Polar, Northern, Central and Southern Urals (Fig. 1). The characteristic feature of the fold belt is a distinct, though disturbed, linearity of tectonic zones. A continental passive margin to the west has been underthrust below units derived from a marine domain juxtaposed against a calc-alkaline palaeo-active margin. Thanks to Late Cenozoic tectonic movements the tectonic zones are all exposed in the Southern Urals. In the north, the easternmost zones are covered by the Mesozoic and Cenozoic sediments of the West Siberian basin.

The Uralian foldbelt is one of the oldest and richest mining regions of Russia. Therefore, it has attracted the attention of many geologists among whom one should mention Murchison (the founder of the Permian system), Karpinsky, (the proponent of the contractionist ideas who suggested that the changes in the Ural's strike were influenced by the outline of the rigid Russian plate) and Shatsky (who established the Riphean system in the Urals, discovered relicts of a Late Proterozoic foldbelt and developed the theory of relationships between geosynclines and platforms).

Understanding the general features of the Uralian structure and history has depended much on the development of general tectonic ideas. Publication of Wegener's 'Die Enstehung der Kontinente' (translated into Russian in 1924) and of the important works of Argand and Staub, led to a mobilist model of the Urals in the Thirties. In the 1940s and up to the 1960s, a fixist paradigm took over and 'charriage' (thrusting) practically became a prohibited word in publications concerning the Urals. Moreover, it is in the Urals that the concept of deep-seated faults was proposed by Peyve (1945). It was initially thought that these faults, such as the Main Uralian Fault, were near-vertical, reached deep into the mantle, controlled magmatism and metallogeny and were intrinsically a proof that continents could not drift over the mantle. For several decades these ideas were foremost in all tectonic interpretations and are still propounded by some researchers (note deep-seated faults in Fig. 2). At the same time, the Urals were regarded as an exemplary geosyncline. In 1972, in light of the introduction of plate tectonics, Peyve and Ivanov proposed that the Urals represent a closed Palaeozoic ocean. The first tectonic map at a scale of $1:10^6$ based on plate tectonic ideas was published in 1977 by Peyve *et al.* Subsequent work has accumulated a considerable volume of information supporting this interpretation (e.g. Kamaletdinov 1974; Perfilyev 1979; Puchkov 1979, 1991, 1993; Ruzhentsev 1986; Ivanov *et al.* 1986; Savelyeva 1987; Yazeva *et al.* 1989; Seravkin *et al.* 1992; Svyazhina *et al.* 1992).

From Burg, J.-P. & Ford, M. (eds), 1997, *Orogeny Through Time*,
Geological Society Special Publication No. 121, pp.201–236.

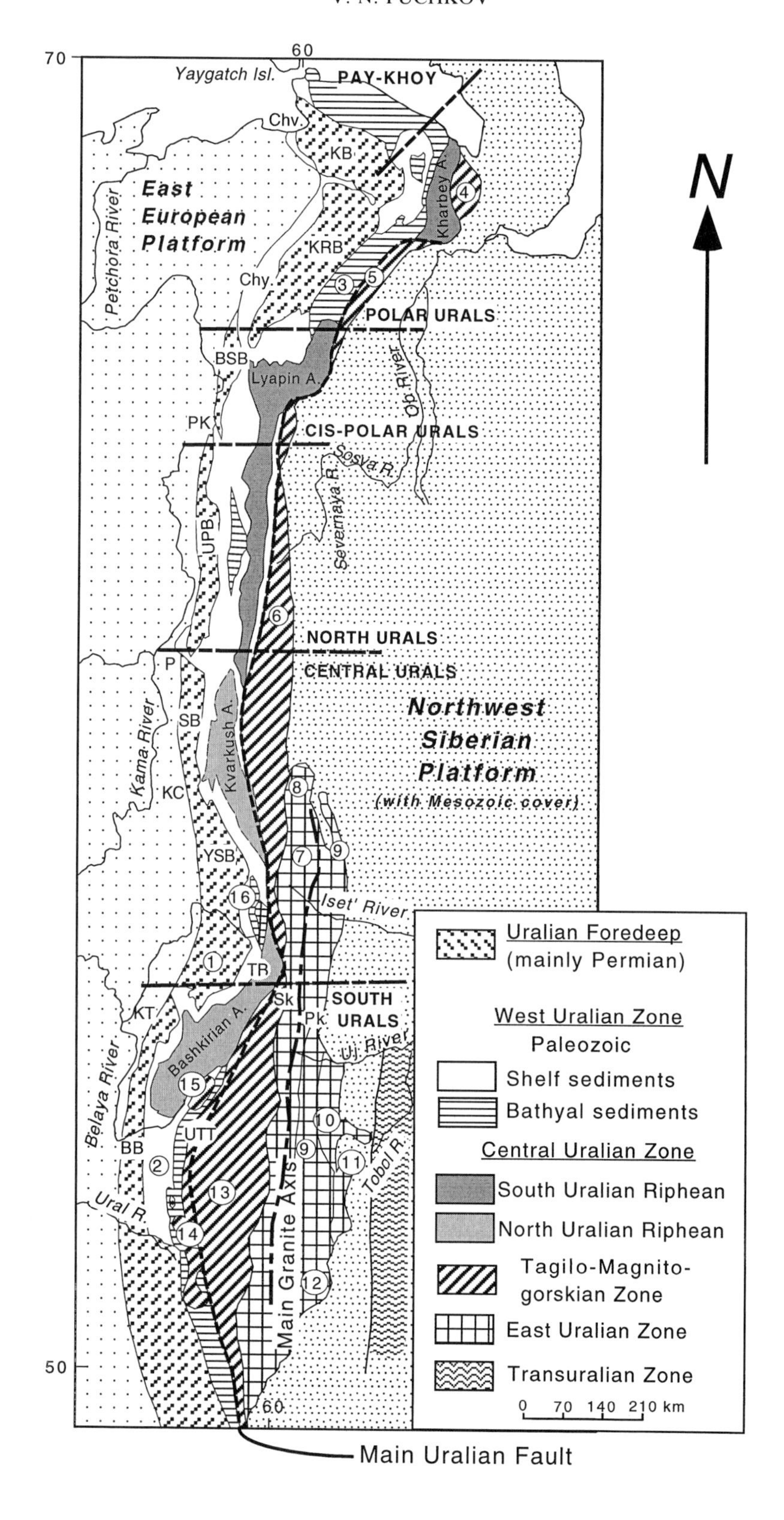
70
60
Yaygatch Isl.
PAY-KHOY
Chv.
KB
Kharbey A.
4
East European Platform
Petchora River
KRB
Chy.
3
5
POLAR URALS
BSB
Lyapin A.
Ob River
PK
CIS-POLAR URALS
Sosva R.
Severnaya R.
UPB
6
NORTH URALS
CENTRAL URALS
P
SB
Kama River
Kvarkush A.
Northwest Siberian Platform
(with Mesozoic cover)
KC
8
9
7
YSB
16
Iset' River
1
TR
Sk
SOUTH URALS
KT
Pk
Bashkirian A.
Uj River
Belaya River
15
10
UTT
BB
9
2
11
Tobol R.
13
Main Granite Axis
Ural R.
14
12
50
60
Main Uralian Fault
N
Uralian Foredeep
(mainly Permian)
West Uralian Zone
Paleozoic
Shelf sediments
Bathyal sediments
Central Uralian Zone
South Uralian Riphean
North Uralian Riphean
Tagilo-Magnito-gorskian Zone
East Uralian Zone
Transuralian Zone
0 70 140 210 km

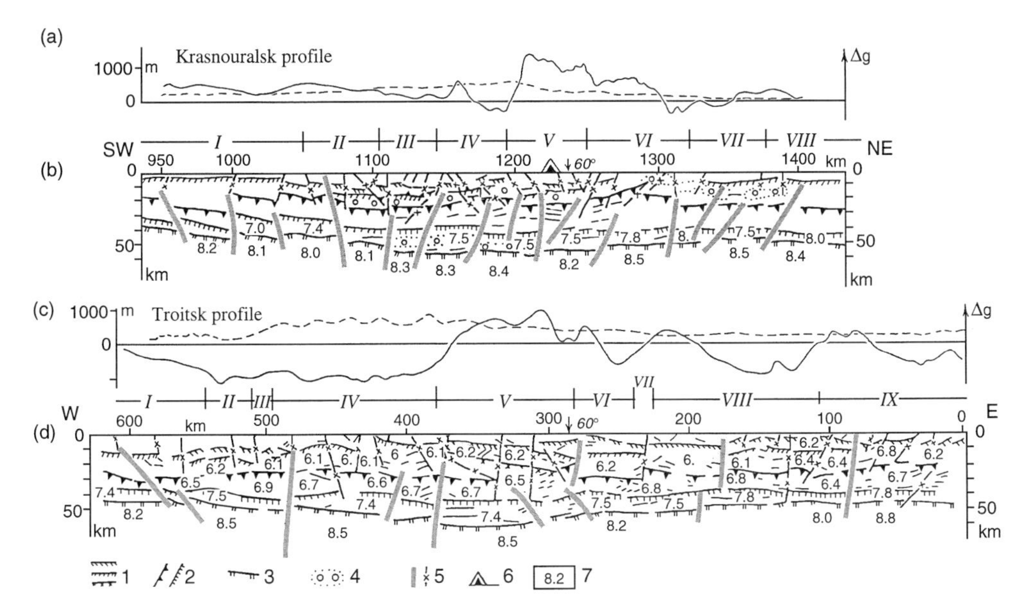

Fig. 2. Geophysical data from two profiles through the Urals (located on Fig. 3). In (**a**) and (**c**) a topographic profile (dashed line) and a gravity anomaly profile (solid line, no scale given) are shown for each profile. (**b**) and (**d**) show deep seismic sounding sections (very generalized after Avtoneev *et al.* 1991). Key: 1 and 2, unspecified boundaries of different, crustal seismic layers; 3, Moho surface; 4, anomalously low velocity zones; 5, hypothetical deep-seated faults of different order; 6, the Uralian Super-deep borehole; 7, layer velocities in km s^{-1}. Structural zones: I, East European platform; II, Uralian foredeep; III, West Uralian Zone; IV, Central Uralian Zone; V, Taglio-Magnitogorskian zone; VI and VII, Central Uralian zone; VIII, Transuralian zone; IX, Kazakhstanian 'Caledonides'.

Our understanding of the deep structure of the Urals has been improved considerably by a combined analysis of magnetic and gravity fields. More than 15 Deep Seismic Sounding (DSS) profiles and more than 20 seismic reflection profiles have been completed (e.g. Necheukhin *et al.* 1986; Avtoneev *et al.* 1991; Sokolov, 1992). This paper gives a summary of the current state of knowledge on the Urals.

Structural zonation of the Urals

The Urals are divided into six sub-longitudinal zones (Fig. 1), that differ both in their structure and stratigraphy. From west to east they are (1) the Uralian Foredeep, (2) the West Uralian, (3) the Central Uralian, (4) the Tagilo-Magnitogorskian, (5) the East Uralian and (6) the Transuralian zones.

Zones 1, 2 and 3 represent the former passive margin of the East European (Euramerian, Laurussia) continent (Puchkov 1979). This margin formed from the Late Cambrian to the Early Ordovician and was stable during Ordovician, Silurian and Devonian times. In the Late Palaeozoic, the platform was deformed to become a part of the Uralian foldbelt. The East European basement comprises granitic and metamorphic rocks of Precambrian age (Gafarov 1970). The crust is 34–42 km thick (Necheukhin *et al.* 1986; Avtoneev *et al.* 1991) (Figs 2 & 3). Zone 3 is bounded to the east by the Main

Fig. 1. Tectonic sketch map of the Urals showing the main tectonic zones and geographic sub-divisions of the orogen. Encircled numbers: 1, Ufimian amphitheatre. 2–13, Subzones: 2, Zilair; 3, Lemva; 4, Schchuchya; 5, Voykar; 6, Tagil; 7, Murzinka–Aduy; 8, Salda; 9, East Uralian volcanics; 10, Troitsk; 11, Denisovka; 12, East Mugodzhary; 13, Magnitogorsk; 14–16, tectonic klippen in the West Uralian zone, containing ophiolites or serpentinitic melange; 14, Sakmara; 15, Kraka; 6, Bardym (Nyazepetrovsk); A, Four anticlinoria of the Central Uralian zone; TR, Taratash complex. Transversal structural elements of the Uralian foredeep. Uplifts: KT, Kara–Tau; KC, Kosva–Chusovaya; P, Polyud; PK, Pechora–Kozhva; Chy, Chernyshov uplift; Chv, Chernov uplift. Basins: BB, Belsk; YSB, Yuryuzan–Sylva; SB,Solikamsk; UPB, Upper Pechora; BSB, Bolshaya Synya; KRB, Kosyu–Rogovaya; KB, Korotaikha.; PK, Poletayevka area; UTT, Ural–Tau metamorphic terranes; SK, Selyankino.

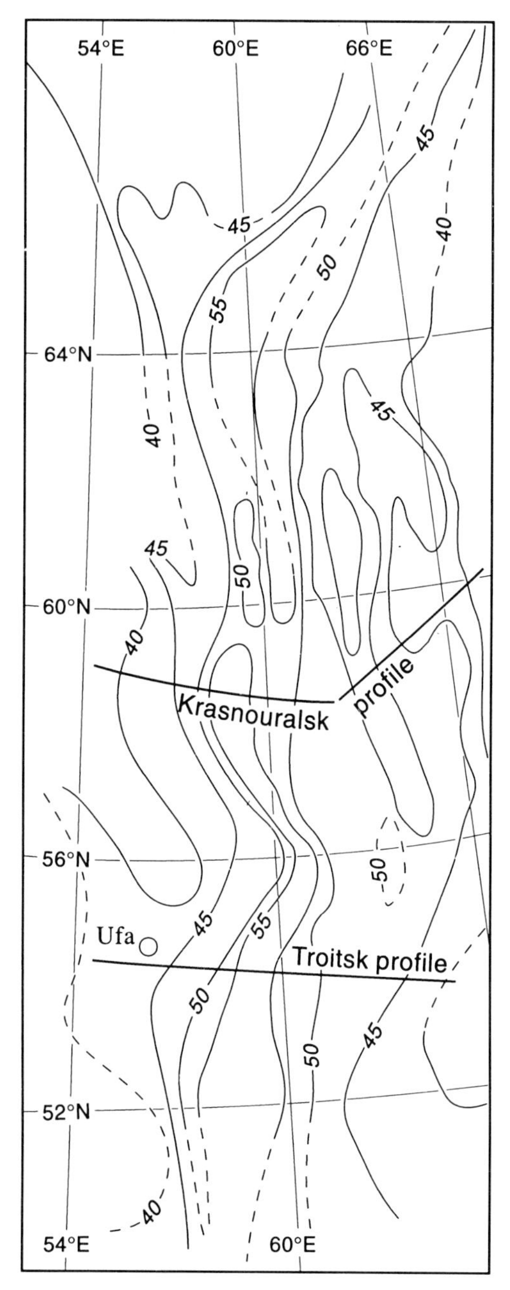

Fig. 3. Thickness of the crust in the Urals and adjacent areas (after Avtoneev *et al.* 1991; isopleths in km). Compiled from gravity anomalies and 16 DSS profiles.

Uralian Fault, the principal suture marked by a serpentinitic mélange and inclined to the east by 20 to 40–50° (locally steeper). It may flatten at depth, as suggested by seismic data (Sokolov 1992; Petrov *et al.* 1994) (Figs 4 & 5).

The Uralian foredeep (Zone 1)

The Uralian foredeep, 50–75 km wide, is filled with Permian flysch and molasse of eastern provenance, up to 6 km thick (Nalivkin 1949; Makedonov 1965; Chuvashov & Dyupina 1973) underlain by 4–7 km thick Ordovician-Carboniferous shelf deposits (Fig. 6), which, in turn, unconformably cover Precambrian sedimentary, metamorphic and magmatic complexes. The western boundary of the foredeep is marked only in the Southern Urals by a chain of barrier reefs of Early Permian age (Asselian to Early Artinskian; Chuvashov & Nairn 1993). The terrigenous facies of the eastern provenance are spread far onto the platform, and the western boundary of the foredeep is usually expressed only by a more or less pronounced downward bend of the top surface of the carbonates underlying the molasse and in a corresponding eastward increase in the molasse thickness.

Facies changes of the foredeep sediments are complex, but the main steps in the foredeep development are the same at all latitudes. The foredeep began with the establishment of a deep-water basin on shelf sediments, west of the orogenic front. The basin was filled by molasse and flysch sediments grading westward into deep-water facies and still further west into reefs and biostromes (Fig. 7). Facies boundaries are diachronous (discussed later in this paper). The basin was filled with depositional evaporites of Kungurian (latest Early Permian) age. In the north the evaporites are replaced by terrigenous sediments with paralic coals. The Late Permian is represented in the foredeep by shallow marine, lacustrine and continental sediments, mostly terrigenous, red-coloured and variegated in the south, grey-coloured and coal-bearing in the north.

The eastern boundary of the foredeep was affected by westward thrusting. To the east the molasse and flysch are preserved in some deep synclines of the West Uralian fold zone. The outer (western) subzone of the foredeep is characterized mostly by smooth, open, non-linear folds typical for platform areas; the inner (eastern) subzone is characterised by thrusts and folds of a thin- to medium-skinned type (Puchkov 1975; Kazantsev 1984; Yudin 1994) (Fig. 4). Most of the structures of the inner foredeep are west-vergent. There are also some subordinate back-thrusts or underthrusts. In the Northern and Polar Urals seismic profiles suggest the presence of wedge-like (klinoform) structures composed mainly of Devonian, Carboniferous and Early Permian carbonates, thrust under the Permian molasse (Sobornov & Bushuev 1992; Sobornov & Tarasov 1992; Fig. 8).

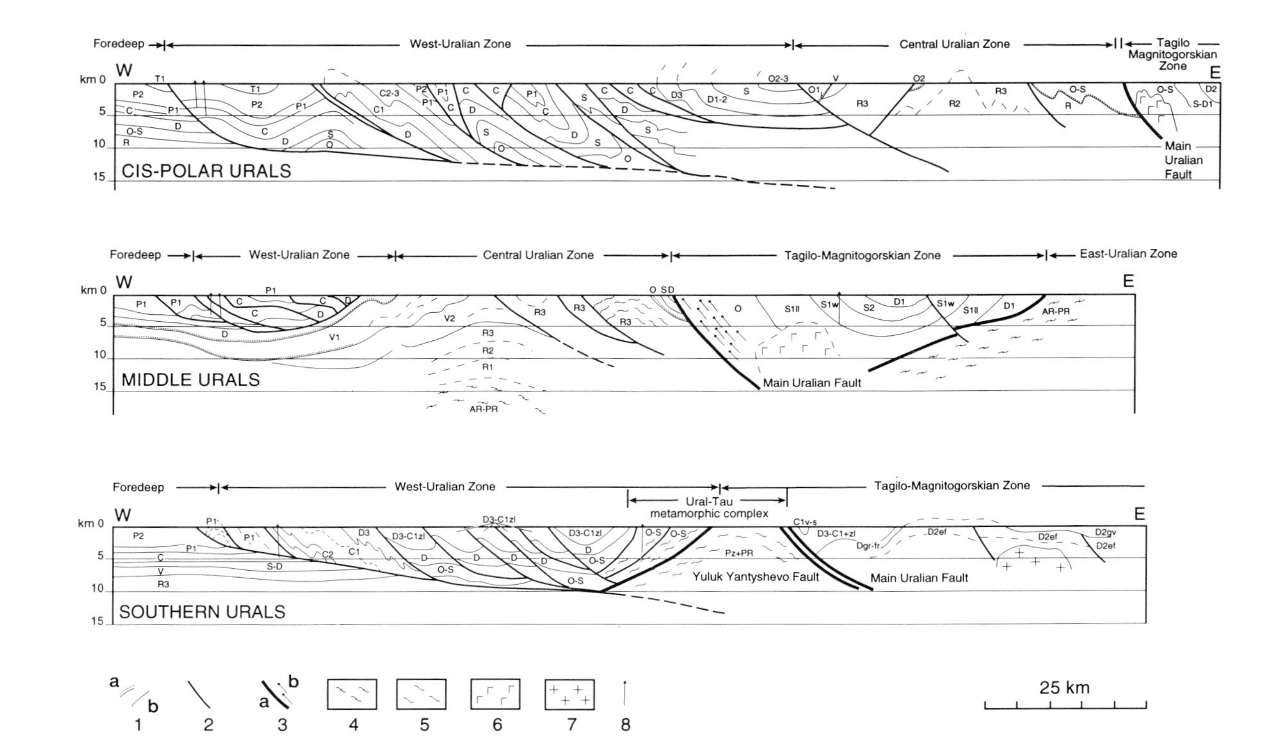

Fig. 4. Geological sections (located in Fig. 23) across the Urals, drawn after the Zilair–Kugarchi and Kushva–Serebryanka CDP seismic profiles, the SG4 near-vertical reflection profile, drilling data and geological observations (Puchkov 1975 and unpublished data; Yudin *et al.* 1983; Ablizin *et al.* 1982; Sokolov 1992; Kazantsev 1984; Seravkin *et al.* 1992). 1a, transgressive boundary; 1b, normal stratigraphic boundary, extrapolated where dashed; 2, faults, extrapolated where dashed; 3, major fault complex with a, serpentinitic mélange or b, blastomylonites; 4, metamorphic complexes derived from Archaean–Proterozoic rocks; 5, metamorphic complexes derived from Riphean and/or Palaeozoic rocks; 6, gabbro of the Platinum-bearing Belt; 7, granites, 8, important wells. Stratigraphic lettering: AR–PR, Archaean–Early Proterozoic; R, Riphean; V, Vendian; O, Ordovician; S, Silurian (Sll, Llandoverian, Sw, Wenlockian); D, Devonian (Def, Eifelian; Dzv, Givetian; Dfr, Frasnian; D3-C1zl; Zilair series (Frasnian–Tournaisian greywacke)), C, Carboniferous; P, Permian. Subscripts 1, 2 and 3 indicate Upper, Middle and Lower respectively.

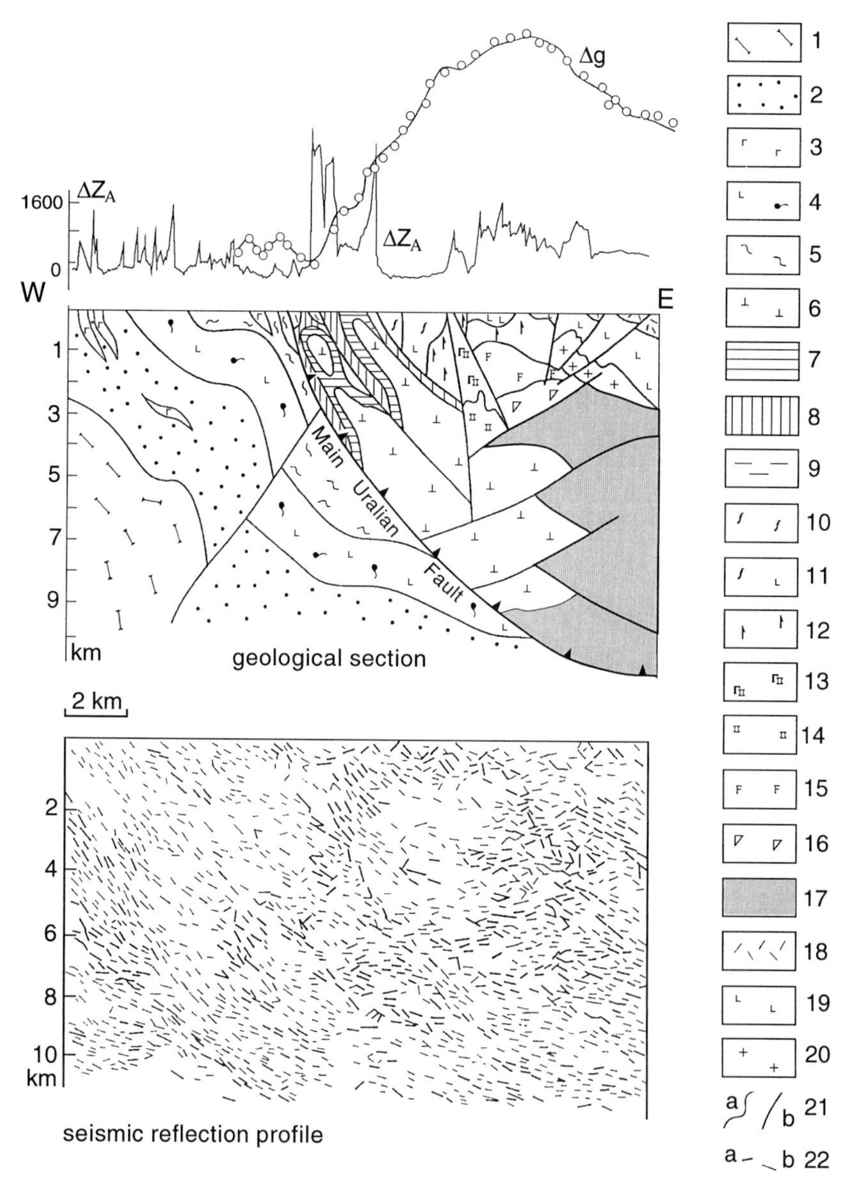

Fig. 5. Geophysical data and geological interpretation along a section across the Main Uralian Fault in the Northern Urals (located in Fig. 23, adapted from Petrov & Puchkov 1994). *1–5, Palaeocontinental sector*: 1, pre-Palaeozoic metamorphic complexes; 2, quartzites and shales; 3, gabbro–diabase; 4, metabasalts intercalated with silty shales; 5, metabasalts intercalated with silty shales. *6–10, Rocks of intensely deformed zone*: 6, dunites and harzburgites; 7, serpentinized peridotites; 8, serpentinites; 9, carbonaceous shales; 10, tuff shales. *11–20, Rocks of less intensely deformed zone*: 11, greenschist metabasalts; 12, sheeted dykes; 13, gabbro-norites; 14, pyroxenites, gabbro-amphibolites; 15, foliated amphibolites; 16, olivine gabbro; 17, mafic and ultramafic rocks undifferentiated; 18, Ordovician rhyolites; 19, Ordovician basalts; 20, plagiogranites. 21: a, Lithological boundaries; b, faults. 22: a, strong and b, weak seismic reflectors. ΔZ_A, intensity of the magnetic field; Δg, relative intensity of the gravity field (no scale given).

Most of the thrust-and-fold structures of the foredeep were formed during the Late Palaeozoic collision. In the southernmost and northern parts of the foredeep where Kungurian evaporites have a sufficient thickness, north–south trending salt ridges and oval domes are traced (Fig. 9).

Structures of the foredeep include not only

Period		Stage	thickness m		Lithology
Permian	Upper	Tatarian	1300		
		Kazanian	400-800		
		Ufimian	800-1000		
	Lower	Kungurian	500		
		Artinskian	1000		
		Sakmarian	70	150	
		Asselian	150		
Carboniferous	Upper	Gzhellian	180	150	
		Kasimovian	150		
	Middle	Moscowian	150		
		Bashkirian	100		
	Lower	Serpukhovian	270		
		Visean	700		
		Tournaisian	300	25	

Period		Stage	thickness m		Lithology
Devonian	Upper	Famennian	500	175	
		Frasnian	25		
	Middle	Givetian	100		
		Eifelian	130		
	Lower	Emsian	280		
		Pragian	220		
		Lohkovian	300		
Silurian	Upper	Pridolian	150		
		Ludlovian	170		
	Lower	Wenlockian	250		
		Llandoverian	650		
Ordovician	Upper	Ashgilian	440		
	Middle	Caradocian	345		
		Llandeilian	350		
		Llanvirnian	265		
	Lower	Arenigian	360		
		Tremadocian	1200		

1 2 3 4 5 6

Fig. 6. Summarized section of the Palaeozoic deposits of the Uralian foredeep and West Uralian zone in the Cis-Polar Urals (Schchugor and Kozhim rivers) compiled after the All-Russian Stratigraphic Committee (1993) and author's personal data. 1, dolomitized limestones; 2, shaly limestones; 3, sandy limestones; 4, dolomitic marls; 5, cherty limestones; 6, carbonate breccia. Other symbols as in Fig. 10.

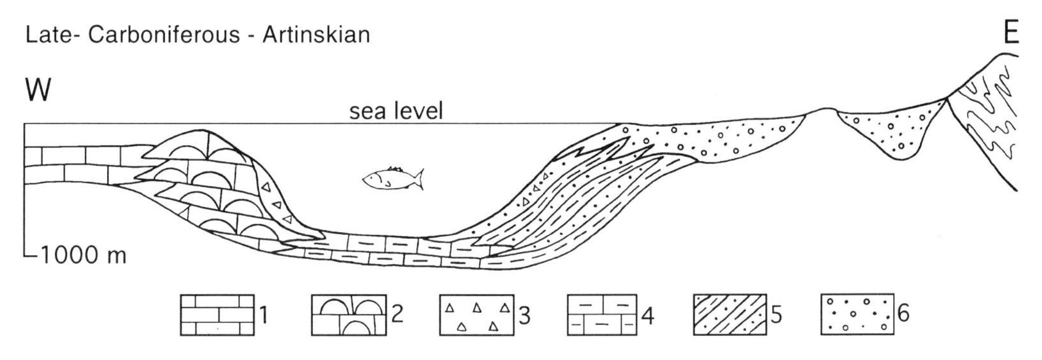

Fig. 7. Cartoon summarizing the principal lithofacies across the Late Carboniferous–Artinskian Uralian foredeep (after Nalivkin 1949 and Chuvashov *et al.* 1984). 1, shallow-water sediments, mostly bedded carbonates; 2, reefal limestones; 3, carbonate olistostrome; 4, basinal shales, cherts, marls; 5, turbidites; 6, molasse.

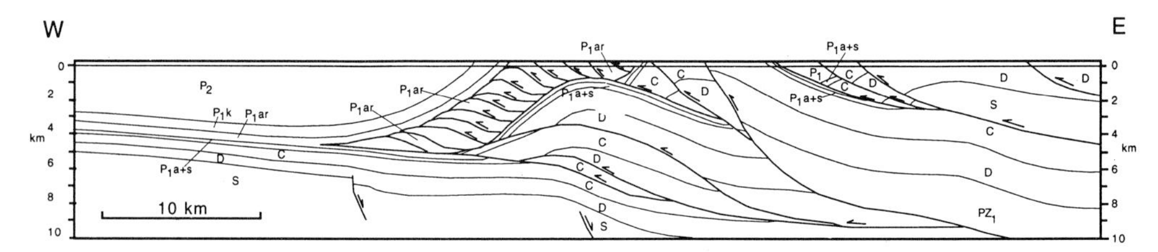

Fig. 8. Geological section across the boundary between the Uralian foredeep and West Uralian zone in the southern part of the Cis-Polar Urals, based on a CDP seismic profile (Sobornov & Bushuyev 1992). Location on Fig. 23, Stratigraphic lettering as on Fig. 4. Permian subunits: P_{1k}, Kungurian ; P_{1ar}, Artinskian; P_{1a+s}, Asselian and Sakmarian.

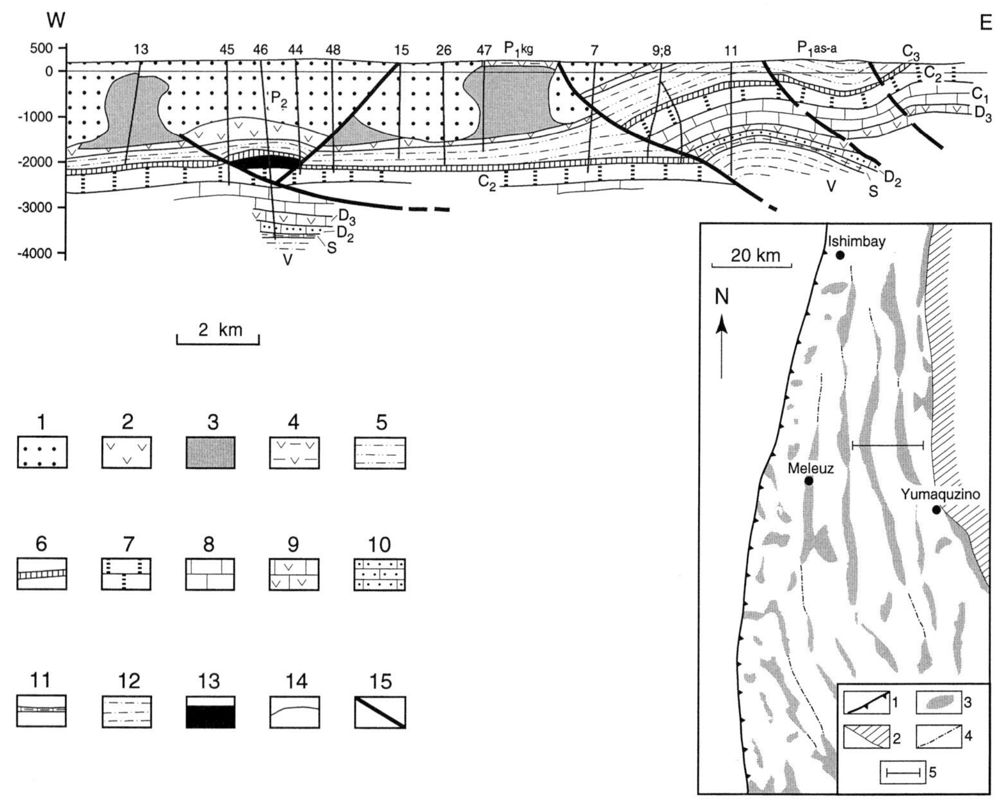

Fig. 9. Geological section across salt ridges (indicated on inset) (from Kamaletdinov *et al.* 1977, modified according to later seismic data). 1, Late Permian terrigenous sediments; 2–4, Kungurian stage (P_{kg}); 2, clays, anhydrites; 3, salt; 4, terrigenous sediments with gypsum; 5, Asselian, Sakmarian and Artinskian (P_{as-a}) terrigenous sediments; 6, Late Carboniferous limestones, marls and siltstones (C_3); 7, Middle Carboniferous limestones and dolomites (C_2); 8, Early Carboniferous limestones and dolomites with shale intercalations (C_1); 9, Late Devonian limestones (D_3); 10, Mid-Devonian limestones, sandstones and shales (D_2); 11, Silurian marls, dolomites and sandstones (S); 12, Vendian terrigenous sediments (V); 13, oil pool; 14, stratigraphic boundaries, 15, thrusts. Numbers near subvertical lines mean wells.
Inset map (located on Fig. 23) showing the salt ridges in the southernmost part of the Pre-Uralian foredeep. 1, western boundary of the foredeep; 2, area where Kungurian deposits are absent; 3, salt ridges; 4, chains of salt ridges; 5, line of section.

longitudinal, but also transverse (NW–SE) elements, partially inheriting their strikes from the structures of the crystalline basement. Among them (from south to north, Fig. 1):

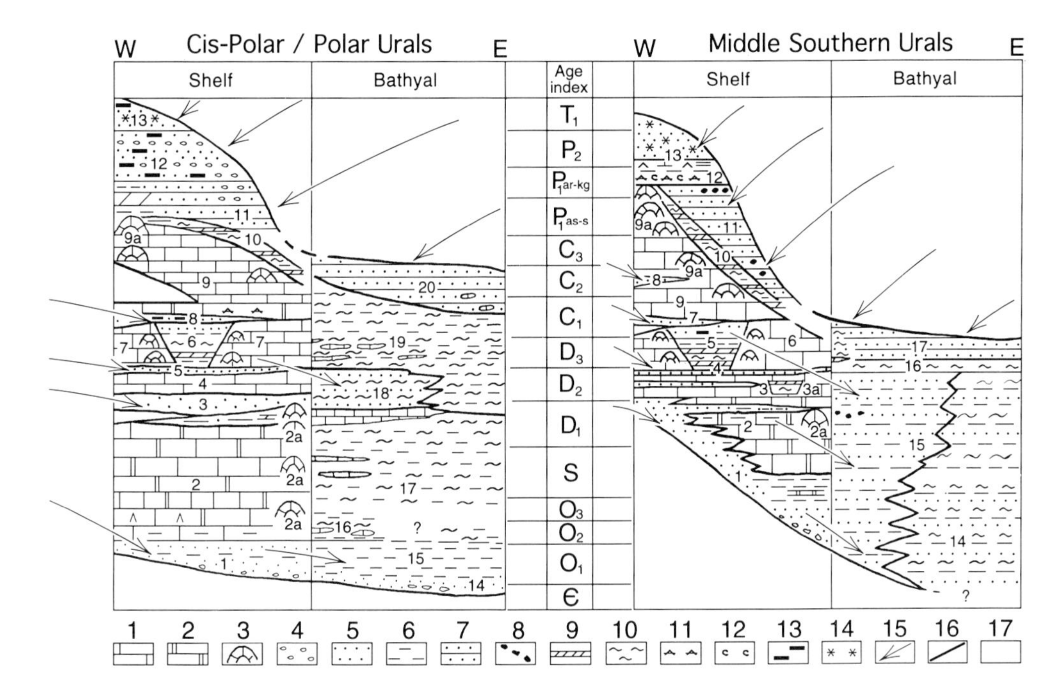

Fig. 10. Schematic correlation of Palaeozoic sedimentary formations belonging to the passive margin of the East-European continent. Symbols in boxes: 1, limestones; 2, dolomites; 3, big biostromes and reefs; 4, conglomerates; 5, sandstones and siltstones; 6, siltstones and shales; 7, flysch; 8, olistostromes; 9, marls; 10, cherts; 11, sulphates; 12, salt; 13, coal; 14, red beds; 15, direction of terrigenous influx; 16, boundaries of the most important formations; 17, sediments eroded or not deposited.
Numbers in the figure refer to the most important formations. *Cis-Polar/Polar Urals*. 1, Telpos terrigenous oligomictic formation, up to 1600 m thick. 2, Kozhimian shallow-water carbonate formation, up to 2300 m thick; 2a, barrier reefs, individual build-ups up to 500 m high. 3, Philippchuk–Takata terrigenous oligomictic formation, up to 400 m thick. 4, Carbonate formation, up to 200 m. 5, Pashiya terrigenous oligomictic formation, up to 26 m. 6, Formations of the Kama–Kinel trough system: Domanik basinal starved formation, up to 300 m thick, overlain by a Famennian–Tournaisian carbonate-terrigenous formation, up to 1000 m thick. 7, Carbonate formation with biostromes and reefs, up to 1300 m thick, aligning the Kama–Kinel trough system. 8, Uglenosnaya (coal-bearing) terrigenous oligomictic formation, up to 200 m thick. 9, Carbonate formation, up to 600 m thick; 9a, barrier reefs and thick biostromes, up to 200 m thick. 10, Basinal starved formation, up to 50 m thick. 11, Flysch of the Uralian foredeep, up to 1000 m thick. 12, Coal-bearing molasse, up to 3300 m thick. 13, Red-coloured and variegated grading upwards to a coal-bearing molasse, 2200 m thick. 14, Pogurey terrigenous oligomictic formation, up to 600 m thick. 15, Grube-Yu fine-grained terrigenous formation, up to 1000 m thick. 16, Kachamylk carbonate–terrigenous formation, up to 1300 m thick. 17, Kharota carbonaceous limestone–cherty–shaly formation, up to 300 m thick. 18, Paga cherty–terrigenous oligomictic formation, up to 600 m thick. 19, Kolokolnya limestone–cherty–shaly formation, up to 500 m thick. 20, Kechpel and Yayu flysch formations, up to 2000 m thick.
Middle Southern Urals. 1, Diachronous terrigenous oligomictic formation, up to 400 m thick. 2, Shallow-water carbonate formation, up to 570 m thick; 2a, Barrier reef, up to 1200 m thick. 3, Carbonate-terrigenous formation, up to 200 m thick; 3a, Infradomanik basinal formation, up to 80 m thick. 4, Pashiya terrigenous oligomictic formation, up to 10 m thick. 5, Kama–Kinel trough system: Domanik basinal formation, up to 350 m thick, overlain by Late Tournaisian–Early Visean terrigenous formation, up to 500 m thick. 6, carbonate formation, with barrier reefs aligning the Kama–Kinel trough system, up to 700 m thick. 7, Uglenosnaya (coal-bearing) terrigenous oligomictic formation, up to 60 m thick. 8, Carbonate–terrigenous oligomictic formation, up to 75 m thick. 9, Carbonate formation, up to 2500 m thick; 9a, barrier reefs at the western margin of the Uralian foredeep, up to 1000 m thick. 10, Basinal formation, up to 100 m thick. 11, Flysch and olistostrome formation, up to 2500 m thick. 12, Evaporite formation, up to 500 m thick. 13, Variegated molasse, up to 3000 m thick. 14, Uzyan cherty-terrigenous oligomictic formation, up to 1000 m thick. 15, Suvanyak terrigenous oligomictic group, up to 4000 m thick. 16, Ibragimovo cherty formation, up to 150 m thick. 17, Zilair flysch formation, up to 2000 m thick.

Kara–Tau, Kosva–Chusovaya, Poliud Range, Pechora–Kozhva, Chernyshov Range and Chernov Range uplifts. These structures differ in their morphology, origin and age. The Kara–Tau and the Poliud Range are combinations of a sinistral strike-slip fault and several folds and thrusts; the Kosva–Chusovaya is a gentle saddle-like uplift under the molasse. Pechora–Kozhva is a Devonian aulacogen inverted in the Late Palaeozoic. Chernyshov and Chernov are Mid-Jurassic, narrow, linear fold-and-thrust zones probably connected to detachments in the Early Palaeozoic strata deep under the basin. These transverse uplifts divide the foredeep into several semi-isolated basins: Belsk, Yuryuzan–Sylva, Solikamsk, Upper Pechora, Bolshaya Synya, Kosyu–Rogovaya and Korotaikha (Fig. 1).

The West Uralian zone (Zone 2)

The West Uralian zone comprises predominantly intensely folded and thrust Lower and Middle Palaeozoic sediments characterizing the former passive margin of the East European continent. There is no conspicuous facies change at the boundary between the Uralian foredeep and the West Uralian zone. Such a change, partially affected by later thrusts, occurs further to the east within the zone, as an abrupt transition from shelf to bathyal sediments (Figs 6 & 10). Sedimentary facies changes occur across the passive margin as well as along its 2000 km length.

Shelf sediments. Figure 10 shows synoptic stratigraphic sections of the shelf sediments in the western slopes of the Urals. The real sections are most complete in the northern parts of the West Uralian zone, where the thickness of the pre-flysch Palaeozoic sediments is up to 6–7 km. In the Middle and Southern Urals these sediments are much thinner (up to 2.5–3 km), and here, in the western sections of the zone, the Ordovician, Silurian and Lower Devonian sediments are absent, so the Middle Devonian strata unconformably overlie Riphean and Vendian sediments.

The complete sections of shelf sediments, most typical for the northern and easternmost parts of the zone (Figs 6 & 10) are composed of transgressive Ordovician quartzites and basal conglomerates grading upsection to shallow water dolomites and limestones of Silurian–Earliest Devonian age. Among the Upper Ordovician and Silurian limestones, a reefal facies formed barriers along the eastern margin of the shelf, mostly in the northern and central areas of the Urals. The Lower Devonian regressive succession of argillaceous limestones, dolomitic marls, siltstones and shales, are followed by Emsian transgressive quartzites, sandstones and siltstones grading upwards mostly into shallow-water open-sea limestones. The longest Lower Devonian barrier reef is traced along the margin of the shelf zone from the Polar to the Southern Urals. In the south Middle Devonian basinal shales, marls and cherts (so-called infradomanik) developed. In the Latest Mid-Devonian a new transgressive series was deposited with thin quartzites and shales in the bottom. Over the area of development of the Upper Givetian terrigenous facies and open-sea limestones, the so-called Kama–Kinel system of deep-water troughs was established. The troughs can be traced from the platform directly into the West Uralian zone as the best evidence of their former unity (Figs 10 & 11). The basinal, 'domanik' facies in the axial parts of the troughs is represented by a starved, condensed unit of marls, cherts and oil shales. The troughs are bordered by reef limestones. This type of sediment distribution persisted through the Famennian and Tournaisian, related to the high stand of sea level across the platform. The regressive Lower Viséan and transgressive Middle Viséan sediments are characterized by the wide development of terrigenous and carbonate-terrigenous facies, including quartzites, shales and siltstones with coal layers. The Kama–Kinel troughs were filled with Early Viséan sediments and ceased to exist. The Early Carboniferous, Late Viséan–Serpukhovian as well as the Mid-Carboniferous Bashkirian stage are represented mostly by pure shallow-water limestones. The next very important transgressive–regressive boundary of the stratigraphic sequences, marked by unconformity and the deposition of terrigenous sediments is within the Moscovian (Middle Carboniferous), but in the West Uralian zone this unconformity is not pronounced due to a general eastward inclination of the shelf. Therefore, the Middle–Upper Carboniferous is represented in the West Uralian zone predominantly by shallow-water limestones, restricted above, in many places, by a stratigraphic unconformity. The Middle Carboniferous of the southern part of the shelf area within the West Uralian zone displays the first signs of the westward terrigenous influx. From this time on, the area of sedimentation influenced by the eastern source of terrigenous material was widening to the west and north and the foredeep started to form above the shelf subzone (see the corresponding section of this paper). Two more major transgressive–regress-

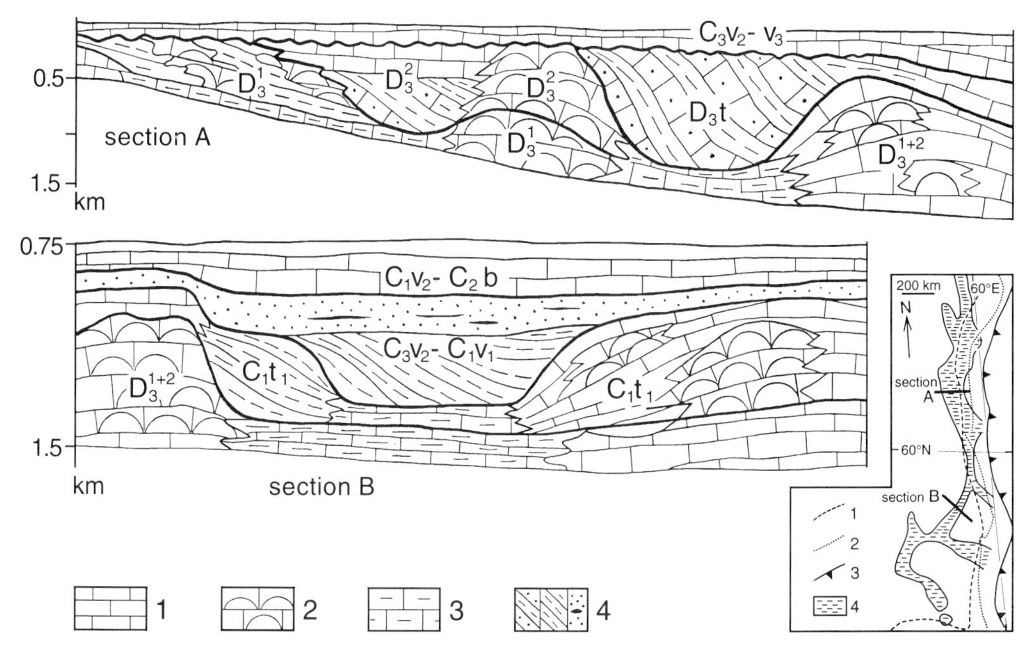

Fig. 11. Principal geological sections across the Kama-Kinel trough system (simplified after Parasyna *et al.* 1989). 1, Shallow-water shelf sediments (predominantly bedded carbonates); 2, reefal massifs and bioherms; 3, basinal (Domanik type) shales and marls marking the axis of the Kama–Kinel system of troughs; 4, terrigenous and carbonate-terrigenous sediments filling the troughs; left, siltstones, shales, oil shales, marls; centre, shales, siltstones; right, sandstones with coal seams. Sections located on inset map, where 1, Uralian foredeep boundary; 2, boundary of the West Uralian zone; 3, boundary of the Tagilo–Magnitogorskian zone; 4, Kama–Kinel troughs.

ive cycles of Early and Late Permian age developed as a background to the formation of the foredeep.

Bathyal sediments. The width of the bathyal zone varies from 10 to 60 km and is widest in the Southern and Polar Urals where the Zilair and Lemva synforms respectively (2 and 3, Fig. 1) occur. The bathyal complexes are best developed there, where the Precambrian rocks of the Central Uralian zone are almost absent. In contrast to the shelf facies, the areas of bathyal sediments are discontinuous, though wide enough for their main features to be studied (Fig. 1). Conodont studies have allowed considerable progress on their stratigraphy, leading to correlation of sections and determination of facies patterns (Puchkov 1979).

The Palaeozoic section of the bathyal complexes (Fig. 10), best studied in the Polar Urals, starts with Uppermost Cambrian sandstones. The Lower Ordovician is also terrigenous, green and red coloured, up to 2–2.5 km thick, siltstones being predominant. The Middle Ordovician in the westernmost bathyal sections, transitional to the shelf facies, comprises shales, limestones and cherts. The Middle Ordovician of the bathyal facies proper is represented by thin cherts in the Southernmost and Polar Urals. Upper Ordovician strata are not known. The Silurian and Early Devonian (Lohkovian) are represented by a very typical condensed unit of carbonaceous cherty shales and limestones, up to 250 m thick, covered by a marker horizon of tentaculite knotty limestones, several metres thick. Only in some areas of the Southern Urals are these cherts and limestones probably replaced by a more than 1000 m thick terrigenous series. The Middle Devonian is usually represented by up to 500 m of quartzites, siltstones, shales and cherts. When the quartzites are absent the section is very condensed. The Upper Devonian is represented by cherts, locally intercalated with 100–300 m thick limestones. In the Southern Urals the cherts are overlain by up to 3000 m of greywacke flysch of Latest Frasnian–Tournaisian age of eastern provenance, polymictic, with fragments of cherts, volcanites and granites, with grains of chromspinel. In the Northern territories the cherty section are also in the Carboniferous, and the first greywackes do not appear earlier than the Late Viséan. The deep-water flysch trough existed here at least until the end of the Late Carboniferous (some additional details will be given later in the paper).

Thrusting is very characteristic for many areas in the West Uralian zone, with detachments in the sedimentary cover and along the cover/basement boundary. The detachment levels may be different depending on the lithologies in the different parts of the zone. In the southernmost and middle Urals the detachments under the Uralian foredeep and the West Uralian zone are situated mostly at the base and top of the Vendian, at depths of 1.5–4 km, and probably down to 10 km in the eastern parts of the zone. In the northern parts of the Urals, where the thickness of the Palaeozoic is greater, the main detachment may be situated at depths of 10 km and more, close to the cover–basement boundary and in Ordovician shales (medium- to thick-skinned tectonic style). Clays and evaporites of some horizons of the Devonian, Carboniferous and probably Permian acted as weaker horizons hosting detachments (Figs 4 & 6; Puchkov 1975; Kazantsev 1984; Yudin 1983, 1994; Dembovsky *et al.* 1990; Sobornov & Bushuev 1992).

As in the foredeep, the general vergence is to the West and there are wedge-like structures which may be regarded as underthrusts or backthrusts. Usually backthrusts are subordinate, with the exception of the westward-dipping Yuluk-Yantyshevo fault (Fig. 4c), the western boundary of the Ural–Tau complex in the Southern Urals.

Ophiolites and island-arc sediments (the Sakmara, Kraka and Nyazepetrovsk-Bardym allochthons of the West Uralian zone: 14, 15 and 16 in Fig. 1) have been thrust westward onto this zone. The oceanic thrust sheets are usually associated with serpentinite mélanges and bathyal complexes of the passive continental margin. The bathyal complexes also form some independent allochthons, with a minimum displacement of 20 km (Puchkov 1979). The allochthons have been folded after their emplacement into 5–10 km wide synforms and antiforms. The best explored is the bottom of the Nyazepetrovsk–Bardym nappe which is penetrated by 5 boreholes to depths of between 500 and 1000 m (Puchkov & Ivanov 1982).

The Central Uralian zone (Zone 3)

The Central Uralian zone (Zone 3), up to 70–75 km wide, is characterized by well exposed Precambrian sedimentary, metamorphic and magmatic rocks which are, in some places, thrust over the rocks of the West Uralian zone (Fig. 4a).

The rocks of the Central Uralian zone are mostly Late Proterozoic in age (i.e. Riphean and Vendian, 1650–570 Ma, Keller & Chumakov 1983; Krasnobayev 1985). The recognition of Early Proterozoic rocks is problematic (they are probably developed in the cores of some thermal domes). Archaean rocks are found only in one place: the Taratash complex which is thought to be a fragment of the basement of the East European platform (TR, Fig. 1, Lennykh & Krasnobayev 1978), represented by polymetamorphic rocks which probably belonged to a greenstone belt (Lennykh 1986). The rocks experienced a granulitic metamorphism by the end of the Archaean, when enderbites, two-pyroxene crystalline schists, gabbro–diorite-gneisses and magnetite quartzites were formed. Later they were subjected to retrograde metamorphism varying from amphibolite to greenschist facies (Lennykh *et al.* 1978). Late Proterozoic complexes are represented in most of the sections by up to 13 km of shallow-water terrigenous sediments with quartz and subarkose sandstones, siltstones, shales and carbonate biostrome horizons. Volcanites are a typical, though far from ubiquitous component of the sections, represented predominantly by sub-alkaline basalts and basalt–rhyolite complexes of ensialic geochemical character.

The Riphean sections of the southern and northern parts of the Urals differ first and foremost in the age and amount of volcanics in the series. In the Southern Urals the volcanites (sub-alkaline basalts and basalt–rhyolite complexes) are not very widely developed, being present at the base of the Early Riphean, in the lower part of the Middle Riphean and also in the Early Vendian (Parnachev 1981; Parnachev *et al.* 1981). In the northern parts of the Urals the maximum volcanic activity took place in the Late Riphean, leading to the formation of a widely developed basalt–rhyolite series and associated granite and gabbro intrusions (Goldin *et al.* 1973). Calc-alkaline volcanites are reported from the Late Riphean of the Polar Urals (Rumyantseva 1984). The Vendian is represented both in the southern and northern sections by a molasse-like series of polymictic conglomerates, sandstones and siltstones (Bekker 1968; Puchkov 1975).

The Late Proterozoic series are affected by a Barrovian-type metamorphism, localized in dome-like structures and dated as Late Vendian

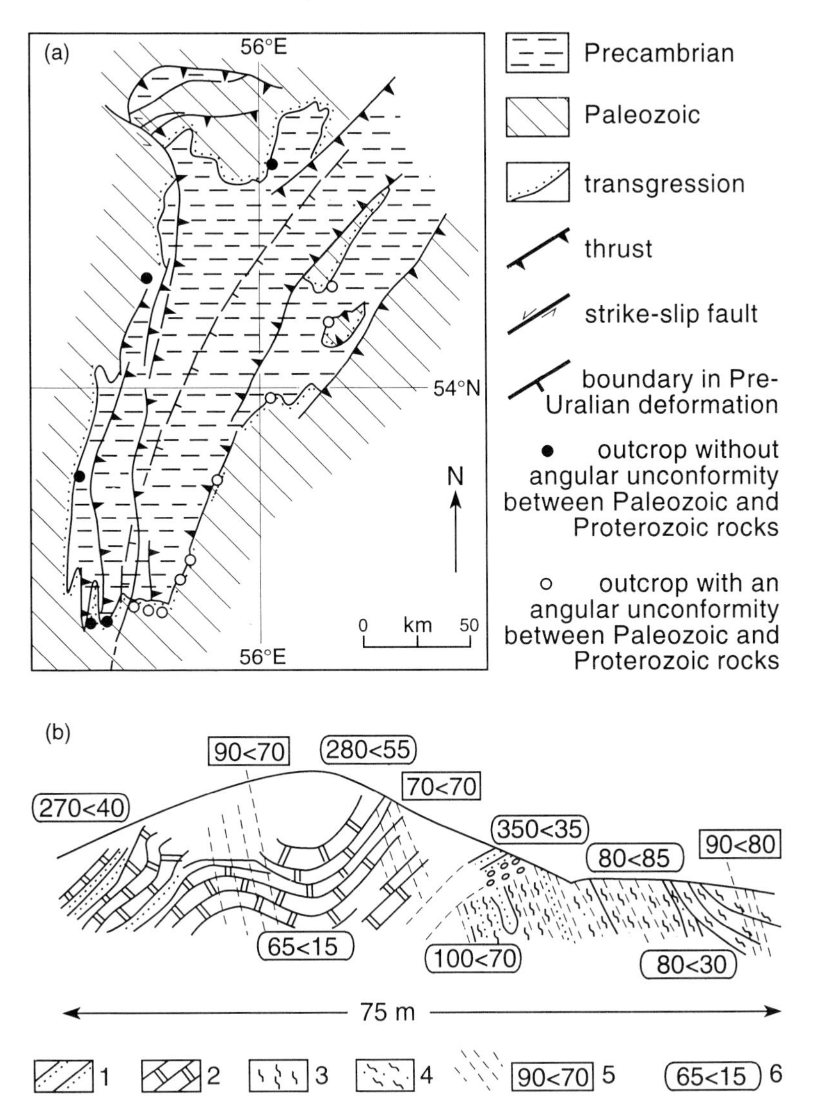

Fig. 12. Relationships between the Precambrian and Paleozoic complexes on the limbs of the Bashkirian anticlinorium (Fig.1). (**a**) Relationships in mapview. (**b**) An example of an outcrop with an angular unconformity between the Vendian and Ordovician. 1 & 2, Ordovician: 1, quartz sandstones; 2, dolomites. 3 & 4, Early Vendian: 3, phyllite; 4, schistose siltstones and quartzites. 5 & 6, Strike and dip of: 5, cleavage; 6, layering. Location of (**a**) is shown in Fig. 23.

(Puchkov 1995). With increasing distance from these complexes to the west, the metamorphic grade of the sediments decreases to anchizone (deep catagenesis in Russian terminology; Anfimov 1986). Therefore, in the Central Uralian zone, relicts of the earlier, Vendian orogen are present. The restored general features of this orogen which differs strongly in structure and strike from the Uralian one, will be discussed later.

The Palaeozoic sediments of the Urals usually cover the Late Proterozoic with angular unconformity. So the structures in the Late Proterozoic are here the result of two orogenic deformations. But in some places, such as the western parts of the Bashkirian and Kvarkush anticlinoria (Fig. 1), the unconformity disappears, as well as the metamorphism, and the linear thrust-and-fold structures affecting the Late Proterozoic sediments are purely Palaeozoic in age (Uralian; Figs 12 & 13). The latter do not necessarily belong to the thin-skinned

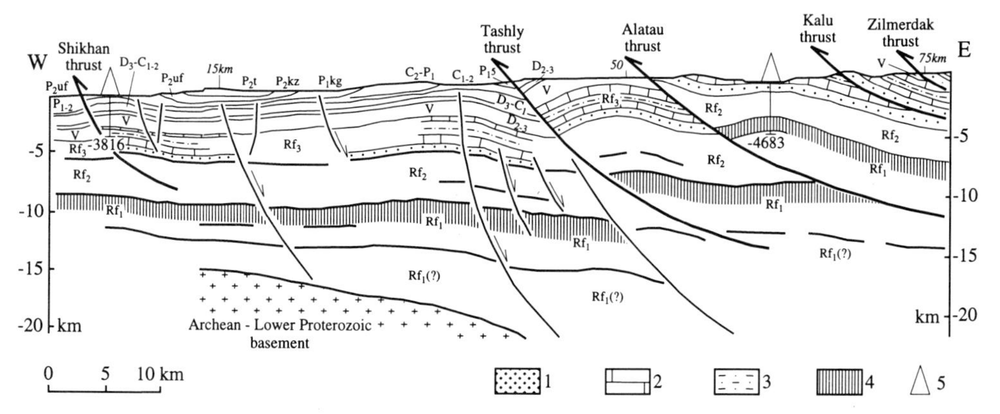

Fig. 13. Geological section across the western slope of the Southern Urals based on a CDP seismic profile (Skripiy & Yunusov 1989). Located on Fig. 23. 1, sandstones of the Upper Riphean Zilmerdak suite; 2, limestones and dolomites of the Upper Riphean Katav and Minyar suites; 3, terrigenous rocks of the Upper Riphean Inzer suite; 4, marker carbonate series of the Lower Riphean Bakal suite; 5, deep wells. Rf_{1-3}: terrigenous–carbonate complexes of the Lower-Upper Riphean, generating strong reflections; V, terrigenous Vendian complex; D_{2-3}, terrigenous-carbonate deposits of the Mid-Late Devonian age ('D' reflector); D_3–C_1 and C_{1-2}, carbonate complexes of the Late Devonian–Early Carboniferous and Early-Mid-Carboniferous, respectively, separated by 'U' reflector; C_2–P1, Mid-Carboniferous–Early Permian carbonate complex, overlying the 'V' reflector. $P_1$5, P_1kg, P_2uf, P_2kz P_2tat: stages of the Permian (respectively, Sakmarian, Kungurian, Ufimian, Kazanian and Tatarian).

deformation predominant in the Palaeozoic cover. In these areas the detachment surfaces could be situated at depths of 10–15 km, probably because of the immense thickness of the Late Proterozoic sediments (Figs 4b & 13).

Structurally, the Central Uralian zone is often described as a chain of anticlinoria (from north to south: Kharbey, Lyapin, Kvarkush and Bashkirian, Fig. 1). Another anticlinorium, the Ural–Tau, is often listed among them (UTT, Fig. 1). But the structures exposed in the Ural–Tau Range present a special problem. Metamorphic terranes of this range, belonging to two complexes, Suvanyak and Maksyutovo, are traditionally attributed to Late Proterozoic. But new data, especially new finds of fauna, make this interpretation very doubtful. The Suvanyak complex, 250 km long and 10–20 km wide, is represented by thick quartzites, siltstones and shales metamorphosed in greenschist facies. It conformably underlies Frasnian cherts of the Ibragimovo suite and contains rare graptolites, brachiopods, conodonts, acritarchs and some other fauna in the range of Ordovician–Mid-Devonian. Zakharov & Puchkov (1994) interpret the Suvanyak complex as a series of Palaeozoic bathyal sediments of the passive continental margin. The Maksyutovo complex, 200 km long and 15 km wide to the east of the Suvanyak complex, is a famous site of the development of HP–LT (eclogite–glaucophane schist) metamorphism (Dobretsov 1974; Valizer & Lennykh 1988; Puchkov 1995). For a long time, the complex was thought to be composed of Riphean sedimentary and magmatic rocks of very contrasting primary ophiolitic and arkose types. The metamorphism was also thought to have occurred in the Riphean. Several recent finds of conodonts in marbles among the ophiolite-type succession (up to Late Silurian–Early Devonian in age according to Zakharov & Mavrinskaya 1994) as well as K–Ar ages for phengite (about 400 Ma, Valizer & Lennykh 1988) and Ar–Ar ages for eclogites (about 370 Ma, Matte *et al.* 1993), make a revised interpretation of this complex possible (Zakharov & Puchkov 1994). These authors proposed that the Maksyutovo metamorphic complex represents subducted and then exhumed accretionary fragments of a Devonian island arc and contains both Palaeozoc and probably Precambrian rocks of different types. During the Late Palaeozoic collision, the Maksyutovo block was folded into an antiform and thrust under the metasedimentary Suvanyak complex of a passive continental margin along the west-dipping Yantyshevo–Yuluk Fault, marked by a serpentinitic mélange (Zakharov & Puchkov 1994; Fig. 4c). In some other places, complexes resembling the Maksyutovo are developed in the footwall of the Main Uralian Fault, mostly in the north of the Urals (Nerka–

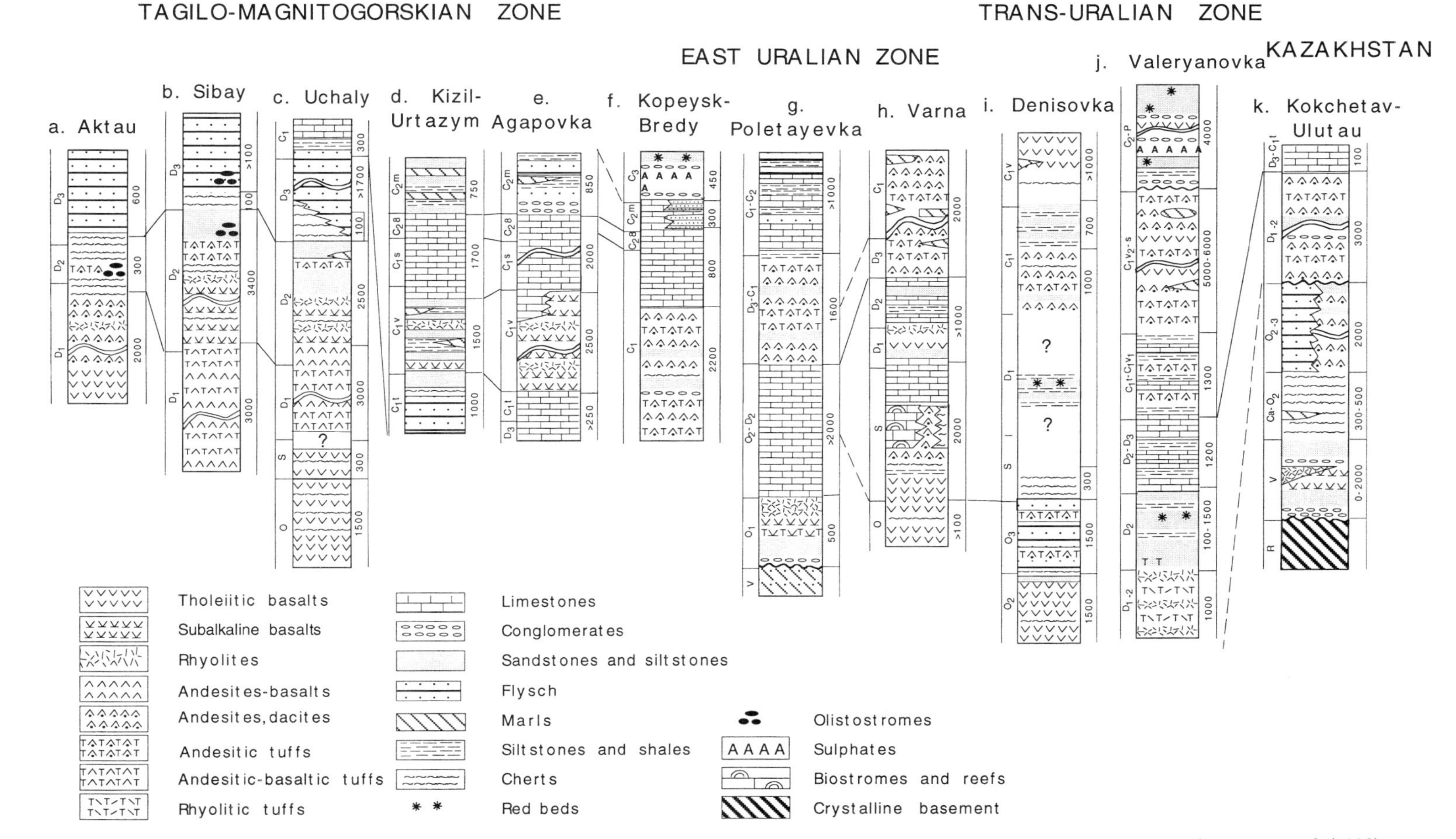

Fig. 14. Schematic correlation of Palaeozoic stratigraphic sections east of the Main Uralian Fault. Compiled after Stratigraficheskiye (1993), Maslov *et al.* (1993) Snachev *et al.* (1994), Puchkov & Ivanov (1985), Chuvashov *et al.* (1984) and Abdulin (1984). The positions of the sections are shown in Fig. 20. Stratigraphic lettering as in Figs 4, 9 and 13.

Yu and Maruun–Keu areas), but more stratigraphic research and isotope dating is needed to make a certain conclusion on their age and nature.

The Tagilo–Magnitogorskian, East Uralian and Transuralian zones belonged to the active margin of the Kazakhstanian (Kazakhstano–Kirgizian) continent (Yazeva *et al.* 1989; Puchkov 1991). They are less uniform compared with the first three zones. What unites them, is a wide development of magmatic complexes, indicators of subduction (Fig. 14). Their geochemical trends show that the subduction zone was dipping beneath the accreting Kazakhstanian continent, to the east in modern coordinates (Yazeva *et al.* 1989; Seravkin *et al.* 1992).

The Main Uralian Fault

The Central Uralian and Tagilo–Magnitogorskian zones are thought to be divided by an east-dipping major suture zone called the Main Uralian Fault (Fig. 4). A considerable part of it is marked by serpentinitic mélanges whose matrix records a combination of intense brittle and low-temperature (greenschist) ductile deformation. In some places the mélange is replaced by zones of blastomylonites of different metamorphic grade. Blocks in the mélanges represent mostly rocks of the hanging wall of the Main Uralian Fault. The most common are ultramafites, gabbros, amphibolites, basalts, cherts and cherty shales of Ordovician, Silurian and Devonian age, Silurian and Devonian limestones, Devonian andesites and dacites, Upper Devonian–Lower Tournaisian greywackes and granitoids of different types. Very typical are low-temperature metasomatic rocks of rodingite type.

The Tagilo–Magnitogorskian zone (Zone 4)

This zone is composed of Palaeozoic complexes of oceanic basins, island arcs, Andean-type belts, flysch troughs covered by shallow-water sediments: limestones and coal-bearing terrigenous sequences (Peyve *et al.* 1977). The ophiolites are widespread (Savelyeva 1987). The volcanic calc-alkaline complexes characteristic of subduction (tuffs, lava basalt and andesite series, as well as their intrusive equivalents) are also well represented (Yazeva *et al.* 1989; Seravkin *et al.* 1992).

The stratigraphy and history of some parts of this zone differ considerably, so that four main subzones, Magnitogorsk, Tagil, Voykar and Schchuchya are established successively (13, 6, 5 and 4 in Fig. 1). The main differences between them are not in the character and composition of the main formations, but mostly in their age. For example, in the Tagil subzone, the ophiolites are Ordovician and the calc-alkaline formations are mostly Silurian. The Devonian is represented by flysch, trachyandesites and shoshonite-type magmatic rocks and the widely developed sedimentary cover of a mature island arc, represented by layered and reef limestones with bauxite deposits. The specific feature of the Tagil subzone is an exemplary development of mafic–ultramafic massifs forming the so-called platinum-bearing belt (Yefimov *et al.* 1993). As for the Magnitogorsk subzone (Fig. 14), the ophiolites range in age from Ordovician to Emsian, the calc-alkaline basalts appear in the Ordovician, but intense calc-alkaline magmatism combined with abundant volcano-terrigenous sedimentation ranges in age from Emsian to Famennian. Flysch sedimentation took place in Famennian–Tournaisian time. Shallow-water limestones and sub-alkaline volcanites of a mature, and probably rifted arc developed in the Viséan-Serpukhovian (Perfilyev 1979; Seravkin *et al.* 1992; Salikhov *et al.* 1993). Massifs of platinum-bearing association are not typical for this subzone.

No Precambrian granitoid and metamorphic rocks outcrop in the Tagilo–Magnitogorskian zone. Their presence at depth is probable as a consequence of a major westward displacement along the Main Uralian Fault, as follows from the interpretations of the geological and geophysical data (Sokolov 1992; Petrov & Puchkov 1994; Figs 4 & 5). The crust of Tagilo–Magnitogorskian zone as a whole is very dense, composed mostly of ultramafic, mafic and intermediate rocks. Derived from the high gravity and magnetic anomalies over it as well as interpretations of many DSS profiles crossing the Urals at different latitudes (Necheukhin *et al.* 1986; Avtoneev *et al.* 1991), crustal thickness is believed to be 45–70 km (Figs 2 & 3). The zone is limited to the east by serpentinite-bearing faults of westward (mostly 20–40°) inclination as is well documented by seismic profiling (Menshikov *et al.* 1983; Sokolov 1992; Fig. 4b). They may be connected with eastward directed backthrusts over the East Uralian zone. The central parts of the Tagilo–Magnitogorskian zone are moderately deformed compared to what can be expected in the centre of a fully fledged foldbelt. There are even very weakly deformed blocks (e.g. area of Podolsk copper-pyritic deposit, Fig. 4c, easternmost part of the profile), the intense dislocations being concentrated in serpentinite mélanges, like those of the Main Uralian Fault.

These mélange zones are typically tens to hundreds of metres thick (up to several km wide), and are in fact megabreccias containing blocks of dismembered ophiolites, calc-alkaline volcanites, granites, cherts and limestones. Their serpentinite matrix is cut by numerous slickensided surfaces. Taking into account the opposite dip of the mélanges across the axis of the Tagilo–Magnitogorskian zone (Fig. 4b; Sokolov 1992) and an evidently considerable tectonic transport along these mélanges, it has been proposed that the Tagilo–Magnitogorskian zone was completely allochthonous in character (Kazantsev 1991). The following arguments seem to make this idea invalid.

(1) It has been shown that on the western slopes of the Urals, in spite of folding and thrusting, the primary facies of the passive continental margin with its shelf and bathyal zones are easily restored (Puchkov 1979). It has been shown that the outer border of the shelf zone was rimmed by a stable band of barrier reefs of Late Ordovician, Mid-Silurian and Early–Mid-Devonian age (Antoshkina & Eliseev 1988; Chuvashov & Shuysky 1990). These observations indicate that the edge of the continent was located in this area and only a small part of it (no more than a few tens of kilometres wide) could be thrust below the Main Uralian Fault. Therefore, the platform cannot continue at depth east of the middle of the Tagilo–Magnitogorskian zone or as a part of the East Uralian zone.

(2) Studies undertaken by Khatyanov (1963), Gafarov (1970) and others show that magnetic lineations reflecting the structures of the Precambrian folded basement of the platform can be traced from the western slope of the Urals up to the Main Uralian Fault. The prolongation of these anomalies is not identified in the east as would be the case if the Tagilo–Magnitogorskian zone were a synformal klippe thrust from somewhere in the east.

(3) All deep seismic sounding profiles as well as the presence of a high gravity anomaly over the Tagilo–Magnitogorskian zone suggest that substantial deep structural and compositional crustal changes occur to the west of the Main Uralian Fault (Necheukhin *et al.* 1986, Figs 2 & 3).

(4) Palaeomagnetic data (Svyazhina *et al.* 1992) give evidence of a significant horizontal movement (about 2000 km) of the East Mugodzhary block relative to the East Europe in Ordovician–Carboniferous time. In addition, CDP profiling and deep drilling in the Central Uralian zone of the Southern Urals have produced no evidence of the autochthonous Paleozoic platform cover at any depth (Skripiy & Yunusov 1989; Fig. 13).

The idea of a completely allochthonous, exotic character for the Tagilo–Magnitogorskian zone can be rejected due to the presence of a high-amplitude thrust on its western limb and a backthrust along its eastern boundary (Puchkov 1991, 1993; Matte 1995).

The East Uralian zone (Zone 5)

This zone is distinguished by the presence of sialic, microcontinental complexes, fragments of Precambrian continental crust (including the East Mugodzhary, Murzinka–Aduy, Salda, Selyankino blocks; Krasnobayev 1985; Krasnobayev *et al.* 1995; successively 12, 7, 8, 3, SK in Fig. 1). They have distinctive Palaeozoic magmatism and metallogeny (e.g. carbonatites connected with the Selyankino block, Levin 1984) and their own Palaeozoic sedimentary cover, although very poorly preserved and difficult to identify. The sediments of the cover are observed in graben-like depressions of the southern part of the East Mugodzhary block (Southernmost Urals), and also in the Poletayevka and Rezh areas (Middle Urals) and are represented mostly by carbonate, terrigenous and cherty sediments (Puchkov 1993). Volcanic formations are also developed.

In the Poletayevka area (Fig. 14g) Precambrian metasediments are overlain by a Lower–Middle Ordovician, 1500 m thick, rhyolite–basalt series. This in turn is overlain by Middle Ordovician–Middle Devonian carbonate deposits, more than 2000 m thick. The Late Devonian–Early Carboniferous is represented by a volcano-sedimentary series that contains andesites and tuffs, up to 1600 m thick. The uppermost part of the section is represented by a Lower–Middle Carboniferous terrigenous-carbonate series (Snachev *et al.* 1994). We interpret this section as a record of a microcontinental block that underwent rifting in the Early–Mid-Ordovician, tectonic quiescence in the Mid-Ordovician–Mid-Devonian and was influenced by a subduction zone in the Late Devonian–Early Carboniferous.

Palaeozoic ophiolitic and island-arc complexes are present in the East Uralian zone as serpentinitic mélanges, tectonic klippen and thrust sheets (the largest of them is probably the Denisovka subzone, though its interpretation as a suture cannot be excluded; Puchkov & Ivanov 1985). The East Uralian volcanogenic belt (Koroteev *et al.* 1979; Seravkin *et al.* 1992) is probably allochthonous: on geological maps its southern end looks like a periclinal synform.

Another specific feature of the East Uralian zone is the abundance of granites that align submeridionally to form the 'Main Granitic Axis of the Urals' (Fig. 1). Two main types of granitoids are recognized (Puchkov *et al.* 1986; Fershtater *et al.* 1994): Upper Devonian–Lower Carboniferous tonalite–granodiorites with a calc-alkaline geochemical affinity and Upper Carboniferous–Permian potassium–sodium granites thought to result from partial melting of the collided continental blocks. The magmatism of the Main Granitic Axis of the Urals was also accompanied by Late Palaeozoic zonal metamorphism, due to a thermal influx and deep burial of the East Uralian zone during the late stages of orogenesis (Puchkov 1996).

The Transuralian zone (Zone 6).

This, the easternmost, most poorly exposed and least studied zone, has a rather controversial eastern boundary separating the Uralides from the Kazakstanides (Caledonides). Only Carboniferous and Devonian rocks of ensialic nature are known. The most important are calc-alkaline volcano-plutonic complexes (Fig. 14j). The Devonian sequence consists of: (1) Lower–Middle Devonian volcanogenic units of tuffs and lavas of liparite and dacite composition, up to 1000 m thick; (2) volcano-sedimentary units (Middle Devonian?), represented by continental red-coloured tuffites, polymictic and volcanomictic siltstones and sandstones, 700–1300 m thick; (3) marine shallow-water carbonate-terrigenous series of Givetian–Famennian age, 1200 m thick. The Carboniferous rocks belong to a Valeryanovka marginal volcano-plutonic belt and are comparatively well studied. Here they consist of three series (Chuvashov *et al.* 1984). (1) A Tournaisian–Early Viséan 1200 m unit of shales, siltstones, limestones with subordinate layers of andesites, andesite–basalt porphyrites and tuffs. (2) A mid-Viséan–Serpukhovian 5000–6000 m thick series of basic and intermediate porphyrites, tuffs and lava breccias, with subordinate limestones. The first two series are mostly open-sea shallow-marine in character. (3) Unconformably overlying these is a red-coloured terrigenous continental series composed of conglomerates, sandstones and siltstones with subordinate porphyrites and anhydrites, dated by spores and pollen complexes as Mid-Carboniferous–Permian (Abdulin 1984) and having a thickness of up to 4 km. The volcanites are penetrated by comagmatic intrusions of gabbro–diorites and diorites.

The pre-Mid-Carboniferous rocks are deformed into a series of narrow linear folds with limb dips between 25 and 60°. The overlying molasse-like Upper Palaeozoic redbeds form more gentle synclines and graben-like structures. The zone is limited to the east by the Urkash fault marked by serpentinites (serpentinitic mélange?) which may be a suture zone between the Uralides and the Kazakhstanides. There is a deficit of hard data to establish the nature of this boundary precisely. But it is worth mentioning that one of the main differences between the eastern zones of Uralides and the western zones of Kazakhstanides lies in the different age of the main rifting episode preceding the formation of oceanic and marginal complexes (Ordovician in the Urals, Vendian in the Kazakhstanides; Fig. 14).

East of the suggested suture zone, a Devonian volcano-plutonic belt occurs. The thick calc-alkaline series is superimposed on terrigenous-cherty, greywacke and andesite rocks of the Vendian–Late Ordovician complexes typical for the Kazakhstanian 'Caledonides'. These volcano-plutonic belts are probably fragments of a longer and wider zone, connected with the Chatkal–Naryn belt of the South Tien-Shan.

The present-day crust of the East Uralian and Transuralian zones is 38–42 km thick and composed of rocks lighter than the Tagilo–Magnitogorskian zone (Necheukhin *et al.* 1986; Avtoneev *et al.* 1991; Figs 2 & 3). The crust of the Urals as a whole records thrusting and stacking of differently composed tectonic units along faults whose inclination changes with depth, creating several levels of listric faults. This is documented by seismic reflection and refraction data and complicates the relationships between deep and shallow structures of the Uralian megazones which involve structural detachments between various deep horizons, density inversions in the crust and close horizontal connections (even interfingering) between adjacent megazones, however contrasting their character at the surface (Puchkov & Svetlakova 1993).

Relicts of the Pre-Uralian orogen

In the Central Uralian zone Upper Proterozoic (Riphean and Vendian) deformed and metamorphosed rocks are exposed. They belong to the 'Douralides' tectonic complex of Kheraskov (1967). The English equivalent of the term can be Pre-Uralides because they form the basement of the Palaeozoic complex of the 'Uralides'. They have also been called Timanides, because the basement of the Timan Range was an important part of this orogen. Crystalline com-

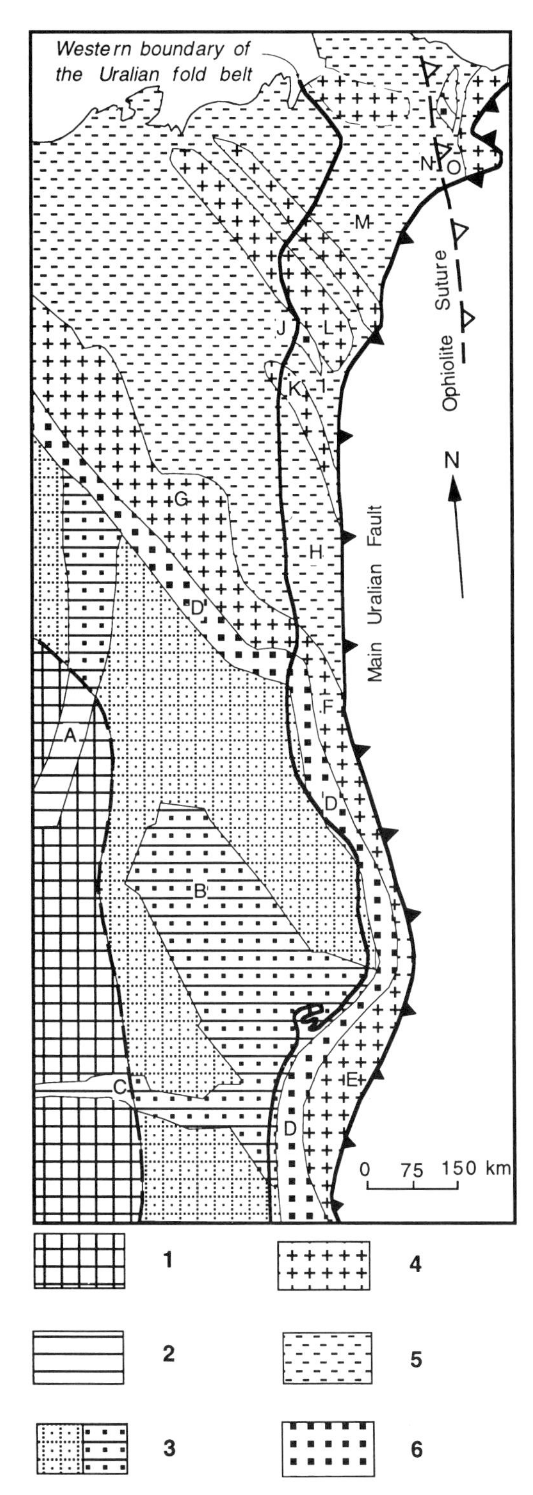

Fig. 15. Schematic tectonic zonation of the Pre-Late Proterozoic basement structures of the East European platform traced to the western slope of the Urals 1–3: Pre-Riphean craton; 1, Uplifted Archaean–Early Proterozoic blocks; 2, Riphean aulacogens; 3, first two structural zones overlain by a Vendian platform cover. 4–7: Pre-Uralides; 4, anticlinoria; 5, synclinoria; 6, foredeep and intermontane depressions. The letters in the figure correspond to those discussed in the text. The position is shown on Fig. 23.

plexes of Archaean–Early Proterozoic age are exposed as small isolated regions among the Pre-Uralides (Taratash complex of the Middle Urals and some other metamorphic terranes of granulite and amphibolite facies), separated from the Late Proterozoic (Riphean) complexes by faults.

General features of the Pre-Riphean structures in the Central Uralian zone cannot be reliably deciphered. However, like the Late Proterozoic structures, they can be traced from the Central Urals to the basement of the northeastern part of the East European platform, using geophysical and geological data (Gafarov 1970; Puchkov 1975; Fig. 15).

By the end of the Proterozoic, these regions belonged to a single continent. In its southern part, an Archaean–Lower Proterozoic craton was subdivided into older cores (Belomorides) and younger foldbelts (Karelides). The cratonic basement of the Volga–Ural region is complicated by three Riphean graben-like structures: the Kazhim , Kaltasy and Sernovodsk–Abdulino aulacogens (A, B, C respectively on Fig. 15). They are up to 8–10 km deep (Romanov & Isherskaya 1994), continue in the western part of the Bashkirian anticlinorium of the Urals, and are overlain by up to 2 km of Vendian terrigenous sediments that form the lowermost member of the platform cover proper.

The intensity of tectonism of the Late Proterozoic rocks increases to the east and north. By Vendian times, a vast collisional orogen had formed at the western slope of the Urals (including the Central Uralian zone), in the Timan Range and Timan–Pechora Basin (Getsen 1991; Puchkov 1993), with the following structural elements shown on Fig. 15 (Puchkov 1975, 1993): a foredeep filled with Late Vendian molasse (D, Fig. 15); marginal anticlinoria: East Bashkirian (E), Kvarkush (F), and Timan (G); Vishera–Ilych–Chiksha (H) and Denisovka–Sablya (I) synclinoria, Man'–Khambo (K) and Kolva–Khobeiz (L) anticlinoria, the Laptopay intermontane depression (J), the Lemyu–

Khoreyver undifferentiated zone (M), the Yengane–Pe ophiolite suture (N), the Oche–Nyrd subduction-related volcanic zone (O) and the Marunkeu–Kharbey anticlinorium (P). The pre-Uralian zones (Fig. 15) have different trends compared to the overlying Uralides (only in the Southern Urals do they coincide to some extent).

The Riphean–Early Vendian history of the Pre-Uralian orogen is characterized by the development of wide troughs on a continental crust due to epicontinental rifting. Extension was accompanied in some places by the formation of sub-alkaline basalts or a rhyolite–basalt series of continental affinities (e.g. Goldin *et al.* 1973; Parnachev 1981; Parnachev *et al.* 1981; Getsen *et al.* 1987; Ivanov 1987). Among the sedimentary complexes of the Pre-Uralides, shallow-water quartzites, subarkoses, arkoses, bioherm algal limestones and dolomites are predominant. Poorly studied formations with ophiolitic and calc-alkaline affinities (Rumyantseva 1984; Puchkov 1993) are found only in the Polar Urals. Some authors (e.g. Ivanov 1987) discard part of these data as well as the presence of Pre-Uralian folding and orogenesis, in contrast to many earlier and some later researchers (e.g. Shatsky 1963; Perfilyev 1968; Gafarov 1970; Puchkov 1975, 1993; Getsen 1991).

The Pre-Uralian orogen was formed as a result of a Late Vendian (approx. 630–570 Ma) continental collision (Puchkov 1993). The orogeny was characterized by (a) deformation reflected in an unconformity at the base of the Ordovician sequence; (b) Barrovian metamorphism and S-type magmatism connected with thermal domes, partially distorted by later deformations, both Late Vendian and Late Palaeozoic; (c) formation of a foredeep filled with Late Vendian polymictic and arkosic terrigenous sediments, including conglomerates and (to the south of the Timan Range) evaporites. Rather small intramontane basins filled with molasse of Vendian age are present in the northern Urals (Puchkov 1975). In addition to lithologies of the Riphean complexes of the Bashkirian anticlinorium (quartzites, crystalline schists, dolomites, granites, syenites), Vendian conglomerates of the Southern Urals also contain pebbles of red jasper which may have originated from Riphean–Early Vendian deep oceanic sediments which are either not now at the surface (existed in overthrusts and now eroded) or are as yet undated.

Correlation of the Pre-Uralides with orogenic events in areas outside the Urals and Timan has not been thoroughly investigated. Of the adjacent areas where an orogeny can be tentatively correlated to the Pre-Uralides, two can be named: (1) Spitzbergen, where a pre-Mid-Ordovician unconformity is described, with metamorphic basement rocks, dated by the K–Ar method at 556, 584 and 621 Ma (Ohta *et al.* 1986) and (2) Taymyr, where collisional granites and amphibolites are dated at 560–625 Ma (Vernikovsky 1995).

The Pre-Uralian orogen was previously erroneously correlated with the Baykalides (e.g. Shatsky 1963). In fact, its origin appears to be unrelated to the Late Precambrian tectonic events of the Baykal area of Siberia. The time of formation of the Pre-Uralides is close to that of the Cadomian foldbelt of northern Europe (Puchkov 1988*a*). Recent global paleotectonic reconstructions for the Vendian and Early Cambrian time (Nance & Murphy 1994; McKerrow 1994) ignore the Pre-Uralides which should be placed very close to fragments of the Cadomian–Avalonian orogen (Fig. 16). It is probable that the Pre-Uralides were an integral part of a great Late Precambrian orogen within the Rodinia supercontinent.

In Early Palaeozoic time Rodinia was affected by intense rifting and one of the branches of the rift system ran parallel to the later Main Uralian Fault. The event led to the formation of the Palaeo-Uralian ocean and the passive margin of the East European continent. The continental rifting in the latest Cambrian and earliest Ordovician (Tremadoc) was accompanied by the formation of 'graben complexes' usually represented by a terrigenous series of variable thickness. They are accompanied by sub-alkaline and alkaline basalts and rhyolites, picrite porphyrites and tuffs (Goldin & Puchkov 1978; Ivanov *et al.* 1986; Dembovskiy *et al.* 1990). The presence of the Early Ordovician (Arenig) and younger (up to Eifelian) ophiolites in the Tagilo–Magnitogorskian zone and in some thrust sheets in the West Uralian zone reliably date the Palaeo-Uralian ocean (Puchkov 1993).

The tectonic nature of the Uralian Cambrian

The Cambrian in the Urals is poorly developed. In some places the latest Cambrian, represented by coarse terrigenous sediments and rift volcanics conformably underlies Lower Ordovician quartzites and quartz sandstones which are widely developed in the western slope of the Urals. The Mid-Cambrian is not known, probably due to general uplift and erosion, which affected the eastern part of the East European platform at this time. Massive, reef limestones

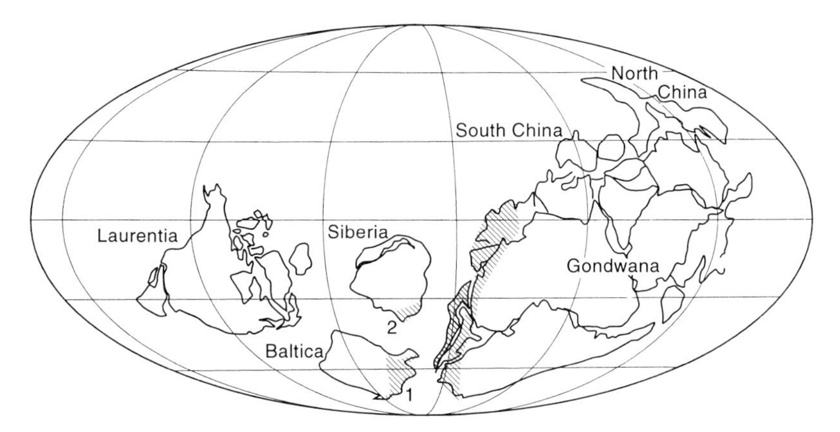

Fig. 16. Palinspastic reconstruction of continents in the Cambrian (McKerrow 1994, modified). Hatched areas are the Late Proterozoic orogens; 1, Pre-Uralides (Timanides), and 2, Taymyr are added to the reconstruction.

with archaeocyatids and algae of Early Cambrian age are known in two restricted areas: the Sakmarian allochthon (14, Fig. 1) and the Troitsk subzone (10, Fig. 1). In the Sakmarian subzone, the limestones constitute part of two 'suites': a lower one consisting of sandstones with subordinate tuffites and basalts and an upper one composed mostly of basalts, volcanic breccias and cherts. The thickness of these units varies between 600 and 1700 m. Limestones are encountered in all parts of the section, and their nature is a subject of continuing discussion. Some geologists believe they are '*in situ*' bioherms, others that they are biohermal olistoliths in sandstones and xenoliths in basalts. As such, they provide only the lower age limit of the suites. The only reliable '*in situ*' occurrence of conodonts in cherts dates the rocks, which enclose Lower Cambrian limestones, as Late Cambrian (Puchkov 1993). The Late Cambrian age and the unusually high Mn content of the volcanic rocks in the section suggests that the sequence records epicontinental rifting, the precursor of the Palaeo-Uralian ocean.

Development of the Palaeo-Uralian ocean

The main stages the Palaeo-Uralian ocean in the Palaeozoic are shown in Fig. 17 (see also Puchkov 1991, 1993). The most important features are: early development of an island arc and accretion against the Kazakhstano–Kirgizian continent, itself a collage of microcontinents and island arcs. The subduction zone was dipping to the east (in modern coordinates), under the Kazakhstano–Kirgizian continent. The sense of dip is deduced from the analysis of geochemical trends of calc-alkaline magmatism (Yazeva *et al.* 1989; Seravkin *et al.* 1992). It is also supported by the fact that the East-European/Laurussia margin of the Palaeo-Uralian ocean was passive (Puchkov 1979; Fig. 10).

The Palaeozoic continental collision in the Urals

The Uralian orogeny resulted from collision of the active and passive margins. Prior to collision, some minor collisions of microcontinent–island arc type and microcontinent–continent type took place throughout the Ordovician, Silurian and Devonian along the margin of the Kazakhstano–Kirgizian continent. The convergence of plates whose margins remain almost parallel for a long time is an exception. Margins of colliding continents often have uneven, indented outlines, which can cause major differences in structural development along orogens. Many researchers have considered the role of plate margin geometry and oblique collision in their analysis of Phanerozoic and Precambrian foldbelts, (e.g. Glazner 1991; Lyberis *et al.* 1992; Russo & Speed 1992; Ryan & Coleman 1992). Observations relevant to this problem have been made in the Urals too, although not in a systematic way.

Palaeomagnetic data (Svyazhina *et al.* 1992) show that beginning in the Ordovician, the East Mugodzhary and Kokchetav continental blocks and intervening Denisovka (primarily oceanic) block (their position is shown on Fig. 19) moved along similar trajectories and the scale of their presumed convergence and collision (Puchkov 1991) was too small to be seen by the palaeomagnetic method. These blocks that belonged to the

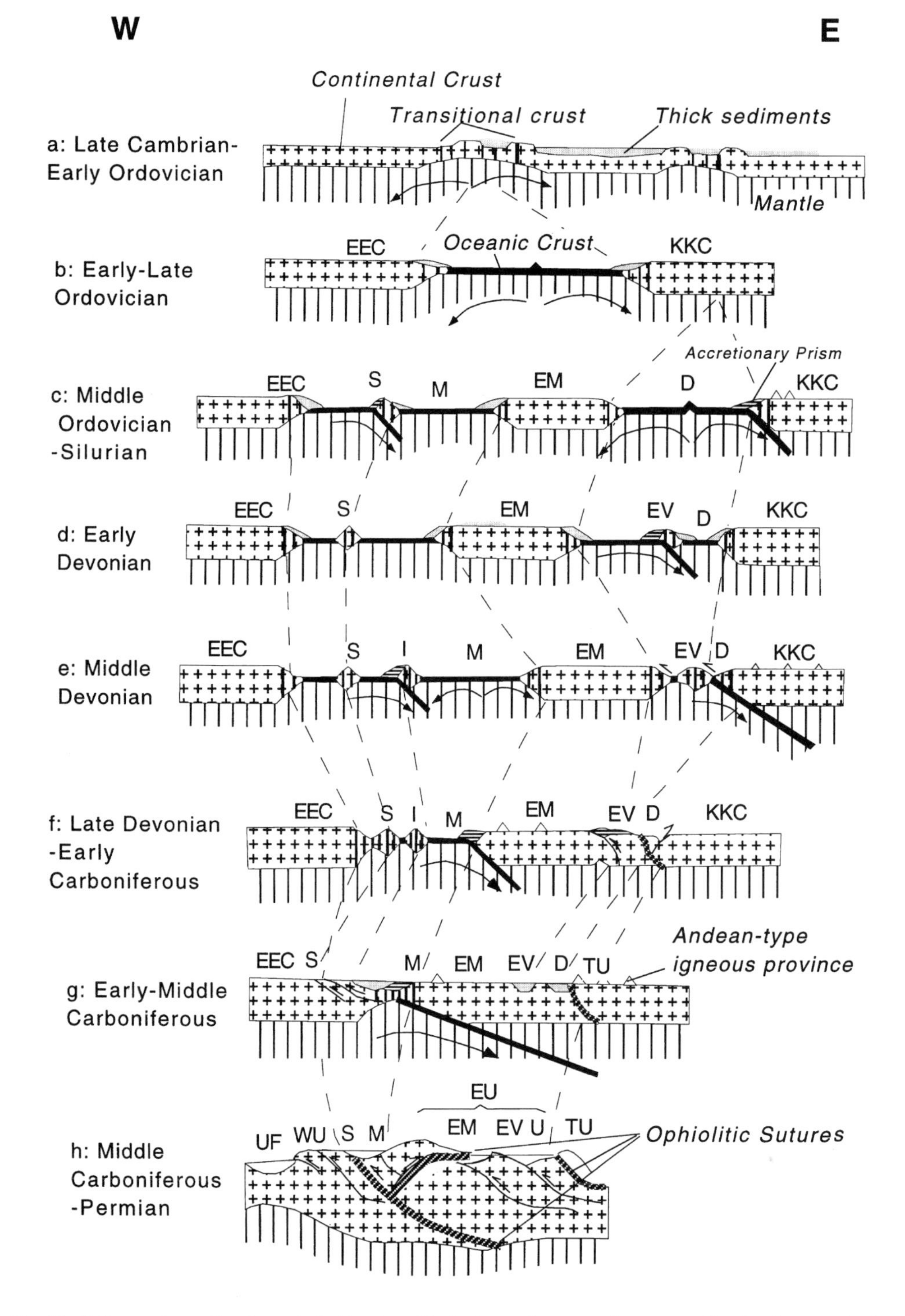

Fig. 17. Palinspastic profiles through the Southern Urals during the successive stages of its development in the Palaeozoic. EEC, East European continent; KKC, Kazakhstano–Kirgizian continent; S, Sakmara island arc; M, Magnitogorsk oceanic basin; EM, East Mugodzhary microcontinent; D, Denisovka oceanic basin; EV,East Uralian volcanic subzone; WU, West Uralian zone; EU East Uralian zone; TU, Transuralian zone.

Palaeozoic Kazakhstano–Kirgizian continent were situated: (1) at the same latitude as the territory of Kola Peninsula in the Ordovician, (2) opposite to the Middle Urals in the Late Carboniferous and (3) in a relative position close to the modern one in the Permian. According to these palaeomagnetic data, we must accept a non-cylindrical convergence of the two continents during their Palaeozoic history, including the period of their collision. Geological data strongly support this point.

The first sign of collision between the Kasakhstanian active margin and the East European passive margin is found in the Upper Frasnian in the Southern Urals. In the Sakmara subzone and the eastern limb of the Zilair synform the Zilair flysch of eastern provenance conformably overlies Eginda and Ibragimovo cherts dated as Frasnian and interpreted as bathyal sediments of the East European passive margin (Puchkov 1979). This flysch was thought to be Famennian–Tournaisian in age, although Chibrikova & Olli (1987) reported Late Frasnian spores and pollens and the Latest Frasnian ages for chert intercalations close to the bottom of the flysch. In any case, the middle of the Late Devonian may be chosen formally as the beginning of collision (*sensu lato*), when the continental margins came into close proximity.

During the Famennian, Early Carboniferous and most of the early Mid-Carboniferous subduction was still active with the formation of andesite, trachyandesite–rhyolite, monzonite–granosyenite (shoshonite), tonalite–granodiorite formations in the east (Ivanov *et al.* 1986; Puchkov *et al.* 1986; Salikhov *et al.* 1993 and others). These processes at the colliding margins can be compared to the modern collision between Australia and Southeastern Asia where the Sunda arc may play the role of a damper before the main collision. According to this model, applied to the Palaeozoic Urals, two stages in the collision can be recognized. The first is the transitional stage, termed a soft collision (Late Devonian–Early Carboniferous), when the continental margins came in contact, but the relict slab of the oceanic lithosphere was still being subducted, undergoing partial melting and producing calc-alkaline volcanites. The type of deformation associated with this stage is difficult to establish. The first thrusts carrying oceanic crust onto the former passive margin (Bardym, Kraka and Sakmara nappes) may have developed by this time. These thrust sheets were subsequently folded, overlie rocks no younger than the Late Devonian Zilair flysch and contain no rocks with proved age younger than Famennian. If so, the crust may have thickened slightly as a result of thrust stacking, but it was not so thick that it produced a high land mass. The second stage, termed a rigid collision (collision *sensu stricto*) is dated as Mid-Carboniferous–Late Permian and produced intense stacking of thrust sheets (involving Carboniferous and Permian rocks), growth of the Uralian foldbelt and a mountain range with a mountain root, generation of anatectic granites of the Main Granitic Axis (Fig. 1; Fershtater *et al.* 1994), and development of the Uralian foredeep and intermontane basins.

The Famennian–Tournaisian flysch trough (Fig. 18a) in the southernmost Western Urals has a distinct deep-water character. The thick Famennian graywacke series graded westward into the condensed Kiya unit of marls, cherts and bituminous shales (Puchkov & Ivanov 1987). The greywacke with eastern provenance did not reach the shelf zone of the passive continental margin until the deep-water trough, the relict of the bathyal zone, was filled. Therefore, the trough acted as a sedimentary trap for the terrigenous sediments from the eastern Uralian Zones. The time when this trap was filled is difficult to define and was certainly different at different latitudes. Keller (1949) has shown that in the northwestern part of the Zilair synform, Viséan greywackes overlie shallow-water shelf sediments of Famennian and Tournaisian age. Therefore the initial deep-water trough was filled no later than Tournaisian. In contrast, in the Ufimian amphitheatre of the Middle Urals (1, Fig. 1), where the passive margin had a distinct ‘promontory’, Famennian greywackes are reported to overlie immediately the Frasnian shelf limestones (Smirnov & Smirnova 1961, 1967; Kamaletdinov 1974). There are two explanations for the absence of bathyal sediments under the greywacke series. Either the collision against the ‘promontory’ was more rigid and the deep-water trough between the margins was rapidly closed, or the graywackes are thrust upon the shelf sediments.

Further to the north, the Famennian–Tournaisian graywacke is present in the Tagilo–Magnitogorskian zone, but nowhere overlies the bathyal sediments of the East European continental margin (Fig. 18a). The disappearance of the initial flysch trough was followed by the formation of the Uralian foredeep which migrated westward across the continental shelf. At the same time a considerable influx of terrigenous material from the east took place. Therefore, the margins of the continents first came in contact only in the south of the Urals.

In the Southern Urals the flysch trough disappeared in the Viséan, and probably earlier

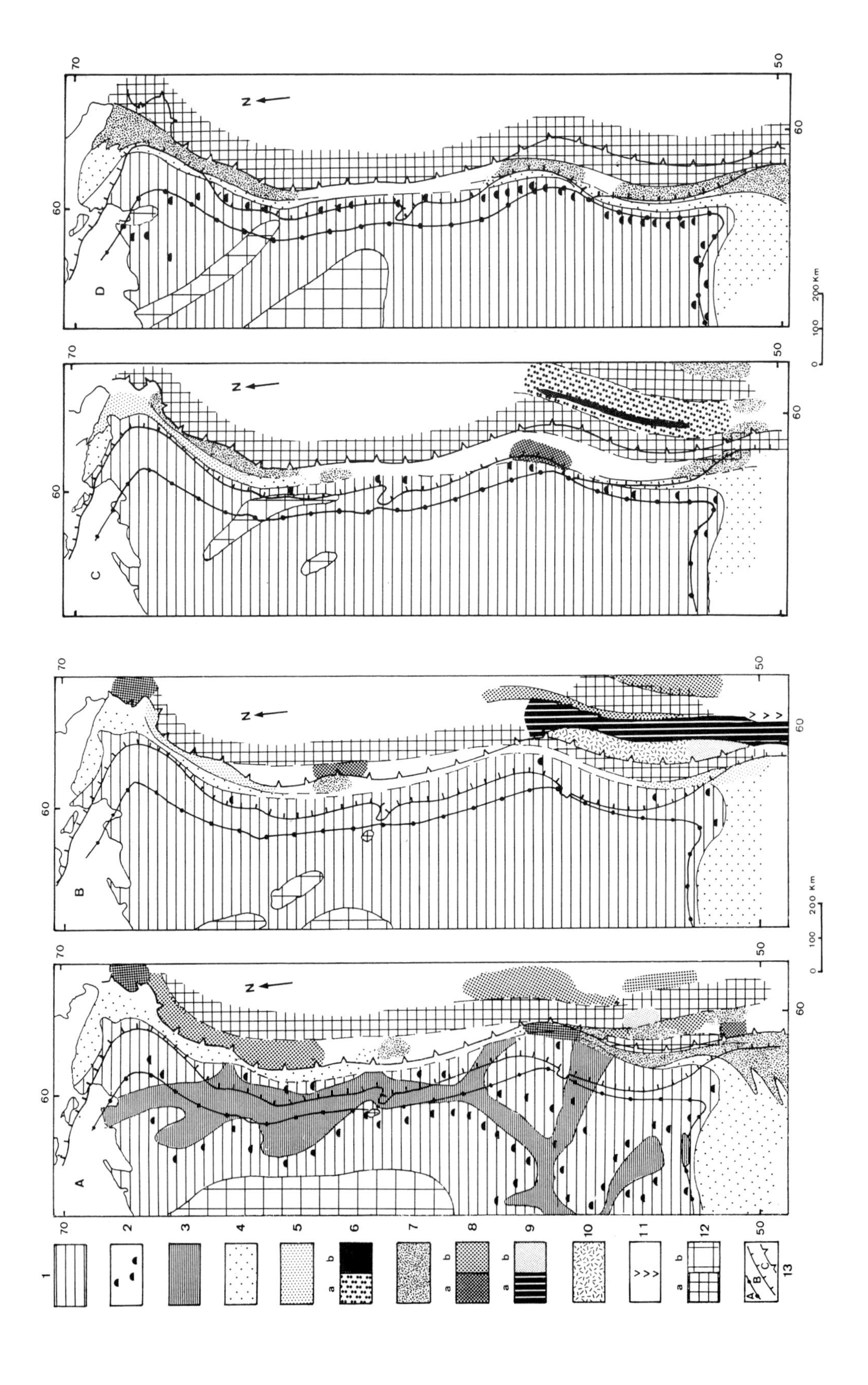

in the Middle Urals. But during most of the Late Tournaisian, Viséan, Serpukhovian and Early Bashkirian the influx of terrigenous material was neither abundant, nor constant, being interrupted by periods of accumulation of pelagic limestones and cherts (Kuruil and Bukharcha suites in the west of the Zilair synform, Keller 1949; modern data on their ages given by Sinitsina *et al.* 1984). Only later in the Mid- and Late Carboniferous when rigid collision started in the Southern Urals, was terrigenous sedimentation intensified. A series of sedimentary facies belts typical of a foredeep was established over the shelf sediments of the former continental margin. Coarse terrigenous sediments identified as molasse (Chuvashov & Nairn 1993), were followed by flysch and olistostromes, then by condensed basinal sediments, carbonate reefs, and finally shallow-water interbedded limestones and dolomites (Fig. 7). Such a facies series existed in the southernmost Urals since the Mid-Carboniferous, in the Middle Urals, probably since the Late Carboniferous. In the north the complete series was established in the Early Permian (Fig. 18b, c & d). The facies belts migrated to the west, so that the older and easternmost facies were reworked by fold-and-thrust belt structures to the rear of the foredeep (Nalivkin 1949 ; Khvorova 1961; Chuvashov & Dyupina 1973). The Uralian foredeep developed in the Southern Urals from the Mid-Carboniferous. The Carboniferous facies, typical for the foredeep, were later deformed and became part of the West Uralian zone, and the structure of the current foredeep was established only in Permian time.

A comparable transition from residual trough in the bathyal zone of the passive continental margin to a foredeep migrating westward onto the shelf of the margin, has been established in the Lemva zone of the Polar Urals, (Yeliseev, 1973; Puchkov, 1979, compare again Fig. 10, a and b) where analogous events took place later than in the Southern Urals. The appearance of greywacke flysch in the eastern side of the residual bathyal trough is dated as Okian–Serpukhovian (latest Early Carboniferous) time; the trough was being filled by flysch during the Mid-and Late Carboniferous; terrigenous sediments of the eastern influx appeared on the shelf only in the Early Permian. Conversely, in Kungurian (latest Early Permian) time the terrigenous influx, reflecting the intensity of collision, was at its highest in the Polar Urals, where a thick terrigenous coal-bearing series was formed. In more southern regions evaporites were at this time predominant, conditioned by both climatic variations and low terrigenous influx.

Such diachronism of events is connected with a shift of the collisional process along the Urals' strike. The shift was gradual and continuous, which can be illustrated by a series of simplified structural-paleogeographic schemes for the Late Devonian–Early Permian period (Fig. 18). One can see that the first appearance of greywacke in the corresponding structural zones of the western slope of the Urals became progressively younger to the north. The facies boundaries cut the main structural boundaries of the western slope of the Urals at a very acute angle (2–4°) and move with time not only from east to west, but from south to north. The terrigenous influx became more intense in the north while simultaneously getting weaker in the south. At the eastern margin of the Pricaspian Basin (Fig. 19a) this influx reaches a maximum intensity in Zilair (Famennian–Early Carboniferous) time (Volozh 1991). It was the time when the South Emba branch of the Variscides (Fig. 19), connecting the Urals with

Fig. 18. Schematic distribution of lithofacies in the Urals and adjacent areas. (**a**) End of the Devonian–beginning of the Carboniferous. (**b**) Early Carboniferous (Viséan and Serpukhovian time). (**c**) Mid- and greater part of the Late Carboniferous. (**d**) End of Carboniferous and beginning of Permian. Symbols: 1, shallow-water shelf sediments (mostly layered carbonates); in the Early–Mid-Viséan. mostly terrigenous sediments with oligomyctic sandstones and coals. 2, reefal massifs and bioherms; 3, comparatively deep-water (domanikoid) basinal facies, mostly shales and marls, developed in the axial parts of the Kama–Kinel trough system; 4, deep-water, bathyal facies: pelagic limestones, cherts, shales; 5, the same, alternating with greywacke or overlain by them. 6, shallow water limestones changing upsection to flysch near uplifts; (a) Late Carboniferous sediments are absent, (b) Late Carboniferous sediments are present; 7, mostly terrigenous sediments of the Uralian provenance (purely terrigenous flysch or tuffaceous-terrigenous turbidites; also molasse for the later stages of development). 8a, terrigenous sediments of the Uralian provenance, alternating with shelf complexes or overlying them; 8b, calc-alkaline volcanics of the subduction zone and accompanying terrigenous rocks; 9a), terrigenous sediments of a local influx intercalated with shallow-water limestones; ab, pure shallow-water limestones. 10-clastic rocks and limestones, associated with trachyrhyolite–basalt volcanics. 11, tholeiitic basalts of the Irgiz zone. 12, uplifts, a, intense, b, weak, subsequently eroded; 13a, western boundary of the Pre-Uralian foredeep prolonged by a northern boundary of the Pricaspian basin; 13b, western boundary of the West Uralian megazone; 13c, Main Uralian Fault.

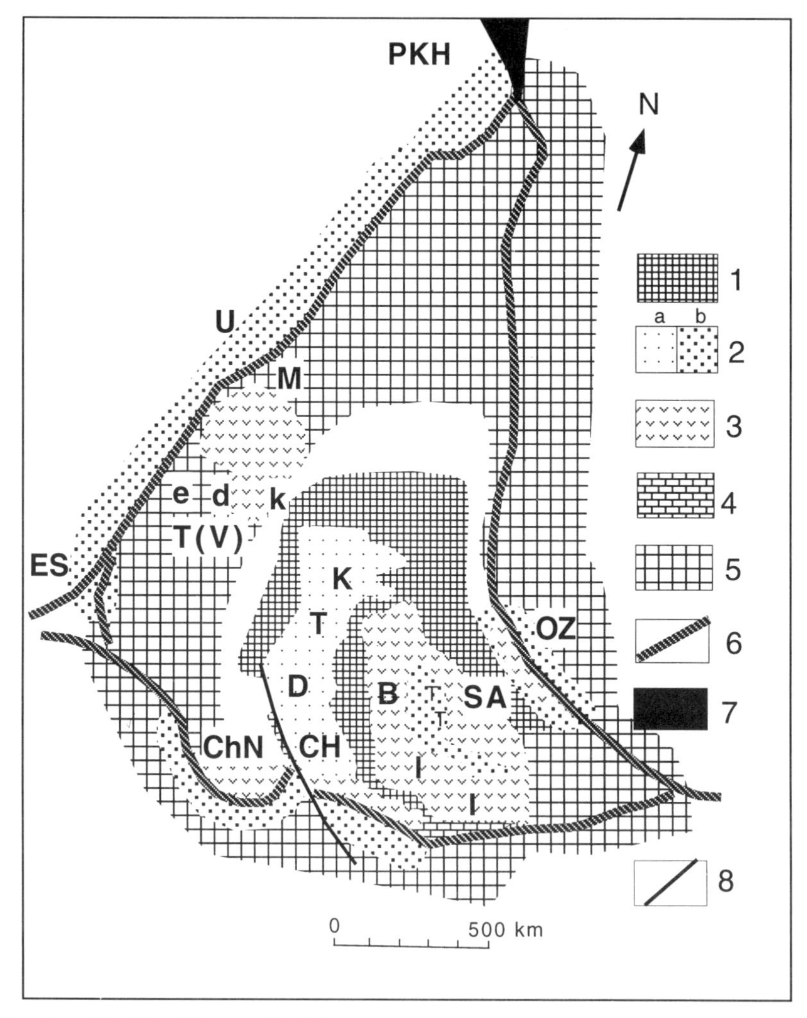

Fig. 19. Zonation of the Kazakhstano–Kirgizian continent in the Late Paleozoic (stage of 'rigid' collision) (Puchkov 1991, modified). 1, Eroded Middle Carboniferous-Permian internal rises; 2, terrigenous sediments, a, flysch and molasse, b, lagoon tuff-terrigenous deposits; 3, Volcano-plutonic complexes; 4, carbonate deposits; 5, Late Carboniferous-Permian orogenic rises connected with collision; 6, suture zones, former boundaries of collided continents/microcontinents; 7, supposed relict oceanic basin; 8, -faults. Structural zones: M, Magnitogorsk; T (V), Turgay (Valerianovka); ChN, Chatkalo–Naryn; K, Karaganda; T, Tengiz; D, Dzhezkazgan; CH, Chuya; B, Balkhash; I, Ili; SA, Sayak; OZ, Ob-Zaysan; PKH, Pay-Khoy; U, Uralian foredeep; ES, South Emba; structural subzones: e, East Mugodzhary; d, Denisovka; k, Kokchetav

Greater Caucasus, was active; in the north the peak activity was in Kungurian time.

This mode of Palaeozoic collision in the Urals was complicated by the uneven outline of the passive margin, with a promontory in the Middle Urals, called the Ufimian amphitheatre (1 in Fig. 1). The promontory may have acted as a pivot around which the Kazakhstanian continent rotated several degrees counter-clockwise. Such a rotation, which probably occurred during waning subduction, could result in the formation of local tensional structures in the upper levels of the lithosphere. In fact, this time (mostly Viséan–Early Bashkirian, immediately preceding rigid collision) was characterized in the Southern Urals by the formation of sedimentary and magmatic complexes atypical of collision or subduction: extensively developed shelf carbonates, mantle-derived gabbro–granite and sub-alkaline trachyrhyolite–basalt magmatic associations, layered gabbro–diabase intrusions of trapp affinities, parallel diabase dykes and

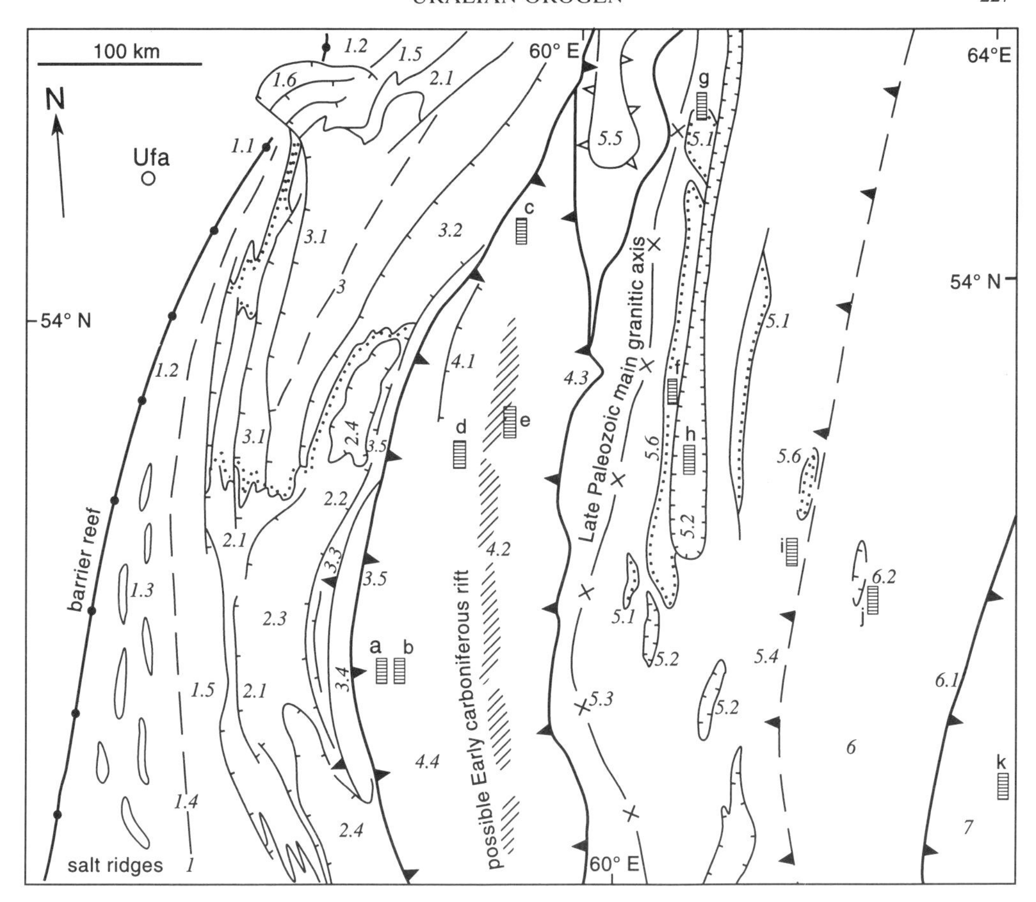

Fig. 20. Most important structures of the Uralian orogen, connected with the Late Paleozoic intracontinental collision (located on Fig. 23). Decorated lines, thrusts; undecorated lines, boundaries of structural subzones. Heavy lines with black triangles, suture zones marked by serpentinitic mélanges. Lines accentuated with points, unconformity between Proterozoic and Palaeozoic complexes. The suggested Early Carboniferous rift zone is marked by mantle-derived basalt–trachyrhyolite and gabbro–granite complexes. a–k, Palaeozoic stratigraphic sections of Fig. 14. Numbers in italic are explained in the text. In 5.5: Late Palaeozoic thermal dome characterized by a Barrovian metamorphism.

tholeiitic pillow basalts of the Irgiz zone (Ivanov *et al.* 1984; Frolova & Burikova 1977; Salikhov *et al.* 1993; Fershtater & Bea 1993).

The onset of a rigid collision in the eastern zones of the Southern Urals can be dated reliably enough as the beginning of the Moscowian. All manifestations of subduction related and rifting volcanism had stopped, and carbonate sedimentation began to be progressively substituted by terrigenous deposition; flysch troughs of NNE strike are thought to have existed east of the Main Uralian Fault by the end of the Middle Carboniferous (Chuvashov *et al.* 1984; Chuvashov & Puchkov 1990). Different opinions have been expressed about the presence of the Late Carboniferous and Permian in the eastern zones of the Urals. According to Chuvashov *et al.* (1984) sediments of this age are practically absent. Howevever, in later publications (Abdulin 1984; All-Russian Stratigraphic Committee 1993) one can recognize some relicts of intermontane basins filled with Late Carboniferous and even Permian molasse (red-coloured conglomerates and sandstones with flora remains and anhydrite layers) preserved in the Southern Urals (Figs 14f, g & 20). We have shown the relicts of these intermontane basins in the east of the Urals (Fig. 20, structures numbered 5.6 and 6.2), but only conditionally, taking into account the existing controversy.

The Uralian collision started early in the south, and its intensity varied much along the length of the orogen until the Late Permian, shifting wave-like to the north. By the beginning

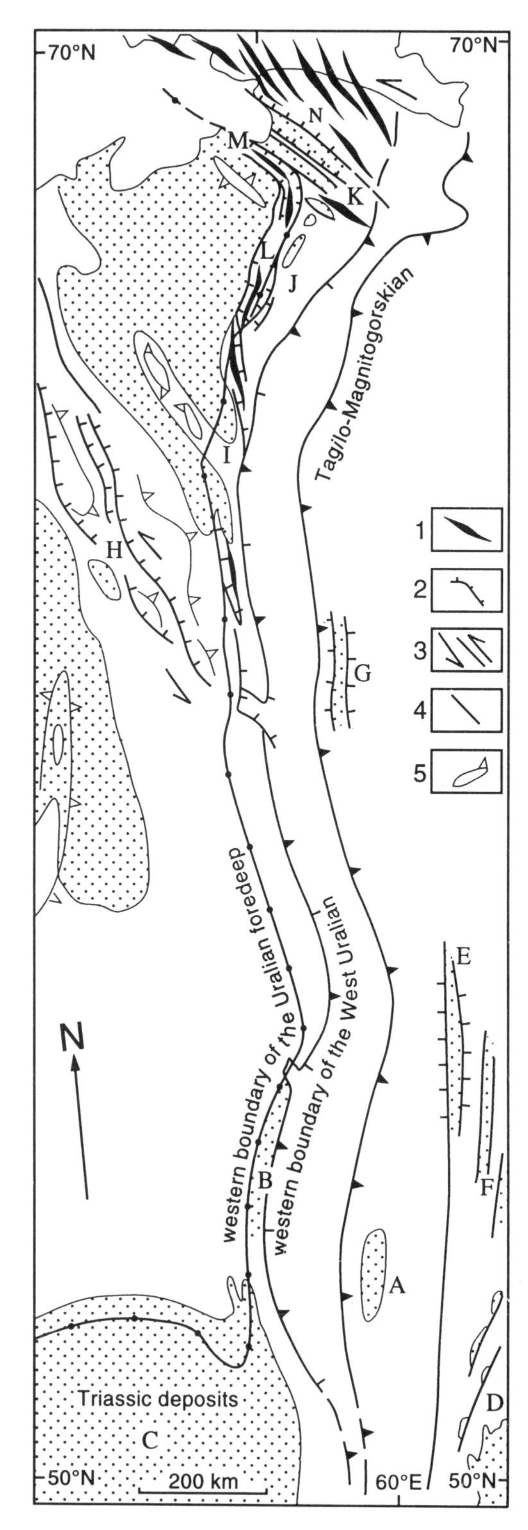

Fig. 21. Old Cimmerian structures in the Urals and adjacent areas compared with the structural boundaries of Uralian Orogen (Variscides). 1, Fold-and-thrust dislocations; 2, overthrusts; 3, wrench faults; 4, undefined faults; 5, gentle, smooth anticlines and flexures of platform type. Letters are explained in the text. Same location as Fig. 18 in Fig. 23.

of the Late Permian the intensity seems to have become more or less equal in all parts of the orogen (oblique collision changed for a time to a cylindrical one). By the end of the Permian, the orogeny was least intense in the east of the Southern Urals, where short ingressions of the Tethys-connected seas were registered (Chuvashov *et al.* 1984).

It is worthwhile noting that the Late Palaeozoic orogenic processes affected areas much wider than the Urals and enclosed the former Kazakhstano–Kirgizian and Siberian continents: wide intermontane basins were formed in Kazakhstan by this time (Fig. 19).

Mesozoic tectonic deformations

The uneven character of the collision was manifested again in the Mid-Jurassic, the Old Cimmerian stage of the region. The Palaeozoic and Old Cimmerian stages were separated by a period of a comparative quiescence with a dissipated rifting episode at the Palaeozoic–Mesozoic boundary, when the territory became a marginal part of the gigantic magmatic trapp province of Siberia.

The Old Cimmerian fold and thrust deformation in the Urals is a comparatively weak, intraplate shortening. Further to the north, in the Pay–Khoy and Novaya Zemlya this deformation was more intense and led to the formation of the Pay–Khoy–Novozemelian foldbelt which did not exist earlier as shown by palaeographic data, pre-Mid-Jurassic structural unconformities and isotopic dates (Chermnykh 1972; Korago *et al.* 1989; Rasulov 1982; Yudin 1994). This foldbelt resulted from direct convergence between the East European and Siberian continents (the structures of the Kazakhstanian continent are not traced so far to the north under the cover of the West Siberian plate, Fig. 19). Although the continents were, by Mesozoic time, integral parts of Pangaea, they were still loose enough to change their relative positions along a system of sinistral and subordinate dextral wrench faults (the directions of movements along these faults are shown in Fig. 21).

Comparison of the structures in areas where Triassic sediments are preserved (Fig. 21), gives

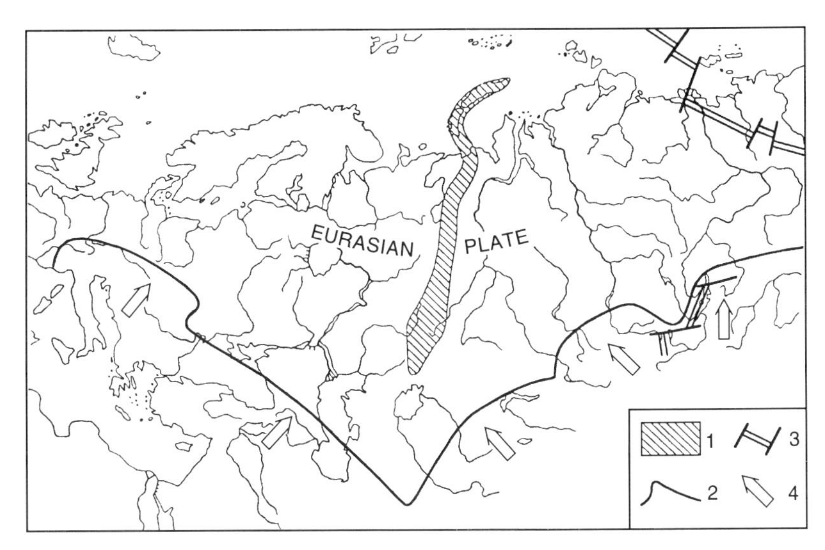

Fig. 22. Position of the Urals mountains in the Eurasian plate with respect to plate margins (Puchkov, 1988). 1, Neotectonic orogen of the Urals mountains; 2 & 3. margins of the Eurasian plate: 2, convergent; 3, divergent. 4, main directions of forces acting at the convergent margin of the Eurasian plate.

some important details concerning the changes in style and intensity of the Old Cimmerian deformations from south to north and west to east. In the Southern Urals Triassic terrigenous coal-bearing sediments of the Orsk basin (Fig. 21a) and coarse terrigenous sediments of the Uralian foredeep (Fig. 21b) are very thin (up to tens of metres) and practically undeformed (Ozhiganov 1964). In the Pricaspian depression (Fig. 21c), where a transition from continental to marine sediments occurred, salt domes are developed, but compressional structures (folds and thrusts) are absent. Immediately to the east of the Southern Urals, in the Turgay depression (Fig. 21d), the Triassic–Early Jurassic sequences are developed in linear graben-like depressions and consist of two series, the lower composed of trapp volcanics, and the upper consisting of terrigenous coal-bearing sediments. Their thickness varies from hundreds of metres to 3 km. Deformation formed gentle folds and normal faults (Abdulin 1984). Further north, in the eastern part of the Middle Urals (Chelyabinsk area, Fig. 21e), as well as in the base of the Mesozoic cover of the West Siberian basin (Fig. 21f) and in the Tagil zone (Fig. 21g), narrow, linear graben-like depressions are filled by Triassic trapp volcanics and terrigenous, partly coal-bearing, mostly alluvial sediments, up to 2–3 km thick. The margins of the depressions are affected by sharp folds and thrusts, usually several km in amplitude, often oppositely vergent (Tuzhikova 1960; Rasulov 1972).

Timan (Fig. 21h) did not experience uplift either in the Carboniferous, or even in the Late Permian, so its structure is not Late Palaeozoic (Raznitsyn 1968). It was covered by a shallow sea during the Kazanian transgression and was a place of terrigenous sedimentation in intracontinental basins during the Tatarian (end of Permian). In Early Triassic time, block faulting led to some local erosion, but Timan was not an obstacle for the transport of polymictic sandstones from the Urals to the central parts of the East European platform. Uplift and deep erosion of the growing Timan Range took place later than the Early Triassic, but before the Mid-Jurassic. The latter lies unconformably over different horizons of Late Palaeozoic age and has basal conglomerates with clasts of metamorphic schists from local uplifts of the basement.

In the western slope of the Cis-Polar (I) and Polar Urals, the Triassic terrigenous sediments (up to 2200 m thick) are weakly coaliferous and partly variegated. They are underlain by a continuous cover of trapp volcanics in the Kosyu–Rogovskaya (Fig. 21j) and Korotaikha (Fig. 21k) depressions of the Uralian foredeep. Subhorizontal over large areas, they possess an intense thrust-and-fold character in local, narrow linear zones such as the Chernyshov (Fig. 21l) and Chernov (Fig. 21m) ranges (Puchkov 1975; Timonin 1976).

Pay-Khoy (Fig. 21n) is a Mesozoic alpine-type foldbelt. According to palaeogeographic, structural and isotopic data, it is superimposed on a deep Permian molasse basin, a former part of

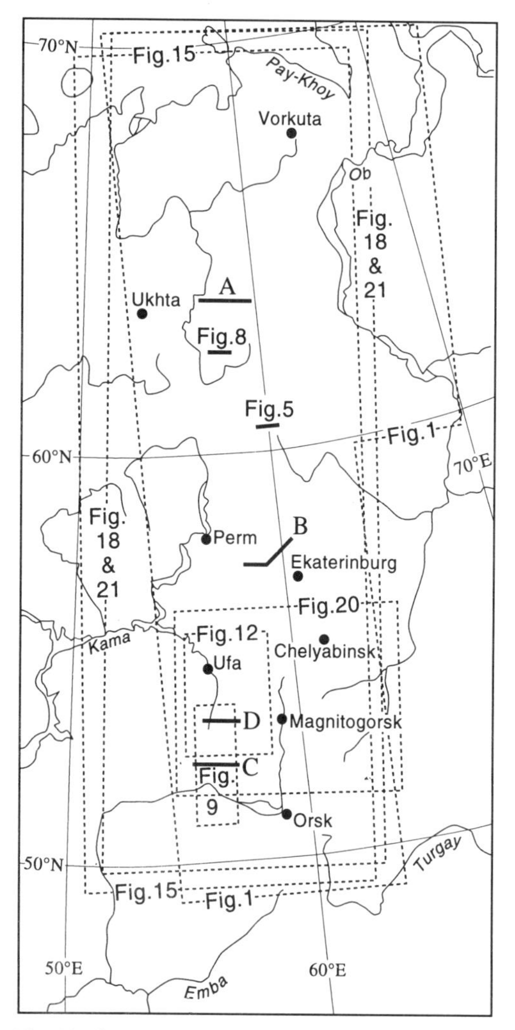

Fig. 23. Geographical location of previous figures. Sections: A, B and C are the Cis-Polar, Middle and Southern Urals sections of Fig. 4, respectively. D: section of Fig. 13.

the Uralian foredeep (Chermnykh 1972; Korago *et al.* 1989; Yudin 1994).

This tentative scheme of the Old Cimmerian deformations allows two important conclusions. First, the western boundary of the area with intense Old Cimmerian deformation has a northeastern strike and crosses at some angle the Uralian structural boundaries. Second, the 'en échelon' character of the deformations, especially the NW-trending, SW-verging structures of Timan and Pay-Khoy, suggests a sinistral, transpressional origin. On this basis, it can be also suggested that the Uralian structure of the Polar Urals continues to the north, and is substituted by a younger, pre-Mid-Jurassic foldbelt immediately to the west of it.

The southern continuation of the Urals was different at different stages in its history. As was pointed out earlier in this paper, the early orogenic uplift in Famennian–Tournaisian time, reflected in the development of the Zilair flysch, can be traced from the Urals to the south and southwest as the South Emba foldbelt (Fig. 19), connecting the Urals with the Variscides of the Greater Caucasus. But before the end of the Early Carboniferous (after the Mid-Viséan) this orogen ceased to develop, as is reflected in the presence of predominantly carbonate facies in the Late Viséan–Late Carboniferous of the southeastern margin of the Pricaspian basin (Volozh 1991). Conversely, the Mid- and Late Carboniferous orogenic movements of the Urals are reflected in the presence of coarse terrigenous facies in the northeastern part of the basin. These and the Permian movements are correlated with the analogous events in the Southern Tien-Shan, which are explained as a result of collision between the Kazakhstano–Kirgizian and Tarimo-Tadzhikian continental masses in the process of closure of Turkestanian ocean and formation of the South Tien-Shan suture zone of Pangaea (Puchkov 1991; Fig. 19).

The origin of the modern Ural mountains

The Late Palaeozoic mountain belt rejuvenated by the Old Cimmerian dislocations existed only for a short time, due to a rapid erosion. By the end of the Jurassic and during the Cretaceous the Urals were low hills and partially a lowland ingressed and covered by seas (Papulov 1974). Only since the Late Oligocene have the Ural mountains started to grow again (Rozhdestvensky 1994). These processes are still active, proved by geodetic and horizontal stress measurements as well as by earthquakes of weak to medium magnitude (Aleynikov *et al.* 1976). This orogeny probably results from an intracontinental deformation that followed favourable directions in the lithosphere of Eurasia. As indicated earlier (Puchkov 1988*b*), the Urals bisect the angle made by the southern, convergent margin of the modern Eurasian lithospheric plate (Fig. 22).

Conclusions

The general features of the Uralian orogen are summarised in a simplified diagram (Fig. 20), representing the Southern Urals, the best exposed area of the orogen. The following typical structural elements can be established from west to east.

The Uralian foredeep: filled with Permian molasse, underlain by Palaeozoic shelf deposits of the East European continent. It is divided into the following units: 1.1, the chain of the Early Permian barrier reefs at the western boundary of the foredeep; 1.2, the outer part of the foredeep, characterized by gentle, platform-type structures or (where Kungurian salt is present) by salt ridges; 1.3, salt ridges of the southern part of the foredeep; 1.4, the frontal line of the west-vergent thrusts and folds of the orogen; 1.5, the internal part of the foredeep, characterized by a thin-skinned to medium-skinned style of thrust-and-fold structure and probably by the presence of wedge-like (underthrust, backthrust) structures at the boundary with the next zone to the east; 1.6, the transverse Kara-Tau uplift, dividing the foredeep into basins.

The West Uralian zone: 2.1, mostly west-vergent thrust-and-fold structures of a thin- to thick-skinned style, affecting Palaeozoic shelf and bathyal deposits of the passive margin of the Palaeozoic East European continent; 2.2, backthrusts characteristic of the eastern limb of the Zilair synform; 2.3, the axial part of the Zilair synform; 2.4, allochthons (klippen) composed of ophiolites and bathyal complexes situated in the axial part of the synform.

The Central Uralian zone: 3.1 and 3.2 – the core of the Bashkirian anticlinorium, an exhumed Pre-cambrian basement of the Palaeozoic continental margin, including crystalline complexes, produced by two or more stages of deformation and metamorphism, sedimentary sequences of Riphean aulacogens and Vendian molasse of the Late Precambrian Pre-Uralian orogen. Structures of the western part of zone 3.1 are Variscan, their morphology being close to those of the West Uralian zone. Structures in the eastern and northern parts of zone 3.2 are a complex result of two or more deformational phases (Variscan and Pre-Variscan); 3.3 – 3.5, the Ural-Tau metamorphic complex, probably Palaeozoic; 3.3, the Yantyshevo–Yuluk backthrust; 3.4, the Maksyutovo metamorphosed accretionary complex of a Devonian island arc(?); 3.5, the Suvanyak metamorphosed bathyal complex of the passive margin of the East European continent; 3.6, the Main Uralian Fault, represented by an east-dipping zone of serpentinitic melange. Elements 3.3–3.6 are conventionally attributed to the Central Uralian zone.

The Tagilo–Magnitogorskian zone: 4.1, Internal mélange zones (thrusts); 4.2, central rift of Early Carboniferous age, marked by intrusions of a mantle-derived gabbro–granitic complex and comagmatic trachyrhyolite–basalt volcanics; 4.3, backthrusts of the eastern boundary of the zone (serpentinitic melanges); 4.4, the axial part of the Magnitogorsk synform.

The East Uralian Zone: This is a collage of microcontinental blocks with relicts of an autochthonous Palaeozoic sedimentary and volcano sedimentary cover (5.1) and allochthonous Palaeozoic ophiolite and island-arc formations (5.2); 5.3, the Main Granitic Axis of the Urals; 5.4; the Denisovka suspected suture zone, with ophiolites and serpentinitic mélanges; 5.5, thermal domes with Barrovian metamorphism and uplifted Early Proterozoic complexes; 5.6, relicts of intermontane depressions, filled with Middle Carboniferous (Moscowian) flysh-like deposits and probably Upper Carboniferous molasse.

The Transuralian Zone: this is a volcano-plutonic belt composed mainly of calc-alkaline magmatites. 6.1, the Urkash fault (suture zone?), a suggested boundary between the Urals and the Kazakhstanides; 6.2, a hypothetical relict of an intermontane basin.

The Kazakhstanides: 7.1, Variscan uplifts; 7.2, Variscan intermontane depressions, filled with Carboniferous and Permian molasse (to the east of the limits of the scheme).

The following are the most important stages of development of the Urals since the latest Precambrian.

(1) Vendian: continental collision and orogeny resulting in the Pre-Uralian (Timanides) foldbelt, which was probably part of the Cadomian orogenic belt within the Rodinia super-continent.
(2) Late Cambrian–Early Ordovician: rifting, break-up of the super continent, formation of the Palaeo-Uralian ocean and the passive margin of the East European continent.
(3) Mid-Ordovician–Mid-Devonian: subduction and accretion along the active margin of the Kazakhstanian continent on the eastern side of the Palaeo-Uralian ocean.
(4) 'Soft' and oblique collision between the passive and active margins, starting in the Southern Urals in the Late Frasnian and in the Northern Urals in the Late Viséan.
(5) 'Rigid' collision between the continents started in the Mid-Carboniferous. The oblique character of the collision was maintained until the Late Permian; with

time the orogenic processes gradually became more intense in the north than in the south.
(6) At the Permian–Triassic boundary: an episode of dissipated (areal) tension accompanied the development of a vast volcanic (trapp) province (including the Urals, Timan, Novaya Zemlya, Taymyr, Western and Eastern Siberia).
(7) Continental and intracontinental collision took place shortly before the Middle Jurassic. It affected the northern and eastern parts of the Urals and created the Timan Range, Chernyshov and Chernov thrust-and-fold zones and the Pay-Khoy–Novozemelian foldbelt.
(8) A long tectonic pause between the end of Jurassic and Late Palaeogene, led to the complete erosion of the Variscan Uralian orogen.
(9) Since the Late Oligocene, a new phase of intracontinental deformation led to the modern Ural mountains.

Future research in the Urals should focus on some of the targets listed below.
(1) The search for more reliable data on Precambrian ophiolites and calc-alkaline volcanics in the Urals with a geodynamic analysis of the Pre-Uralian (Timanides) foldbelt.
(2) Reliable palaeomagnetic determination of ancient pole positions of continents and terranes enclosed in the Urals. In particular Riphean, Vendian, Silurian and Devonian palaeolatitudes are needed.
(3) Creation of improved paleocontinental reconstruction, especially for the Vendian time.
(4) Research on the stratigraphy and tectonic nature of the Uralian Cambrian.
(5) Geological and isotopic age study of the areas affected by HP–LT and other types of metamorphism. The finds of Palaeozoic fauna in the Maksyutovo complex of the Southern Urals makes one anticipate analogous finds in some other metamorphic terranes which are now thought to be Precambrian.
(6) A thorough basin analysis must be applied to the Proterozoic and Palaeozoic sections of the Western slope of the Urals and adjacent area of the East European platform, as former parts of a single basin.
(7) Determination of directions of tectonic transport for many thrusts (e.g. checking the backthrust concept along the eastern limb of the Magnitogorskian synform). Very important are structural studies aimed at palinspastic reconstruction of the orogen. Also very important are structural studies of the metamorphic terranes of the Urals.
(8) Deeper study of the easternmost structures of the Urals and better definition of their boundary with the Kazakhstanides.
(9) Geological interpretation of the international geophysical profile URSEIS-95 Sterlitamak–Novonikolayevka (Southern Urals) which is close to completion while this article is being edited. There is hope that the profile will provide more detailed and reliable knowledge of the deep structure of the Urals. Combined with the other seismic profiles and applied to geological data, it gives a basis for creation of a 4D model of the Urals.
(10) Use of structural and geodynamic data to better understand the metallogeny of the Urals.
(11) Transformation of geological data of the Urals into digital form, using geoinformation systems (GIS).

This work was completed under the sponsorship of the Russian Foundation for Fundamental Investigations (Ref.no: RFFI 93–05–14033) and INTAS (Ref.no: INTAS–94–1857). The author expresses his deep gratitude to J.-P. Burg, M. Ford, J. Knapp and anonymous reviewers for corrections and useful suggestions they have made while evaluating and editing the manuscript. Author's e-mail address: puchkov@giras.baskiria.su.

References

ABDULIN A. A. (ed.) 1984. [*The geology and raw materials of the SE part of the Turgay basin and the Northern Ulytau*]. Nauka, Alma-Ata [in Russian].

ABLIZIN, B. D., KLYUZHINA, M. L., KURBATSKAYA, F. A. & KURBATSKY, A. M. 1982. [*The Late Riphean and Vendian of the western slope of the Urals*]. Nauka, Moscow [in Russian].

ALEYNIKOV, A. L., BELLAVIN, O. V., BUGAYLO, V. A. *et al.* 1976. [*Some probems of the Ural's geodynamics*]. *In*: [*Structure and development of the Earth's crust and formation of the ore fields of the Urals according to geophysical data*]. UNTs AN SSSR, Sverdlovsk, 88–90 [in Russian].

ALL-RUSSIAN STRATIGRAPHIC COMMITTEE 1993. [*Stratidraphic schemes of the Urals: Precambrian and Paleozoic*]. Ekaterinburg, 151 scheme. [in Russian].

ANFIMOV L. V. 1986. [*Nature of the post-diagenetic changes of the sedimentary rocks of the Bashkirian anticlinorium.*] Yezhegodnikø-1985. Sverdlovsk, IGG UNTs AN SSSR. 24–26. [in Russian].

ANTOSHKINA, A. I. & ELISEEV, A. I. 1988. [*The Paleozoic reefs in the north of the Urals and adjacent territories*]. *In*: [*Lithology of carbonate rocks of the North Urals, Pay-Khoy and Timan*]. Institute of Geology, Komi Scientific Centre, Syktyvkar, 5–18 [in Russian].

AVTONEEV, S. V., ANANYEVA, YE. M., BASHTA, K. G. & 21 authors. 1991. [*Deep structure of the Urals according to geophysical data*]. *In*: [*Deep structure of the USSR territory*]. Nauka, Moscow [in Russian].

BEKKER YU. R. 1968. [*Late Precambrian molasse of the Southern Urals*] L., Nedra [in Russian].

CHERMNYKH, V. A. (ed.) 1972. [*Atlas of lithologic-paleogeographic maps of the Paleozoic and Mezozoic of the North Cis-Urals*]. Nauka, Leningrad, 49 maps [in Russian].

CHIBRIKOVA YE. V. & OLLI V. A. 1987. [*New data on the stratigraphy of the northern part of Kazakhian Urals*]. Ufa, IG UFAN SSSR [in Russian].

CHUVASHOV B. I. & DYUPINA G. V. 1973. [*Late Paleozoic terrigenous sediments of the western slope of the Middle Urals*]. Nauka, Moscow [in Russian].

—— & NAIRN, A. E. M. (eds) 1993. *Permian system: Guides to geological Excursions in the Uralian Type Localities*. Occasional Publications ESRI, University of South Carolina, New Ser. **10**.

—— & PUCHKOV, V. N. 1990. [Geological history of the Urals in the Carboniferous]. *In*: [*New data on the geology of the Urals, Western Siberia and the Kazakhstan*]. Institute of Geology and Geochemistry, Sverdlovsk, 11–16 [in Russian].

—— & SHUYSKY, V. P. 1990. [Evolution and tectonic position of the Paleozoic reefs in the Urals]. *In*: [*New data on the geology of the Urals, Western Siberia and the Kazakhstan*]. Institute of Geology and Geochemistry, Sverdlovsk, 3–10 [in Russian].

——, IVANOVA, R. M. & KOLCHINA, A. N. 1984. [*Late Paleozoic of the eastern slope of the Urals*]. UNTs AN SSSR. Sverdlovsk, [in Russian].

DEMBOVSKY, B. YA., DEMBOVSKAYA, Z. P., NASSEDKINA, V. A. & KLYUZHINA, M. L. 1990. [*The Ordovician of the Cis-Polar Urals.V: Geology, lithology, stratigraphy*]. *In*: PUCHKOV, V. (ed.) UrO RAN SSSR, Sverdlovsk [in Russian].

DOBRETSOV N. L. 1974. [*Glaucophane-schist and eclogite-glaucophane-schist complexes of the USSR*]. Novosibirsk [in Russian].

FERSHTATER, G. B. & BEA, F. 1993. [Geochemical features of Uralian granitoids of different magmatic derivation]. *Geokhimiya*, **11**, 1579–1599. [in Russian].

——, BORODINA, N. S., RAPOPORT, M. S., OSIPOVA, T. A., SMIRNOV, V. N. & LEVIN, V. YA. 1994. [*Orogenic magmatism of the Urals*]. Miass, IGG RAN [in Russian].

FROLOVA T. I. & BURIKOVA I. A. 1977. [*Geosynclinal volcanism at the example of the eastern slope of the Urals*]. Moscow,Mosc. State Univ. [in Russian].

GAFAROV, P. A. 1976. [*Comparative tectonics of the basements of ancient platforms and types of their magnetic fields*]. Nauka, Moscow [in Russian].

GAFAROV, R. A. 1970. [On the deep structure of the basement in the conjugation zone between the East-European platform and the Urals]. *Izvestiya AN SSSR, ser. geol.*, **N8**, 3–14 [in Russian].

GETSEN, V. G. 1991. [Geodynamical reconstruction of the development of the N-E. European part of the USSR at the Late Proterozoic stage]. *Geotektonika*, **5**, 26–37 [in Russian].

——, DEDEEV, V. A., AKIMOVA, G. N, ANDREICHEV, V. L., BASHILOV, V. I., BELYAKOVA, L. T., GORNOSTAY, B. A. & DEMBOVSKY, B. YA. 1987. [*The Riphean and Vendian of the European Nortn of the USSR*]. Inst. Geol., Komi Filial AN SSSR., Syktyvkar [in Russian].

GLAZNER, A. F. 1991. Plutonism, oblique subduction and continental growth: An example from the Mesozoic of California. *Geology*, **19**, 784–786.

GOLDIN, B. A. & PUCHKOV, V. N. 1978. [Early Paleozoic (rift) magmatism of the western slope of the Urals]. *In*: [*Precambrian and Lower Paleozoic of the Urals*]. Institute of Geology and Geochemistry, Sverdlovsk, 63–71 [in Russian].

——, FISHMAN, M. V., KALININ, YE. P. & DAVYDOV V. A. 1973. [*Volcanic complexes in the North of the Urals*]. Nauka, Leningrad [in Russian].

IVANOV, S. N. 1987. [*On the baykalides of the Urals and nature of metamorphic complexes flanking eugeosynclines*] Institute of Geology and Geochemistry, Sverdlovsk [in Russian].

——, SEMENOV, I. V. & CHERVYAKOVSKY, G. F. 1984. [Magmatism of a precontinental stage of the Ural's development]. *Doklady AN SSSR*, **274(2)**, 387–391 [in Russian].

——, PUCHKOV, V. N., IVANOV, K. S., SAMARKIN, G. I., SEMENOV, I. V., PUMPYANSKY, A. I., DYMKIN, A. M., POLTAVETS, YU. A., RUSIN, A. I. & KRASNOBAYEV, A. A. 1986. [*The formation of the Uralian earth's crust*]. Nauka, Moscow [in Russian].

KAMALETDINOV, M. A. 1974. [*The nappe structures of the Urals*]. Nauka, Moscow [in Russian].

——, KAMALETDINOV, M. A., KAZANTSEV, YU. V., KAZANTSEVA, T. T. & FATTAKHUTDINOV, S. G., 1984. [*The main tectonic features and regularities of localization of the oil and gas deposits in the Uralian foredeep*]. Chast 1. Ufa, Inst. Geol. Bashkirian. Filial AN SSSR [in Russian].

KARPINSKY, A. P. 1919. [On the tectonics of the European Russia]. *Izvesyia of the Academy of Sciences*. *In*: *Collected works of A. P. Karpinsky*, **2**, USSR, Moscow-Leningrad [in Russian].

KAZANTSEV, YU. V. 1984. [*Structural geology of the Uralian foredeep*]. Nauka, Moscow [in Russian].

—— 1991. [Synforms of the Magnitogorsk synclinorium]. *Doklady AN USSR*, **316(5)**, 1183–1188 [in Russian].

KELLER, B. M. 1949. [*Paleozoic flysh formation in the Zilair synclinorium of the Southern Urals and comparable complexes*]. Moskva,Trudy Instituta Geologicneskikh Nauk, **104**, [in Russian].

—— & CHUMAKOV, N. M. (eds) 1983. [*The stratotype of the Riphean. Stratigraphy, geochronology*]. Trudy Geologicheskogo Instituta AN SSSR, **337**, Nauka, Moscow [in Russian].

KHATYANOV, F. I. 1963. [Division of the Uralian foldbelt into a platform and geosynclinal parts as it follows from geophysical data]. *Doklady AN SSSR*, **150**, N5, 1325–1329 Russian].

KHERASKOV, N. P. 1967. [*Tectonics and formations*]. Izbrannye trudy. Nauka, Moscow.

KHVOROVA, N. V. 1961. [*Flysh and lower molasse formations of the Southern Urals*]. Nauka, Moscow [in Russian].

KORAGO, YE. A., KOVALEVA, G. N. & TRUFANOV, G. N. 1989. [Formations, tectonics and geological history of the Novozemelian Cimmerides]. *Geotektonika*, **6**, 40–61 [in Russian].

KOROTEEV, V. A., DIANOVA, T. V. & KABANOVA, L. YA. 1979. [*Middle Paleozoic volcanism of the Eastern zone of the Urals*]. Nauka, Moscow [in Russian].

KRASNOBAYEV, A. A. 1985. [Problems of the Proterozoic geochronology in the Urals]. *IGG UNTs AN SSSR. Yezhegodnik. Sverdlovsk*, 26–30 [in Russian].

——, KUZNETSOV, G. P. *et al.* 1995. [Uranium-Lead age of zircons from gneisses of the Chelyabinsk complex]. *Yezhegodnik-1994 Institute of Geology and Geochemistry, Yekaterinburg*, 34–37 [in Russian].

LENNYKH, V. I. 1986. [Metakomatiites of the Taratash complex]. *In*: [*Precambrian volcano-sedimentary complexes of the Urals*]. UNTs AN SSSR, Sverdlovsk, 70–73 [in Russian].

—— & KRASNOBAYEV, A. A. 1978. [On isotopic ages of metamorphic rocks]. *In*: [*Precambrian and Lower Paleozoic of the Urals*]. UNTs AN SSSR, Sverdlovsk, 69–76 [in Russian].

——, PANKOV, YU. D. & PETROV, V. I. 1978. [Petrology and metamorphism of the migmatite complex]. *In*: [*Petrology and iron ore deposits of the Taratash complex*]. UNTs AN SSSR, Sverdlovsk, 3–45 [in Russian].

LEVIN, V. YA. 1994. [Geological and structural position,internal structure and composition of carbonatite complexes of the Urals]. *In*: [*Magmatism and metamorphism of the conjugation zone of the Urals and the East-European platform*]. UNTs AN SSSR, Sverdlovsk, 3–18 [in Russian].

LYBERIS, N., YURUR, T., CHOROVICH, J., KASAPOGLU, E. & GUNDOGLU, N. 1992. The East Anatolian Fault: an oblique collisional belt. *Tectonophysics*, **204**, 1–15.

MCKERROW, W. S. 1994. Terrane assembly in the Variscan belt of Europe. *Europrobe news*, **N5**, 4–5.

MAKEDONOV, A. V. (ed.) 1965. [The coal-bearing formation and its main features]. *In*: [*The history of the coal accumulation in the Pechora basin*]. Nauka, Leningrad, 47–134 [in Russian].

MASLOV, V. A., CHERKASOV, V. L., TISCHCHENKO, V. T., SMIRNOVA, I. A., ARTYUSHKOVA, O. V. & PAVLOV, V. V. 1993. [*On stratigraphy and correlation of the Middle Paleozoic complexes of main copper-pyritic areas of the Southrn Urals*] Ufimsky Nauch. Tsentr,Ufa [in Russian].

MATTE, PH. 1995. Southern Uralides and Variscides:comparison of their anatomies and evolutions. *Geologie en Minjbouw*, **74**, 151–166.

——, MALUSKI, H., NICOLAS, A., KEPEZHINSKAS, P. & SOBOLEV, S. 1993. Geodynamic model and $^{39}Ar/^{40}Ar$ dating for generation and emplacement of the High Pressure metamorphic rocks in SW Urals. *Comptes Rendus de l'Academie des Sciences, Paris*, **317**, sér. II, 1667–1674.

MENSHIKOV, YU. P., KUZNETSOVA, N. V., SHEBUKHOVA, S. V. & NIKISHEVA, G. N. 1983. [*Earth's crust faults and methods of their study*]. UNTs AN SSSR. Sverdlovsk, 65–78 [in Russian].

NALIVKIN, V. D. 1949. [*Facies and geological history of the Ufimian plateau and Yurezan-Sylva depression*]. Moskva,Trudy VNIGRI (Oil Research Institute), nov. ser., **46** [in Russian].

NANCE, R. D. & MURPHY, J. B. 1994. Contrasting basement isotopic signatures and the palinspastic restoration of peripherial orogens: example from the Neoproterozoic Avalonian-Cadomian belt. *Geology*, **22**, 612–620.

NECHEUKHIN, V. M., BERLYAND, N. G., PUCHKOV, V. N. & SOKOLOV, V. B. 1986. [*Deep structure, tectonics and metallogeny of the Urals*]. Institute of Geology and Geochemistry, Sverdlovsk [in Russian].

OHTA, Y., HIRAJIMA, T. & HIROI, Y. 1986. Caledonian high-pressure metamorphism in Central Western Spitsbergen. *In*: EVANS, B. W. & BROWN, E. H. (eds) *Blueschists and Eclogites*. Geological Society of America Memoirs, **164**, 205–216.

OZHIGANOV, D. G. (ed.) 1964. [*The Geology of the USSR, v.XIII; (Bashkiria,Orenburg district)*]. Nedra, Moscow [in Russian].

PAPULOV, G. N. 1974. [*The Cretaceous sediments of the Urals*]. Nauka, Moscow [in Russian].

PARASYNA, V. S., SOLOMATIN, A. V. & SHLEZINGER, A. YE. 1989. [The Pechora Late Devonian-Early Carboniferous deep-water basin]. *Geotektonika*, **5**, 82–92 [in Russian].

PARNACHEV, V. P. 1981. [Volcanic complexes and tectonic regime of the western slope of the Urals in the Late Precambrian]. *In*: [*Ancient volcanism of the Southern Urals*]. Uralian Scientific Centre, AN SSSR, 18–30 [in Russian].

——, KOZLOV, V. I. & TITUNINA, I. V. 1981. [*New data on the structure, composition and origin of the Arsha metavolcanic complex of the Southern Urals (Late Precambrian)*]. Sverdlovsk, 69–86 [in Russian].

PERFILYEV, A. S. 1968. [*Tectonics of the North of the Urals*]. Nauka, Moscow. [in Russian].

—— 1979. [*On the formation of the earth's crust of the Uralian eugeosyncline*]. Trudy Geol. Inst. AN SSSR, **328** [in Russian].

PETROV, G. A. & PUCHKOV, V. N. 1994. [Tectonics of the Main Uralian Fault zone] *Geotektonika*, **1**, 35–47 [in Russian].

PEYVE, A. V. 1945. [Deep-seated faults in geosynclines]. *Izvestia of the Academy of Sciences of USSR, series geology*, **5**, 3–19 [in Russian].

——, IVANOV, S. N., NECHEUKHIN, V. M., PERFILYEV, A. S. & PUCHKOV, V. N. 1977. [*Tectonics of the Urals. The explanatory notes for the 1:1 000 000-scale tectonic map*]. Nauka, Moscow [in Russian].

PUCHKOV, V. N. 1975. [*The structural connections*

between the Cis-Polar Urals and flanking part of the Russian platform]. Nauka, Leningrad [in Russian].
—— 1979. [*Bathyal complexes of the passive margins of geosynclines*]. Nauka, Moscow [in Russian].
—— 1988*a*. Correlation and geodynamic features of Pre-Alpine tectonic movements throughout and around the Alpine Orogen. *In*: *Studia Geologica Polonica*, **91**, Warszawa, 77–92.
—— 1988*b*. [Intraplate events in the geological history of mobile belts,at the example of the Urals]. *In*: [*Intraplate phenomena in the Earth's crust*]. Nauka, Moscow, 167–175 [in Russian].
—— 1991. *The Paleozoic of the Uralo-Mongolian fold system*. Occasional Publications ESRI, University of South Carolina, New Series, N7, part II.
—— 1993. [Paleooceanic structures of the Urals]. *Geotektonika*, **3**, 18–33 [in Russian].
—— 1996. [Geodynamic control of the regional metamorphism of the Urals]. *Geotektonika*, **2**, 16–35. [in Russian].
—— & Ivanov, K. S. 1982. [*Geology of the allochtonous complexes of the Ufimian amphitheatre*]. Sverdlovsk, UNTs AN SSSR [in Russian].
—— & —— 1985. [First data on the volcanic -cherty Ordovician series in the East of the Urals]. *Doklady AN SSSR*, **285**, N4, 966–970 [in Russian].
—— & —— 1987. [On the stratigraphy of the Late Devonian-Early Carboniferous series of the Sakmara zone]. *In*: [*New data on geology of the Urals*]. UNTs AN SSSR, Sverdlovsk, 84–93 [in Russian].
—— & Sokolov, V. B. 1992. Deep crustal structure of the Urals sccording to geological and geophysical data (abstr). *29th International Geological Congress, Kyoto*, **1**, 154.
—— & Svetlakova, A. N. 1993. [Structure of the Southern Urals at the cross-section of the Troitsk profile]. *Doklady RAN*, **333**, N3,348–351 [in Russian].
——, Rapoport, M. B. *et al.* 1986. [Tectonic control of the Paleozoic granitoid magmatism in the eastern slope of the Urals]. *In*: [*Studies of geology and metallogeny of the Urals*]. UNTs AN SSSR, Sverdlovsk, 85–95 [in Russian].
Rozhdestvensky, A. P. 1994. [Main features of the Urals neotectonics]. *Yezhegodnik-1993, Inst. Geol. UNTs RAN, Ufa*, 63–66 [in Russian].
Rumyantseva, N. A. 1984. [Pre-ordovician volcanic formations of the Oche-Nyrd uplift]. *In*: [*Magmatism and metamorphism of the conjugation zone of the Urals and the East-European platform*]. UNTs USSR, Sverdlovsk, 19–35 [in Russian].
Rasulov, A. T. 1982. [*Tectonics of Early Mezozoic depressions of the eastern slope of the Urals*]. UNTs AN SSSR, Sverdlovsk, [in Russian].
Raznitsyn, V. A. 1968. [*Tectonics of the Middle Timans*]. Nauka, Leningrado [in Russian].
Romanov, V. A. & Isherskaya, M. V. 1994. [*On the Riphean series of the Western Bashkiria*]. Ufa, Inst. Geologii, [in Russian].
Russo, R. M. & Speed, R. C. 1992. Oblique collision and tectonic wedging of the South American continent and Caribbean terranes. *Geology*, **20**, 447–450.
Ruzhentsev, S. V. 1986. [*The marginal ophiolite allochtons: their tectonic nature and structural position*]. Trudy Geol. Inst. AN SSSR, **283** [in Russian].
Ryan, H. P. & Coleman, P. J. 1992. Composite transform- convergent plate boundaries: Description and discussion. *Marine and Petroleum Geology*, **9**, 89–97.
Salikhov, D. N., Yusupov, S. Sh. & Mitrofanov, V. A. 1993. [*Early collisional stage of development of the Southern Urals and its metallogeny*]. Ufa [in Russian].
Savelyeva, G. N. 1987. [*Gabbro-ultramafic complexes of the Uralian ophiolites and their analogues in the modern oceanic crust*]. Nauka, Moscow, [in Russian].
Seravkin, I. B., Kosarev, A. M., Salikhov, D. N., Znamensky, S. Ye., Rykus, A. M. & Rodicheva, Z. I. 1992. [*Volcanism of the Southern Urals*]. Nauka, Moscow, [in Russian].
Shatsky, N. S. 1963. [The Riphean era and Baykalian folding phase]. *In*: [*Academician Shatsky: Selected works*], **1**, AN-SSSR, Moscow, 600–619 [in Russian].
Skripiy, A. A. & Yunusov, N. K. 1989. [The tension structures in the conjugation zone between the Southern Urals and the East -European platform]. *Geotektonika*, **6**, 62–71 [in Russian].
Sinitsina, Z. A., Sinitsin, I. I. & Shamov, D. F. 1984. [*A concise stratigraphic description of the Late Paleozoic of the Southern Urals*]. Guidebook of excursion 047 k 27mu MGK. Nauka, Moscow, 9–19 [in Russian].
Smirnov, G. A. & Smirnova, T. A. 1961. [*Materials for the paleogeography of the Urals. Essay III, Famennian time*]. UFAN SSSR [in Russian].
—— & —— 1967. [*Materials for the Paleogeography of the Urals. Essay IV. Tournaisian time*]. UFAN SSSR [in Russian].
Snachev, V. N., Kuznetsov, N. S., Rachev, P. I. & Kovalev, S. G. 1994. [*Magmatism and metallogeny of the northern part of the East-Uralian rift system*]. Ufa, Geol. Inst. [in Russian].
Sobornov, K. O. & Bushuev, A. C. 1992. [Kinematics of the conjugation zone between the Northern Urals and Verkhnepechorskaya basin]. *Geotektonika*, **4**, 39–51 [in Russian].
—— & Tarasov, P. P. 1992. [Two-layer subthrust structure of the southern part of the folded limb of the Kosyu-Rogovskaya basin]. *In*: [*Regional studies and new exploration targets for oil and gas*]. Institute of Geology and Exploitation of Combustible Raw Materials, Moscow, 59–66 [in Russian].
Sokolov, V. B. 1992. [Structure of the Uralian Earth's crust]. *Geotektonika*, **5**, 3–19 [in Russian].
Svyazhina, I. V., Puchkov, V. N. & Ivanov, K. S. 1992. [Reconstruction of the Ordovician Uralian ocean based on paleomagnetic data]. *Geologiya i Geofyzika*, **4**, 17–22 [in Russian].
Timonin, N. I. 1975. [*Tectonics of the Chernyschov Range*]. Nauka, Leningrad [in Russian].
Tuzhikova, V. I. 1960. [*Geotectonic conditions of formation of Early Mezozoic coal-bearing deposits of the eastern slope of the Middle Urals and*

Cis-Urals]. Trudy GGI UF AN SSSR, **46** [in Russian].
VALIZER, P. M. & LENNYKH, V. I. 1988. [*The amphiboles of the blueschists of the Urals*]. Nauka, Moscow [in Russian].
VERNIKOVSKY, V. A. 1995. [Characteristic features of formation of the metamorphic complexes of the Taymyr foldbelt in the Riphean and Paleozoic]. *Petrologiya/Petrology*, **3(1)**, 64–83 [in Russian].
VOLOZH, YU. A. 1991. [*Sedimentary basins of the western Kazakhstan based on seismostratigraphic analysis*]. Doctoral Dissertation, Moscow, Geol. Inst. AN SSSR [in Russian].
YEFIMOV, A. A., YEFIMOVA, L. P. & MAYEGOV V. I. 1993. [*Tectonics of the Platinum-bearing belt of the Urals:relationships of material complexes of the Urals and a mechanism of structure formation*]. *Geotektonika*, **3**, 34–46 [in Russian].
YAZEVA, R. G., PUCHKOV, V. N. & BOCHKAREV, V. V. 1989. [*Relics of the active continental margin in the Urals*]. *Geotektonika*, **3**, 76–85 [in Russian].
YELISEEV, A. I. 1973. [*The Carboniferous of the Lemva zone in the Urals*]. Nauka, Moscow [in Russian].
YUDIN, V. V. 1983. [*The Variscides of the Northern Urals*]. Nauka, Leningrad [in Russian].
—— 1994. [*The orogeny of the Northern Urals and Pay-Khoy*]. Nauka, Ekaterinburg [in Russian].
ZAKHAROV, O. A. & MAVRINSKAYA, T. M. 1994. [New paleontological data on the age of the protolith of the Ural-Tau metamorphic rocks]. *Yezhegodnik–93, IG UNTs RAN, Ufa*, 19–20 [in Russian].
—— & PUCHKOV, V. N. 1994. [*On the tectonic nature of the Maksyutovo metamorphic complex in the Southern Urals*]. Ufa, UNTs RAN [in Russian].

Cenozoic evolution of the Central Andes in Bolivia and northern Chile

SIMON LAMB, LEONORE HOKE, LORCAN KENNAN & JOHN DEWEY
Department of Earth Sciences, University of Oxford, Parks Road, Oxford OX1 3PR, UK

Abstract: The Central Andes in Bolivia and northern Chile form part of a wide and obliquely convergent plate-boundary zone where the oceanic Nazca plate is being subducted beneath the continental South American plate. In the latest Cretaceous and Palaeocene, this part of the Central Andes formed a volcanic arc along what is today the forearc region of northern Chile, with a wide zone of subsidence, as much as 400 km wide, at or close to sea level behind the arc. In the Eocene, the central part of the behind-arc basin was inverted to form a zone of uplift (proto-cordillera), about 100 km wide and along what is today the western margin of the Eastern Cordillera of Bolivia. The Altiplano basin and an early foreland basin were initiated at this time, receiving sediment from the Eocene proto-cordillera. Subsequently, the proto-cordillera widened, as the rate of deformation increased and deformation spread westwards into the early Altiplano basin, and also eastwards towards the Brazilian Shield. In the Late Miocene, deformation essentially ceased in the Altiplano and Eastern Cordillera. An intense zone of shortening was initiated in what is today the Subandean Zone on the eastern margin of the Central Andes, deforming the Oligo-Miocene foreland basin. Shortening in the Subandean Zone accommodated both underthrusting of the Brazilian Shield and also bending of the entire mountain belt about a vertical axis. It is suggested that much of the distinctive Cenozoic tectonic evolution of this part of the Andes is related to pre-Andean strength inhomogeneities in the South American lithosphere.

The Andes are one of the largest active plate-boundary zones, forming a mountainous region which extends for over 5000 km along the western margin of South America (Fig. 1a) as a result of the subduction since the Cretaceous of the oceanic Nazca (or formerly Farallon) plate beneath the South American plate (Dewey & Bird 1970; Pardo-Casas & Molnar 1987). They show marked variation in tectonic style and evolution along their length, as well as several major changes in trend.

The Andes are highest and widest in the Central Andes of northern Chile and Bolivia, where the present-day relative plate convergence is roughly ENE–WSW at *c.* 85 mm a^{-1} (Fig. 1, DeMets *et al.* 1990). Here, there is a pronounced bend (Arica bend) in both the structural and topographic trends, which swing round from *c.* NW–SE, north of the bend, to *c.* N–S further south (Fig. 1a). An active volcanic arc follows the western margin of the high Andes, and there is a zone of Cenozoic magmatism east of the arc, up to several hundred kilometres wide. Also, in this part of the Andes, there are thick continental sedimentary sequences, deposited during the Cenozoic tectonic evolution. The region is well populated and easily accessible. For all these reasons, this is an extremely good place to study the development of wide zones of continental deformation.

We describe here the tectonic evolution since the Cretaceous of the Central Andes in northern Chile and Bolivia, based mainly on our own extensive field work, unpublished oil company data, and new geochronological, geochemical and palaeomagnetic data.

Physiographic and geological provinces

The Andes between 16°S and 23°S form a wide region up to 800 km wide and reaching elevations over 6000 m, bounded in the east by the Peru–Chile trench and in the west by the Amazon basin and Chaco plains (Fig. 1). Topographic cross-sections (Fig. 1b) show that in detail there are several distinctive physiographic provinces. These provide a convenient way of describing the geology, because the tectonic evolution of each province has been generally distinct (Figs 1 & 2). Crustal-scale cross-sections through the Bolivian Andes are shown in Figures 1c and 3.

From Burg, J.-P. & Ford, M. (eds), 1997, *Orogeny Through Time,*
Geological Society Special Publication No. 121, pp. 237–264.

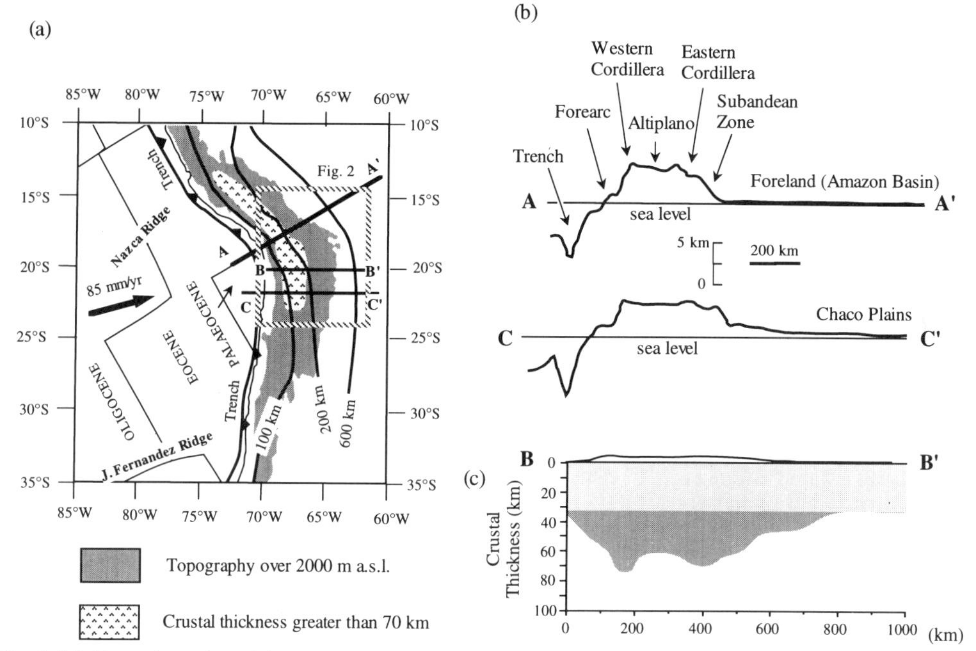

Fig. 1. (**a**) Tectonic setting of the Central Andes on the western margin of South America where the oceanic Nazca plate is being subducted beneath the South American plate in a roughly ENE-WSW direction at *c.* 85 mm a^{-1} (large arrow, DeMets *et al.* 1990). The position of the trench and depth contours of the Benioff zone are also shown (bold curves). The oceanic Nazca plate is oldest in the bend region of the trench with a Palaeocene age. The region of topography over 2000 m a.s.l. and also crustal thickness greater than 70 km in the Andes are shaded. Topographic cross-sections AA' and CC' are shown in (b). The box outline defines the region shown in Fig. 2. (**b**) Topographic profiles AA' and CC' (see (a) for location) define a mountainous zone up to 900 km wide, ranging from *c.* −6000 m in the trench to *c.* 4000 m in the Altiplano. Several physiographic provinces (Forearc, Western Cordillera, Altiplano, Eastern Cordillera, Subandean Zone, Foreland) can be recognized. (**c**) Crustal scale cross-section through the Bolivian Andes for an E–W transect at 20°S (BB^1 in (a))based on seismic data (after Beck *et al.* in press). Light shading defines original crustal thickness (*c.* 35 km) and dark shading shows the crustal thickening.

Foreland and Subandes

The foreland region along the eastern margin of the Central Andes, in the Amazonian basin and Chaco plains, is generally at an altitude of a few hundred metres above sea level (Fig. 1b). This is underlain by the Brazilian Shield, which has been a stable nucleus of South America since the Proterozoic (Litherland *et al.* 1986), and overlain by up to 5 km of Neogene sediments immediately east of the Andes in a foreland basin about 200 km wide (Fig. 3, unpublished oil company data). The topography increases towards the west in the foothills of the Andes, referred to as the Subandean Zone, which is about 100 km wide and reaches altitudes over 1500 m, dominated by major valleys and ridges which follow the general structural grain (Fig. 1b). The valleys follow synclines of Cenozoic sediment, separated by faulted anticlinal ridges of older Mesozoic and Palaeozoic sequences (Figs 2 and 3). Seismic refraction (Wigger *et al.* 1993) and reflection data (unpublished oil company data) shows that the Subandes are part of a thin-skinned fold and thrust belt accommodating shortening above a basal décollement at a depth of 7–10 km and dipping at 2–3° towards the west and southwest (Fig. 3). The Brazilian Shield has been underthrust beneath the Subandean Zone and Eastern Cordillera.

Eastern Cordillera

West of the Subandean Zone, the mountains rise progressively to altitudes over 4000 m, in a region referred to as the Eastern Cordillera. This is up to 200 km wide and made up mainly of thick (up to 10 km) sequences of Palaeozoic flysch-like deposits, with thin (<3 km) infolded Cretaceous and Cenozoic sequences. The most westerly parts of the Eastern Cordillera form a high spine, referred to in the north as the

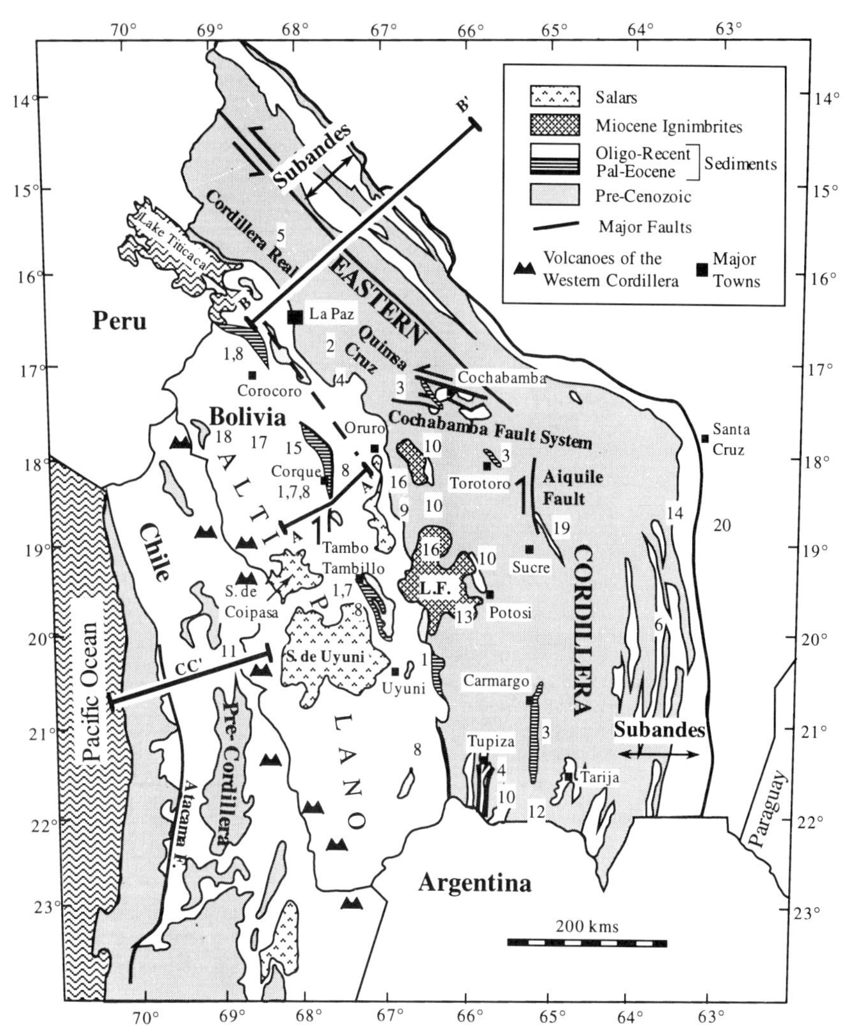

Fig. 2. Map of the Central Andes in northern Chile and Bolivia (see Fig. 1), showing the distribution of Cenozoic and pre-Cenozoic sedimentary sequences and major structures. Major physiographic features are also shown, including the large salars in the Altiplano and Lake Titicaca. AA' and BB' show the line of a composite geological cross-section in Fig. 3. Line CC' shows the line of a schematic cross-section in Fig. 4a. Numbers refer to localities where various formations or rock units, mentioned in the text (see Figs 9 & 11), are defined: (1) Potoco, Huayllamarca, Turco, Tihuanacu Formations; (2) Cordillera Real graniotoids; (3) Camargo, Torotoro, Morachata/Viloma Formations; (4) Salla, Tupiza Formations; (5) Quimsa Cruz granites (Mina Viloco and Mina Argentina bodies); (6) Petaca Formation; (7) Azurita Formation; (8) Totora, Huayllapucara, Caquiaviri, Tambillo Formations; (9) Challapata ignimbrite; (10) Mondragon, Bolivar, Parotani, Nazareno Formations; (11) Chilean precordillera ignimbrites; (12) Quebrada Honda Formation; (13) pre-Los Frailes ignimbrites; (14) Yecua, Tariquia Formations; (15) Crucero, Pomata Formations; (16) Los Frailes, Morococalla ignimbrites; (17) Umala Formation, Turco ignimbrite; (18) Perez ignimbrite; (19) Sucre-Tarabuco tuffs; (20) Emborozu Formation and recent sedimentation.

Cordillera Real and Quimsa Cruz (Fig. 2), rising to over 6000 m where Palaeozoic, Triassic and possibly Cenozoic intrusive granitoid bodies outcrop. There is a marked change in vergence of structures east and west of the central part of the Eastern Cordillera. In the west, vergence is mainly towards the west, whereas in the east vergence is towards the east (Fig. 3).

In the bend region, at *c.* 17.5°S, a series of Plio-Pleistocene basins occur within the Eastern Cordillera. These are bounded by ESE-trending normal faults with a sinistral strike-slip com-

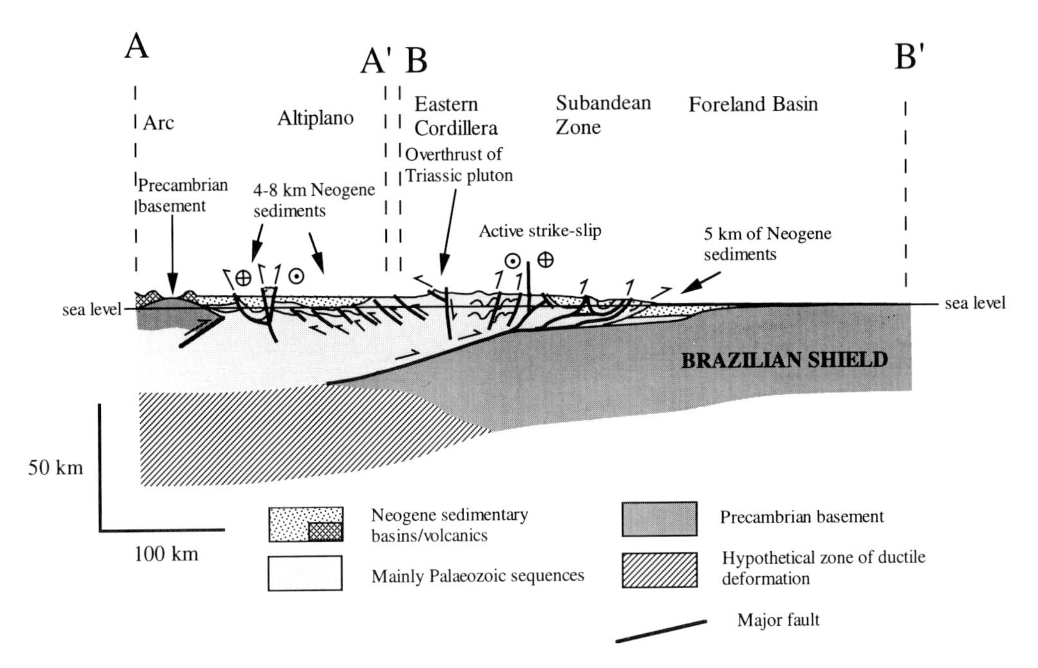

Fig. 3. Composite crustal-scale geological cross-section through the Bolivian Andes (see Fig. 2), based mainly on unpublished oil company data. The principal features are the presence of thick Neogene sedimentary basins in both the Subandes and Foreland, and also the Altiplano region. Note the marked changed in vergence of structures east and west of the central part of the Eastern Cordillera. Precambrian basement has been underthrust beneath the Subandean Zone and Eastern Cordillera, and also outcrops on the western margin of the Altiplano (Troeng *et al.* 1994). A zone of distributed ductile deformation may occur at depth beneath the Altiplano.

ponent which comprise the Cochabamba Fault System (Fig. 2, Dewey & Lamb 1992; Kennan 1994; Kennan *et al.* 1995).

Altiplano

The Altiplano forms a *c.* 200 km wide region of subdued relief, west of the Eastern Cordillera, at an average altitude of *c.* 3800 m (Figs 1b & 2). It is the second largest high plateau region, after Tibet, on Earth, and is essentially a region of internal drainage. Near La Paz the drainage has locally broken through to the Amazon basin. The vast salars of Uyuni and Coipasa, and also Lake Poopo, are the remnants of once extensive Pleistocene lakes (Servant & Fontes 1978). The Altiplano has been an important locus of sedimentation, where thick sequences of red-beds have accumulated in the Cenozoic (Figs 2 and 3).

Western Cordillera and Forearc

The Western Cordillera is the active volcanic arc along the international border between Bolivia and Chile, and consists of spaced Miocene and Quaternary andesitic volcanoes and small volcanic centres which have erupted through a poorly known sequence of Cenozoic, Cretaceous and older rocks (Figs 2 and 3). Volcanic cones rise over 2000 m above the general land surface, reaching elevations over 6000 m.

The western margin of the high Andes comprises the Precordillera of northern Chile, which consists mainly of Precambrian basement rocks and Mesozoic sedimentary sequences and Cenozoic intrusive and extrusive rocks. A major fault system, which extends for hundreds of kilometres along the Precordillera and further south (Cordillera Domeyko), appears to have accommodated both sinistral and dextral stike-slip during the Cenozoic (Mpodozis *et al.* 1993; Reutter *et al.* 1993). The Precordillera slopes from *c.* 4000 m to an altitude of *c.* 1000 m, in a region mantled by Miocene ignimbrites (Figs 2 & 4a). The foot of this ignimbrite slope is a flat region of younger Cenozoic deposits in the central depression of northern Chile (Figs 2 & 4a). The Atacama fault, which extends for hundreds of kilometres along the coastal parts of

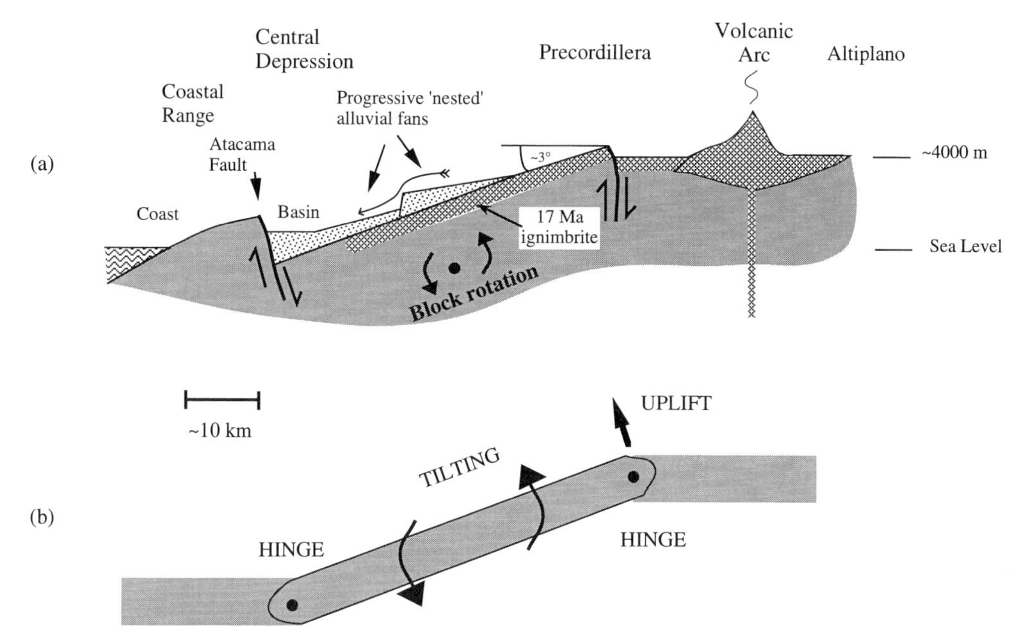

Fig. 4. Diagrams illustrating the Plio-Pleistocene kinematics of the western margin of the Central Andes in the forearc region of northern Chile. See section CC' in Fig. 2 for general location. (**a**) Early Miocene ignimbrites slope off from the Altiplano and Precordillera, dipping gently towards the Atacama Fault (active in the Plio-Pleistocene principally as a normal fault). A series of younger nested alluvial fans above the ignimbrite suggest progressive block tilting and down-cutting. (**b**) Block tilting of the western margin of the Central Andes appears to have accommodated uplift of the Altiplano relative to the coastal regions. The normal displacements on the Atacama and other faults may be second-order accommodation structures related to this tilting with overall convergence between the coast and the volcanic arc.

northern Chile as a high angle fault with a normal dip-slip component of motion, forms the western margin of this depression, separating it from the narrow coastal ranges which rise up locally to over 2000 m (Figs 2 & 4a). A zone of Cenozoic thrusting occurs south of 22°S in the Salar de Atacama region (Jolley *et al*. 1990, Fig. 2).

West of the coastal ranges, the topography slopes below sea level in the Pacific Ocean down to depths of 6 km in the Peru–Chile trench about 75 km off-shore, forming a deformed crustal prism (Fig. 1b).

Lithospheric structure

The lithospheric structure of the Central Andes can be constrained by both geophysical data (seismicity and gravity), and also geochemical evidence. In particular, a study of the helium isotopic signature of geothermal areas places constraints on processes near the base of the lithosphere.

Seismicity and gravity

Figure 5a shows the epicentres of all earthquakes between 1964 and 1992 in the Central Andes between 10°S and 35°S, shallower than 70 km and magnitude $M_b > 4.0$. The clear coincidence between the shallow seismicity and the topographic expression of the Central Andes, shows that the Central Andes are an actively deforming mountain belt. A projection of both shallow and deep focus earthquakes on to an east–west vertical cross-section in the vicinity of 20°S (Fig. 5b) shows that the bulk of the seismicity defines a Benioff zone which dips below the western margin of South America at *c*. 30°, reaching depths greater than 600 km (Figs 1 & 5b). However, south of 27°S and north of 15°S, the Benioff zone flattens markedly, associated with gaps in the volcanic arc.

Studies of seismicity, gravity and topography in the Bolivian Andes show that the crustal thickness beneath the Altiplano is up to 75 km thick (Fig. 1a and c, James 1971; Wigger *et al*.

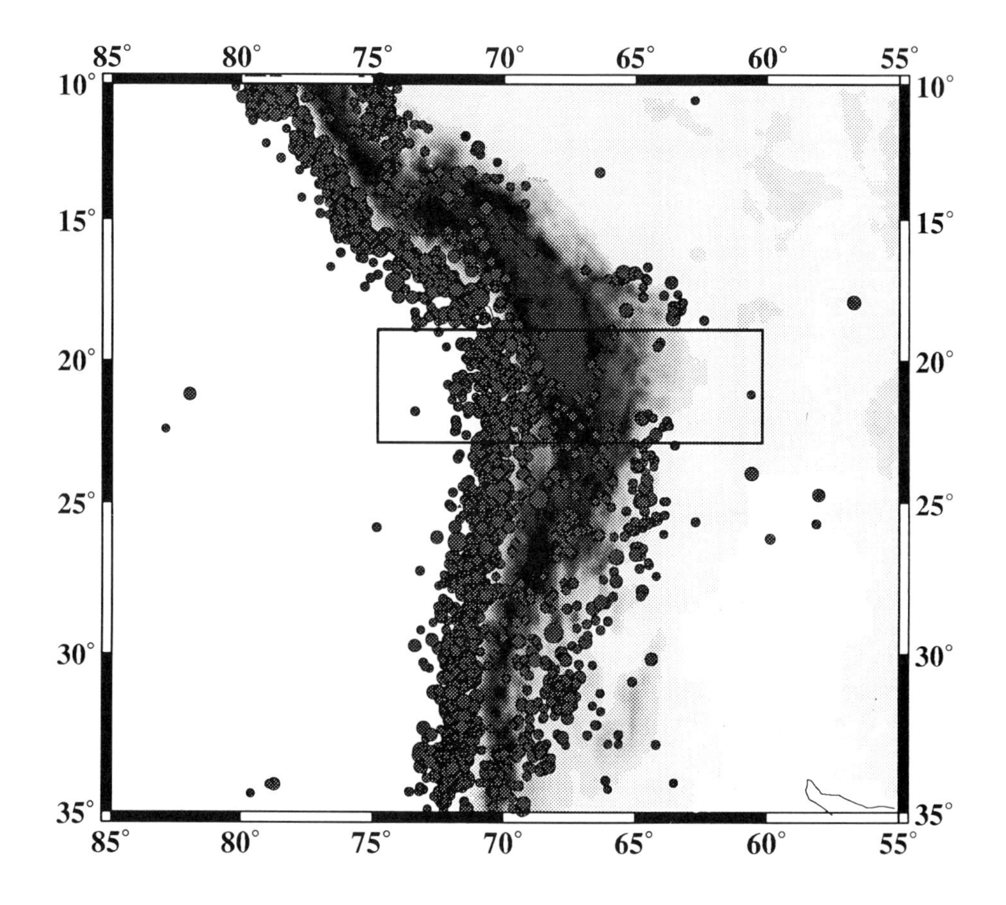
85°
80°
75°
70°
65°
60°
55°
10°
15°
20°
25°
30°
35°

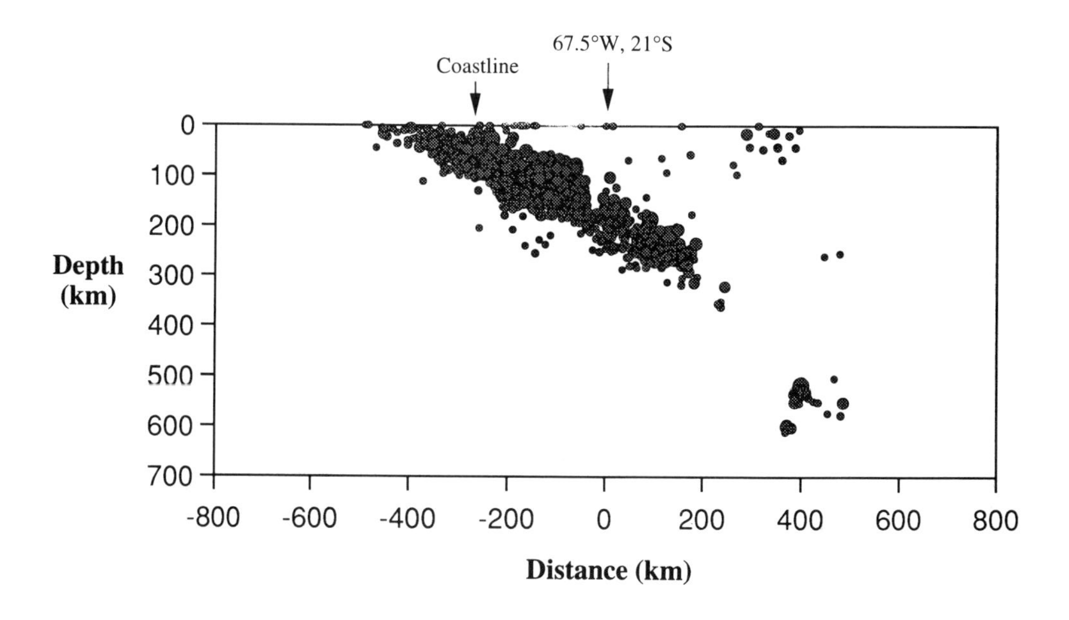
67.5°W, 21°S
Coastline
0
100
200
300
400
500
600
700
Depth (km)
-800
-600
-400
-200
0
200
400
600
800
Distance (km)

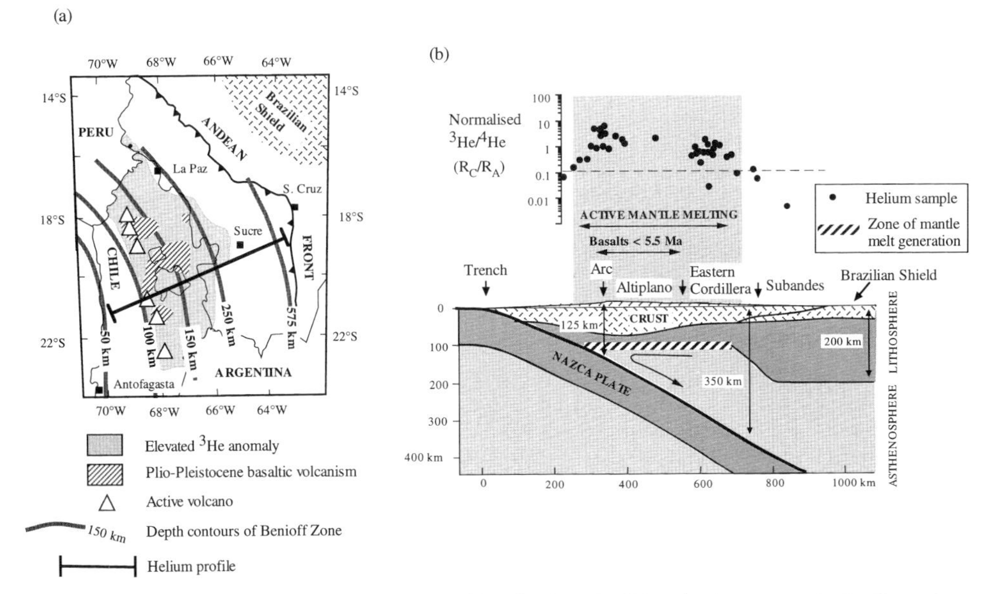

Fig. 6. Diagram illustrating helium isotopic data and implications for lithospheric structure in the Central Andes of northern Bolivia and Chile, after Hoke *et al.* (1994*b*). (**a**) Map showing the region characterized by hot springs, mineral waters and sulfataras with an anomalously high $^3He/^4He$ ratio (normalized $^3He/^4He$ with respect to air >0.1). Also shown are the general regions of outcrop of monogenetic basaltic Plio-Pleistocene volcanism, the location of the active volcanic arc, and depth contours of the Benioff zone. Basaltic volcanism and elevated $^3He/^4He$ ratios occur in a wide region up to 300 km behind the arc. (**b**) Lithospheric-scale cross-section across the Central Andes (see (a) for location), showing the position of the subducted slab and structure of the overlying South American plate. The thickness of the South American lithosphere is partly based on an interpretation of the helium data and topography. The elevated $^3He/^4He$ ratios and basaltic volcanism are interpreted to be the consequence of behind-arc mantle melting. The broad consistancy of the helium data suggests a 'thinned lithosphere', *c.* 100 km thick, extending behind the volcanic arc and beneath the Altiplano and part of the Eastern Cordillera , with a zone of mantle melting at its base. The drop-off in the $^3He/^4He$ ratio further east is interpreted to mark the presence of the Brazilian Shield and 'thick' old lithosphere (see text).

1993; Dorbath *et al.* 1993; Zandt *et al.* 1994; Beck *et al.* 1996), and thins both east and west towards regions of normal continental crustal thickness of *c.* 35 km in the Chilean forearc region and the Brazilian shield. However, there are also marked isostatic gravity anomalies. A negative isostatic anomaly east of the Subandean Zone suggests an excess of low density material, whereas a positive anomaly in the Subandean Zone and Eastern Cordillera suggests an excess of high density material (Lyon-Caen *et al.* 1985; Watts *et al.* 1995). This positive – negative coupling of isostatic gravity anomalies is characteristic of regions where the lithosphere has marked flexural rigidity. Detailed modelling suggests a flexural rigidity for the lithosphere equivalent to an elastic thickness greater than 50 km in the bend region, decreasing to less than 25 km both north and south (Watts *et al.* 1995; Whitman 1995). Thus, the region of flexurally strongest lithosphere coincides with the old and 'cold' Brazilian Shield, where the base of the lithosphere might be expected to be at depths greater than 200 km (*c.* 1200°C isotherm). The

Fig. 5. (**a**) Map of the Central Andes showing the location of earthquake events between 1964 and 1992 shallower than 70 km and with magnitudes $M_b > 4.0$ (data taken from the IS catalogue). Shading defines general topography, becoming darker in higher ground. The bulk of the shallow seismicity occurs off-shore near the trench. Most of the remaining earthquake moment release occurs much further east in the Subandean Zone and Foreland region. (**b**) Seismicity between 1964 and 1992 in an east–west vertical cross-section between latitudes 19°S and longitude 23°W for events with $M_b > 1.0$ (see box in (a) for location). The section is centred on latitude 21°S and longitude 67.5°S. The bulk of the seismicity defines a Benioff zone, dipping at *c.* 30° to the east. The remaining moment release defines more diffuse deformation in the overlying plate.

simplest interpretation is that this shield area has been underthrust beneath the Subandean Zone in the bend region of the Bolivian Andes and is providing flexural support for the topographic load of the Andes in the Subandean Zone and at least part of the Eastern Cordillera. Flexing of the lithosphere has also created a foreland basin up to 5 km deep and up to two hundred kilometres wide (unpublished oil company data) along the eastern margin of the Central Andes (Fig. 3).

Helium isotope studies

Information about the lithospheric structure also comes from a study of helium emitted in geothermal springs. A recent helium isotope study right across the Central Andes in Bolivia and northern Chile (Hoke *et al.* 1993, 1994*b*) has documented an elevated ^{3}He signature, greater than that typical for crustally derived helium. This zone extends for up to 400 km from the Precordillera in the west, across the active volcanic arc and the Altiplano high plateau region to the western part of the Eastern Cordillera (Fig. 6a). In the absence of significant crustal sources of ^{3}He, the most likely sources for the ^{3}He are mantle melts present at depth. Therefore, the elevated ^{3}He signature maps an extensive area of active mantle melt generation and concomitant subsurface basalt addition to the Andean crust (Hoke *et al.* 1993, 1994*b*), in a region where the crust is up to 75 km thick and the subducting slab is at depths of 100–350 km (Fig. 6b). Mantle melting is also suggested by the rare presence of small young olivine-bearing basaltic andesite volcanic centres throughout the Altiplano (Fig. 6a), with ages less than *c.* 5 Ma (Hoke & Lamb, unpublished data).

Hoke *et al.* (1994*b*) interpreted the elevated ^{3}He signature in the volcanic arc to be a consequence of melting in the convecting asthenospheric mantle wedge between the subducting slab and overlying lithosphere, where dehydration of the downgoing slab may hydrate the overlying mantle and cause melting if temperatures are greater than 1100°C (see also Davies & Bickle 1991). This model explains sustained arc volcanism over tens of million years and implies that the base of the lithosphere beneath the arc is much shallower than the depth to the Benioff zone here (<120 km). A different mechanism of melting to that for arc volcanism is required to explain the presence of mantle melts at depth beneath both the Altiplano and western parts of the Eastern Cordillera (see discussion below). However, as both the observed ^{3}He signature and topography are broadly constant right across the width of the Altiplano and into the Eastern Cordillera (Fig. 6b), it is suggested here that the zone of thin lithosphere beneath the arc (<120 km thick) extends beneath the Altiplano and western part of the Eastern Cordillera where there is a significant mantle-derived contribution to the helium isotope signal (Hoke *et al.* 1994*b*; Hoke & Lamb 1994; Fig. 6b).

We interpret the transition to pure crustal helium isotope ratios in the Eastern Cordillera to mark the change to a thicker lithosphere (>200 km) and the western limit of underthrusting of the flexurally-strong Brazilian Shield beneath the Eastern Cordillera, defined by the gravity data (Fig. 6b). Thus, our proposed lithospheric structure for the Bolivian Andes, with a wide zone of thinned lithosphere beneath the Altiplano and western part of the Eastern Cordillera, is similar to that previously proposed for the Bolivian Altiplano (Isacks 1988), but differs from the model of Whitman *et al.* (1992) based on seismic data.

Cenozoic basin evolution and deformation

Since the Cretaceous, the Andes in northern Chile and Bolivia have essentially formed part of a convergent plate margin, with the subduction of the oceanic Nazca plate (or Farallon plate) beneath the South American continent (Pardo-Casas & Molnar 1987). It is likely that this type of plate margin existed prior to the Cretaceous as well (Coira *et al.* 1982; Flint *et al.* 1993).

The stratigraphy and distribution of both Cretaceous and Cenozoic sedimentary basins, as well as Cenozoic intrusive and extrusive bodies, place constraints on processes occurring at the surface and deeper in the lithosphere. A composite crustal cross-section through the northern part of the Bolivian Andes shows the overall style of deformation and the location of the two principal regions of Cenozoic sedimentation in the Subandean and foreland regions and the Altiplano (Fig. 3). In the following sections, the structure and stratigraphy of sedimentary rocks are used to reconstruct the Cenozoic evolution of sedimentary basins and related deformation. The distribution of deformation and sedimentation, both in time and space, for a generalized east–west transect across the Andes in northern Chile and Bolivia at the latitudes 16°S to 22°S, is summarized in Fig. 7. The general history is one of a widening zone of deformation which commenced in what is today the Eastern Cordillera in the early Cenozoic, when most of the Bolivian Andes was near or at sea level. The extensive Altiplano basin formed as a region trapped between the

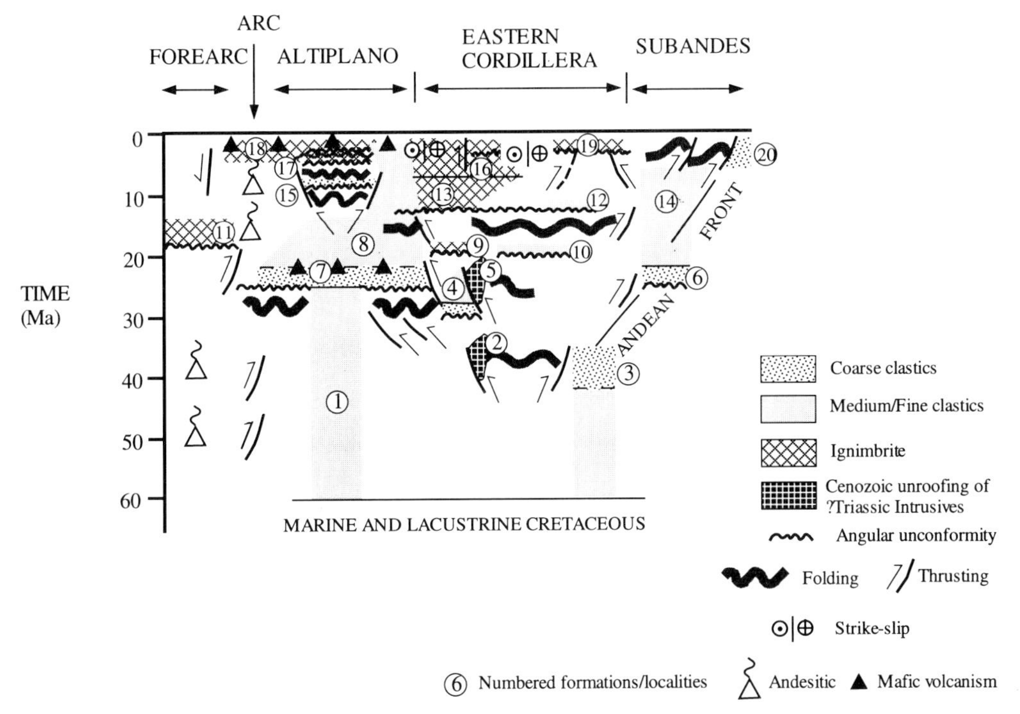

Fig. 7. Diagram summarizing the distribution and timing of deformation and sedimentation in the Central Andes for a generalized east–west transect between the latitudes 16°S and 22°S. Numbers in circles refer to geological formations or localities, defined in Fig. 2 (see also Fig. 9).

uplifting Eastern Cordillera and the volcanic arc and received sediment for most of the Cenozoic. Also, Cenozoic sequences were deposited in local fault-bounded compressional basins within the Eastern Cordillera, and in a series of foreland basins further east. Formation names used in the following sections are those adopted by the Bolivian Geological Survey in their 1 : 100 000 and 1 : 250 000 geological map series.

Pre-Cretaceous, Cretaceous and early Palaeocene

The backbone of most of the Bolivian Andes, especially in the Eastern Cordillera, consists of sequences of deformed Ordovician and Silurian flysch deposits, many kilometres thick. There is evidence for significant pre-Cretaceous deformation, but the nature and tectonic setting of this deformation is not well understood.

The volcanic arc in the Cretaceous was in the present-day forearc of northern Chile (Fig. 8; Campusano 1990). There is a marked angular unconformity at the base of the Cretaceous in the Eastern Cordillera and Altiplano in Bolivia. Open to tight folding, with limb dips up to 50°, a weak axial planar cleavage and extensive quartz veining are found in Palaeozoic flysch deposits beneath the Cretaceous throughout the Eastern Cordillera (for instance in the Cochabamba, Potosi and Camargo regions). They are truncated by Cretaceous conglomerates (La Puerta and Angostura Formations). The metamorphic grade of pre-Cretaceous rocks suggest that at least 5 km of pre-Cretaceous rocks were stripped off by erosion prior to Cretaceous deposition.

From fossil evidence (summarized in Riccardi 1988), most of the Cretaceous in Bolivia above the basal conglomeratic beds is Cenomanian and younger. Isopach maps of the Cretaceous sequences in Bolivia define a general depocentre near the western margin of the Eastern Cordillera (Fig. 8; unpublished oil company data). In detail, there are two sub-basins, referred to as the Sevaruyo and Maragua basins, where Cretaceous sequences are up to 3 km thick (Riccardi 1988; Rouchy *et al.* 1993; unpublished oil company data), separated by a local 'high' that coincides with the southern extension of the Cordillera Real along the western margin of the Eastern Cordillera (Fig. 8). Local mafic volcanism, including pillow lavas, could be associ-

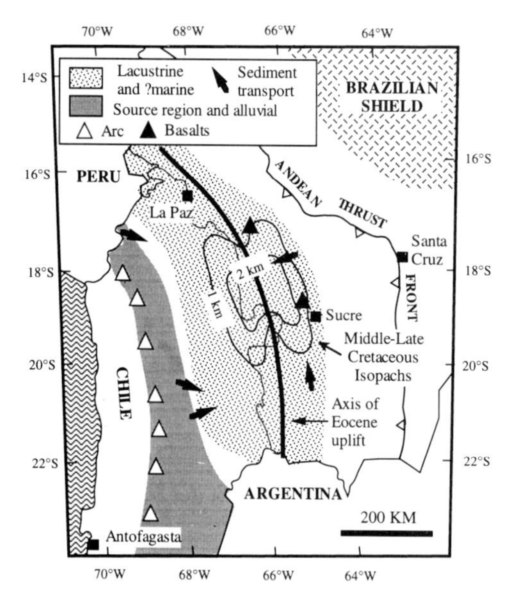

Fig. 8. Generalized palaeogeographic map for the Central Andes in northern Chile and Bolivia in the latest Cretaceous and earliest Cenozoic (*c.* 60 Ma). The position of the volcanic arc and main sediment source regions and basins and thrust front are plotted in their present location and are not palinspastically restored. Sediment transport directions are shown in their present-day orientations. Note general position of the axis of the Cretaceous basin, with palaeoflow towards this depocentre. The axis more-or-less coincides with the zone of Eocene uplift (see Figs 10 and 16).

ated with a small amount of lithospheric extension concomitant with subsidence of the Cretaceous basin (Soler & Sempere 1993). There is evidence for a marine trangression in the lower part of the Cretaceous sequence, where distinctive grey limestones occur (Miraflores Formation). Maastrichtian sequences progressively onlap the margins of the sub-basins and here the Cretaceous sequences are much thinner (usually ≪1 km). The lateral continuity and extensive outcrop of the Maastrichtian and Palaeocene lacustrine and possibly marine limestones and shales (Fig. 9, El Molino and Santa Lucia Formations, Gayet *et al.* 1991; Rouchy *et al.* 1993), preserved throughout the Eastern Cordillera and Altiplano, demonstrate that most of the Bolivian Andes was a region of very subdued topography at or near sea level at this time. The active plate margin was a relatively narrow zone much further west, where thick Late Cretaceous to Palaeocene clastic sequences were deposited in what is today the forearc region of northern Chile (Hartley *et al.* 1992; Flint *et al.* 1993).

Interestingly, the general axis of the Cretaceous basins in the Eastern Cordillera coincides with the region of most intense pre-Cretaceous shortening, where angular truncations at the Cretaceous unconformity are greatest.

Palaeocene to Oligocene

The volcanic arc in the Palaeocene and Eocene followed what is today the Precordillera in northern Chile (Figs 9 and 10, Campusano 1990). However, the apparent lack of arc volcanics in the age range 25 Ma to 35 Ma (Lahsen 1982; Campusano 1990), suggests that arc volcanism may have been limited in the Oligocene. This region also appears to have been the locus of deformation (Hartley *et al.* 1992). North–south horizontal shear has been described from the Cordillera Domeyko (Mpodozis *et al.* 1993) and the Chilean Precordillera further north (West Fissure System, Reutter *et al.* 1993) in what appears to be a major arc-parallel fault zone extending for hundreds of kilometres. However, the large-scale kinematic significance of this fault zone is unclear, and Reutter *et al.* (1993) describe Eocene dextral strike-slip and Oligocene sinistral strike-slip motion, whereas Mpodozis *et al.* (1993) describe the converse with Eocene sinistral strike-slip and Oligocene dextral strike-slip motion. The total Cenozoic displacements across this fault zone have yet to be quantified.

There is evidence for Eocene deformation in a narrow zone much further east along what is today the western margin of the Eastern Cordillera. Here, in the Cordillera Real, both K/Ar mica ages (McBride *et al.* 1987; Farrar *et al.* 1988) and fission track apatite and zircon ages (Benjamin *et al.* 1987) indicate rapid cooling at ca. 40 Ma. Also, Farrar *et al.* (1990) document evidence for folding of the Cordillera Real between the youngest K/Ar resetting age of muscovite (*c.* 60 Ma) and biotite (*c.* 39 Ma). A fission track zircon age of 38 Ma from the Quimsa Cruz batholith, at the southern end of the Cordillera Real, also suggests Eocene cooling (Lamb & Hurford unpublished data). Marocco *et al.* (1987) describe folding on the western margin of the southern Eastern Cordillera which may be Palaeocene in age, though the field relations and dating are not conclusive.

Evidence for uplift at this time also comes from the Cenozoic sedimentary record in both the Altiplano region and Eastern Cordillera.

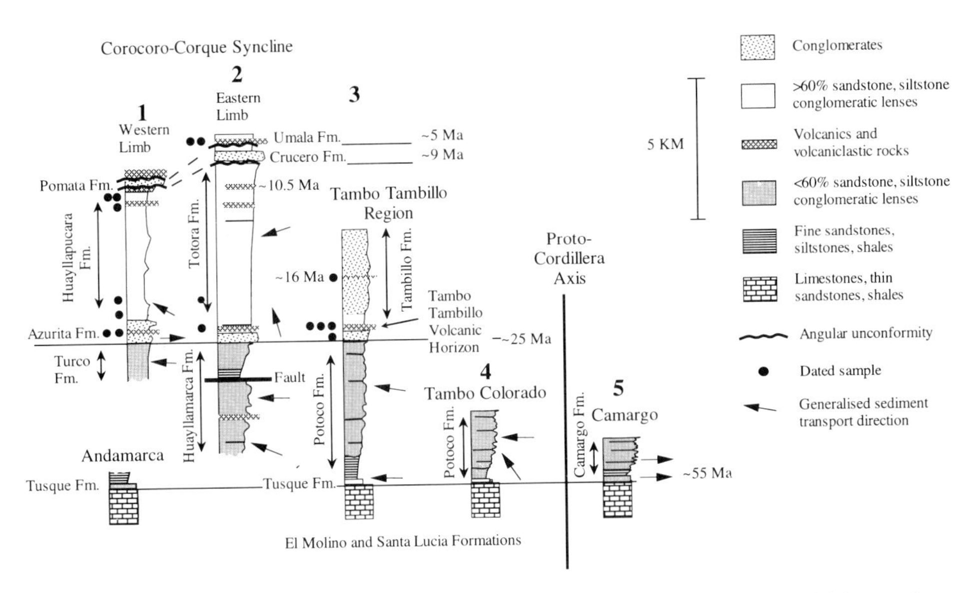

Fig. 9. Diagram showing schematic Cenozoic stratigraphic sections for five localities in the Altiplano and Eastern Cordillera, defining the principal Cenozoic formations and the localities of dated volcanic horizons (Kennan *et al.* 1995; Lamb unpublished data). See Fig. 10 for locations of numbered sections. Generalized sediment transport directions (present-day orientation) are also shown, each based on numerous (15–100) measurements of rib and furrow or trough cross-bedding orientation. Note the change in sediment transport directions between sections 4 and 5. The latter defines the axis of Eocene uplift which resulted in a divergence of palaeoflow, either into the Altiplano basin (section 4) or early foreland region (section 5).

Fossiliferous Cretaceous limestones (El Molino Formation, Gayet *et al.* 1991) and Palaeocene lacustrine deposits (Santa Lucia Formation, Gayet *et al.* 1991; Marshall & Sempere 1991) pass conformably into thick red-bed sequences (Fig. 9, Cayara, Tusque, Camargo and Potoco Formations). For instance, in the Camargo area of southern Bolivia, this transition can be traced along strike for over a hundred kilometres and is perfectly conformable. The basal part of the red-bed sequence (Santa Lucia Formation) consists of red siltstones with thin medium sandstone interbeds. Only *c.* 150 m stratigraphically above the Maastrichtian limestones, conglomerates and thick coarse sandstones are well developed (Fig. 9, Camargo Formation). Similar transitions can be found further north near Torotoro, and also near Cochabamba (Kennan 1994), and also in the Altiplano near Corque, Tambo Tambillo and Tambo Colorado (Figs 2, 9 & 10).

The sediment source regions for these red-bed sequences can be defined with some precision from the pattern of sediment transport directions. The centre of the Cretaceous to Palaeocene basin was inverted, so that the deepest part formed a narrow uplifting region (proto-cordillera) in the Eocene (Figs 8 & 10), which shed sediment both to the east and west. The eastern sedimentary basins, outcropping between Cochabamba and Camargo, and further south, formed the foreland basin for the Bolivian Andes at this time, accumulating up to a kilometre of sediment derived from the west (Fig. 9). The proto-cordillera developed as a narrow isolated range in what is today the western part of the Eastern Cordillera, separated from the active volcanic arc by a region several hundred kilometres wide in Bolivia (Fig. 10). Thus, the intervening region formed a large intermontane basin which is now preserved in the Altiplano region of the Bolivian Andes. Near Tambo Tambillo and Corque in the central Bolivian Altiplano, up to 5 km of Early Cenozoic continental sediments (Fig. 9, Tusque, Potoco, Tihuanaco, Huayllamarca and Turco Formations), transported partly from the Eastern Cordillera, were deposited in what appears to be the main depocentre of the Altiplano basin.

Late Oligocene to Early Miocene

Undeformed Early Miocene ignimbrites (Baker & Francis 1978) show that north of the latitude 22°S, significant shortening deformation in

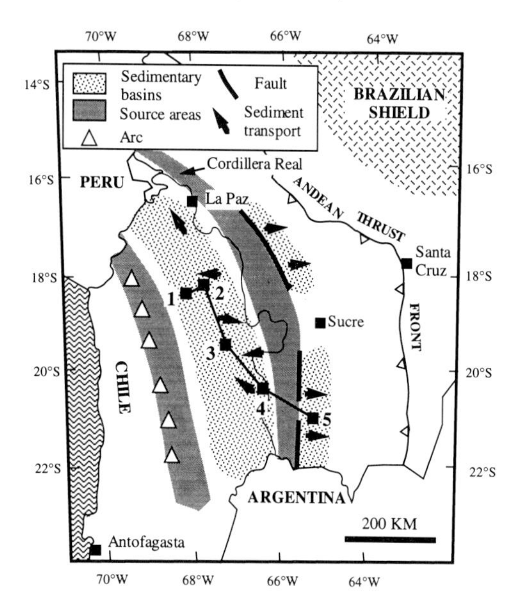

Fig. 10. Generalized palaeogeographic map for the Central Andes in northern Chile and Bolivia in the Eocene (35–55 Ma), with the same conventions as Fig. 8. The location of numbered stratigraphic sections, illustrated in Fig. 9, are also shown. Note the development of two basins at this time, separated by a general axis of uplift on the western margin of the Eastern Cordillera. The western basin represents the early stages of the Altiplano basin. Palaeoflow is both towards the Altiplano basin, where several kilometres of early Cenozoic red-beds have accumulated, and an early foreland basin which presently outcrops in the Cochabamba, Torotoro and Camargo regions (see Fig. 2).

northern Chile ceased in the Early Miocene, though relative uplift of the Western Cordillera continued. South of 22°S, these ignimbrites have been internally shortened in a thrust belt (Jolley *et al.* 1990).

The first appearance of thick coarse conglomeratic sequences in the Altiplano region occurred at *c.* 25 Ma (Kennan *et al.* 1995; Lamb unpublished data). These conglomerates (Azurita Formation) can be traced throughout the central Altiplano in a region extending at least 200 km in a N–S direction (Fig. 9). However, there is marked variation in provenance for these conglomerates. The eastern outcrops, such as those near Corque, consist of Palaeozoic and Mesozoic clasts from the Eastern Cordillera. Further west, the clasts in the conglomerates consist mainly of metamorphic and igneous clasts derived from the west. The metamorphic clasts are similar to Precambrian basement rocks, currently outcropping in the Western Cordillera and Precordillera of northern Chile (Troeng *et al.* 1994). In the centre of the Altiplano basin, near Corque and Tambo Tambillo, these conglomerates are conformable with the underlying early Cenozoic succession. However, both to the east and west, there is evidence for a marked angular unconformity. The angular unconformity, overlain by conglomerates rich in metamorphic clasts, is exposed between the Uyuni and Coipasa salars (Eduardo Soria-Escalante pers. comm. 1993). Further east, near Lake Poopo, an angular unconformity is imaged in seismic reflection profiles (unpublished oil company information). The prominent reflector immediately above this unconformity may be equivalent to or very slightly younger than the *c.* 25 Ma conglomeratic horizons outcropping further west. Beneath the unconformity, sequences are folded and thrusted (Fig. 3).

The prominent angular unconformity and deposition of thick conglomeratic sequences, described above, indicates significant shortening throughout much of the Altiplano in the latest Oligocene and Early Miocene, while deposition continued in the central part of the basin. This coincides with fission track evidence for rapid cooling in the Quimsa Cruz pluton, on the western margin of the Eastern Cordillera between 27 and 22 Ma (Lamb & Hurford, unpublished data). The distribution of Oligo-Miocene sedimentary sequences in the Bolivian Eastern Cordillera shows that local fault-bounded compressional basins also formed at this time, which locally lie with angular unconformity on older sequences. For instance, the red-bed sequences in the Salla basin on the western margin of the Eastern Cordillera were deposited between *c.* 30 Ma and 22 Ma (Marshall & Sempere 1991), and sequences near Tupiza are this age too (Fig. 2, Tupiza Formation, Herail *et al.* 1993*b*).

An important event in the Altiplano at *c.* 24 Ma (Hoke *et al.* 1994*a*; Kennan *et al.* 1995; Lamb unpublished data) was the emplacement of a laterally extensive extrusive-intrusive basaltic complex (Tambo Tambillo volcanic complex, Hoke *et al.* 1994*a*) which forms sills and volcanic centres with oceanic island basalt characteristics (Davidson & de Silva 1992) in a region of at least 1000 km^2 (Figs 9 & 10). The significance of this mafic volcanism is discussed below.

Mid-Miocene

In the Early to Mid-Miocene, medium to coarse red-bed sequences were deposited right across

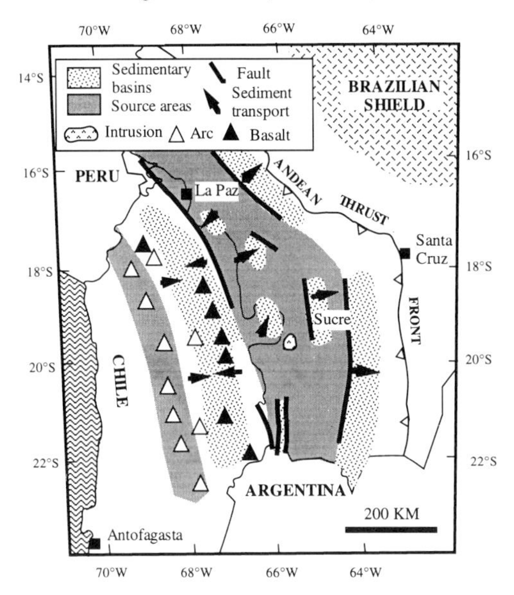

Fig. 11. Generalized palaeogeographic map for the Central Andes in northern Chile and Bolivia in the Oligo-Miocene (15–30 Ma), with the same conventions as Fig. 8. Note the widening of the uplifting regions in the Eastern Cordillera, compared to that in the Eocene, with the development of numerous small intramontane compressional basins accumulating several hundred metres of sediment. Also, the locus of sedimentation in the foreland region had migrated significantly further east. An intense phase of crustal shortening occurred in the Altiplano prior to 25 Ma, with widespread erosion in all but the centre of the basin. Subsequently, the Altiplano basin was rejuvenated with the accumulation of several kilometres of red-beds, derived both from the Western and Eastern Cordilleras. At this time, there is evidence for both silicic magmatism on the western margin of the Eastern Cordillera, and widespread basaltic volcanism in the Altiplano basin.

the Altiplano basin (Fig. 9, Totora, Huayllapucara, Corniri and Caquiaviri Formations), resulting in the rejuvenation of the Altiplano basin and nearly a doubling in the sedimentation rate compared to that in the Early Cenozoic (Figs 9 & 10). Eight kilometres of red-beds were deposited west of Corque, and several kilometres were deposited in the Lake Poopo area between 25 and 9 Ma. Synsedimentary deformation on the eastern margin of the Altiplano basin resulted in progressive folding of these sequences, with relative uplift of the regions even further east (Herail *et al.* 1993*a*; unpublished oil company data). At this time, a very coarse conglomeratic sequence, with clasts of Cretaceous limestone up to a metre in diameter (Fig. 9, Tambillo Formation), was deposited in the Tambo Tambillo region and most likely derived from uplifted regions to the east.

In the Eastern Cordillera, younger sandy and conglomeratic sequences, dated at 17–20 Ma, rest with marked angular unconformity on folded Cretaceous and Palaeozoic sequences in the Eastern Cordillera (Fig. 11), for instance near Cochabamba (Bolivar and Parotani Formations, Gubbels *et al.* 1993; Kennan *et al.* 1995) and Potosi (Mondragon Formation, Sempere *et al.* 1989; Kennan *et al.* 1995), and Tupiza (Nazareno Formation, Herail *et al.* 1993*b*). The angular unconformity indicates shortening throughout the Eastern Cordillera prior to *c.* 20 Ma, concomitant with rapid cooling of granitoid bodies on the western margin of the Eastern Cordillera, and also the rapid deposition in the Altiplano basin. Shortening continued after deposition of the 20 Ma sequences, as they are themselves gently folded.

High level shortening in the Eastern Cordillera appears to have a concertina-style, accommodated by kilometre- to tens of kilometre-scale folding and relatively high angle reverse faulting, with changes in vergence. No evidence for nappe-style deformation has been observed. Our examination of the so-called 'Calazaya Nappe' (Sempere *et al.* 1991; Baby *et al.* 1992*a*) on the western margin of the Eastern Cordillera near Uyuni, shows that this structure is an illusion created by the confusing interference of different trends of folding, young strike-slip faults and a pronounced angular unconformity at the base of the Cretaceous sequences.

The nature of the eastern front of the Andes in the Middle Miocene is not clear, though the deposition of Oligo-Miocene coarse conglomerates (Fig. 11, Petaca Formation, Sanjines & Jimenez 1976; Marshall *et al.* 1993) in the Subandean Zone suggests some sort of a foreland basin existed.

Late Miocene to Pliocene

In the forearc region and Precordillera of northern Chile, at *c.* 20°S–22°S, Early Miocene ignimbrites define a regional dip slope tilted less than 5° towards the west (Fig. 4a). The only evidence for significant younger deformation in this region is along the Atacama fault, which is a high angle normal fault which offsets Pliocene tuffs (Fig. 4a, Naranjo 1987; Dewey & Lamb 1992), and also on the western margin of the volcanic arc, where high angle normal faults also offset Late Miocene and younger ignimbrites

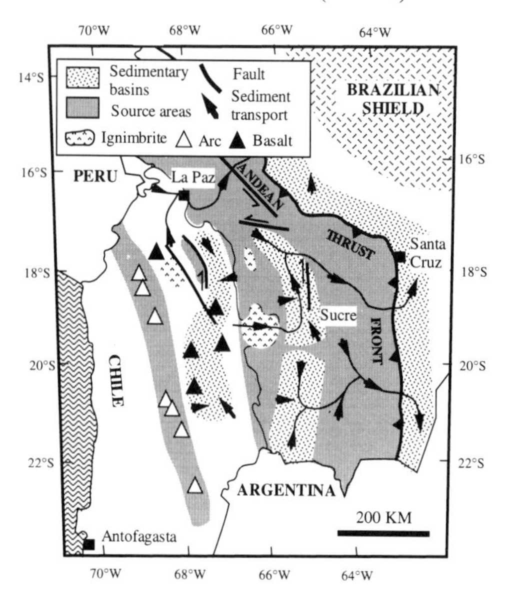

Fig. 12. Generalized palaeogeographic map for the Central Andes in northern Chile and Bolivia in the Late Miocene and Pliocene (2–8 Ma), with the same conventions as Fig. 8. The palaeogeographic reconstruction spans two main periods. Between 5 and 8 Ma, deformation in the Altiplano (see Fig. 16) resulted in uplift, erosion and breaching of the Altiplano basin, so that sediment was transported from the Altiplano region into the foreland basin. At the same time, extensive low relief peneplains developed in the Eastern Cordillera, which also ultimately drained into the foreland (San Juan del Oro and other surfaces). There were also widespread eruptions of ignimbrites on the western margin of the Eastern Cordillera (Los Frailes and satellite volcanic fields). Between 5 and 2 Ma, the Altiplano again became an internal drainage basin with the accumulation of generally thin fluvial and lacustrine sequences ($\ll$500 m), with scattered basaltic volcanism. Deformation in the Subandean Zone may have increased river gradients in the Eastern Cordillera, resulting in downcutting and dissection of the San Juan de Oro and other peneplain surfaces (schematic drainage system shown).

and lavas (Fig. 4a). However, south of 22°S, Late Miocene ignimbrites west of the volcanic arc, have been deformed in a thrust belt in the Salar de Atacama region (Jolley *et al.* 1990), though the shortening is probably small ($\ll$5 km).

The central part of the Altiplano basin started to shorten internally in the Late Miocene, during a period of a few million years. An angular unconformity, imaged in seismic reflection profiles and exposed near Corque, is overlain by conglomeratic sequences which contain *c.* 9 Ma volcanic ashfall deposits (Crucero and Pomata Formations, Lamb unpublished data). However, the most intense phase of internal deformation occurred between *c.* 9 and 5 Ma, when Mid- to Late Miocene sequences (Fig. 9, Totora, Huayllapucara and Mauri Formations) were tightly folded and cut by thrust faults with displacements of up to several kilometres (Kennan *et al.* 1995; Lamb unpublished data; unpublished oil company data). Deformed sequences as young as 9 Ma are overlain with marked angular unconformity by *c.* 5 Ma ignimbrites (Umala Formation) which are only very gently folded, and *c.* 3 Ma ignimbrites (Perez ignimbrite, Marshall *et al.* 1992) which are flat lying. The period of intense shortening prior to *c.* 5 Ma in the Altiplano marks the termination of the main phase of accumulation of sediments in the Altiplano basin. Subsequently, only a few hundred metres of Plio-Pleistocene sediments and volcanics accummulated, except locally in the region west of Oruro where more than a 1000 m may have been deposited (Fig. 12; unpublished oil company data).

In the Eastern Cordillera, regional flat-lying peneplain surfaces (Juan del Oro surface and others) developed since the Middle Miocene, with the local accumulation of sedimentary sequences as old as 12 Ma (McFadden *et al.* 1990). These are currently at altitudes between 2 and 4 km (Fig. 12, Servant *et al.* 1989; Gubbels *et al.* 1993; Kennan *et al.* 1995, 1997). Also, 12 Ma and younger ignimbrites are essentially flat lying, including the *c.* 7 Ma extensive ignimbrites of the Los Frailes and satellite massifs in the western parts of the Eastern Cordillera, and *c.* 3 Ma tuffs which mantle the remains of peneplain surfaces throughout the Eastern Cordillera (Gubbels *et al.* 1993; Kennan *et al.* 1995). Dissection of these surfaces occurred in the last 3 Ma, when the present drainage pattern was established. However, there is evidence for conjugate strike-slip faulting in the Eastern Cordillera and Altiplano, which offset Pliocene and younger strata (Dewey & Lamb 1992; Kennan 1994; Kennan *et al.* 1995; Lamb unpublished data). These faults are generally spaced at tens of kilometres with individual offsets of much less than 1 km and have accommodated shortening in an approximately east–west direction, with concomitant north–south extension.

In the Subandean Zone, Miocene to Recent sequences up to 5 km thick (unpublished oil company data), deposited in a foreland basin related to deformation in both the Eastern Cordillera and Subandes itself, are intensely folded and faulted in an active thin-skinned fold and thrust belt which has accommodated *c.* 140 km of shortening in the Bolivian bend region (Figs 3, 12, Baby *et al.* 1992*b*, 1993). The

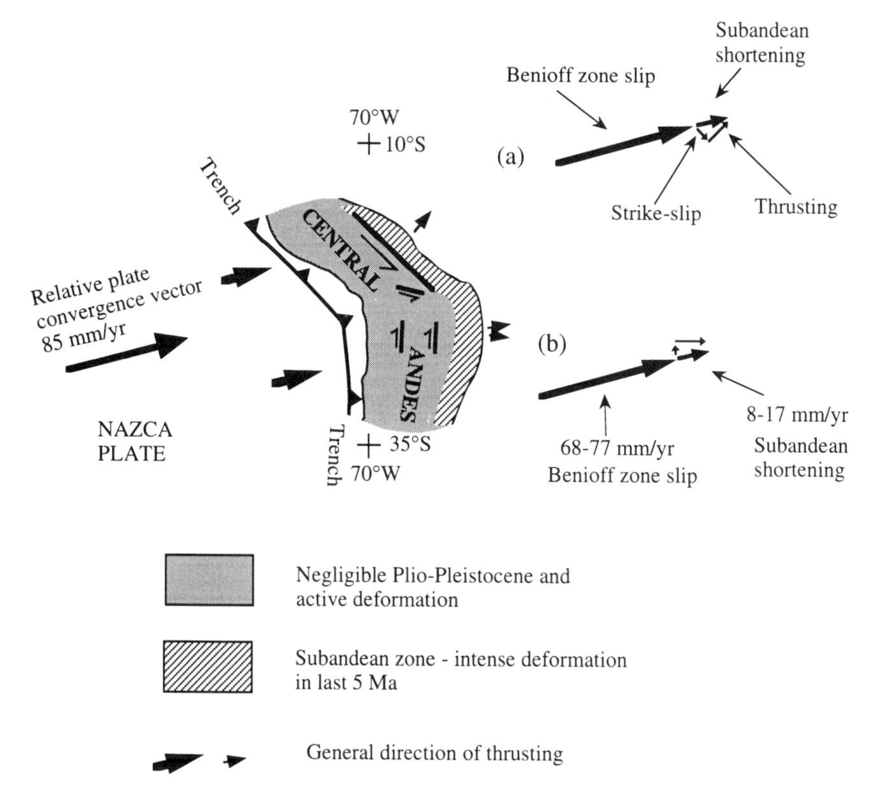

Fig. 13. Diagrams illustrating the partitioning of the relative plate convergence between the Nazca and South American plates at the latitudes of the Bolivian Andes. The total plate convergence vector for the Nazca plate relative to the South American plate is $85 \pm 5\,mm\,a^{-1}$ at $076 \pm 5°$ (DeMets *et al.* 1990). The active and Plio-Pleistocene deformation is mainly partitioned into Benioff zone slip parallel to the plate convergence vector at 68–77 mm a^{-1}, and deformation in the Subandes. (**a**) North of the bend, Subandean deformation appears to be partitioned into sinistral strike-slip at the back of the Subandean Zone and parallel to the general structural trend, and shortening in the Subandean Zone perpendicular to the structural trend (see text). (**b**) South of the bend, shortening in the Subandes may be more-or-less parallel to the plate convergence direction, though there is some evidence for dextral strike-slip within the Eastern Cordillera. The net deformation in the Subandean Zone and Eastern Cordillera results in shortening in a direction subparallel to the plate convergence vector at 8–17 mm a^{-1}.

exact timing of the inception of deformation in the Subandean Zone is not clear, but it is certainly likely to have been most active in the last 5 Ma, when shortening elsewhere in the Andes of northern Chile and Bolivia was very small.

Plio-Pleistocene and active kinematics

Earthquake moment release

Diffuse seismicity at depths less than 70 km is found throughout the Central Andes as predominantly thrust events (Fig. 5a, b) The total moment release for crustal events since 1960 accounts for less than 20% of the instantaneous relative plate convergence between the Nazca and South American plates (Dewey & Lamb 1992). This suggests that the bulk of the relative plate convergence on this short time scale is taken up in the brittle crust as elastic and anelastic strain. However, Dewey & Lamb (1992) suggest that the distribution of moment release might reflect the longer term part of active deformation on a time scale of tens of thousands to hundreds of thousands of years (cf. Jackson *et al.* 1995). The pattern of earthquake moment release suggests that the relative plate motion has been accommodated in two principal zones, along the plate interface offshore and as more distributed deformation much further east in the Andes. The net shortening direction in both zones is approximately parallel to the relative plate convergence vector (Dewey & Lamb 1992). The bulk of shallow seismicity is found offshore near the plate interface (Fig. 5a, b), suggesting that 80–90% (68–77 mm a^{-1}) of

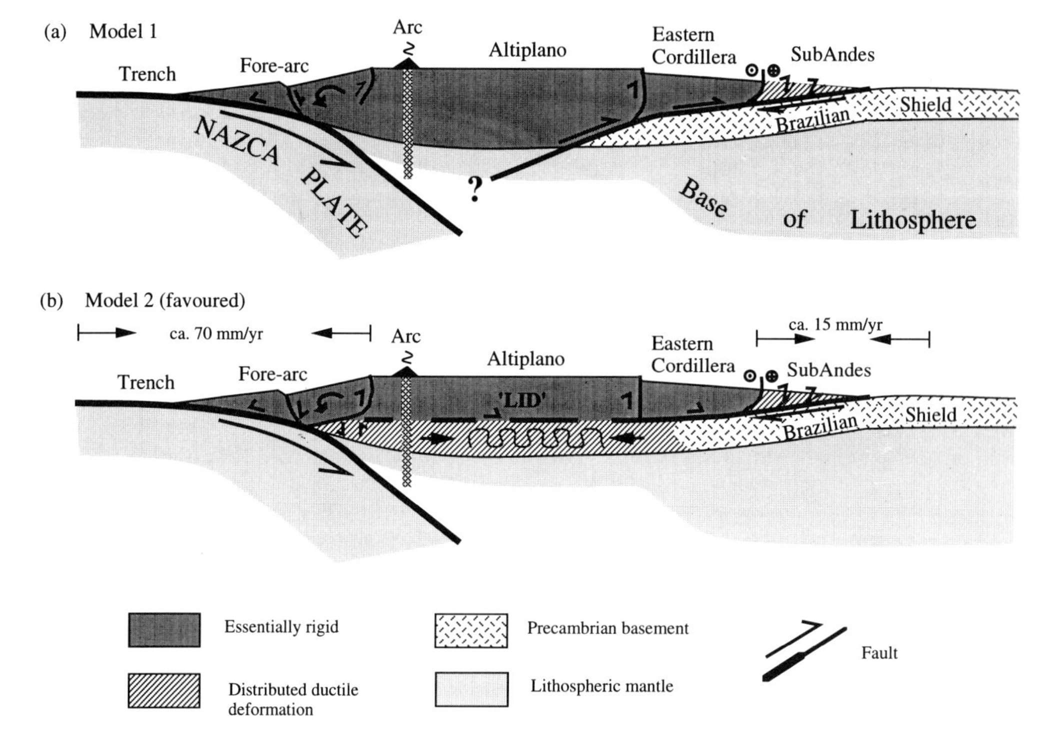

Fig. 14. Cartoon lithospheric-scale cross-sections through the Central Andes in the vicinity of the Bolivian 'bend' illustrating the 2D Plio-Pleistocene and active kinematics. Surface deformation is partitioned principally between eastwards underthrusting of the Nazca plate at the trench and thin-skinned folding and thrusting in the Subandean Zone. The main uncertainty is how deformation in the Subandean Zone connects with Benioff zone slip to accommodate the overall plate convergence (see text). (**a**) In model 1, thin-skinned deformation in the Subandes connects with a deeper thrust which cuts the whole lithosphere, accommodating underthusting of the whole Brazilian Shield lithosphere. (**b**) In model 2, favoured in this paper, underthrusting of the Brazilian Shield in the Subandes is accommodated at depth beneath the western parts of the Eastern Cordillera and Altiplano by a zone of distributed ductile deformation. In this model, the Altiplano forms an essentially rigid lid.

the active relative plate convergence is absorbed along the plate interface during large events with slip parallel to the relative plate convergence vector (Fig. 6, DeMets *et al.* 1990; Dewey & Lamb 1992). The remaining *c.* 10–20% of the relative plate convergence or *c.* 8 to 17 mm a^{-1} of shortening is absorbed in the Andes (Fig. 13).

Horizontal kinematics

A distinct zone of shallow thrust earthquakes follows the eastern foothills of the Andes in the Subandean Zone (Fig. 5a). The geometry of ramp structures on major thrusts (Roeder 1988), and also the direction of thrusting during earthquakes suggests that thrusting in the northern part of the Bolivian Subandes is nearly orthogonal to the trend of the Subandean Zone (Vega & Buforn 1991; Fig. 13). Further south (south of the Arica bend), thrust earthquakes have slip directions ranging between orthogonal to the Subandean Zone to essentially parallel to the plate convergence direction (Vega & Buforn 1991; Dewey & Lamb 1992). In this case, the marked bend in the Subandean Zone in Bolivia implies a divergence in the shortening directions on the eastern margin of this part of the Andes (Fig. 13). This divergence appears to be accommodated within the Andes by normal faulting, block rotations about vertical axes, and conjugate strike-slip faulting in the Eastern Cordillera such as the Cochabamba Fault System (Fig. 2; Dewey & Lamb 1992; Kennan 1994; Lamb unpublished data).

Satellite images of the southwestern margin of the Bolivian Subandes, north of the bend where the Andes trend *c.* NW, show a pronounced NW-trending lineament which appears to offset rivers (Figs 2 & 13; Kennan 1994). This suggests young sinistral strike-slip motion, parallel to the general structural trend and at the back of the Subandean Zone in this region. This and fault plane solutions (Vega & Buforn 1991; Dewey & Lamb 1992) suggest that in detail the deformation in the Subandean Zone is partitioned into strike-parallel shear (sinistral strike-slip north of the bend) west of the purely compressional fold and thrust belt.

Vertical kinematics

If deformation in the Subandean Zone absorbs a significant fraction of the total convergence between the Nazca and South American plates, then this deformation must be kinematically linked to deformation throughout the thickness of the South American lithosphere. There are two end-member models (Fig. 14): (1) the basal décollement in the Subandean Zone extends into a narrow shear zone at depth which cuts the whole thickness of the lithosphere, and the Brazilian shield is effectively being subducted into the mantle beneath the Eastern Cordillera (Fig. 14a); or (2) the underthrusting of the Brazilian shield is absorbed by a zone of distributed shortening beneath the undeforming 'lid' of the Eastern Cordillera and Altiplano (Fig. 14b, Isacks 1988; Gubbels *et al.* 1993).

If the first model is correct, the Brazilian shield has been subducted into the underlying mantle and one might expect a west dipping Benioff zone beneath the Eastern Cordillera, for which there is no seismic evidence (Fig. 5b). Also, motion on such a lithosphere-scale thrust might be expected to result in regional tilting of the hanging wall in the Eastern Cordillera. The presence of regional and flat-lying Late Miocene peneplains in this region (Gubbels *et al.* 1993; Kennan *et al.* 1995, 1997) suggest that such tilting has not occurred. For these reasons, we favour the second model (Fig. 14b). In this case, 140 km of shortening in the Subandean Zone could generate *c.* 2 km of surface uplift in this region since the Late Miocene (Kennan *et al.* 1997).

Tectonic rotation

Cenozoic deformation in the Bolivian Andes has been responsible for substantial rotations about a vertical axis of crustal blocks, relative to the South American plate. Palaeomagnetic work (MacFadden *et al.* 1990; Scanlon & Turner 1992; Macedo-Sanchez *et al.* 1992; Roperch *et al.* 1993; MacFadden *et al.* 1995; Lamb unpublished data) has defined three principal domains which can be characterized by particular regional rotations about a vertical axis (Fig. 15a). Local small block rotations complicate the regional pattern of tectonic rotations. The rotations are observed in Cretaceous to Miocene sedimentary and volcanic rocks, suggesting that all the rotation occurred in the last 20 Ma (Fig. 15b).

In the north, the Bolivian Andes are characterized by small regional anticlockwise rotations up to 20°, observed in Cretaceous to Late Miocene sediments and volcanics, right across the width of the Andes (Fig. 15a, Domain 1a). Further south, a domain can be defined, also extending right across the Andes, in which regional clockwise rotations up to 20° are typical, observed in Cretaceous to Early Miocene sediments (Fig. 15a, Domain 2). Even further south, there is a domain, which again extends right across the width of the Andes, characterized by regional clockwise rotations up to 40°, observed in Cretaceous to Late Miocene sediments and volcanics (Fig. 15a, Domain 3a).

These three principal domains can be ascribed to along-strike gradients in the shortening in the Subandean Zone, which has resulted in 'bending' of almost the entire width of the Central Andes since the Late Miocene. Shortening in the Subandean Zone at a latitude of *c.* 17°S in the bend could be as high as 140 km (Baby *et al.* 1993), while shortening further south near the Argentinian border may be *c.* 70 km (Baby *et al.* 1993). Thus, the average shortening gradient along the length of the Subandean Zone could accommodate a *c.* 15° clockwise regional rotation of the Bolivian Andes further east. Shortening gradients in the Subandean Zone, north of the bend, are poorly documented, but overall there seems to be a reduction in Subandean shortening in Peru, which could accommodate an anticlockwise rotation of the Andes further east.

In the Cochabamba area, in a region *c.* 100 km by 100 km, local clockwise rotations, between 20° and 45°, have been detected in Cretaceous sediments, within the general area of anticlockwise rotations (Fig. 15a, Domain 3b). These may be related to block rotation accommodated by sinistral strike-slip on ESE-trending faults, active in the Miocene-Pliocene and accommodating the divergence in shortening round the pronounced bend in the Bolivian Andes (Fig. 2, Cochabamba Fault System).

In general, zones of tectonic rotation can be correlated with the general structural trend. However, regions of anomalous strike orientation, for instance in the Otavi syncline near Potosi, are not always associated with significant Cenozoic tectonic rotation and probably reflect the reactivation of pre-Cretaceous trends, which

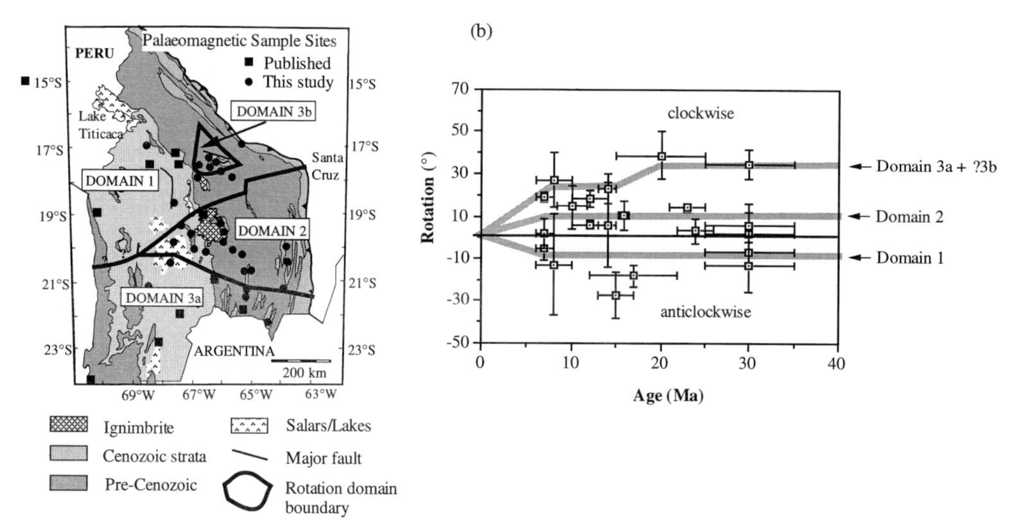

Fig. 15. Diagram illustrating the distribution and timing of tectonic rotations about a vertical axis in the Central Andes of northern Chile and Bolivia. (**a**) Simplified geological map of the Central Andes in northern Chile and Bolivia showing the location of palaeomagnetic samples analysed for tectonic rotations in this study (Lamb unpublished data) and previous work (MacFadden *et al.* 1990, 1995, Macedo-Sanchez *et al.* 1992; Roperch *et al.* 1993). Results can be used to define three principal domains. (**b**) Graph showing the regional rotation history for samples from various domains, shown in (a). Local small block rotation may perturb the general pattern. Domain 1 is characterized by *anticlockwise* tectonic rotations up to 20° in Miocene sedimentary rocks. Domain 2 is characterised by *clockwise* rotations up to 20° in Cretaceous and Miocene sedimentary and volcanic rocks. Domain 3a is characterized by *clockwise* rotations up to 40° for Miocene sedimentary and volcanic rocks. Domain 3b is characterized by *clockwise* rotations up to 40° in Cretaceous and younger sedimentary rocks.

differ from prevailing Andean trends (Lamb *et al.* 1993).

Magmatism

Cenozoic magmatism in the Andes of northern Chile and Bolivia is not just confined to the volcanic arc, but occurs in a region several hundred kilometres wide, behind the arc. However, in contrast to the andesitic-rhyolitic volcanism along the volcanic arc, with its well-defined tectonic setting related to subduction of the oceanic Nazca plate, the origin of the magmatism in a wide region to the east of the arc is unclear.

Volcanic arc

The narrow region of active andesitic arc volcanism follows the 100–150 km depth contours of the Benioff zone and coincides with the anticipated melting zone at *c.* 120 km depth where the hydrated mantle wedge above the subducting slab is expected to undergo the transition from amphibolite to eclogite facies metamorphism (Davies & Bickle 1991). The precise location of volcanoes in the arc at different times during the Cenozoic has shifted from the Precordillera in the Early Cenozoic to locally up to 70 km east of the volcanic arc in the Miocene. This movement of the arc may reflect both shortening in the forearc and small changes in the geometry (e.g. flattening) of the Benioff zone through time.

Widespread ignimbritic eruptions within the volcanic arc are probably the result of crustal melting, triggered by the introduction of arc-related mantle melts at high levels in the crust.

Altiplano and Eastern Cordillera

Scattered Plio-Pleistocene mafic and potassium-rich volcanic centres are found throughout the Altiplano in a region up to 200 km wide east of the volcanic arc (Fig. 6, Davidson & DeSilva 1992; Hoke *et al.* 1993, 1994*a*, *b*; Hoke & Lamb unpublished data). Also, a major basaltic extrusive–intrusive volcanic complex of Early

Miocene age outcrops throughout the central and eastern Altiplano.

Cenozoic magmatism east of the volcanic arc is also characterized by high-level dioritic and granitoid intrusions and ignimbrites, emplaced during the Late Oligocene to Pleistocene (Everden *et al.* 1977; Grant *et al.* 1977; Schneider 1985; McBride *et al.* 1987; Miller 1988; Lavenu *et al.* 1989; Kennan *et al.* 1995; Lamb unpublished data) and found mainly in the western part of the Eastern Cordillera, and also the eastern margin of the Altiplano. Many of the large granitoid bodies in the Cordillera Real and Quimsa Cruz were once thought to be Cenozoic (McBride *et al.* 1987). However, recent geochronological and field studies suggest that most, and possibly all of these, have a Carboniferous to Triassic age (McBride *et al.* 1987; Miller 1988; Kennan *et al.* 1995; Jones and Lamb unpublished data).

The origin of the Cenozoic intrusive and volcanic (basalts and ignimbrites) magmatism in the Eastern Cordillera and Altiplano is unclear. We consider two processes: (1) crustal melting due to heating during phases of Cenozoic crustal shortening (shear heating) and/or thermal relaxation after crustal shortening with a tens of million year time lag (England & Thompson 1984); (2) mantle melting. We believe that a combination of these two processes explains the observations best, though further ongoing work will test these preliminary conclusions. The timing of deformation and magmatism provides some support for the first model. For instance, a major phase of ignimbritic activity at *c.* 7 Ma (Los Frailes Formation), may be the result of crustal melting during thermal equilibration after the major phase of shortening in the Eastern Cordillera around 25 Ma, with a 10–20 Ma time lag. However, we postulate that the injection of mantle melts into the crust should be considered as an additional factor which triggered crustal melting and ignimbrite eruptions on the western margin of the Eastern Cordillera, as well as generating widespread mafic volcanism in the Altiplano. A mechanism for mantle melting is discussed below.

Convective removal of the basal part of the lithosphere

The lithospheric structure of the Andes in Bolivia and northern Chile, inferred from the helium isotopic studies (Fig. 6b), with a zone of 'thin' lithosphere beneath the Altiplano and part of the Eastern Cordillera, requires a mechanism of lithospheric thinning. We believe that this thinning is a direct consequence of the convective removal of the basal part of the lithosphere, driven by the gravitational instability associated with a growing lithospheric root above a convecting asthenosphere (Houseman *et al.* 1981; Molnar *et al.* 1993; Platt & England 1993). In effect, the thickening of the lithosphere, as a consequence of lithospheric shortening, is counteracted by removal and thermal 'erosion' of the base of the lithosphere. However, it is not clear whether this has occurred gradually or catastrophically.

If a volatile-rich part of the lithospheric mantle (McKenzie 1989) was brought into contact with the convecting asthenosphere as a consequence of convective removal of the base of the lithosphere, a temperature increase sufficient to cause melting could be attained fairly rapidly by heat conduction from the adjacent 'hot' asthenosphere (McKenzie 1989; Platt & England 1993). This suggests the following possible scenario for the Central Andes. Shortening in the Bolivian Altiplano, between 10 Ma and *c.* 5 Ma, accelerated convective removal of the basal part of the lithosphere in this region. This resulted in heating of the mantle metasomatic layer as the basal part of the lithosphere was replaced by 'hot' asthenosphere, triggering the generation of mantle melts at *c.* 5 Ma and the release of mantle volatiles, including helium, in a wide region. We speculate that an earlier phase of widespread mafic volcanism in the central Bolivian Altiplano at *c.* 24 Ma (Figs. 7, 9 & 11; Hoke *et al.* 1993, 1994*a*, *b*; Hoke & Lamb 1994; Kennan *et al.* 1995; Lamb unpublished data) may be related to an earlier phase of convective removal of the basal part of the lithosphere.

The model of convective removal of the base of the lithosphere will also cause surface uplift in isostatic response to the detachment of the dense lithospheric root and its replacement with buoyant hot asthenosphere (Molnar *et al.* 1993; Platt & England 1993), though as yet we cannot quantify the uplift expected in the Bolivian Altiplano.

Cenozoic crustal thickening

It is clear that there is substantial crustal thickening beneath the Central Andes, compared to the crust further east in the foreland (Fig. 1a and c; James 1971; Wigger *et al.* 1993; Dorbath *et al.* 1993; Zandt *et al.* 1994; Beck *et al.* 1996). The history of crustal shortening and magmatism in the Central Andes, outlined in the previous sections, suggests that there has been a long term history of surface uplift in this region since the early Cenozoic from essentially sea level to average elevations of nearly 4000 m

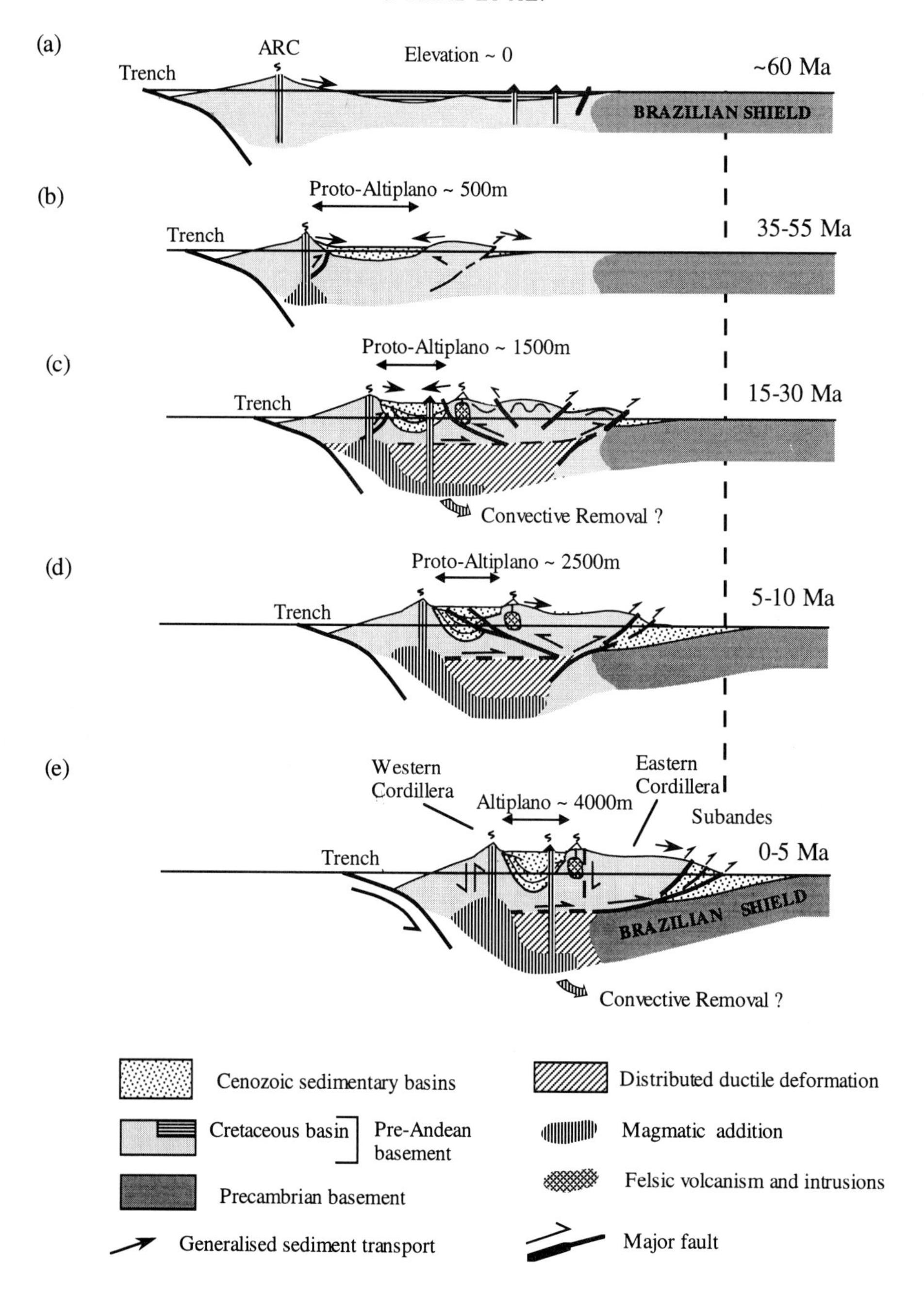
(a)
ARC
Trench
Elevation ~ 0
~60 Ma
BRAZILIAN SHIELD
(b)
Proto-Altiplano ~ 500m
Trench
35-55 Ma
(c)
Proto-Altiplano ~ 1500m
Trench
15-30 Ma
Convective Removal ?
(d)
Proto-Altiplano ~ 2500m
Trench
5-10 Ma
(e)
Western Cordillera
Altiplano ~ 4000m
Eastern Cordillera
Subandes
Trench
0-5 Ma
BRAZILIAN SHIELD
Convective Removal ?
Cenozoic sedimentary basins
Cretaceous basin
Pre-Andean basement
Precambrian basement
Generalised sediment transport
Distributed ductile deformation
Magmatic addition
Felsic volcanism and intrusions
Major fault

today (see Fig. 16). The approximate crustal cross-sectional area for an east-west section through the Central Andes at 20°S is shown in Fig. 1c. If the crustal thickness throughout the Bolivian Andes at the end of the Cretaceous was similar to that in the Brazilian shield today (*c.* 34 km), then the area of the topography and crustal root is *c.* 1.7×10^4 km^2 (Fig. 1c).

The existence of both the Altiplano basin and a substantial foreland basin shows that the shape of the crustal cross-section is partly a consequence of erosion and sedimentation within the Central Andes. The widespread preservation of Cretaceous sequences suggests that Cenozoic erosion has not stripped off more than a few kilometres thickness from the Eastern Cordillera, though this may be greater north of the Arica bend. K/Ar cooling ages on micas and zircon fission track ages (McBride *et al.* 1987; Benjamin *et al.* 1987; Kennan *et al.* 1995; Lamb unpublished data) from Triassic granitoid intrusives suggest that up to 10 km may have been locally stripped off from the western margin of the Eastern Cordillera. All this suggests that less than 0.1×10^4 km^2 has been removed by erosion from the E–W crustal profile at 20°S (Fig. 1c).

We consider two mechanisms of crustal thickening and their relative roles in contributing to the Andean crustal cross-section: (1) crustal shortening; (2) magmatic addition.

Crustal shortening

If the mass of the Andean crust is preserved, then crustal shortening will result in crustal thickening. If all the crustal root is a consequence of crustal shortening, then for an initial uniform crustal thickness of 34 km, the crustal cross-section at 20°S (Fig. 1c) suggests that *c.* 500 km of east–west Cenozoic crustal shortening are required. This calculation takes into account erosion but ignores the effects of density changes or crustal material transported into or out of the plane of the cross-section. Also, the crust beneath the Western Cordillera may have been much thicker at the beginning of the Cenozoic as a consequence of its Mesozoic evolution. If we just consider crustal thickening to the east of the Western Cordillera, then this part of the crustal root in Fig. 1c could be generated by *c.* 375 km of shortening in this region.

Estimates of E–W Cenozoic shortening at *c.* 20°S, based on structural studies, are about 140 km in the Subandean Zone (Baby *et al.* 1993) and 140±20 km in the Altiplano (this study; unpublished oil company data). Sheffels (1990) estimates a minimum shortening in the Bolivian Eastern Cordillera at roughly 18°S of about 140 km based on structural interpretations of the geology. However, Sheffels (1990) ignored the effects of a marked angular unconformity at the base of Cretaceous sequences in this region, so it is open to question whether this a minimum value. If we assume that Cenozoic tilting of Cretaceous strata throughout the Eastern Cordillera is on average *c.* 35°, then this tilting alone could accommodate a shortening of *c.* 30 km. Also, the general spacing of major reverse faults with displacements of a few kilometres and truncating Cretaceous sequences, is *c.* 10 km. Thus, reverse faulting would be expected to accommodate at the very least 20 km of shortening across the Eastern Cordillera. In this case, Cenozoic shortening in the Eastern Cordillera is at least 50 km.

If we accept all the above shortening estimates at face value, this suggests a total shortening of about 320 km across the Altiplano, Eastern Cordillera and Subandean Zone in the Cenozoic

Fig. 16. Cartoon cross-sections illustrating the generalised crustal evolution of the Central Andes in Bolivia and northern Chile in the last 60 Ma. Estimates of the average surface elevation for the proto-Altiplano region are also given. (**a**) In the latest Cretaceous, most of the Bolivian Andes formed a wide back-arc region of subsidence near sea level, associated with a minor amount of crustal thinning, bounded to the east by the rigid Brazilian Shield. (**b**) A narrow zone of uplift developed in the centre of the old Cretaceous basin, dividing the early Altiplano basin from an early foreland basin. (**c**) The zone of 'thick-skinned' deformation widened, resulting in thrusting of the Eastern Cordillera over the eastern margin of the Altiplano basin, and also towards the east in the foreland region. Numerous local fault-bounded compressional basins developed, and the foreland basin migrated further east. Widespread behind-arc basaltic volcanism in the Altiplano region may be associated with convective removal of part of the lithospheric mantle. (**d**) The Brazilian Shield began to impinge on the eastern margin of the Andes, with the initiation of a well-developed thin-skinned fold and thrust belt in the Subandean Zone. Shortening was also concentrated in the Altiplano, with a marked difference in vergence. (**e**) Subsequently, shortening has been concentrated in the thin-skinned Subandean Zone, accommodating underthrusting of the Brazilian Shield. Uplift of the Altiplano may be partly associated with crustal thickening in a ductile zone of deformation at depth. Widespread behind-arc Plio-Pleistocene volcanism may be associated with convective removal of the lower part of the lithosphere.

at *c.* 20°S, or about 85% of that deduced from simple crustal thickness calculations. Cenozoic shortening in the Western Cordillera and forearc has not been quantified, but is very small (<20 km) since the Early Miocene, and is likely to be much less than 100 km since the early Cenozoic. Thus, the previous simple estimates suggest that E–W crustal shortening in the Andes at 20°S may not be sufficient to explain all the observed crustal thickness and additional significant mechanisms of crustal thickening cannot be ruled out by.

Magmatism

There has been widespread magmatic activity throughout much of the Central Andes during the Cenozoic. Extrapolating estimates of the volcanic arc production rate for the last 10 Ma in the Central Andes (Francis & Hawkesworth 1994), a possible volume of 80 km^3 km^{-1} arc length of mantle melts may have been added to the surface in the volcanic arc during the Cenozoic. Mantle melt addition may be occurring in a wide region behind the arc (Hoke *et al.* 1993, 1994*b*). The model of convective removal of the basal part of the lithosphere, described previously (Hoke *et al.* 1993, 1994*b*; Hoke & Lamb 1994) suggests that basaltic addition to the crust is the result of melting of the lithospheric mantle. This may have happened twice in the Cenozoic, once during the widespread eruption of mafic volcanics in the central Altiplano at *c.* 24 Ma, and also during the Plio-Pleistocene.

The surface volcanism may only be a small fraction of the total melt addition to the crust. However, the small volume of surface volcanism in the region to the east of the volcanic arc, beneath the Altiplano and part of the Eastern Cordillera of Bolivia, suggests that any subsurface melt additions to the crust here are likely to be less than that beneath the volcanic arc. If we assume that as little as 10% of the total melt addition reaches the surface (Aitcheson *et al.* 1993), then up to 800 km^3 km^{-1} arc length of mantle melt may have been added to the crust beneath the arc during the Cenozoic, but less than this in the region further east. This would contribute a crustal root equivalent to a maximum of 50 km of crustal shortening, for an initial crustal thickness of *c.* 34 km.

The above discussion suggests that mantle melt additions are not the most important mechanism of crustal thickening on a regional scale, probably contributing less than 15% to the total crustal thickening. However, locally beneath the volcanic arc, this melt addition could be important, possibly accounting for as much as 40% of the crustal thickening here.

Summary of Cenozoic tectonic evolution

The Cenozoic tectonic development of the Central Andes in northern Chile and Bolivia is summarised here, based on the previous descriptions of the nature, timing and distribution of sedimentary basins, deformation and magmatism. Five main stages are recognized in what is viewed overall as a continuous evolution of this part of the Andes. These are illustrated by the cartoon diagrams in Fig. 16.

(1) In the latest Cretaceous (Maastrichtian) and Palaeocene, the Central Andes at the latitudes of northern Chile and Bolivia formed a volcanic arc along what is today the Precordillera in the forearc of northern Chile (Fig. 16a). Behind the arc, there was a zone of subsidence up to 400 km wide which was at or close to sea level. This formed a wide basin, receiving sediment both from the volcanic arc and also regions in the east. The general axis of this zone of subsidence more-or-less coincides with the region of greatest erosion prior to the Cretaceous, where pre-Cretaceous deformation was most intense and Ordovician and Silurian flysch-type deposits are thickest. This suggests a complex history of alternating pre-Cenozoic basin formation and inversion.

(2) In the Eocene, the Cretaceous backarc basin was inverted and formed a narrow zone of uplift (<100 km wide and referred to as the proto-cordillera), along what is today the western margin of the Eastern Cordillera and several hundred kilometres east of the volcanic arc at that time (Fig. 16b). This created a basin in what is today the Altiplano, which began to rapidly accummulate sediment derived partly from the uplifting proto-cordillera. Sediment was also shed to the east of the proto-cordillera into a narrow foreland basin within what is today the Eastern Cordillera. It appears that strike-slip motion was accommodated in a laterally extensive fault system along the volcanic arc, though the sense of shear is uncertain.

Total crustal shortening in the Eocene is probably of the order of a few tens of kilometres, with average elevations in the proto-cordillera of *c.* 500 m (Fig. 16b).

(3) By the late Oligocene, the uplifting proto-cordillera had widened. Deformation spread westwards into the Altiplano basin as a west-verging fold and thrust belt, so that deposition was terminated in all but the centre of the Altiplano basin (Fig. 16c). Deformation also spread eastwards with the development of eastward verging structures (Fig. 16c). Deformation appears to be thick-skinned involving upright folding and relatively high angle reverse faults. Uplift and rapid unroofing of both the

Western and Eastern Cordilleras caused a rejuvenation of the Altiplano basin, with the widespread development of conglomerates derived from both sides of the basin at *c.* 25 Ma. Deformation in the Eastern Cordillera also spread eastwards with widespread internal folding and the development of a new foreland basin in what is today the Subandean Zone.

At *c.* 24 Ma, there was widespread mantle melting beneath the Altiplano, possibly as a consequence of a phase of convective removal of the basal part of the lithosphere, which had been thickened to a critical point during earlier lithospheric shortening (Fig. 16c). Also, local intrusions and ignimbrites were emplaced on the western margin of the Eastern Cordillera. By 20 Ma, a number of local fault-bounded compressional basins existed throughout the Eastern Cordillera, accumulating sediments which rest with angular unconformity on older deformed Cretaceous and Palaeozoic sequences. However, the main vergence of deformation appears to have been towards the west.

Total crustal shortening by the Middle Miocene was probably of the order of 200 km, with average surface elevations in the proto-Altiplano of *c.* 2000 to 2500 m (Fig. 16c).

(4) In the Late Miocene, the pattern of deformation in the Bolivian Andes changed markedly. Deformation essentially terminated in the Eastern Cordillera and was concentrated in the Altiplano region which started to shorten internally, probably around 9 Ma. The intense phase of shortening had ceased by 5 Ma. Thermal re-equilibration after Oligo-Miocene crustal thickening on the western margin of the Eastern Cordillera may have triggered a major phase of ignimbritic activity at *c.* 7 Ma.

Total crustal shortening by the latest Miocene was probably of the order of 250 km (Fig. 16d).

(5) The cessation, between 9 and 5 Ma, of intense deformation in the Altiplano region may have coincided with the initiation of significant underthrusting of the Brazilian Shield and deformation in the Subandean Zone. Palaeomagnetic data suggest that the main phase of regional rotation about a vertical axis in the Bolivian Andes is younger than *c.* 8 Ma. Thus, regional shortening gradients in the Subandean Zone since *c.* 8 Ma accommodated a *c.* 10° clockwise rotation of the Eastern Cordillera and Altiplano south of the bend, and a *c.* 10° anticlockwise rotation north of the bend. A local increase in shortening gradients in the Subandean Zone in southern Bolivia seems to have accommodated an even larger clockwise rotation, up to 20°, in the same period.

The initiation of important shortening in the Subandean Zone may mark an important change in vergence of deformation in the Bolivian Andes from a pronounced W-verging component prior to 5 Ma to essentially E-verging subsequently. At this time, the second phase of mafic volcanism in the Altiplano commenced, possibly related to a second phase of convective removal of the basal part of the lithosphere as a consequence of Mid- to Late Miocene thickening. Shortening in the Subandean Zone during this period of *c.* 140 km brings the estimated total crustal shortening to around 400 km, with average surface heights in the Altiplano of *c.* 4000 m (Fig. 16e).

Rates of deformation, sedimentation and erosion

The previous description can be used to estimate rates of deformation in the Central Andes, though these estimates have large uncertainties which are difficult to quantify. The general picture is one of increasing rates of deformation towards the present, but on average only a small proportion ($<15\%$) of the relative plate convergence has been absorbed by shortening in the Andes. The bulk of the relative plate convergence has been accommodated by shear near the plate interface. Rates are comparable with those in other wide plate-boundary zones, such as the Central Asian collision zone (Lamb 1994).

The time-averaged rate of shortening over the entire Cenozoic history of Andean deformation is *c.* $10\,\mathrm{mm\,a^{-1}}$. Shortening strain rates were probably mainly low (*c.* $2 \times 10^{-8}\,\mathrm{a^{-1}}$) for much of the early Andean deformation history, though increasing by an order of magnitude to *c.* $2 \times 10^{-7}\,\mathrm{a^{-1}}$ in the last 10 Ma when deformation was concentrated in the Subandean Zone. Palaeomagnetic data suggests that the bulk of the bending about a vertical axis of the Bolivian Andes also occurred in the last 10 Ma, with rotation rates of 1–4° $\mathrm{Ma^{-1}}$.

Fission track age data suggests that average denudation rates have been locally as high as $0.2\,\mathrm{mm\,a^{-1}}$ on the western margin of the Eastern Cordillera. Likewise, average sedimentation rates in the Neogene range between 0.2 and 0.5 $\mathrm{mm\,a^{1}}$ in the Altiplano and foreland basins. Surface uplift rates in the Eastern Cordillera and Altiplano are difficult to estimate, but have probably been in the range 0.05–0.2 $\mathrm{mm\,a^{-1}}$ during the Cenozoic.

Discussion

The overall picture of the Cenozoic tectonic evolution of the Andes in northern Chile and Bolivia is one of progressive shortening and crustal thickening on the western margin of the South American plate, above the subducting

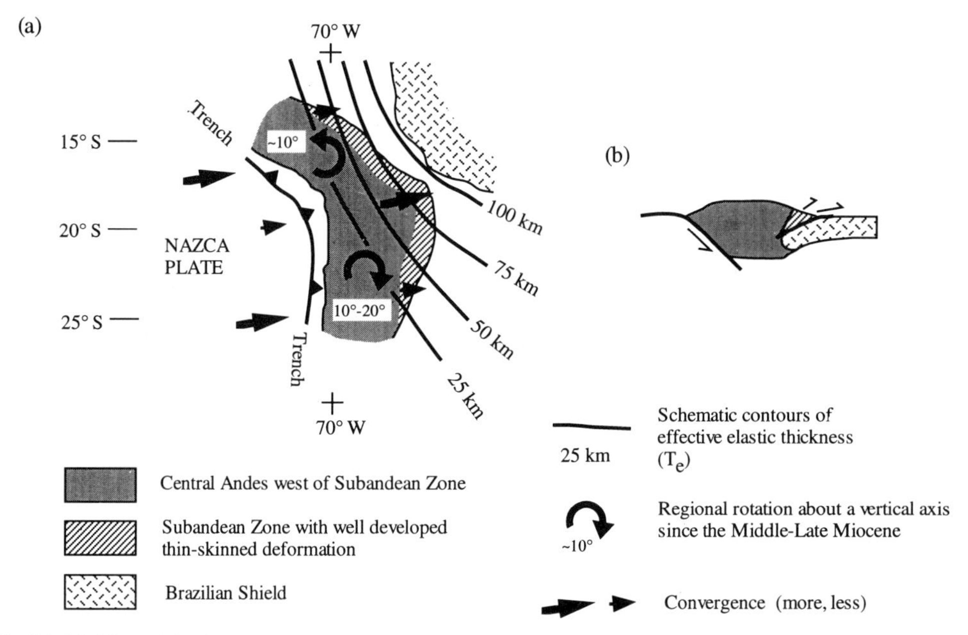

Fig. 17. (**a**) Map of the Central Andes (after Watts *et al.* 1995) illustrating the possible relationship between bending, foreland shortening and lithospheric flexure. The impingement of the flexurally strong Brazilian Shield since the Late Miocene, illustrated by contours of effective elastic thickness, favoured the development of a thin-skinned fold and thrust belt in the bend region of the Bolivian Andes in the Subandean Zone, and the underthrusting of the Brazilian Shield. A larger proportion of the overall Andean shortening budget has been absorbed in this region, compared to the foreland regions further north and south, where deformation is more diffuse and involves basement. This has resulted in substantial bending of the whole Central Andes since the Late Miocene. (**b**) Schematic E–W cross-section through the Central Andes at ~18°S, showing the partitioning of Andean shortening since the Late Miocene into slip near the plate interface and underthrusting of the Brazilian Shield much further east.

Nazca plate. However, the causes of the way this has occurred are unclear, and any of the following may have played a role: the rate and direction of relative plate motion; nature of the subducted plate; mantle convection; strength inhomogeneities in the overlying South American plate; body forces related to variations in crustal thickness; the distribution of pre-Andean sedimentary basins.

The apparent coincidence between the marked acceleration in relative plate convergence rate in the Eocene (Pardo-Casas & Molnar 1987) and the onset of deformation in the Bolivian Andes suggests a cause and effect relationship, especially if more rapid subduction results in an increase in average shear stresses on the plate interface. Likewise, another increase in relative plate convergence in the Late Oligocene may be related to the marked increase in tectonic activity in the Bolivian Andes at that time (Pardo-Casas & Molnar 1987; Sempere *et al.* 1990). The weak correlation between the long term position of features of the subducted plate (aseismic ridges or age of oceanic crust) and Andean segmentation, suggests that these features do not play a major role in controlling Andean tectonics (Dewey & Lamb 1992).

We believe that strength inhomogeneties in the overlying plate, which are ultimately related to the thermal history of the South American lithosphere, have also played an important role in determining the tectonic evolution of this part of the Andes. For instance, the development of the proto-cordillera in the Eocene along the axis of the Cretaceous back-arc basin suggests that shortening was localised in a weak zone where the lithosphere was thinnest and possibly warmest, as a consequence of the Cretaceous back-arc extension history. It is also interesting that the depocentre of the Cretaceous basin coincides with the region of most intense pre-Cretaceous folding and erosion.

Watts *et al.* (1995) have shown that there are marked spatial variations in the flexural strength of the lithosphere. They suggest that the most plausible distribution of flexural strength can be described by contours of effective elastic thickness that parallel and increase towards the region of outcrop of the Brazilian Shield and that this variation is a pre-Andean feature (Fig. 17). In this case, only part of the Bolivian Andes in the 'bend' region rests on flexurally strong lithosphere. The tectonic evolution of this part of the Andes, summarised in Figure 16, suggests that only in the last 10 Ma have the flexurally strongest parts of the South American lithosphere (elastic thickness >25 km) been underthrust beneath the Central Andes, and it is in this period that the distinctive thin-skinned fold and thrust belt of the Subandean Zone developed (Fig. 17).

We suggest that the impingement of flexurally strong lithosphere in the Brazilian Shield on the eastern margin of the Bolivian Andes in the Late Miocene is causally related to the development of a thin-skinned zone of deformation, though the presence of thick Palaeozoic shale sequences in this region may also have helped to provide a weak décollement. Prior to the Late Miocene, deformation in what is today the Eastern Cordillera may have been thick-skinned with concomitant deformation of the underlying lower parts of the lithosphere. The development of a major thin-skinned zone of deformation on the eastern margin of the Bolivian Andes in the Subandean Zone also coincides with a marked increase in the rates of shortening in the Andes and also regional bending of the Central Andes about a vertical axis. Thus, it seems that once a thin-skinned fold and thrust belt developed, it absorbed more of the overall shortening budget in the plate-boundary zone, compared to the foreland regions further north and south where deformation is thick-skinned, involving basement (Fig. 17).

Finally, the mechanism of convective removal of the basal part of the lithosphere may have played a role in controlling the style of Andean deformation. Thus, the style of Plio-Pleistocene and acive kinematics proposed here for the Bolivian Andes (Fig. 14b), in which the Altiplano appears to have acted as a rigid 'lid' above a deforming lower lithosphere, may be a consequence of thermal weakening of the lower part of the lithosphere as it was replaced by 'hot' asthenosphere during convective removal. Also, surface uplift as a consequence of convective removal would tend to increase body forces on the eastern margin of the Andes, promoting deformation in the Subandean Zone (cf. Molnar *et al.* 1993). These effects have yet to be quantified.

References

AITCHESON, S., FORREST, A. & ENTENMANN, J. 1993. Recharge-Assimilation-Fraction-Tapping ('RAFT') processes and magma enrichment in the Central Andes. *In*: *Extended Abstracts*: *2nd International Symposium, Oxford 1993, ORSTOM*, 327–330.

BABY, P., SEMPERE, T., OLLER, J. & HERAIL, G. 1992*a*. Evidence for major shortening on the eastern edge of the Bolivian Altiplano: the Calasaya nappe. *Tectonophysics*, **205**, 155–169.

——, HERAIL, G., SALINAS, R. & SEMPERE, T. 1992*b*. Geometry and kinematic evolution of passive roof duplexes deduced from cross-section balancing: example from the foreland thrust system of the southern Bolivian Subandean Zone. *Tectonics*, **11**, 523–536.

——, GUILLIER, B., OLLER, J., HERAIL, G., MONTEMURRO, D. & SPECHT, M. 1993. Structural synthesis of the Bolivian Subandean Zone. *In*: *Extended Abstracts*: *2nd International Symposium, Oxford 1993, ORSTOM*, 159–162.

BAKER, M. & FRANCIS, P. 1978. Upper Cenozoic volcanism in the Central Andes – ages and volumes. *Earth and Planetary Science Letters*, **41**, 175–187.

BECK, S., ZANDT, G., MYERS, S., WALLACE, T., SILVER, P. & DRAKE, L. 1996. Crustal thickness variations in the Central Andes. *Geology*.

BENJAMIN, M., JOHNSON, N. & NAESER, C. 1987. Recent rapid uplift in the Bolivian Andes: Evidence from fission-track dating. *Geology*, **15**, 680–683.

CAMPUSANO, T. B. 1990. *Kontinentale Sedimentation der Kreide und des Alttertiaers im Umfeld des subduktionsbedingten Magmatismus in der chilenischen Praekordillere (21°S–23°S)*. Berliner Geowissenschaftliche Abhandlungen, Reihe A, **123**, Berlin.

COIRA, B., DAVIDSON, J., MPODOZIS, C. & RAMOS, V. 1982. Tectonic and magmatic evolution of the Andes of northern Argentina and Chile. *Earth Science Reviews*, **18**, 303–332.

DAVIDSON, J. P. & DE SILVA, S. L. 1992. Volcanic rocks from the Bolivian Altiplano: Insights into crustal structure, contamination, and magma genesis in the central Andes. *Geology*, **20**, 1127–1130.

DAVIES, J. & BICKLE, M. 1991. A physical model for the volume and composition of melt produced by hydrous fluxing above subduction zones. *Philosophical Transactions of the Royal Society of London*, **A335**, 355–364.

DEMETS, C., GORDON, R. G., ARGUS, D. F. & STEIN, S. 1990. Current plate motions. *Geophysical Journal International*, **101**, 425–478.

DEWEY J. & BIRD, J. 1970. Mountain belts and the new global tectonics. *Journal of Geophysical Research*, **75**, 2625–2647.

—— & LAMB, S. H. 1992. Active tectonics of the Andes. *Tectonophysics*, **205**, 79–95.

DORBATH, C., GRANET, M., POUPINET, G. & MARTINEZ, C. 1993. A seismic study of the Altiplano and the Eastern Cordillera in Northern Bolivia:

new constraints on a lithospheric model. *In*: *Extended Abstracts*: *2nd International Symposium, Oxford 1993, ORSTOM*, 7–10.

ENGLAND, P. & THOMPSON, A. 1984. Pressure-Temperature-Time Paths of Regional Metamorphism I. Heat Transfer during the Evolution of Regions of Thickened Continental Crust. *Journal of Petrology*, **25**, 894–928.

EVERNDEN, J. F., KRIZ, S. J. & CHERRONI, C. 1977. Potassium-Argon ages of some Bolivian rocks. *Economic Geology*, **72**, 1042–1061.

FARRAR, E., CLARK, A., KONTAK, D. & ARCHIBALD, D. 1988. Zongo-San Gaban Zone: Eocene foreland boundary of the Central Andean orogen, northwest Bolivia and southeast Peru, *Geology*, **16**, 55–58.

——, CLARK, A. & HEINRICH, S. 1990. The age of the Zongo pluton and the tectono-thermal evolution of the Zongo San-Gaban zone in the Cordillera Real, Bolivia. *In*: *Abstracts, 1st International Symposium on Andean Geodynamics, Grenoble 1990, ORSTOM*, 171–174.

FRANCIS, P. & HAWKESWORTH, C. 1994. Late Cenozoic rates of magmatic activity in the Central Andes and their relationships to continental crust formation and thickening. *Journal of the Geological Society, London*, **151**, 845–854.

FLINT, S., TURNER, P., JOLLEY, E. & HARTLEY, A. 1993. Extensional tectonics in convergent margin basins: An example from the Salar de Atacama, Chilean Andes. *Geological Society of America Bulletin*, **105**, 603–617.

GAYET, M., MARSHALL, L. & SEMPERE, T. 1991. The Mesozoic and Palaeocene vertebrates of Bolivia and their stratigraphic context: a review. *In*: SUAREZ, R. (ed.) *Fossiles y facies de Bolivia*, **1**. Revista Tecnica de Yacimientos Petroliferos Fiscales Bolivianos, **12**, 393–433.

GRANT, J. N., HALLS, C., AVILA SALINAS, W. & SNELLING, N.J. 1979. K-Ar ages of igneous rocks and mineralisation in part of the Bolivian tin belt. *Economic Geology*, **74**, 838–851.

GUBBELS, T. L., ISACKS, B. L. & FARRAR, E. 1993. High-level surfaces, plateau uplift, and foreland basin development, Bolivian central Andes. *Geology*, **21**, 695–698.

HARTLEY, A., FLINT, S., TURNER, P. & JOLLY, E. 1992. Tectonic controls on the development of a semi-arid alluvial basin as reflected in the stratigraphy of the Purilactis Group (Upper Cretaceous – Eocene) northern Chile. *Journal of South American Earth Sciences*, **5**, 275–296.

HERAIL, G., SOLER. P., BONHOMME, M. & LIZECA, J. 1993*a*. Evolution geodynamique de la transition entre l'Altiplano et la Cordillere Orientale au nord d'Oruro (Bolivie) – Implications sur le deroulement de l'orogenese andine. *Comptes Rendus de l'Academie des Sciences de Paris, Serie II*, **317**, 515–522.

——, OLLER, J., BABY, P., BLANCO, J., BONHOMME, M. G. & SOLER, P. 1993*b*. The Tupiza, Nazareno and Estarca basins (Bolivia): strike-slip faulting and thrusting during the Cenozoic evolution of the southern branch of the Bolivian Orocline. *In*: *Extended Abstracts*: *2nd International Symposium, Oxford 1993, ORSTOM*, 191–194.

HOKE, L, HILTON, D., LAMB, S., HAMMERSCHMIDT, K. & FRIEDRICHSON H. 1993. 3He evidence for a wide zone of active mantle melting beneath the Central Andes. *In*: *Extended Abstracts*: *2nd International Symposium, Oxford 1993, ORSTOM*, 371–372.

—— & LAMB, S. 1994. Lithospheric structure of the Central Andes, constrained by ^{3}He and geological data. *In*: *Extended abstracts*: *Annual Conference of the Geological Society of Chile, Concepcion, Chile.*

——, LAMB, S. & ENTENMANN, J. 1994*a*. Comment on Davidson, J. & DE SILVA, S. *In*: *Geology*, **20**, 1127–1130 (1993). *Geology*, **21**, 1147–1149.

——, HILTON, D., LAMB, S., HAMMERSCHMIDT, K. & FRIEDRICHSON, H. 1994b. ^{3}He evidence for a wide zone of active mantle melting beneath the Central Andes. *Earth and Planetary Science Letters*, **128**, 341–355.

HOUSEMAN, G. A., MCKENZIE, D. P. & MOLNER, P. 1981. Convective instability of a thickened boundary layer and its relevance for the thermal evolution of continental convergent belts. *Journal of Geophysical Research*, **86**, 6115–6132.

ISACKS, B. 1988. Uplift of the Central Andean Plateau and Bending of the Bolivian Orocline. *Journal of Geophysical Research*, **93**, 3211–3231.

JACKSON, J., HAINES, J. & HOLT, W. 1995. The accommodation of Arabia-Eurasia plate convergence in Iran. *Journal of Geophysical Research*, **100**, 15,205–15,219.

JAMES, D. E. 1971. Andean crustal and upper mantle structure. *Journal of Geophysical Research*, **76**, 3246–3271.

JOLLEY, E., TURNER, P., WILLIAMS, G., HARTLEY, A. & FLINT, S. 1990. Sedimentological responses of an alluvial system to Neogene thrust tectonics. *Journal of the Geological Society, London*, **147**, 769–784.

KENNAN, L. 1994. *Cenozoic tectonics of the Bolivian Andes*. PhD thesis, University of Oxford.

——, LAMB, S. & RUNDLE, C. 1995. K-Ar dates from the Altiplano and Cordillera Oriental of Bolivia: stratigraphic and tectonic implications. *Journal of South American Earth Sciences*, **8**, 163–186.

——, —— & HOKE, L. 1997. High-altitude palaeosurfaces in the Bolivian Andes: evidence for late Cenozoic surface uplift. *In*: WIDDOWSON, M. (ed.) *Palaeosurfaces: Recognition, Reconstruction and Palaeoenvironmental Intepretation*. Geological Society, London, Special Publications, **120**, 307–324.

LAHSEN, A. 1982. Upper Cenozoic volcanism and tectonism in the Andes of northern Chile. *Earth Science Reviews*, **18**, 285–302.

LAMB, S. 1994. A simple method for estimating the horizontal velocity field in wide zones of active deformation – II. Examples from New Zealand, Central Asia and Chile. *Geophysical Journal International*, **119**, 313–337.

——, HOKE., L. & KENNAN, L. 1993. Tectonic evolution of the Central Andes since the Cretaceous. *In*: *Extended Abstracts*: *2nd Inter-*

national Symposium, Oxford 1993, ORSTOM, 207–210.

LAVENU, A., BONHOMME M. G., VATIN-PERIGNON N. & DE PACHTERE, P. 1989. Neogene magmatism in the Bolivian Andes between 16°S and 18°S: Stratigraphy and K/Ar geochronology. *Journal of South American Earth Sciences*, **2**, 35–47.]

LITHERLAND, M., ANNELLS, R. N., APPLETON, J. D., BERRANGE, J. P., BLOOMFIELD, K., BURTON, C. C. J., DARBYSHIRE, D. P. F., FLETCHER, C. J. N., HAWKINS, M. P., KLINCK, B. A., LLANOS, A., MITCHELL, W. I., O'CONNOR, E. A., PITFIELD, P. E. J., POWER, G. & WEBB, B. C. 1986. *The geology and mineral resources of the Bolivian Precambriam shield*. British Geological Survey Overseas Memoir, **9**.

LYON-CAEN, H., MOLNAR, P. & SUAREZ, G. 1985. Gravity anomalies and flexure of the Brazilian shield beneath the Bolivian Andes. *Earth and Planetary Science Letters*, **75**, 81–92.

MCBRIDE, S., CLARK, A., FARRAR, E. & ARCHIBALD, D. 1987. Delimitation of a cryptic Eocene tectono-thermal domain in the Eastern Cordillera of the Bolivian Andes through K-Ar dating and ^{40}Ar–^{39}Ar step-heating. *Journal of the Geological Society, London*, **144**, 243–255.

MCKENZIE, D. P. 1989. Some remarks on the movement of small melt fractions in the mantle. *Earth and Planetary Science Letters*, **95**, 53–72.

MACEDO-SANCHEZ, O., SURMONT, J., KISSEL, C. & LAJ, C. 1992. New temporal constraints on the rotation of the Peruvian Central Andes obtained from palaeomagnetism. *Geophysical Research Letters*, **19**, 1875–1878.

MACFADDEN, B. J., ANAYA, F., PEREZ, H., NAESER, C. W., ZEITLER, P. K. & CAMPBELL, K. E. 1990. Late Cenozoic paleomagnetism and chronology of Andean basins of Bolivia: evidence for possible oroclinal bending. *Journal of Geology*, **98**, 541–555.

——, ANAYA, F. & SWISHER III, C. 1995. Neogene paleomagnetism and oroclinal bending of the central Andes of Bolivia. *Journal of Geophysical Research*, **100**, 8153–8167.

MAROCCO, R., SEMPERE, T., CIRBIAN, M. & OLLER, J. 1987. Mise en evidence d'une deformation paleocene en Bolivie du Sud. *Comptes Rendus de l'Academie des Sciences de Paris*, **304**, II, 18, 1139–1143.

MARSHALL, L. G. & SEMPERE, T. 1991. The Eocene to Pleistocene vertebrates of Bolivia and their stratigraphic context: a review. *In*: SUAREZ, R. (ed.) *Fossiles y facies de Bolivia*, **1**. Revista Tecnica de Yacimientos Petroliferos Fiscales Bolivianos, **12**, 631–652.

——, SWISHER III, C., LAVENU, A., HOFFSTETTER, R. & CURTIS, G. H. 1992. Geochronology of the mammal-bearing late Cenozoic on the northern Altiplano, Bolivia. *Journal of South American Earth Sciences*, **5**, 1–19.

——, SEMPERE, T. & GAYET, M. 1993. The Petaca (Late Oligocene – Middle Miocene) and Yecua (Late Miocene) Formations of the Subandean – Chaco Basin, Bolivia, and their tectonic significance. *Documents du Laboratoire de Geologie de Lyon, France*, **125**, 291–301.

MILLER, J. F. 1988. *Granite Petrogenesis in the Cordillera Real, Bolivia and crustal evolution in the Central Andes*. PhD thesis, Open University, Milton Keynes.

MOLNAR, P., ENGLAND, P. C. & MARTINOD, M. 1993. Mantle dynamics, the uplift of the Tibetan Plateau, and the Indian monsoon. *Reviews in Geophysics*, **31**, 357–396.

MPODOZIS, C., MARINOVIC, N. & SMOJE, I. 1993. Eocene left lateral strike-slip faulting and clockwise block rotations in the Cordillera de Domeyko, west of Salar de Atacama, northern Chile. *In*: *Extended Abstracts*: *2nd International Symposium, Oxford 1993, ORSTOM*, 225–228.

NARANJO, J. 1987. Interpretation de la actividad Cenozoica superior a lo largo de la zona de Falla Atacama, Norte de Chile. *Revista Geologica de Chile*, **31**, 43–55.

PARDO-CASAS, F. & MOLNAR, P. 1987. Relative motion of the Nazca (Farallon) and South American plates since Late Cretaceous time. *Tectonics*, **6**, 233–248.

PLATT, J. & ENGLAND, P. 1993. Convective removal of the lithosphere beneath mountain belts: thermal and mechanical consequences. *American Journal of Science*, **294**, 307–336.

REUTTER, K., CHONG, G. & SCHEUBER, E. 1993. The 'West Fissure' and the cordilleran fault system of northern Chile. *In*: *Extended Abstracts*: *2nd International Symposium, Oxford 1993, ORSTOM*, 237–240.

RICCARDI, A. 1988. *The Cretaceous System of Southern South America*. Memoir of the Geological Society of America, **168**.

ROEDER, T. 1988. Andean-age structure of Eastern Cordillera in the province of La Paz, Bolivia. *Tectonics*, **7**, 23–39.

ROPERCH, P., FORNARI, M. & HERAIL, G. 1993. A paleomagnetic study of the Altiplano. *In*: *Extended Abstracts*: *2nd International Symposium, Oxford 1993, ORSTOM*, 241–244.

ROUCHY, J. M., CAMOIN, G., CASANOVA, J. & DECONINCK, J. F. 1993. The central palaeo-Andean basin of Bolivia (Potosi area) during the late Cretaceous and early Tertiary: reconstruction of ancient saline lakes using sedimentological, paleoecological and stable isotope records. *Palaeogeography, Palaeoclimatology, Palaeoecology*, **105**, 179–198.

SANJINES, G. & JIMENEZ, F. 1976. Communicacion preliminar acerca de la presencia de fossiles vertebrados en la formacion Petaca del area de Santa Cruz. *Revista Tecnica de Yacimientos Petroliferos Fiscales Bolivianos*, **4**, 147–156.

SCANLON, P. & TURNER, P. 1992. Structural constraints on palaeomagnetic rotations south of the Arica bend, northern Chile: Implications for the Bolivian Orocline. *Tectonophysics*, **205**, 141–154.

SCHNEIDER, A. 1985. *Eruptive processes, mineralisation and isotopic evolution of the Los Frailes – KariKari region, Bolivia*. DIC thesis. Imperial College, London.

SEMPERE, T., CHAVEZ, H. & PEREZ, M. 1989. Estudio sedimentologico preliminar de la parte superior de la formacion Mondragon en las cercanias de Lenas (Departamento de Potosi): consecuencias geodinamicas. *Actas del VIII Congreso Geologico de Bolivia, La Paz*, 273–278.

——, HERAIL, G., OLLER, J. & BONHOMME, M. 1990. Late Oligocene – Early Miocene major tectonic crisis and related basins in Bolivia. *Geology*, **18**, 496–949.

——, BABY, P., OLLER, J. & HERAIL, G. 1991. La nappe de Calazaya: une preuve de raccourcissements majeur gouvernes par des elements paleostructuraux dans les Andes boliviennes. *Comptes Rendus de l'Academie des Sciences de Paris, Serie II*, **312**, 77–83.

SERVANT, M. & FONTES, J. C. 1978. Les lacs quaternaires des hautes plateaux des Andes Boliviennes. Premierès interprétations paléoclimatiques. *Cahiers ORSTOM. Série Geologie*, **10**, 9–23.

——, SEMPERE, T., ARGOLLO, J., BERNAT, M., FERAUD, G. & LO BELLO, P. 1989. Morphogenese et soulevement de la Cordillere Orientale des Andes de Bolivie au Cenozoique. *Comptes Rendus de l'Academie des Sciences, Paris*, **309**, 417–22.

SHEFFELS, B. 1990. Lower bound on the amount of crustal shortening in the central Bolivian Andes. *Geology*, **18**, 812–815.

SOLER, P. & SEMPERE, T. 1993. Stratigraphie, geochemie et signification paleotectonique des roches volcaniques basiques mesozoiques des Andes boliviennes. *Comptes Rendus de l'Academie des Sciences de Paris, Serie II*, **316**, 777–784.

TROENG, B., SORIA-ESCALANTE, E., CLAURE, H., MOBAREC, R. & MURILLO, F. 1994. Descubrimiento de basamento precambrico en la Cordillera Occidental Altiplano de los Andes Bolivianos. *Memorias del XI congreso geologico de Bolivia*, 231–237.

VEGA, A. & BUFORN, E. 1991. Focal mechanisms of intraplate earthquakes in Bolivia, South America. *PAGEOPH*, **136**, 449–458.

WATTS, A., LAMB, S., FAIRHEAD, J. & DEWEY, J. F. 1995. Lithospheric flexure and bending of the Central Andes. *Earth and Planetary Science Letters*, **134**, 9–21.

WHITMAN, D. 1995. Moho geometry beneath the eastern margin of the Andes, northwest Argentina: constraints on the elastic thickness of the Andean foreland. *Journal of Geophysical Research*, **99**, 15 277–15 289.

——, ISACKS, B., CHATELAIN, J-L., CHIU, J-M. & PEREZ, A. 1992. Attenuation of high frequency seismic waves beneath the Central Andean plateau. *Journal of Geophysical Research*, **97**, 19 929–19 947.

WIGGER, P., SCHMITZ, M., ARANEDA, M., ASCH, G., BALDZUHN, S., GIESE, P., HEINSOHN, W-D, MARTINEZ, E., RICALDI, E., ROWER, P. & VIRAMONTE, J. 1993. Variation in the crustal structure of the southern Central Andes deduced from seismic refraction investigations. *In*: REUTTER K-J., SCHEUBER, E. & WIGGER, P. (eds) *Tectonics of the Southern Central Andes*. Springer, Berlin, 23–48.

ZANDT, G., VELASCO A. & BECK S., 1994. Composition and thickness of the southern Altiplano crust, Bolivia. *Geology*, **22**, 1003–1006.

Index